£34-90
27/7

AF850985

WITHDRAWN

Napier Polytechnic Library
Spinning, extruding, and processing of f
MLen 620.88 8 10700680 1
3 8042 00015 5149

SPINNING, EXTRUDING AND PROCESSING OF FIBERS

SPINNING, EXTRUDING AND PROCESSING OF FIBERS

Recent Advances

Edited by J.S. Robinson

NOYES DATA CORPORATION
Park Ridge, New Jersey, U.S.A.
1980

Sole distribution in the UK by:
Gothard House Publications
Gothard House
Henley-on-Thames, Oxon
RG9 1AJ Tel. 049 12 3602

Copyright © 1980 by Noyes Data Corporation
No part of this book may be reproduced in any form without permission in writing from the Publisher.
Library of Congress Catalog Card Number: 80-13137
ISBN: 0-8155-0801-8
Printed in the United States

Published in the United States of America by
Noyes Data Corporation
Noyes Building, Park Ridge, New Jersey 07656

NAPIER COLLEGE OF COMMERCE AND TECHNOLOGY
AC 81/070068 /01
RC 184 000698
MERCHISTON 677.4 ROB
CAT
CON LC80013137
PR TH 34-90

Library of Congress Cataloging in Publication Data

Main entry under title:

Spinning, extruding, and processing of fibers.

(Chemical technology review ; no. 159)
Includes indexes.
1. Textile fibers--Patents. 2. Spinning--Patents. 3. Plastics--Extrusion--Patents.
I. Robinson, J. S., 1936- II. Series.
TS1449.S645 677'.4 80-13137
ISBN 0-8155-0801-8

Foreword

The detailed, descriptive information in this book is based on U.S. patents issued since January 1977 that deal with spinning, extruding and processing of fibers. This is the second volume of an industrial trilogy on polymer, fiber and fabrics manufacture, the first volume being entitled *Fiber-Forming Polymers, Recent Advances,* and the third to be entitled *Manufacture of Yarns and Fabrics from Synthetic Fibers.*

This book is a data-based publication, providing information retrieved and made available from the U.S. patent literature. It thus serves a double purpose in that it supplies detailed technical information and can be used as a guide to the patent literature in this field. By indicating all the information that is significant, and eliminating legal jargon and juristic phraseology, this book presents an advanced commercially oriented review of recent advances in the spinning, extruding and processing of fibers.

The U.S. patent literature is the largest and most comprehensive collection of technical information in the world. There is more practical, commercial, timely process information assembled here than is available from any other source. The technical information obtained from a patent is extremely reliable and comprehensive; sufficient information must be included to avoid rejection for "insufficient disclosure." These patents include practically all of those issued on the subject in the United States during the period under review; there has been no bias in the selection of patents for inclusion.

The patent literature covers a substantial amount of information not available in the journal literature. The patent literature is a prime source of basic commercially useful information. This information is overlooked by those who rely primarily on the periodical journal literature. It is realized that there is a lag between a patent application on a new process development and the granting of a patent, but it is felt that this may roughly parallel or even anticipate the lag in putting that development into commercial practice.

Many of these patents are being utilized commercially. Whether used or not, they offer opportunities for technological transfer. Also, a major purpose of this book is to describe the number of technical possibilities available, which may open up profitable areas of research and development. The information contained in this book will allow you to establish a sound background before launching into research in this field.

Advanced composition and production methods developed by Noyes Data are employed to bring these durably bound books to you in a minimum of time. Special techniques are used to close the gap between "manuscript" and "completed book." Industrial technology is progressing so rapidly that time-honored, conventional typesetting, binding and shipping methods are no longer suitable. We have bypassed the delays in the conventional book publishing cycle and provide the user with an effective and convenient means of reviewing up-to-date information in depth.

The table of contents is organized in such a way as to serve as a subject index. Other indexes by company, inventor and patent number help in providing easy access to the information contained in this book.

15 Reasons Why the U.S. Patent Office Literature Is Important to You

1. The U.S. patent literature is the largest and most comprehensive collection of technical information in the world. There is more practical commercial process information assembled here than is available from any other source.
2. The technical information obtained from the patent literature is extremely comprehensive; sufficient information must be included to avoid rejection for "insufficient disclosure."
3. The patent literature is a prime source of basic commercially utilizable information. This information is overlooked by those who rely primarily on the periodical journal literature.
4. An important feature of the patent literature is that it can serve to avoid duplication of research and development.
5. Patents, unlike periodical literature, are bound by definition to contain new information, data and ideas.
6. It can serve as a source of new ideas in a different but related field, and may be outside the patent protection offered the original invention.
7. Since claims are narrowly defined, much valuable information is included that may be outside the legal protection afforded by the claims.
8. Patents discuss the difficulties associated with previous research, development or production techniques, and offer a specific method of overcoming problems. This gives clues to current process information that has not been published in periodicals or books.
9. Can aid in process design by providing a selection of alternate techniques. A powerful research and engineering tool.
10. Obtain licenses–many U.S. chemical patents have not been developed commercially.
11. Patents provide an excellent starting point for the next investigator.
12. Frequently, innovations derived from research are first disclosed in the patent literature, prior to coverage in the periodical literature.
13. Patents offer a most valuable method of keeping abreast of latest technologies, serving an individual's own "current awareness" program.
14. Copies of U.S. patents are easily obtained from the U.S. Patent Office at 50¢ a copy.
15. It is a creative source of ideas for those with imagination.

Contents and Subject Index

Introduction

This book deals with the spinning and extrusion of polymers, resulting in filaments, films, or shaped articles. Dry spinning, wet spinning, and melt spinning are used to produce filaments and fibers. Extrusion, of which melt spinning may be considered an example, is also used to produce films and shaped articles including hollow fibers. Melt spinning is accomplished by extruding the polymer in a molten condition, but at melt temperatures some polymers tend to degrade and revert to monomer. In dry spinning, a solution of polymer is extruded into a heated zone in which the solvent evaporates and from which the filaments are collected. In wet spinning, a solution of polymer is extruded into a liquid bath in which the solvent is at least partially removed from the filament. Wet spinning is carried out at much lower temperatures than melt spinning and, normally, at substantially lower temperatures than dry spinning. Improvements in these spinning and extrusion processes are discussed in the book's early chapters.

Production of glass fibers, carbon and metal carbide fibers, fibrils such as those used in papermaking and in other nonwoven applications, and core yarns and other composite fibers which allow for the combining of attributes of two or more materials are all covered in detail.

In addition to the spinning or extruding step, the filaments are usually subjected to further processing steps. Drawing, stretching, and heat setting are discussed in later chapters. Yarns and fabrics made from synthetic filaments will be the subject of a forthcoming Noyes publication.

The final chapter deals with polymer recovery, which is of increasing concern as petrochemical supply dwindles and ecological awareness increases.

Dry Spinning

In the wet spinning process, covered in the next chapter, a polymer is dissolved in a suitable solvent and, after passing through a spinneret, the resulting fibers are passed through a liquid which extracts the solvent contained in the fibers. In dry spinning, the solvent is removed from the fibers by heat and vaporization.

In both the dry and wet spinning processes, the fibers are stretched and otherwise processed after spinning and solvent removal.

ACRYLICS AND MODACRYLICS

Solvent System for Polyacrylonitrile

D.B. Parkinson and N. Fishman; U.S. Patent 4,172,065; October 23, 1979; assigned to Trian Company provide a solvent which dissolves acrylonitrile polymers (and other highly polar, difficultly-soluble polymers, such as polyimides) in sufficient concentration to be practicable for casting films, spinning fibers and coating applications, which can be removed in its entirety by application of heat at a temperature below the point of degradation of the end product, and which does not leave an objectionable residue in the end product.

An advantageous solvent for polymers of this character is provided by a solution containing nitromethane, water and 2,2,2-trifluoroethanol (TFE). The proportions of the components of this solvent may vary within limits as follows, percentages being by weight: nitromethane, 75 to 95%; water, 3 to 20%; and TFE, 2 to 10%.

A preferred solvent is as follows: nitromethane, 84.8%; water, 10.1%; and TFE, 5.1%.

The dry polymer to be spun into fibers by the dry process, to be cast as a film or to be applied as a coating is dispersed in this solvent at or above room temperature. Then it is dissolved by vigorous mixing in a high speed homogenizer, or similar dissolving apparatus, at 90° to 100°C for a suitable period of time.

After a solution is obtained, it should be maintained at 60° to 80°C until a stage of fiber spinning or film casting process is reached at which coagulation is desired. Solutions prepared in this manner are water white and have a sufficiently low viscosity to be spun into fibers through spinnerets conventionally used in the dry spinning process. Pilot runs with effective dissolution procedures are able to produce solutions with 30% or higher polymer concentrations.

Bifilar Acrylic Fibers

C. Pieper, K. Feltgen and A. Nogaj; U.S. Patent 4,124,673; November 7, 1978; assigned to Bayer AG, Germany provide acrylic fibers and filaments with a permanent crimp, and a process for their production by spinning two solutions of the same acrylonitrile polymer.

The polymer used is generally a copolymer which contains at least 50% by weight of acrylonitrile and up to 50% by weight of one or more ethylenically unsaturated monomers copolymerizable with acrylonitrile or a mixture of two or more of the abovementioned polymers. However, it is particularly preferred to use polymers which contain at least 85% by weight of acrylonitrile.

Particularly preferred comonomers are acrylic acid or methacrylic acid esters; for example, methyl acrylate and ethyl acrylate; vinyl esters, for example, vinyl acetate; and comonomers which increase affinity for dyes, for example, allyl and methallyl sulfonic acid, styrene sulfonic acid and their salts.

The solutions may be dry or wet spun, preferably dry spun, by any of the known processes in such a way that substantially equal parts of the two components are present in the filament. In order to optimize the permanent crimp, it is best for the ratio of the polymer components in the filament to differ from 1:1, although it should not exceed a ratio of about 1:3.

Temperature differences of about 30°C have proved to be adequate. However, larger temperature differences, for example, from 40° to 120°C, are particularly preferred; the average absolute temperature being of the order normally encountered in this art.

In principle, it is, of course, possible to divide the solution stream and to establish the requisite temperature difference at any given point in front of the orifice of the spinneret. However, it is particularly favorable to carry out these measures as near to the orifice of the spinneret as possible. It is possible, in this way, to eliminate the need for a double pipe system for the solution stream.

The two fiber components are spun by known methods in such a way that they are present in the resulting bicomponent filament in an eccentric arrangement to one another. This eccentric arrangement may be a core/cover arrangement, although it is preferably a side-by-side arrangement.

The bicomponent fibers are then aftertreated by the methods normally adopted for acrylic fibers. Preferably, the filaments are washed, stretched, optionally mechanically crimped, heat treated, dried and optionally cut into fibers.

Once it has been developed, the crimp is permanent and represents the condition of minimal energy for the fibers. It is also permanent and elastic when subjected to deformation. If the crimp is pulled out by mechanical deformation even to the point where the fiber breaks, it springs back into the fiber after a heat treatment.

Example: An acrylonitrile copolymer of 93.6% acrylonitrile, 5.7% methacrylate and 0.7% sodium methallyl sulfonate was dissolved in dimethyl formamide at 90°C. A solution with a polymer content of 29.5% was prepared. The solution was delivered to a bifilar spinneret. The solution stream was divided into two parts. One of the two solution streams (A) was heated to 130°C and the other (B) cooled to 40°C. The solutions had a ball drop time of 24 seconds (A) and 300 seconds (B). (The ball drop time is a measure of solution viscosity. It is the time which a ball with a diameter of 3.175 mm takes to travel a distance of 7 cm in the solution at the temperature indicated.)

The solutions were introduced into a bifilar spinneret in a ratio of A to B of 1.2:1 and were spun side by side. Several bifilar spinnerets were arranged on a die base and air which had been heated to 380°C flowed past the base. The filaments were run off at 300 m/min. The duct situated below the spinneret had a wall temperature of 170°C. After spinning, the spun material was washed with water for 30 seconds at 100°C. The material was then stretched 1:4 in water at 100°C. The filaments were then dried for 2 minutes at 100°C in the absence of tension.

The resulting crimp in the filaments was intensified by boiling the filaments for 10 minutes in water, followed by tempering for 20 minutes at 130°C. The fibers had a denier of 3.3 dtex and contained 4 crimp arcs per centimeter. The number of crimp arcs per centimeter is derived from the quotient of the number of crimp arcs and the length of the fiber from which the crimp has been just pulled out by elongation.

Spinning and aftertreatment under the same conditions as above, but with the stretching ratio altered to 1:2.5, produced a fiber with a denier of 5.8 dtex and 10 crimps per centimeter.

Modacrylics with Improved Coloristic Properties

R. Miessen, G. Blankenstein, S. Korte and C. Süling; U.S. Patent 4,014,958; March 29, 1977; assigned to Bayer AG, Germany describe dry spun modacrylic filaments with improved coloristic properties based on a polymer mixture which comprises at least one acrylonitrile vinyl chloride copolymer and a chlorine containing copolymer containing sulfonic acid ester groups or sulfuric acid ester groups.

Where modacrylic filaments are normally obtained by spinning solutions of the starting materials in acetone, cyclohexanone or acetonitrile, the process is more suitable for the production of modacrylic fibers and modacrylic filaments from solutions of the polymer mixtures in solvents of the type suitable for the production of acrylic filaments. Solvents of this kind are dimethylformamide, dimethylacetamide and dimethylsulfoxide. Particularly favorable results are obtained by dissolving the polymer mixtures in dimethylformamide or dimethylacetamide and spinning the resulting solutions by a dry spinning process.

The concentration of the polymer mixture in the spinning solutions is generally between 25 and 40% by weight. Depending on the average molecular weight of the components, favorable spinning results are obtained either with the fairly high or fairly low concentration of the spinning solutions; although it is preferred to use spinning solutions with concentrations above 30% by weight. In most cases, it is best to spin 35 to 38% solutions. The spinning solutions should have viscosities from 700 to 6,000 poises, preferably from 2,500 to 4,000 poises, as measured at 20°C.

The spinning solution is transported from the particular dissolution units used to the spinnerets through metering units, for example metering pumps, and heating units. In this connection, it can be advantageous to insure, optionally by secondary heating, that the temperature of the spinning solution is regulated to prevent the spinning solutions from gelling. Favorable results are obtained by keeping the spinning solutions at temperatures from 20° to 50°C, depending upon their concentration and viscosity.

The spinning solution does not undergo degradation or gelation even in the event of a prolonged residence time. This surprising effect is a particular advantage of the solvents preferably used, dimethylacetamide and dimethylformamide, over the solvents normally used for dissolving halogen-containing copolymers, such as cyclohexanone, acetonitrile and acetone.

Gear pumps have proved to be particularly effective metering units. The spinning solution is pumped by way of heating units working from 20° to 140°C through spinnerets; for example, in the form of cap dies or ring dies, arranged in heated spinning ducts. The heated spinning ducts are kept at temperatures from 130° to 190°C, preferably from 150° to 170°C. Most of the solvent is removed from the filaments with a hot gas stream, preferably air, with a temperature of 150° to 220°C. Particularly advantageous spinning results are obtained when the temperature of the air used is selected in such a way that temperatures from 170° to 200°C prevail in the immediate vicinity of the spinneret. Tows with a spun denier of 400 to 5,000 dtex are obtained according to the number of bores in the spinneret, the diameter of the bores, which should be between 150 and

300 μ and the quantity of spinning solution delivered per unit of time. The resulting spun material may still contain solvent and is stretched at elevated temperature by methods known and subsequently washed, dried, optionally crimped and cut. Dyeing with cationic dyes is carried out by known methods at elevated temperature in a weakly acid medium. The quantities of dye taken up may readily be quantitatively determined by colorimetric measurement.

Example 1: *Production of a Polymeric Dye Receptive Additive of Acrylonitrile, Vinylidene Chloride and N-Acryloyldimethyltaurine* – The following solutions and mixtures are used:

(1) 4,325 g of dimethylformamide (DMF) and 175 g of deionized water;

(2) 1,000 g of dimethylformamide, 375 g of N-acryloyldimethyltaurine (ADT) and 132.5 g of 2-methylaminoethanol;

(3) 1,250 g of acrylonitrile and 875 g of vinylidene chloride;

(4) 300 g of dimethylformamide and 6 g of ammonium peroxydisulfate; and

(5) 300 g of dimethylformamide and 7 g of oxalic acid.

The DMF/water mixture (1) is initially introduced into a 10 liters glass reaction vessel, equipped with a reflux condenser cooled with iced water, a stirrer and a gas inlet pipe, followed by rinsing with nitrogen at 50°C. The ADT-ammonium salt solution (2) obtained by neutralization at 0°C, and the monomer mixture (3) are then combined with the DMF/water mixture (1). After the reaction temperature of 50°C has been adjusted, polymerization is initiated by the addition of the initiator solutions (4) and (5).

After 15 hours a solids content of 24.0% by weight is reached in the solution, corresponding to a conversion of 75%. Following dilution with water, a polymer of the following composition is recovered from the polymer solution with a conversion of 75% by the addition of electrolyte:

47.0% by weight of acrylonitrile;

32.0% by weight of vinylidene chloride; and

21.0% by weight of N-acryloyldimethyltaurine (in the form of the ammonium salt of 2-methylaminoethanol) (ADT-salt).

The polymer yield was 1,970 g and it had a k value of 72.

Example 2 (Comparative): A copolymer of acrylonitrile and vinyl chloride, containing the comonomers in a ratio of 42:58 and which has an intrinsic viscosity in DMF of 0.99 at 25°C, is made into a paste with dimethylformamide at room temperature, and dissolved over a period of 6.5 hours at 40°C in a stirrer equipped vessel.

In order to remove gas bubbles, the vessel is evacuated for 30 minutes, resulting in the formation of a spinning solution with a viscosity of 2,000 poises at 40°C. This solution is delivered by a gear pump acting as a metering unit through a spinneret in the form of an annular die which is arranged in a vertical spinning duct and comprises 120 bores with a diameter of 0.15 mm.

The solution is pumped at a rate of 38.4 cc/min, and the filaments are taken off at a rate of 100 m/min. The spinning process is carried out at a duct temperature of 145° to 150°C and at an air temperature of 180°C; the draft applied in the spinning duct amounting to 1:5.5. The resulting filaments are stretched to five times their original length in boiling water, after which the material is treated at 140°C resulting in 25% shrinkage.

The modacrylic filaments had a tensile strength of 2.02 poises/dtex and elongation at break of 40.3%. A fiber saturation value S_F of 1.2 and an absorption rate V of 0.76 are obtained during dyeing with basic dyes.

Example 3: The procedure is as in Example 2, except that the vinyl chloride-acrylonitrile copolymer used is replaced by a mixture of that copolymer and the acrylonitrile-vinylidene chloride-acryloyldimethyltaurine terpolymer according to Example 1 in a mixing ratio of 92% by weight of the copolymer and 8% by weight of the terpolymer. Dry spinning was carried out under the following conditions:

Spinning duct temperature	151°C
Air temperature	182°C
Spinneret temperature	100°C
Delivery rate	39.4 cc soln/min
Take-off rate	100 m/min
In-duct draft	1:5.2

The further aftertreatment was carried out in the same way as in Example 2. The filaments obtained had a tensile strength of 1.94 poises/dtex and elongation at break of 38%. A fiber saturation value number S_F of 3.3 and an absorption rate V of 2.07 are obtained during dyeing with basic dyes.

The values characteristic of coloristic behavior are significantly improved in relation to Example 2.

Removal of Residual Solvent

C. Pieper, H. Uhlemann, K. Feltgen, K. Nickel, A. Nogaj and G. Lorenz; U.S. Patent 4,124,674; November 7, 1978; assigned to Bayer AG, Germany describe a process for continuously removing residual solvent from filaments of dry spun acrylonitrile polymers to values of below 2% by weight of solvent by a multi-stage hot wash with low quantities of water.

A primary objective of the process is to minimize the consumption of water during washing by comparison with conventional processes. A low water consumption is important for economic reasons because all the water left after washing has to be evaporated in order to recover the solvent. The filament material may be demoistened between individual washing stages.

Example: An acrylonitrile polymer consisting of 93.6% acrylonitrile, 5.7% methylacrylate and 0.7% sodium methallyl sulfonate was dissolved in dimethylformamide (DMF) at 90°C. The solution was dry spun by conventional methods so that the spun material had a DMF content of 22%. The spun material was cold washed in a cascade of nine stages. The washing liquid was thoroughly mixed in each stage. Between each stage, the washing solution was removed from the tow by means of squeezing rolls. The tow, which had a weight of about 105 g/m, entrained approximately 45% of moisture (based on the fiber weight) from one stage to the next.

The washing water had a temperature of 15°C and flowed in countercurrent to the tow from one stage to the next. In the ninth stage, 0.6 part of washing liquid (water with a DMF content of 0.1%) was added per part of polyacrylonitrile (PAN). After this treatment, the tow had a DMF content of 7%. Within the accuracy of measurement, this DMF content was independent of the speed of travel of the tow, which was varied between 10 and 60 m/min. Even an increase in temperature in the washing liquid to 50°C produced hardly any change in the DMF content. Upon completion of washing, the washing liquid had a DMF content of 33%. The tow then entered another cascade of seven stages. This cascade was encased so that neither DMF nor steam was emitted. In this cascade, the washing liquid had a temperature of 99°C.

The speed of travel of the tow was adjusted in such a way that, in each of the first six stages of the cascade (hot water), the residence time amounted to 10 seconds. Between each stage, the washing liquid was removed from the tow by squeezing out to such an extent that 45% of washing liquid adhered to it. In this cascade, too, the washing liquid flowed in countercurrent to the tow. In the last stage, 0.9 part of water was added per part of PAN. Upon completion of hot washing the washing liquid had a DMF content of about 6.8%.

In the last stage the tow was drawn in a ratio of 1:5. The tow had a DMF content before drawing of less than 1%. The tow was then prepared at 80°C, dried for 1 minute at 140°C with allowance for shrinkage, and then crimped. The fibers had an individual denier of 3.3 dtex and a strength of about 3.7 poises/dtex for an elongation at break of about 39%. The tow was split while still hot and processed into yarns. The yarns had a strength of 0.9 poise/dtex for an elongation of 16%. The fibers contained very few vacuoles, had a stable luster and a density of 1.18 g/cc.

CELLULOSICS

Manufacture of Viscose Filaments

P. Monzie, S. Chaunis, P. Goullioud and P. Lainé; U.S. Patent 4,126,656;

November 21, 1978; assigned to Centre Technique de L'Industrie des Papiers, Cartons et Celluloses et Technique Papetiere Avancee, France disclose a method for manufacturing viscose filaments from a cellulosic starting material. The viscose is extruded by passing it through a spinneret to form continuous thin strands, the extruded strands are passed continuously through a gaseous medium which contains at least one volatile agent of such a nature, and, under such conditions, a coagulating action on the strands results.

The strands which have thus been coagulated are then brought into contact with an acid medium which causes a preregeneration of the initial cellulose, so that the strands are converted to continuous filaments. The filaments are passed through at least one other acid liquid medium which causes ultimate regeneration of the cellulose.

Filaments manufactured by the method have characteristics which allow them to be distinguished from known filaments and exhibit the following features.

Little or no morphological difference between the periphery and the core of the filaments is observed under polarized light, by phase contrast, by examination with a scanning electron microscope, or by staining sections with Victoria Blue (the sample being embedded in methacrylate). The material constituting the new filaments thus appears to be particularly homogeneous.

Little or no difference in the degree of orientation of the cellulosic macromolecules has been detectable on examination by electron diffraction, either on longitudinal sections or on transverse sections. This degree of orientation is medium throughout the filament, in contrast to the conventional filaments which can be either more or less oriented or be highly oriented throughout their length.

A characteristic behavior towards the dyestuff called "Brilliant Dyestuff Blue FF Ciba" of the filaments is observed. In fact, staining takes place during a first stage in the peripheral zone of the filament while thereafter, in a second stage, it reaches the central part so as to give a very intense coloration through the whole fiber.

Finally, an excellent ability of the filaments to swell in water and in sodium hydroxide solution, with a difference in degree of swelling between an internal zone and an external zone, has been noted.

The filaments can be prepared in the same forms as the usual viscose filaments; for instance, as continuous filaments (rayon), and as discontinuous staple fibers as tows, slivers, threads, double yarns, waddings, flocks, and nonwovens.

No particular description or illustration of the means used to prepare viscose from the cellulosic pulp need be given since such is well known.

Figure 1.1 shows an apparatus to be located downstream of such means for preparing viscose. Such apparatus essentially comprises a vertical cylindrical reservoir **1** which is fed continuously with viscose. The viscose, under the effect of a nitrogen pressure exerted by means of an inlet tube **2** provided with a manometer **3**, is extruded through the orifice of a spinneret **4** connected to the small end of the truncated cone shaped bottom part **1a** of the reservoir. The spinneret is advantageously made of glass, though it could be stainless steel or any other appropriate metal, or even plastics, and consists of a cylindrical nozzle merging into a capillary tube, the diameter of which is 600 μ.

A cylindrical column **7** having an internal diameter of 7.5 cm and a length of 125 cm, is located on the truncated cone bottom part of the reservoir and is axially aligned therewith.

Figure 1.1: Viscose Spinning Apparatus

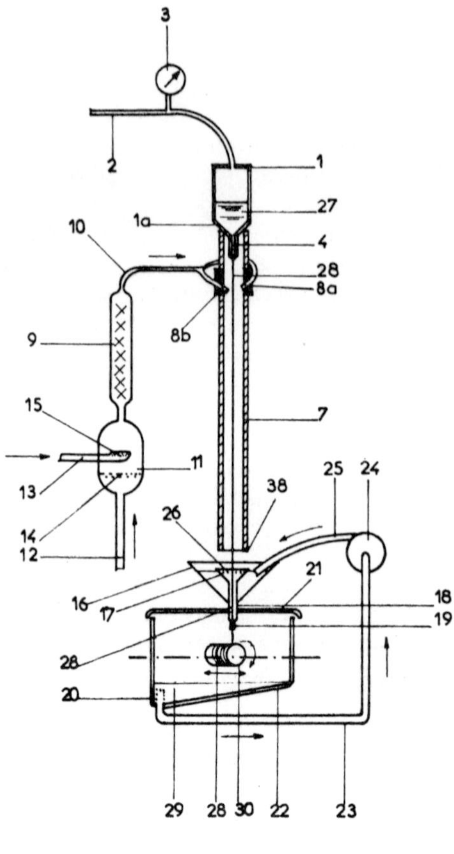

Source: U.S. Patent 4,126,656

At a distance of about 25 cm from the top of the column **7**, there opens into the column, at diametrically opposed points **8a** and **8b**, an inlet for a gaseous mixture fed via a homogenizer **9** and a tube **10**, from a source **11**. This source receives compressed air via a pipeline **12** and hydrogen chloride gas through a pipeline **13**. At **14** and **15** are shown sintered glass distributing devices which ensure good distribution of the mixture.

The base **38** of the column is just above a funnel comprising two coaxial concentric bodies **16** and **17** joined at a point **18** and aligned axially with the column, the body **17** extending below the point to a constriction **19** at which it opens into a tank **20** having a cover **21** and an inclined bottom **22**. At the lower point of this bottom is a tube **23** connected to a pump **24** which, via another tube **25**, can pass liquid **29** contained in the bottom of the tank into the truncated cone annulus between two truncated conical parts of the bodies.

By suitably regulating the flow rate, a part of the liquid continuously remains between the two truncated conical parts and passes into the inner conical part through orifices **26** formed on its walls at the same horizontal level, and spreads as a thin layer over the inner truncated cone surface of the body until it fills the constriction.

Vertically below the combination of reservoir **1**, column **7**, funnels **16** and **17**, point **18** and constriction **19**, and in the tank **20**, is located a wind up cylinder **30** which is rotatable about a horizontal axis and can be subjected to a slow reciprocating movement along this axis by means of a conventional control device. This cylinder is so arranged that a plane tangential to it includes the axis of the funnels and of the point so that a filament **28** which has issued vertically from the constriction engages tangentially with the turns of the coil which it forms on the cylinder **30**. In use, the cylinder is sprinkled with the liquid from the constriction. The liquid is cycled in a closed circuit through these devices.

In the course of the cycle, the content of reagents in the liquid medium can be controlled and continuously adjusted.

The filaments will be subjected to a final treatment with sulfuric acid under such conditions that the ultimate regeneration of the cellulose from any xanthate not completely decomposed during the preregenerating operation takes place, and will also be subjected to various other treatments, such as washing and desulfurization.

Example: A viscose **27**, located in the reservoir, is extruded at ambient temperature under the effect of nitrogen pressure at a speed of 1 to 4 cm/sec, through the capillary of the spinneret **4**, in the form of an initial strand. This strand passes down the column, firstly for a distance of 25 cm through air and then for 100 cm through another gaseous medium consisting of air/hydrochloric acid (feed rate: 0.5 to 1 g HCl/hr; duration of travel of the strand in chamber **7**: 0.1 to 1 sec). During this travel the strand undergoes stretching under the double action of gravity and the mechanical effect of the drawoff device. While in the region of the acid gaseous medium it coagulates acquiring the consistency of a gel.

The acid liquid **29** which travels in a closed circuit below the spinning apparatus continues the preregeneration of the cellulose of the strand when the latter comes into contact with it in the constriction **19**, where it is seized and driven along axially to become a viscose filament **28**, consisting of viscose in the nascent state.

This acid liquid has 40 g/l sulfuric acid, 60 g/l neutral sodium sulfate, 1 g/l zinc sulfate and water to 1 liter.

To achieve ultimate regeneration, the filaments are then subjected to various treatments which, in particular, insure the complete regeneration of the cellulose and its definitive presentation. These various treatments, with their essential details, are as follows.

Steeping–in a bath containing 60 g of sulfuric acid per liter at 50° to 60°C for 10 minutes;

Washing–in softened water at ambient temperature for 5 minutes at a pH increasing firstly from 4.5 to 6.5 and then to about 8;

Desulfurization–in a bath containing 2 g/l of Na_2CO_3 and 10 g/l of Na_2SO_3 at 90° to 95°C for 20 minutes;

Washing–in softened water at ambient temperature initially at pH 9.5 and then down to pH 8 or 7; and

Drying–in a ventilated oven at 105°C for 5 to 8 hours.

Cellulose Spun into Ammonia Atmosphere

A process for producing shaped articles of cellulose is disclosed by *H. Rodier; U.S. Patent 4,145,391; March 20, 1979; assigned to Rhone Poulenc Textile, France* where a solution of cellulose in dimethyl sulfoxide (DMSO) and formaldehyde, at a formaldehyde/cellulose weight ratio of 0.2 to 2, and at a cellulose concentration of at least 6% by weight, based on the volume of DMSO, is spun into a gaseous atmosphere containing ammonia.

The process permits the cellulosic solution to be formed into shaped articles in a very rapid and economical manner. The process is especially useful in the production of filaments which can be used in the textile field, but other shaped articles, such as fibers, threads, films and sheets, can also be made from the cellulosic solution.

Example: 90 g of sulfate wood pulp, viscose grade, having a DP of 450 and a moisture content of 5.5%, were dried to a moisture content of less than 1% and introduced into 1,000 cm^3 of DMSO containing 600 ppm of water and 106 g of 96% strength paraformaldehyde providing a paraformaldehyde/cellulose weight ratio of 1.20. The temperature of the mixture, which was initially at ambient temperature, was raised to 130°C in 1 hour with slow stirring and was maintained at 130°C for 3 hours with continued stirring.

The temperature of the solution was then lowered to 120°C and maintained at that temperature while a stream of nitrogen, which was also at a temperature of 120°C, was bubbled through the solution to remove a part of the formaldehyde. When the formaldehyde/cellulose weight ratio had been reduced to 0.25, the bubbling of the stream of nitrogen was ceased, and the solution was readjusted to compensate for DMSO losses. The resulting solution had a viscosity of 1,100 poises at 20°C, and the dissolution of the cellulose was complete as observed under a polarizing microscope.

The solution was extruded downward at a temperature of 25°C through a spinneret having 48 orifices, each of 0.25 mm diameter, into a 7 m high-spinning chamber having two sections; the upper section being 2 m high, and the lower section being 5 m high. The walls of the upper section were maintained at a temperature of 120°C, and the walls of the lower section were maintained at a temperature of 245°C.

A gaseous mixture, which had been heated to a temperature of 105°C, and consisting of a mixture of nitrogen in which a 28% strength aqueous solution of ammonia had been previously vaporized to an ammonia content in the gaseous mixture, under normal conditions (that is, STP), of 0.10 g/l, was injected into the upper section of the spinning chamber.

Pure nitrogen, at a temperature of 245°C, was injected into the lower section of the spinning chamber.

The gas mixture in the spinning chamber was withdrawn from the lower section of the chamber, and the resulting yarns were withdrawn from the lower section of the spinning chamber at a speed of 140 m/min, and wound at that speed onto perforated metal bobbins. Filled bobbins so obtained were washed with water at ambient temperature to remove residual solvent, and, thereafter, the yarns were sized and dried. The resulting yarn exhibited the following physical characteristics (where the condition state refers to maintaining the filaments until equilibrium at 20°±2°C, and 65±2% relative humidity): 150 dtex/48 filaments gauge; 17 g/tex tenacity after conditioning; 20% elongation after conditioning; 12 g/tex wet tenacity; and 27% wet elongation.

OTHER POLYMERS

Polypyrrolidone

According to *A.C. Tanquary; U.S. Patent 4,094,945; June 13, 1978; assigned to Chevron Research Company,* polypyrrolidone solutions having a bulk viscosity of 100 to 10,000 poises and containing a solvent, such as formic acid, and a volatile diluent, such as methylene chloride, are spun into filaments by extrusion through a spinneret. The solution contains, for example, about 5 to 40 weight percent of polypyrrolidone, about 30 to 90 weight percent of formic acid and about 5 to 60 weight percent of methylene chloride.

By this process, high molecular weight polypyrrolidone is easily spun into filaments without degradation of its molecular weight.

Example 1: Polypyrrolidone of 110,000 average molecular weight was mixed with a 50/50 weight percent solvent/diluent mixture of formic acid/methylene chloride. The filament-forming solution had a solids level of 23% and a bulk viscosity of about 2,000 poises. A portion of the solution was charged to a 130 ml feed reservoir and forced by nitrogen pressure (100 psi) through a screen pack (40 and 250 mesh screens) and a spinneret (single orifice of 10 mils diameter). From the spinneret the monofilament passed in front of a 1,100 W radiant heater for about 2 feet. The temperature along the filament drying zone was about 120°C. No difficulties were encountered in collecting monofilaments of about 350 denier.

Example 2: *Gap Spinning of Nylon 4* – A 500 ml reservoir was charged with the filament-forming solution containing 15.5 weight percent polypyrrolidone (295,000 molecular weight), 38% formic acid and 46.5% methylene chloride. The solution was forced by nitrogen at 140 psi from the reservoir to a Zenith gear pump. The solution was pumped through a stainless steel screen pack containing 40 and 250 mesh screens.

A monofilament was extruded through a spinneret with a 20 mils orifice and then allowed to pass vertically near two sets of 250 W infrared lights. The lights were positioned about 2.5" from the fiber path. The temperature in the fiber path was 120°C. 24" below the spinneret the monofilament entered a 45" trough containing ethyl acetate. The filament passed through the bath, around a glass godet, over a first pair of unheated godets, and finally took several turns around a pair of heated metal godets at 100°C. Fiber samples were collected on a Leesona winder.

The yarn packages were dried 12 hours under vacuum to remove residual formic acid and ethyl acetate. The tensile properties of the fibers are shown below. The drawn fibers show excellent tenacity, initial modulus and tensile factor (square root of percent elongation at break times tenacity).

Tensile Properties of Gap-Spun Fibers

Draw at 175°C	undrawn	3X	undrawn	2X
Denier	80	27	76	45
Tenacity, g/den	0.87	3.40	0.88	1.84
Elongation at break, %	220	27	243	82
Tensile factor	13.0	16.5	13.7	16.7
Initial modulus, g/den	15.5	27.0	9.8	17.2

Halogenated Aromatic Polyesters

According to *W.C. Gilbert; U.S. Patent 4,128,614; December 5, 1978; assigned to Celanese Corporation* certain halogenated aromatic polyester fibers having reduced spin line static generation when prepared via dry spinning from a methylene chloride solution are provided by admixing an esterified hindered phenol prior to dry spinning. A preferred esterified hindered phenol which may be utilized to reduce spin line static is tetrakis[methylene-3-(3',5'-di-tert-butyl-4'-hydroxyphenyl)propionate] methane.

Example: *Preparation of the Polyester* – The brominated aromatic polyester is formed by reacting 201.7 pbw tetrabromobisphenol A and a mixture of 46 pbw isophthaloyl chloride and 30.8 pbw terephthaloyl chloride. The resulting brominated aromatic polyester possesses the structural formula:

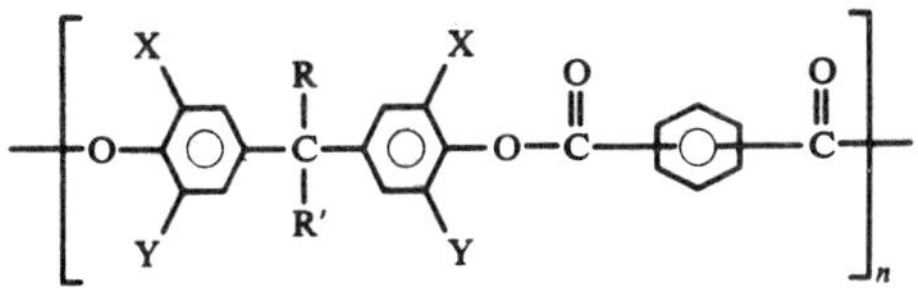

where X and Y are bromine, R and R' are methyl groups, and n is about 50. The brominated aromatic polyester has a bromine content of about 48% by weight, and exhibits an inherent viscosity of about 0.75 dl/g determined at a concentration of 0.1% by weight in a solvent which is a mixture of 10 pbw of phenol and 7 pbw of trichlorophenol.

Preparation of Esterified Hindered Phenol – Tetrakis[methylene-3-(3',5'-di-tert-butyl-4'-hydroxyphenyl)propionate] methane is prepared as described in U.S. Patent 3,330,859.

Dry Spinning – The solution containing the brominated aromatic polyester is then prepared. Thus, 100 pbw of tetrabromobisphenol A is mixed with 0.5 pbw of tetrakis[methylene-3-(3',5'-di-tert-butyl-4'-hydroxyphenyl)propionate] methane previously prepared, and the mixture is added to 372 pbw methylene chloride.

The solution is then separately dry spun into fibers under the conditions shown below.

The spin line static or static charge developed on the fibers is measured by an electrostatic voltmeter by positioning the probe of the instrument about ½" from the moving fibers where they exit from the spinning cabinet. The resulting spin line static charge is also shown below.

Dope solids, % polyester and phenol by weight of solution	21.3
Dope viscosity, poises	1,600
Jet (number of holes, hole size μ)	10/42
Jet face temperature, °C	80
Column temperature, °C, top	60
Column temperature, °C, bottom	50
Take up, m/min	100
Column length, ft	25
Column air	down draft

The spin line static charge is then determined and found to be 0 volts.

Comparative Example: The same procedure of the example is repeated except that the esterified hindered phenol is omitted. The spin line static charge is found to be +45 volts for the brominated aromatic polyester.

As may be seen from the test results, the esterified hindered phenol substantially reduces the generation of spin line static charge of the fibers into which it is incorporated during the dry-spinning process.

Flame Retardant Melamine

Y. Nihongi and N. Yasuhira; U.S. Patent 4,088,620; May 9, 1978; assigned to Kuraray Co., Ltd., Japan describe flame retardant and infusible fibers having a water-swelling degree lower than 2.0, obtained by spinning precondensates of melamine-aldehyde resins and thereafter curing the resulting fibers.

The preferred process for preparing the fiber is a so called "dry reaction spinning" process. Essentially, the spinning solution comprising the precondensate dissolved in a solvent (the preferred solvent being water) is spun through a nozzle into an air or inert gas (e.g., nitrogen) atmosphere, heated to a temperature of from 170° to 320°C.

This has the effect of simultaneously evaporating the spinning solution solvent and curing the precondensate to form the higher molecular weight crosslinked polymer having a degree of water-swelling of less than 2.0, preferably less than 1.5. The particular temperature selected depends upon many factors, such as the speed of spinning, the resin constituting the fiber, the degree of water-swelling, etc. For those fibers consisting only of a melamine-aldehyde resin, a temperature of 170° to 280°C is preferred for spinning speeds of from 10 to 20 m/min.

Since the cured melamine-aldehyde resin forming the fiber of the process is not a linear polymer, it is impossible to determine the molecular weight thereof. However, the extent to which a three-dimensional infusible and insoluble network polymer is formed may be represented by the degree of crosslinking which has occurred during the curing operation, this degree being determined from the degree of swelling of the polymer. Specifically, a polymer having a low degree of swelling has a high three dimensional network structure and, correspondingly, a polymer having a high degree of swelling (by a suitable solvent) does not have an extensive formation of a three dimensional network structure, and can be said to be insufficiently cured according to the process.

The degree of swelling is defined as W/W_0, where W is the weight of the fiber determined after immersing the dry fiber in a solvent for a fixed period of time and W_0 is the weight of the same absolute dry fiber. The degree of swelling decreases with the progress of the crosslinking reaction, with the minimum, of course, being 1. A polymer which is not crosslinked has an infinite degree of swelling since the polymer would be dissolved in the solvent.

To determine the presence of a sufficient cure for the resins of the process, the degree of swelling of the fiber in water must be lower than 2.0, preferably less than 1.5. Water is a solvent for the melamine-aldehyde precondensate. The technique used to determine this degree of swelling is as follows.

A dry fiber having a weight of W_0 is immersed in water at 20°C for 16 hours and then dehydrated in a centrifugal separator operating at 3,000 rpm and a centrifugal acceleration of 1,000 *g* for 5 minutes. The resulting weight of this fiber is termed "W", and the ratio of W/W_0 provides the degree of swelling which must be less than 2.0, preferably less than 1.5. When the degree of swelling is higher then 2.0, the fiber has been insufficiently cured with the result that its tensile strength and flexibility may be too small to provide a fiber having the desired properties.

In addition to having a degree of water-swelling of less than 2.0, as above pointed out, the birefringence (Δn) thereof is less than 0.02. Generally, the value of Δn of a polymeric fiber corresponds to the degree of extent of orientation of the polymer constituting the fiber, and it may generally be observed that as the drawing or stretching ratio (during the manufacture of the fiber) increases, the larger the values of Δn and tenacity of the fiber will become. This is because the higher the drawing ratio, the higher the degree of orientation generally.

Contrary to these facts, with the fiber of this process, although the value of Δn is small, below 0.02, the fiber has a tenacity of from 1.6 to 5.0 g/den and sufficient flexibility. In addition, the fiber has excellent flame retardant properties and is thermally infusible. The fiber is especially superior compared to prior art fibers with regard to the latter property, which may be clearly exhibited by the degree of shrinkage of the fiber at high temperature. The degree of shrinkage is the observed value of shrinkage, represented in percentages based on its original length, of a sample fiber 20 cm in length after being exposed to an atmosphere of air at 300°C for 10 minutes.

Although almost all organic synthetic fibers are completely decomposed (i.e., carbonized) at 300°C or exhibit a shrinkage of more than 50%, the melamine-aldehyde resin fibers exhibit a degree of shrinkage of smaller than 10%, thereby indicating a superior infusibility. Unlike most organic synthetic fibers, these fibers, in addition to exhibiting a superior infusibility, retain their original fibrous state when exposed to fire (in addition to being nonflammable) and do not gradually decompose to carbon and, in addition, generate only a very small amount of smoke when exposed to heat while maintaining a considerable part of their original strength.

Example: An aqueous solution of N-methylol melamine having a 70% resin concentration was prepared by dissolving a commercially available N-methylol melamine (Sumirez Resin 607; the resin having a degree of methylolation of 3.0) in water at 60°C with gentle agitation for 1 hour. This solution was aged at 25°C for a period of 4 days until the viscosity thereof was 450 poises, determined as above indicated.

Reaction spinning of this solution was performed (as explained above) by spinning the solution through a nozzle into an atmosphere of air maintained at a temperature of 210°C. The holes in the nozzle had a diameter of 0.25 mm. A fiber was obtained which was 100% melamine-formaldehyde resin. Its whiteness value was 0.85, it had a denier of 16.5, a tenacity of 2.3 g/den, an elongation of 11% and a degree of swelling (in water as discussed above) of 1.22.

The value of Δn was 0.002, and it was observed by x-ray analysis that the fiber was completely amorphous. After testing, it was determined that this fiber was not combustible and, upon being contacted with a flame, a gentle glowing was observed, and only an extremely small amount of smoke was generated. Upon removal of the flame, the glow ceased instantly with no smolder being observed. No tendency to fuse was observed, and the fiber showed a tendency to decompose gradually only under strong heating to a temperature of 500°C. The thermal shrinkage of the fiber at 300°C in air was 6%, and the dyeability of the fiber was satisfactory. Specifically, the absorption of an acid dye in a dyeing bath at 98°C was 100%. The light fastness of the fiber dyed with the dye was observed as grade 6, measured according to JIS-L0843-71.

The moisture absorption of the fiber at 20°C and 65% relative humidity was 6.2%; a value which is near to that of cotton. On the other hand, the fiber exhibited no shrinkage when placed in hot water at 120°C.

Protein Fibers

J. Visser, R.H.J. Oosthoek, J.W. Groeneweg and H. Dijkstra; U.S. Patent 4,118,520; October 3, 1978; assigned to Thomas J. Lipton, Inc. disclose fibers containing casein and at least one heat settable protein which are prepared by dry spinning an aqueous protein mixture containing calcium ions and phosphate ions at a temperature below the gelling point of the protein mixture and drying the fibers obtained.

In the case where soy protein is used as the heat settable protein, a spinning temperature ranging from room temperature up to about 85°C, and preferably between 40° to 70°C, can suitably be applied.

The dry spinning operation is effected by pumping the spinning dope and pressing it through spinneret orifices of 0.05 to 2.00 mm diameter. The continuous filaments of mixed protein emerging from the spinneret are dried and wound or transported by a reel.

In the case where thin fibers are required, the continuous filaments emerging from the spinneret are conducted through a conditioning zone, preferably air, with a wet bulb temperature which is equal to or lower than the spinning temperature, preferably at least 5°C below the spinning temperature. The drying temperature applied for a relatively quick drying operation is from 100° to 120°C. Drying to a moisture content of less than 11%, and preferably from 0 to 9%, led to filaments or fibers of reasonably good stability.

Before using these fibers in the preparation of meat analogues, the fibers are rehydrated in water at a temperature above the heat setting temperature of the protein, and preferably above 90°C. The stability to boiling water and to sodium chloride environments was increased by including in the rehydrating medium 3 to 10% calcium chloride or by using ordinary tap water, which, in some cases, contains a sufficient amount of calcium ions.

Example 1: A spinning mixture was prepared starting from the following ingredients: 160 g of a soy protein isolate (50% protein) obtained by extracting defatted soybean meal with water and lowering the pH to the isoelectric point to precipitate the protein; 320 g calcium caseinate which contains 0.5 mmol calcium ions per gram of casein; and 520 g of an aqueous solution containing 8.8 g $NaH_2PO_4 \cdot H_2O$. The pH of the mixture was 5.9.

The mixture was intensively mixed, transferred to a sigma kneader, heated to 60°C and deaerated under reduced pressure.

The spinning mixture obtained was brought into the reservoir of a dry spinning device which had been brought to 60°C. Via a gear pump the mass was pressed through a spinneret with orifices of 0.5 mm diameter into a space with conditioned air. Here the wet bulb temperature was about 55°C. The emerging continuous filaments passed a drying zone of about 100°C where they were dried to a moisture content of about 20% and were taken up by a transporting reel and allowed to dry further to a moisture content of 11%.

Stretch ratios of about 40 were reached and fibers of 100 μm diameter were obtained. The fibers were rehydrated before use in boiling water containing 4% calcium chloride.

Example 2: *Preparation of a Chicken Analogue* – A bundle of rehydrated casein/soya (4:1) fibers was cut into small pieces and mixed with a binder using a Hobart mixer.

40 parts of rehydrated fibers were mixed with 40 parts of an egg white binder consisting of 15% dried egg albumin, 2% flavor, 10% fat and 73% water; and

20 parts of an aqueous meat mix containing 10% raw chicken meat, salt and spices. The mix obtained was formed in the shape of a chicken breast and was heat set during 15 minutes at 85°C. After heat setting the product was dipped in batter and stored in deep frozen condition.

ASSOCIATED APPARATUS

Dry Spinning Pack Assembly

P. Klaver and J.G. Steenhof de Jong; U.S. Patent 4,123,208; October 31, 1978; assigned to E.I. Du Pont de Nemours and Company provide an improved apparatus for dry spinning of filaments adapted to minimize the occurrence of oversized filaments during spinning.

Referring now to Figure 1.2a, the embodiment chosen for purposes of illustration includes a housing **10** which forms a chamber **12** supplied with a heated inert gas through inlet conduit **14** which is formed in the upper portion of the sidewall **11** of the housing. The heated gas is uniformly distributed by a screen **13** attached to sidewall **11** below the inlet conduit. A spinning pack **16** is positioned centrally within the housing and attached to a fixed inlet pipe **18** through flange connector **20** adjoining the spinning pack and the pipe. The inlet pipe connects a source of spinning solution (not shown) to the spinning pack.

A circular baffle **15** extends from the top portion of the housing to surround the upper portion of the spinning pack. The baffle directs the gas flow as indicated by the arrows downwardly in the chamber and keeps it from flowing directly onto the upper portion of the spinning pack. The spinning pack includes as component parts a distribution plate **22** sandwiched between spinneret **24** and flange connector **20** and held together by clamp ring **17** which is the peripheral surface of the pack. The spinneret which is the bottom surface of the spinning pack has a plurality of orifices **26** for extruding filaments **28** from the spinning solution supplied to the pack.

As best seen in Figure 1.2b, the orifices **26** through which filaments are extruded are arranged in concentric rings surrounding a central portion **29** of the spinneret. Within the central portion is a circle of holes **30** which lead to passages **32** that extend through the peripheral surface **17** of the spinning pack which is exposed to the gas in chamber **12**.

The structure of spinning pack **16** which embodies the improvement is best understood by referring to Figure 1.2b-e. The circular distribution plate **22** as viewed from the solution inlet is seen to include a recessed cavity **23** in its upper surface and an annular cavity **25** in its lower surface connected by passages **27** which are arranged in a circular pattern from the outside edge of cavity **23** to the outside edge of cavity **25**. Interspersed in a horizontal radial fashion between passages **27** in the distribution plate are passages **32** which at one end are in communication with holes **34** in clamp ring **17** around the periphery of the pack **16** and at the other end joined to holes **30** in the central portion **29** of the spinneret.

Figure 1.2: Dry Spinning Pack Assembly

a.

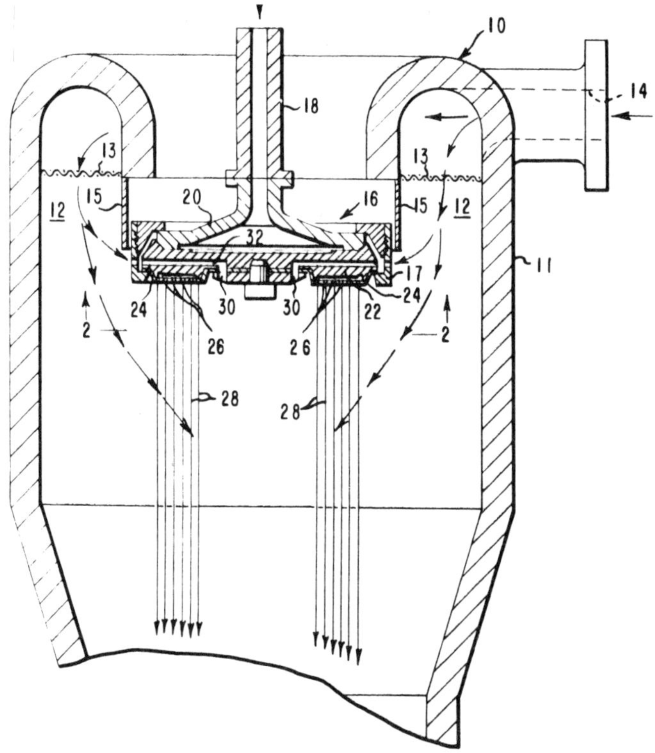

b.

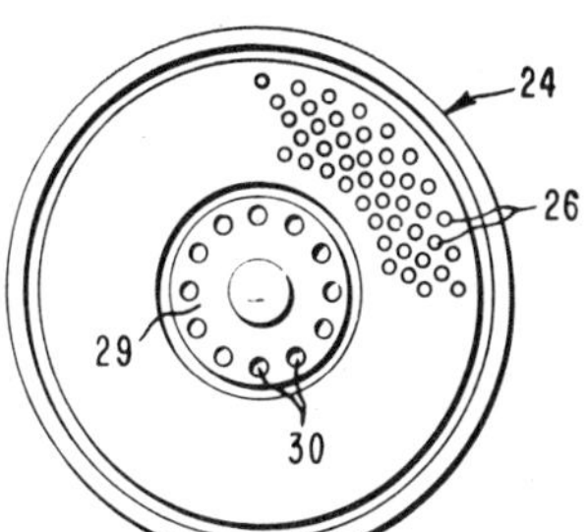

c.

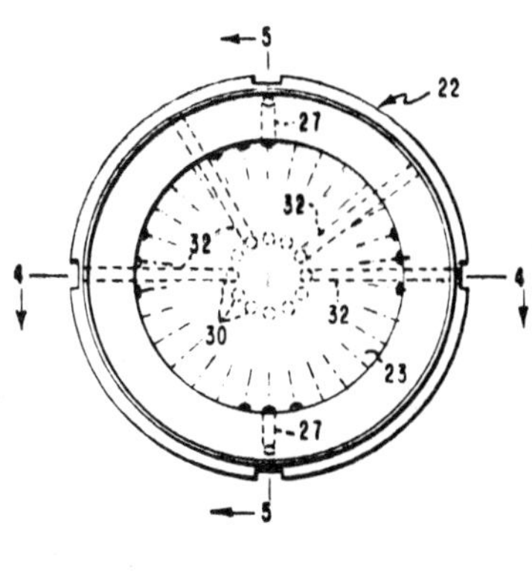

(continued)

Figure 1.2: (continued)

d.

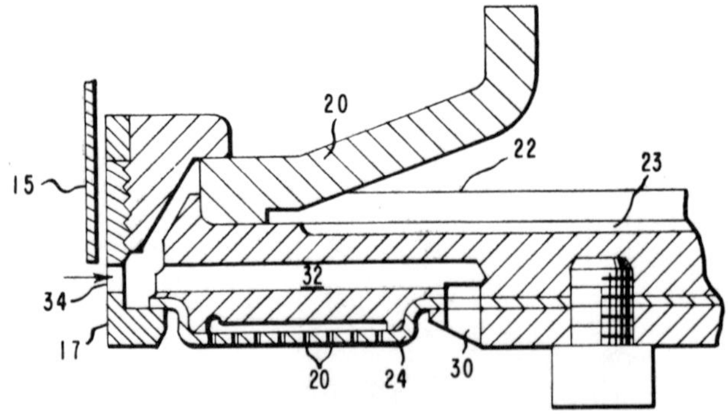

e.

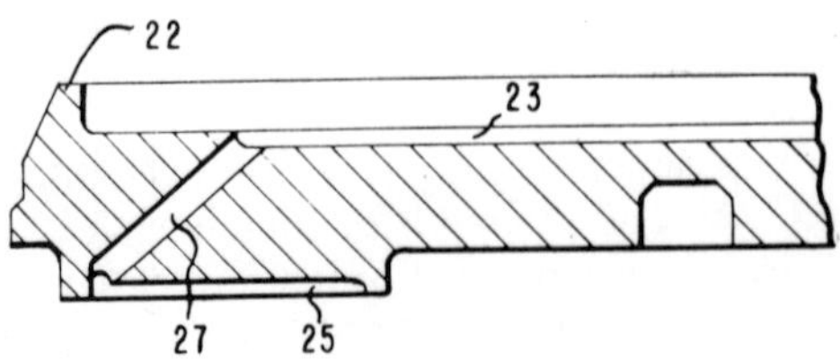

(a) Diagrammatic sectional elevation view showing the spinning pack mounted within a chamber supplied with heated gas;

(b) Bottom view of the spinning pack taken along line **2–2** of (a);

(c) Plan view of the distribution plate which forms a part of the spinning pack;

(d) Cross-sectional side elevation of a portion of the spinning pack taken through line **4–4** of (c); and

(e) Cross-sectional side elevation of a portion of the distribution plate of the spinning pack taken along the line **5–5** of (c).

Source: U.S. Patent 4,123,208

In operation, a spinning solution is metered at a constant rate through inlet conduit **18** into spinning pack **16** where it passes through distribution plate **22** and is extruded from spinneret orifices **26** to form concentric rings of filaments. The constant downward motion of the large number of extruded filaments **28** creates a zone of reduced pressure in the central space within the concentric rings, and the resulting suction assists streams of hot gas in flowing through the passages **32** in the spinning pack from annular gas supply chamber **12** to the central portion **29** of the spinneret. Hot gas also flows in the normal manner through the rings of extruded filaments from the outer portion of the spinning pack to the central spade. However, because part of the gas flows directly to the central space through the passages **32** in the distribution plate **22**, a constant supply of fresh heated gas which contains no evaporated spinning solvent is provided to the innermost rings of extruded filaments.

Thus by distributing some of the hot gas from chamber **12** directly to the center of the spinneret through the passages **32** in the spinning pack a smaller amount of the hot gas flows through the outermost filaments, and the distribution of the hot gas is more uniform resulting in a spinning operation of increased continuity.

Static Mixing Apparatus

A process developed by *D. Brauner, H.J. Kaluza and E. Muschelknautz; U.S. Patent 4,134,954; January 16, 1979; assigned to Bayer AG, Germany* for the homogenization of the spinning fluid in spinning devices, packing material is arranged before the nozzle plate in the free spaces of a static mixing apparatus.

The effect of this is that, in addition to the locally limited fine distribution of the packing material, a desired coarse distribution is achieved by means of the fittings of the static mixing apparatus. It has surprisingly proved that, even for coarse mixing only, very small mixing sections are necessary so that the spinning device is also ideal with respect to maintaining temperature by virtue of its compact construction.

As shown in Figure 1.3a, the spinning device comprises a spinning pump **1**, from which a conduit **3** leads into a housing **4** through an upper frame plate **2**. A static mixing apparatus **5** is arranged within it, whose free spaces are filled with packing material **6**. Below the housing there is located a nozzle plate **7**, which is held by a lower frame plate **8**. Clamping bolts **9** serve to connect the upper frame plate with the lower frame plate. The static mixing apparatus consists of individual inserts **10** according to Figure 1.3b. Each insert is constructed from a connecting element from which in two planes crosspieces **11** extend in all four directions and between them slots **12** are arranged. A plurality of such inserts are interconnected so that in each case crosspieces and slots interpenetrate.

Example: The spinning device according to Figure 1.3 is used. The length of the mixing section is 60 mm; the internal housing diameter 15 mm. The static mixing device consists of pairs of plates with crosspieces and slots each 1.7 mm in width.

A spinning fluid having a viscosity of approximately 1,000 poises and a density of approximately 1 g/cm^3 is to be homogenized as well as possible in respect to residence time and temperature. If the abovedescribed device is used without packing material, then in a sample having a volume of 2 mm^3, a standard deviation of 7% is obtained. However, if the static mixing device is filled with a packing material consisting of sand having an average grain size of 400 to 500 μm, then surprisingly for a sample having a volume of 2 mm^3, a standard deviation of approximately 0.5% can be achieved. With this arrangement for a throughput of 0.3 kg/hr a pressure drop of approximately 165 bars occurs.

If a packing material layer 60 mm in height is used without a static mixing device, then a standard deviation of only between 20 and 40% is obtained.

Figure 1.3: Static Mixing Apparatus

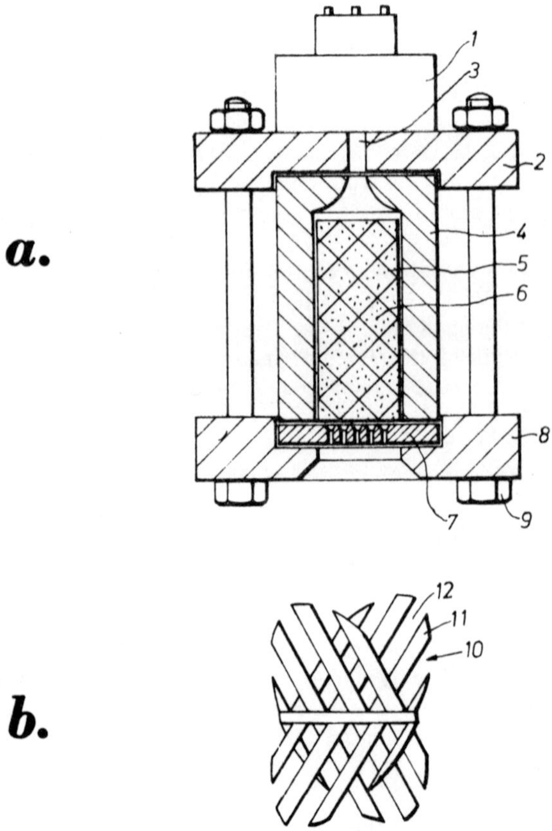

(a) Spinning device with a static mixing apparatus
(b) Single pair of intersecting plates.

Source: U.S. Patent 4,134,954

Wet Spinning

ACRYLICS AND MODACRYLICS

Reduction of Voids in Wet-Spun Acrylic Fibers

An improved process for the wet-spinning of acrylic polymers is disclosed by *L. Patron, P.L. Mazzuco, F. Veneto, E. Trevisan and E. Niero; U.S. Patent 4,001,485; January 4, 1977; assigned to Montefibre SpA, Italy*, wherein a spinning solution of an acrylic polymer in an organic solvent, such as dimethyl acetamide or dimethyl formamide, is extruded through a spinneret into a coagulating bath consisting essentially of water and an organic solvent in a ratio between 30:70 and 60:40 by weight; and wherein the filaments thus obtained are washed, stretched and heat treated; and wherein the coagulating bath contains from 0.03 to 3% by weight of acid corresponding to the organic solvent and has a pH value greater than 5 and preferably between 6.5 and 8. The spinning solvent is preferably dimethyl acetamide containing free acetic acid to an extent of less than 0.1% by weight, if necessary the acetic acid being added to the coagulating bath to bring its concentration up to values between 0.03 and 3% by weight. The pH value greater than 5 is obtained by the addition to the coagulating bath of ammonium hydroxide, an alkali metal hydroxide, or dimethylamine.

The process results in more compact filaments or threads without prejudicing the rheological behavior of the spinning solution at the moment of extrusion.

Example: The polymer used in this example was a copolymer consisting of 93% of acrylonitrile and 7% of vinyl acetate, having a specific viscosity of 0.140 and 35 μeq/g of end acid groups. This copolymer was obtained by polymerization in an aqueous suspension in the presence of a catalytic system consisting of potassium persulfate and sulfur dioxide.

265 g of this copolymer were dissolved in 735 g of dimethyl acetamide with a free acetic acid content of 0.3%, to which had previously been admixed 0.132 g of oxalic acid as a stabilizer and 1.3 g of titanium dioxide as a dulling agent.

The solution thus obtained was fed through a gear pump, at a flow rate of 15.7 cc/min, to a spinneret with 500 holes of 75 μ diameter each. The filaments extruded through the spinneret were coagulated in a coagulating bath containing a mixture consisting of 45% dimethyl acetamide (acetic acid content x 0.3%) and 55% of water, and maintained at a temperature of 45°C. The pH value of the coagulating bath was 4.2 and the concentration of acetic acid therein was 4.2.

The filaments coming out of the coagulating bath were then gathered on rollers and stretched to a stretching ratio of 5.54 in order to obtain filaments with a titer of 2 denier. Upon microscopic examination (at an enlargement of 15) of a sample drawn at the outlet of the coagulating bath and consisting of 20 filaments, there were statistically measured to be about 200 voids/mm of single filament.

In order to determine the spinnability of the spinning solution, the maximum winding-up rate before breakage of the filaments at the outlet of the coagulating bath was measured. The speed proved to be 11 m/min.

By repeating the example, but this time using a coagulating bath with a pH value of 7.5 obtained by the addition of ammonium hydroxide, the maximum winding-up speed remained the same, but the voids present in the filaments leaving the coagulating bath statistically proved to be 80/mm of single filament.

Acrylic Fibers Free from Delustering

According to *K. Kozuka, S. Kurioka, T. Yasumoto, S. Kobayashi, A. Kubota and N. Otoshi; U.S. Patent 4,002,809; January 11, 1977; assigned to Kanegafuchi Kagaku Kogyo KK, Japan*, a method of manufacturing acrylic fibers free from delustering in boiling water comprises the steps of forming a spinning solution by dissolving in a solvent of acetone, acetonitrile, dimethyl formamide and mixtures thereof:

(a) a copolymer of (1) 30 to 80% acrylonitrile and (2) 70 to 20% vinyl chloride or vinylidene chloride; or

(b) a tripolymer consisting of both (1) and (2), and (3) 3.0% or less by weight of ethylenically unsaturated monomer having hydrophilic groups.

One or more compounds selected from the group of homopolymers or copolymers of glycidyl methacrylate and specific organic tin compounds are added. The spinning solution is spun into a first bath consisting of 60% or less of the organic solvent-water.

The yarn is then led into a second bath consisting of 61 to 85% of the organic solvent-water and then in at least a third bath consisting of less than 60% of the organic solvent-water. The yarn is then dried, drawn out and annealed by ordinary methods.

Example: A spinning solution was prepared consisting of an acrylic resin (specific viscosity 0.161 at 30°C of a 2.0 g/l cyclohexanone solution) containing 32.0% acrylonitrile, 67.0% vinyl chloride and 1.0% sodium p-styrenesulfonate dissolved in acetone so as to be 24.0% in resin concentration. A homopolymer of glycidyl methacrylate (specific viscosity 0.091 at 30°C of a 4.0 g/l acetone solution) was used as an additive to the spinning solution at the amount of 3.0% to the acrylic resin.

The spinning solution thus prepared was spun by use of a nozzle having 300 holes, each 0.10 mm in diameter, in a spinning bath consisting of a first bath of a 10% aqueous solution of acetone, a second bath of a 75% aqueous solution of acetone, a third bath of a 50% aqueous solution of acetone, and a fourth bath of water. It was dried at 120°C, further thermally drawn out 300% and heat treated at 140°C for 5 minutes to obtain fibers.

The acrylic fibers thus obtained underwent little or no delustering even when treated in boiling water and were excellent also in heat and light resistance.

In contrast, ordinary acrylic fibers were produced from a spinning solution prepared from the above acrylic resin by adding 3.0% epoxy-based Epon 834 (glycidyl ether of bisphenol A having a molecular weight of 624) thereto. After ordinary spinning conditions consisting of a first bath of a 40% aqueous solution of acetone, a second bath of a 20% aqueous solution of acetone, and a third bath of water, the fibers obtained as well as ordinary acrylic fibers spun by the same spinning method without an additive were readily delustered by boiling-water treatment and were inferior in quality.

Improved Hot/Wet Properties

K. Takeya, H. Suzuki and N. Yamawaki; U.S. Patent 4,012,459; March 15, 1977; assigned to American Cyanamid Company disclose a process for preparing acrylic fibers of improved hot-wet properties and improved elastic recovery by finely dispersing in the spin dope prior to spinning a solution of a polycarbonate resin in a solvent immiscible with the spin dope. The acrylic fibers so produced contain elongated fibrils of the polycarbonate resin of 0.05 to 5.0 microns in diameter and up to several centimeters long.

Example: (All parts and percentages are by weight unless otherwise specified.) A fiber-forming spinning solution was prepared by dissolving 11 parts of a copolymer of 90% acrylonitrile and 10% methyl acrylate in 89 parts of an aqueous solution of 60% sodium thiocyanate. A solution of a polycarbonate obtained by interfacial polymerization of bisphenol A with phosgene having a number average molecular weight of 120,000 was prepared by dissolving 18 parts of the polycarbonate in 82 parts of tetrachloroethane.

The polycarbonate solution was dispersed in the spinning solution using a propeller-type agitator. The polycarbonate solution was added in sufficient amount to provide a ratio of 85:15 of acrylic polymer:polycarbonate and mixing was continued until a uniform dispersion was obtained. The resulting composition was deaerated and spun into an aqueous solution of sodium thiocyanate maintained at -3°C using a spinneret having 50 orifices each of a diameter of 0.1 mm.

The coagulated filaments were subjected to stretching at a stretch ratio of 2 in conjunction with cold-water washing and then subjected to stretching at a stretch ratio of 5 in boiling water. After drying to collapse the wet gel structure, the fiber was heat relaxed at 125°C to obtain an acrylic fiber free from devitrification and fibrillation. The fiber, upon microscopic examination, was found to contain numerous fibrils of polycarbonate discontinuously distributed therethrough, the fibrils having diameters ranging from 0.5 to 2 microns and a length-to-diameter ratio of 10 and greater. The following properties relative to fiber spun from the same acrylic polymer without the polycarbonate additive were obtained.

Properties, in 95°C Water	With Polycarbonate	Without Polycarbonate
Young's modulus, g/den	2.37	0.88
3% elongation elasticity, %	80.4	78.6

Flame-Retardant Acrylics

H. Suzuki, S. Sawanishi and T. Kusunose; U.S. Patent 4,091,066; May 23, 1978; assigned to Japan Exlan Company Limited, Japan describe a process for producing a flame-retardant acrylic fiber by wet-spinning a halogen-containing acrylic spinning solution containing antimony oxide particles dispersed therein. The acrylic fiber having a high degree of flame retardancy and improved fiber properties such as transparency, luster and brightness of colors of dyed products is produced in an industrially advantageous manner by regulating the water content in the hydrogel fiber obtained by hot-stretching after wet-spinning within a range of 50 to 130% based on the fiber-forming polymer.

Example: Particles of antimony oxide having an average particle diameter of 14 mμ were continuously introduced into a 50% aqueous sodium thiocyanate solution of an acrylonitrile copolymer (10% polymer concentration) consisting of 82.7% acrylonitrile, 9% vinylidene chloride, 8% methyl acrylate and 0.3% sodium methallylsulfonate so that the particles constitute 5.3% based on the weight of the copolymer.

The mixture was then blended by stirring with a Pipeline Homomixer, whereby an acrylic spinning solution containing finely and uniformly dispersed antimony oxide was obtained continuously and stably.

The spinning solution thus obtained was extruded into a coagulating bath of a 12% aqueous sodium thiocyanate solution through a spinneret having 5,028 orifices, each 0.14 mm in diameter. The coagulated filaments were stretched 2.5 times their length at room temperature and then washed with water. The filaments in a hydrogel state were stretched 4.8 times their length in boiling water. Subsequently, the filaments were subjected to drying for fiber compacting at a temperature of 125°C and to wet-heat relaxing treatment in saturated steam at 128°C to produce the final fiber.

The percent of water content within the fiber after hot stretching was measured and found to be 64% based on the fiber-forming polymer. The final fiber was substantially free from voids and, therefore, no substantial devitrification (loss of transparency) was observed. Moreover, it had a good luster and did not show any deterioration in color development of dyed products.

A comparative test was carried out in which the same antimony oxide particles as above were continuously introduced into a 50% aqueous sodium thiocyanate solution of an acrylonitrile copolymer (10% polymer concentration) consisting of 83.5% acrylonitrile, 7.3% methyl acrylate and 8.8% vinylidene chloride (thus containing no copolymer component having hydrophilic groups). Thereafter, by the same procedure and under the same conditions as above, wet-spinning, cold-stretching, hot-stretching, drying for fiber compacting, and relaxing heat treatment were carried out to produce the final fiber. The water content percent within the gel acrylic fiber by this procedure and fiber properties of the thus-obtained flame-retardant acrylic fiber are shown below.

	Example	Comparative Example
Water content with gel fiber, %	64	48
Degree of transparency, %	24	18
Δk/s ratio, %	4.5	11

The percent water content within the gel fiber (1), degree of transparency (2), and the degree of deterioration in color development (Δk/s ratio) (3), are calculated from the following measurements.

(1) Acrylic filaments in a gel state are charged into a centrifuge and dehydrated at 3,000 rpm for 2 min. Thereafter, the percent water remaining in the gel filaments is measured by the method of weight loss on drying. The value after deduction of a definite value of 10% (residual water which does not participate in the percent water content) from the abovementioned percent water remaining in the gel filaments is employed.

(2) Finally produced acrylic filaments are cut in 2-mm lengths. 105 mg of the cut filaments are dispersed in a solution adjusted to a refractive index of 1.506 obtained by mixing dimethyl phthalate (1.512), acetophenone (1.529) and ethanol (1.358). A definite quantity of the dispersion is placed in a measuring cell (1 x 1 x 5 cm). A beam of light having a wavelength of 420 mμ is projected at right angles against the surface of the cell. From the percent transmission observed, the degree of transparency is evaluated.

(3) The fiber for measurement is made to completely absorb 0.5% owf (on the dry weight of the fiber) of Aizen Catilon Blue K-2 GLH (a cationic dye). The dyed product is divided into two parts. One is dried at 60°C for 60 min and the other is dried at 120°C for 15 min. The reflection density (k/s value) of each of the two dried dyed products is measured with a Hunter-type reflectometer and the Δk/s ratio is calculated by the following formula

$$\Delta k/s\ \text{Ratio} = \frac{(k/s)_{120°C} - (k/s)_{60°C}}{5} \times 100\%$$

The value of the denominator in the above formula represents the reflection density, obtained according to the above procedure, of the ordinary acrylic fiber. The smaller the Δk/s ratio, the smaller is the degree of deterioration in color development of the final fiber.

As apparent from the results shown in the table on the preceding page, by changing only the content of the hydrophilic groups introduced into the fiber-forming polymer so as to regulate the percent water content in the hydrogel acrylic fiber within the range recommended in the process, a flame-retardant acrylic fiber was obtained which had a reduced loss of transparency and a good luster and was substantially free from deterioration in color development.

Modacryl Filaments with Permanent Brilliance and Transparence

E.-A. Albers, W. Fester and B. Sassenrath; U.S. Patent 4,056,516; November 1, 1977; and E.-A. Albers and B. Sassenrath; U.S. Patent 4,056,517; November 1, 1977; both assigned to Hoechst AG, Germany describe modacryl filaments and fibers made from copolymers insoluble in acetone of acrylonitrile and from 20 to 45 wt % of vinylidene chloride, containing from 0 to 1.5 wt % of unsaturated organic sulfonic acids or the salts thereof, and from 0 to 10 wt % of further copolymerizable compounds, which copolymers have a K value according to Fikentscher of from about 60 to 90, and a process for the manufacture thereof. These modacryl filaments and fibers are especially distinquished by their permanent brilliance and permanent transparence; these properties are maintained even on contact with water having a temperature above 80°C.

The stability to boiling is examined by two different methods. First a comparison is carried out by a microscopic 120-fold magnification of samples boiled in water and dried at 60°C and untreated samples. In the case of filaments or fibers stable to boiling, there is no difference before and after boiling. In the case of products not stable to boiling, the filaments or fibers treated with boiling water look grey-to-black in transmitted light. Second, the remission power at 4200 A of samples is compared, in which samples are dyed for 2 hours at boiling temperature with 5% Remacryl Blue RL (C.I. Basic Blue 22), one sample being dried at 60°C and the other at 125°C.

Fibers and filaments are considered stable to boiling if the relative difference in their remission power does not exceed 3%. In the case of products not stable to boiling, the sample dried at 125°C has less remission power than that dried at 60°C, since the opaqueness of the material is removed by the drying at 125°C, while it is maintained when the material is dried at 60°C.

Example 1: (In the example all percentages indicated are by weight.) A 23% spinning solution is prepared using a copolymer of 60% acrylonitrile and 40% vinylidene chloride having a K value of 86, prepared by suspension-precipitation polymerization, and a solvent mixture of 92% dimethyl formamide and 8% water.

The degased spinning solution is forced through a nozzle having 100 holes, of 70 μm diameter, at a rate of 3.3 cc/min into a first coagulation bath having a length of 60 cm and containing a mixture of 65% dimethyl formamide and 35% water having a temperature of 25°C. Without an intermediate conveying device, the spinning tow obtained is passed on to a second coagulation bath (also 60 cm long, containing 40% dimethyl formamide and 60% water of 25°C). The spinning tow is then drawn off the second coagulation bath by means of a draw-off roller at a speed of 4 m/min; passed on to a third bath containing 20% dimethyl formamide and 80% water of a temperature of 60°C; drawn three-times its length by means of a further draw-off roller; washed with water of 90°C; dried and finally drawn two-times its length at 120°C on a pair of rollers.

The filament so obtained is steamed for 15 minutes in steam of 110°C with free shrinkage, and it then has the following properties: titer, 363 dtex; tensile strength, 3.5 g/dtex; and, elongation at break, 25%.

The filaments obtained have a very good whiteness degree, high brilliance and complete transparence. At a 120-fold magnification under the microscope, a material boiled in water for 1 hour shows no difference as compared to an unboiled material. Both samples are transparent and free from cavities. A product dyed with Remacryl Blue RL at boiling temperature and dried at 60°C has a remission power at 4200 A of 33.3%; and a product dried at 125°C has a remission power of 33.1% under these conditions. The filaments obtained are therefore stable to boiling.

Example 2: As a comparative example, a spinning solution according to Example 1 is spun under the same conditions into a coagulation bath containing 65% dimethyl formamide and 35% water, but after 20 cm of immersion in the bath it is drawn off this coagulation bath and drawn in a drawing bath containing 65% dimethyl formamide and 35% water. All further conditions are as in Example 1.

The filament obtained has the following properties: titer, 367 dtex; tensile strength, 2.5 g/dtex; and, elongation at break, 19%. The filaments are transparent and brilliant; at a 120-fold magnification under the microscope no cavities are observed. On boiling in water, these filaments, however, become dull and opaque. Under the microscope, a filament so treated looks black in transmitted light. Samples dyed with Remacryl Blue RL and dried at 60°C and 125°C give remission values of 44.0% and 36.3%, respectively. The thus-obtained filaments are not stable to boiling.

CELLULOSE AND STARCH

Rayon Fibers Containing Starch

Rayon fibers are made by *F.R. Smith; U.S. Patent 4,144,079; March 13, 1979; assigned to Avtex Fibers Inc.* by spinning a viscose containing dissolved starch. Starch grains may be slurried in water, then made alkaline with NaOH to form a solution and then added to viscose.

Example 1: (All proportions are by weight unless otherwise indicated.) Alkaline starch solution was prepared by mixing a slurry of cornstarch grains in water with an 18% NaOH aqueous solution at about 20° to 25°C, to give a translucent viscous solution comprising 13% starch and 4% NaOH.

In a conventional viscose mixer, viscose containing 9.2% cellulose, 6.2% NaOH, 32% CS_2 based on the weight of the cellulose (boc), and about 0.5% TiO_2 boc was prepared by dissolving xanthated alkali cellulose in aqueous NaOH and mixing for about 2 hours. A quantity of the alkaline starch solution described was then added to the viscose in an amount such that the resulting viscose-starch blend contained 10% starch based on the weight of the cellulose.

Mixing was continued for 1 hour and the solution aged for about 24 hours at about 19°C (including a period of about 12 hours for vacuum deaeration). The solution was filtered both before deaeration and after and directly pumped (e.g., within a half-hour) through the spinneret. At the spinneret, the ball fall viscosity of the viscose-starch blend was about 90 and its salt test value was about 8. The solution was spun through 12,000 circular spinneret holes 0.0025 inch in diameter, into an aqueous spin bath containing 7 to 8% H_2SO_4, about 1.5% $ZnSO_4$, and about 21% Na_2SO_4 at 55°C.

The tow formed in the spin bath was passed around a driven roll and then pulled by a second driven roll through a stretch bath containing 3% H_2SO_4 aqueous solution at about 90°C. The stretch bath is continuously replenished by spin bath carried into it by the tow and by additions of water from time to time. The exit speed, i.e., the speed at the surface of the second driven roll, was 60 m/min and the speed ratio of the first and second driven rolls was such that the tow was stretched about 60 to 75% in the stretch bath.

The length of travel of the tow in the spin bath was 0.5 m and in the stretch bath about 2 m. After leaving the driven roll, the tow dropped into a cutter and the resulting cut fibers dropped into flowing hot water (85° to 90°C) where relaxation and crimping occurred. The fibers were taken up as a blanket, washed with hot water and desulfurized with a conventional solution of sodium polysulfide, and rewashed. A conventional staple finish solution made from Red Oil was applied and the fibers were then dried in hot air, e.g., at about 90°C.

Forty-one samples, each comprising 10 single fibers, were tested for tensile properties. The averaged results are set forth below. Additional tests showed crimps ranging from 9.4 to 12.6 per inch, for an average of 10.95. The denier per filament of the fibers was about 1.5.

Property	Conditioned at 75°F, 57% RH	Wet
Tenacity, g/den	2.84	1.55
Elongation, %	19.15	23.83
Breaking energy, g-cm/cm/den	0.33	0.19

Example 2: This example illustrates the manufacture of a disposable diaper comprising a cover sheet of a nonwoven fabric made of the starch-containing fibers. A mass of the staple fibers (containing 10% starch boc) is opened by conventional means and is then fed to a standard carding machine, or its equivalent, to produce an oriented web of fibers (a card web). A plurality of such webs (e.g., three) is brought together, one on the other, to build up a composite highly porous web weighing about 5 to 8 g/m^2.

A binder is then applied to this composite web to anchor the fibers together. Typically, a latex is employed, such as a latex of an elastomer (e.g., an acrylic latex). A typical binder is Rhoplex HA-8. This is a 45% solids emulsion of a self-crosslinking acrylic copolymer based on ethyl acrylate and a minor amount of an acrylamide crosslinker, which may be crosslinked on drying under the action of a catalyst (such as ammonium chloride which may be incorporated in the aqueous latex before it is applied to the composite web).

The latex may be applied to the web by saturation, e.g., by passing the web through a bath of the latex and squeezing out excess latex, before drying under heat so as to apply some 10 to 50% of the solid binder (based on the weight of the fiber). Preferably, the amount of binder is approximately one-fourth of the total weight of the bonded web, i.e., about one-third of the weight of the fiber. The bonded web is used as the cover stock for making disposable diapers.

Continuous Process for Viscose Yarn

A.T. Serkov, G.A. Budnitsky, B.M. Sokolovsky and V.K. Lezhnev; U.S. Patent 4,045,532; August 30, 1977 describe a process of producing viscose yarn which comprises the steps of continuously feeding a viscose solution through spinnerets into a spinning bath for spinning yarn, with subsequent final regeneration thereof, washing, avivage treatment and drying. A cooling of the yarn during the final regeneration, to a temperature below the boiling point of carbon disulfide, with subsequent washing and avivage treatment of the cooled yarn, and drying, is conducted at a temperature sufficient for concurrent removal of water and carbon bisulfide from the yarn. The term "avivage treatment" as used herein means the treatment of fibers with aqueous solutions or emulsions of fattylike chemical reagents for reducing their brittleness and stiffness and imparting good passability to them in textile processing.

To produce viscose yarn, conventional equipment is used which is installed in an order defined by the production method of this process, as illustrated in the figure below.

Figure 2.1: Apparatus for Production of Viscose Yarn

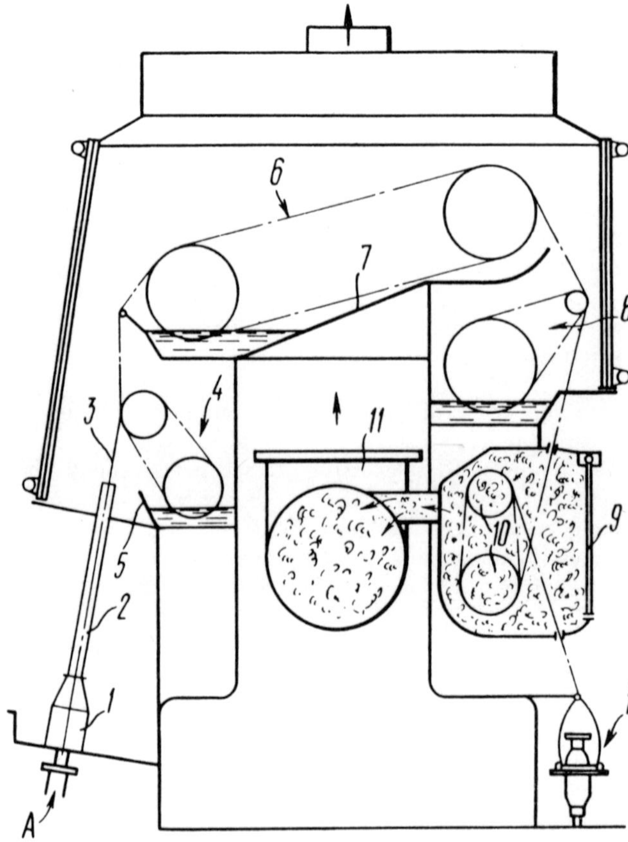

Source: U.S. Patent 4,045,532

The equipment comprises a spinneret **1**, a pipe **2** in which a yarn **3** is spun, a pair of rollers **4**, a bath **5**, a pair of rollers **6**, a bath **7**, an avivage treatment apparatus **8**, a chamber **9** for drying yarn having rolls **10** and an exhaust blower **11**.

A solution of viscose is fed in the direction indicated by arrow **A** through the spinneret into a spinning bath which is formed by the pipe which contains a spinning solution at a temperature below 60°C. In the spinning bath the yarn is spun and subsequently fed into the pipe and wound in two or three turns on the pair of rollers **4** where the yarn is completely regenerated and cooled, one of the rollers of the pair being partially immersed in the bath **5** containing a weak acid solution at a temperature of 15° to 20°C below the boiling point of carbon bisulfide which is formed in the yarn during its formation, carbon bisulfide having a boiling point of 46.3°C.

Subsequently, the cooled yarn is wound in a few turns on the pair of rollers **6**, one of which is partially immersed in bath **7** containing water at a temperature below the boiling point of carbon bisulfide. The yarn is washed on these rollers and is then fed in the pure state into an avivage treatment apparatus, wherein a brightening avivage treatment is conducted also at a temperature below the boiling point of carbon bisulfide. It should be noted that a major part of carbon bisulfide, in an amount of 65% of a specific total content thereof, remains in the yarn in the form of an emulsion due to its slight solubility, in spite of the fact that all of the soluble impurities on the yarn have already been washed off. It is assumed that, generally, 35% of the carbon bisulfide has been released from the yarn in the pipe. The yarn thus treated is then fed for drying into chamber **9**.

The yarn is dried by a known contact method on rolls **10** heated with steam with the surface temperature at 100°C, that is at a temperature above the boiling point of carbon bisulfide. During the drying, carbon bisulfide is removed from the yarn along with water and is discharged from the chamber by means of the exhaust blower for recovery, and the yarn is fed into a receiving winder **12** for winding into packages.

Example: A solution of viscose containing 7.5% by wt of α-cellulose and 6.3% alkali with a ball viscosity of 40 sec was fed through a spinneret die having 30 apertures of 0.08 mm diameter to spin a 13-tex yarn into a spinning bath with a specific gravity of 1.260 and a composition including 145 g/l of sulfuric acid, 18 g/l of zinc sulfate, the temperature of the spinning bath being 50°C. Upon leaving the spinning bath, the yarn was wound in two or three turns on rollers immersed in a weak acid solution at 30°C.

The yarn was then washed and subjected to avivage treatment, with the temperature of the treatment solution being 21°C. After the avivage treatment, the yarn contained 45% of that amount of carbon bisulfide specified for xanthation. The yarn, with the abovementioned content of carbon bisulfide, was fed into a chamber for drying.

During the drying, conducted at 100°C, carbon bisulfide was removed from the yarn along with water. The dried yarn was then wound into packages, and the resulting air-vapor mixture was removed for recovery.

Water-Insensitive Starch Fibers

Water-insensitive starch fibers having an amylopectin content of about 55 to 100% by weight are prepared by *H.R. Hernandez, D.S. Greif, A.N. Barna and*

D.S. Thornton; U.S. Patent 4,139,699; February 13, 1979; assigned to National Starch and Chemical Corporation by extruding a colloidal starch dispersion in threadlike form into a moving coagulating bath. Water-insoluble additives may be incorporated into the starch dispersion so as to produce a fiber containing the additive in encapsulated form.

Example: A slurry was prepared using an unmodified waxy maize starch containing essentially 100% amylopectin in water at a 15% solids level. The slurry was then placed in a boiling-water bath and cooked at 96°C with mechanical agitation for a period of 30 minutes. After cooking, the resulting starch dispersion was cooled to 22°C, and its Brookfield viscosity was found to be 5,000 cp at 20 rpm.

The starch dispersion was then extruded at 703.08 g/cm^2 pressure from a stainless steel spinneret containing 100 apertures, each of which had a diameter of 204.2 μ. The dispersion was extruded at an angle of approximately 90° into an agitated aqueous coagulating bath consisting of a 44% by weight aqueous solution of ammonium sulfate maintained at room temperature. The extrusion process was continued for a period of 30 minutes and the resultant discontinuous fibers were agitated in the salt solution for an additional hour.

Thereafter, the fibers were recovered from the salt solution by collecting them on a 100-mesh stainless steel screen and washed free of salt with water. The fibers at this point may be introduced directly into a papermaking process or consolidated into wet mat form at approximately 50% solids.

Alternatively, the fibers may be reclaimed in dry form after recovery from the salt solution by introducing them into a solution of ethyl alcohol and mixing for a period of 10 minutes. The fibers may then be recovered from the alcohol solution by using screen filtration techniques and either air or oven dried.

The discontinuous fibrous products formed by these techniques were found to possess a cross-sectional diameter averaging approximately 100 μ and a length distribution between 500 and 3,000 μ. The procedure produced a satisfactory starch fiber product, i.e., the fibers were water-insensitive and, after drying, were readily redispersible in water while retaining their original structure and configuration.

POLYAMIDES AND OTHER NITROGEN-CONTAINING POLYMERS

Production of Arylamides with Recovery of Amide Solvent

K. Kouzai, Y. Tabe, E. Masunaga and K. Matsuda; U.S. Patent 4,073,837; February 14, 1978; assigned to Teijin Limited, Japan disclose a process for producing wholly aromatic polyamide fibers by extruding a spinning solution of a wholly aromatic polyamide in an amide solvent into an aqueous coagulating bath containing inorganic salts.

The process comprises feeding an aqueous coagulating liquid containing calcium chloride, having an amide solvent concentration of not more than 3% by weight, to the spinneret; flowing the coagulating liquid in the bath at an average velocity not more than 0.1 times the velocity of yarn take-up in the yarn advancing direction; discharging the coagulating liquid having an amide solvent concentration of not less than 4% by weight from the bath; treating the coagulating bath with an organic solvent consisting substantially of methylene chloride; and, extracting and recovering the amide solvent from the coagulating liquid. The process is illustrated in Figure 2.2, on the following page.

In Figure 2.2a, the reference number **1** represents a spinneret, **2** a coagulating bath, **3** filaments, **4** a take-up roller, **5** a feed inlet for the coagulating liquid, **6** a trough for receiving the coagulating liquid that has overflowed, **7** a coagulating liquid exhaust port, **8** and **13** reservoirs, **9** a cooling coil, **10, 12** and **14** pumps, **11** an extraction tower, and **15** a water washing bath.

A spinning solution obtained by dissolving a wholly aromatic polyamide in an amide solvent is extruded into the coagulating bath through the spinneret. The coagulated filaments are taken up at a certain take-up velocity by the take-up roller, washed with water in the water bath and then fed to the subsequent drawing step.

The coagulating liquid is fed from the feed inlet rearward of the spinneret, and flows within the coagulating bath in the direction shown by arrows in Figure 2.2b. It is then caused to overflow from the yarn take-up side of the coagulating bath and is received on the trough and thence the coagulating liquid is led into reservoir **8** through the exhaust port. It is cooled to about 50°C by the cooling coil through which water flows and supplied to the upper part of the extracting tower by pump **10**.

Methylene chloride is fed into the lower part of the extracting tower, and within this tower the amide solvent as the coagulating liquid is extracted and transferred into the methylene chloride. The coagulating liquid after extraction is taken out by pump **12** from the bottom of the extraction cell and received by reservoir **13**, from which the coagulating liquid is supplied to the coagulating bath by pump **14**.

The methylene chloride containing the amide solvent is taken out from the upper part of the extraction tower and led to an evaporation device where the methylene chloride is evaporated and separated from the amide solvent. The amide solvent is further distilled under reduced pressure to purify and recover it. The amide solvent is reused for the preparation of the spinning solution, and methylene chloride is reused as an extracting agent.

According to this process, wholly aromatic polyamide fibers can be produced in very good spinning condition, and the formation of agglutinized filaments or abnormal filaments is reduced. Furthermore, the solvent can be recovered economically and efficiently. The process makes it possible to produce filaments in various denier sizes, from fine to heavy denier.

Figure 2.2: Arylamide Process

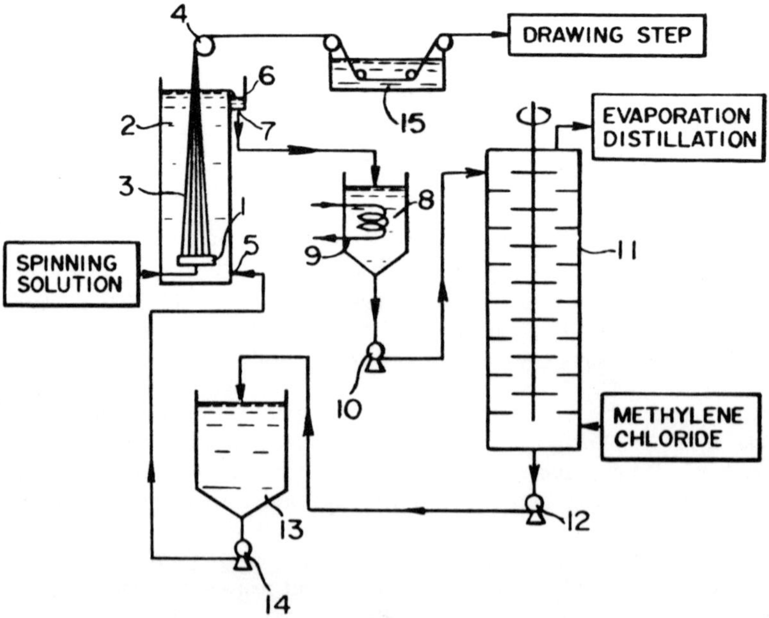

a.

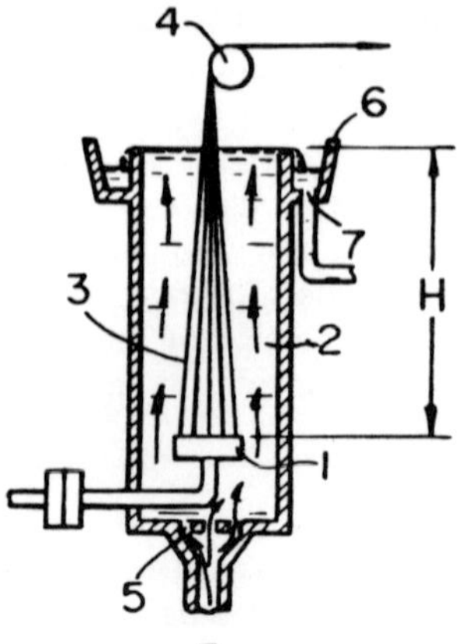

b.

(a) Flow diagram
(b) Vertical-type coagulating bath

Source: U.S. Patent 4,073,837

The various properties of the filaments were measured in accordance with the following methods.

Rate of tenacity fluctuation: This shows the extent of the dispersion in the measured values of a single filament tenacity, and is determined by the following equation:

$$\text{Rate of tenacity fluctuation} = \frac{\text{Dispersion } [\sigma] \text{ of single filament tenacity}}{\text{Average value } [\bar{x}] \text{ of single filament tenacity}} \times 100\%$$

The tenacity of the single filament was measured with respect to 20 samples.

Amount of abnormal fibers: 50 g of raw staples were spread by hand and foreign matters such as dope scum, undrawn filaments or agglutinized filaments were taken out and their weight was measured. The amount obtained is expressed by percent by weight.

Example 1: 20.5 parts of a wholly aromatic polyamide having an inherent viscosity of 1.9 and obtained by copolymerization of m-phenylene diamine, isophthaloyl chloride and terephthaloyl chloride (copolymerization mol ratio 98:2) was dissolved in 79.5 parts of N-methyl-2-pyrrolidone and the solution was defoamed to prepare a spinning solution held at 70°C. The solution was extruded at the rate of 234 cc/min into a coagulating bath through a spinneret having 6,000 holes, each with a diameter of 0.12 mm, secured to the bottom of the coagulating bath substantially horizontal thereto, as shown in Figure 2.2b. It was coagulated in the filament form in the coagulating bath and then taken up at a rate of 2.1 m/min.

The coagulating bath contained a coagulating liquid held at 95°C and consisting of an aqueous solution containing N-methyl-2-pyrrolidone in a concentration of 1.7% and calcium chloride in a concentration of 42%. The coagulating liquid was allowed to flow in the coagulating bath at an average velocity of 0.17 m/min (the yarn take-up velocity multiplied by 0.08), and the coagulating liquid having an N-methyl-2-pyrrolidone concentration of 6.2% and a specific gravity at 60°C of 1.375 to 1.380, was discharged from the upper part of the coagulating bath.

The filaments which had been taken up from the coagulating liquid were washed thoroughly with water to remove the solvent, drawn to 2.75 times the original length in boiling water held at 95°C, dried, drawn to 1.4 times their length on a hot plate held at 310°C, then heat-treated. The filaments were then crimped using a stuffer crimper, and the crimped tow was cut to a fiber length of 76 mm to form staple fibers.

The spinning and drawing conditions were good, and there was no such trouble as agglutinization, opacification or breaking of the filaments. The number of wrapping during the spinning process and the properties of the fibers obtained were as follows:

Number of wrapping per hour	0-0.01
Denier size	9.97
Tenacity, g/den	4.95
Elongation, %	45
Rate of tenacity fluctuation, %	9.1
Amount of abnormal fibers, % by wt	1×10^{-4}

The coagulating liquid which was taken out from the coagulating bath was supplied to the reservoir as shown in Figure 2.2a, and after its temperature was adjusted to 50°C, it was fed to an extraction tower where it was treated with 1.3 times its volume of methylene chloride to extract and recover the N-methyl-2-pyrrolidone present in the coagulating liquid. By this treatment, a calcium chloride aqueous solution having an N-methyl-2-pyrrolidone concentration of 1.7% was continuously taken out from the extraction tower, and recycled to the coagulating bath.

Example 2: As a comparative example, Example 1 was repeated except that the amount of the coagulating liquid fed was decreased, and the average velocity of the coagulating liquid was adjusted to 0.03 times the yarn take-up velocity. The concentration of N-methyl-2-pyrrolidone in the coagulating liquid to be fed to the coagulating bath gradually increased, and the spinnability of the spinning solution became poor. At the same time, the N-methyl-2-pyrrolidone concentration of the coagulating liquid increased and the specific gravity of the coagulating liquid decreased to below 1.35. Finally, it became impossible to extract and recover N-methyl-2-pyrrolidone using methylene chloride in a countercurrent extracting tower.

Air Gage Arylamide Spinning Process

According to *G.B. Lewis; U.S. Patent 4,078,034; March 7, 1978; assigned to E.I. DuPont de Nemours and Company*, the strength of aromatic polyamide filaments spun at high speed downwardly through a noncoagulating fluid into a liquid coagulating bath and subsequently through a spin tube containing the freshly spun filaments and coagulating liquid is improved when the filaments are abruptly deflected away from the walls of the spin tube a distance of 0.025 to 0.15 times the inside diameter of the spin tube.

Referring to the figure on the following page, in Figure 2.3a, spinning solution is supplied through transfer line **1** to spinning block **2** then to spinneret **3** located above vessel **4** open to the atmosphere **7** containing a liquid coagulating bath **5** supplied from pipe **6**. Extruded filaments **9** pass from the spinneret through the air gap **17** into the coagulating liquid and into the spin tube **8**, downward through the relatively larger diameter tube **11** into the atmosphere to change-of-direction means **10** from where the filaments are pulled by driven roll **12** and its associated separator roll **13** and are washed and neutralized while on the rolls. The bulk of liquid **19** falls from the filaments at means **10**. The yarn then passes to heated rolls **14** and **15** for drying and to windup on roll **16**.

Figure 2.3: Air Gage Arylamide Spinning Process

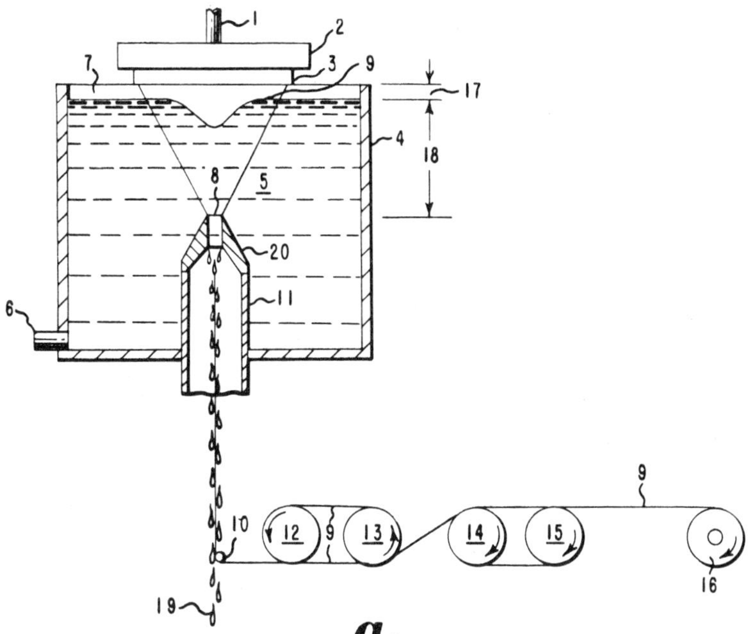

a.

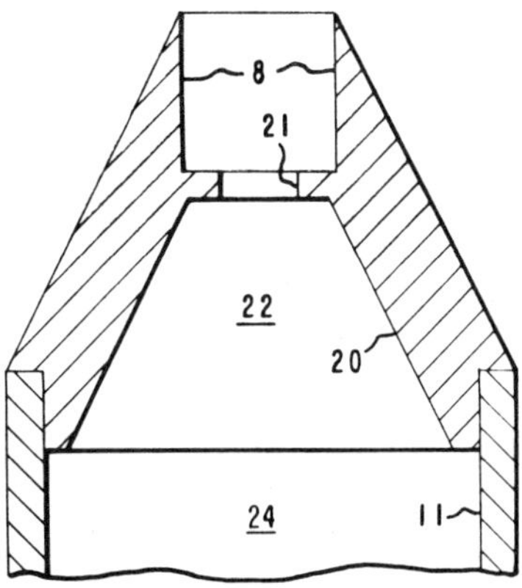

b.

(a) Spinning scheme
(b) Spin tube

Source: U.S. Patent 4,078,034

The vertical distance **17** from the face of the spinneret to the upper level of the coagulating liquid is termed the "air-gap." The vertical distance **18** from the top of the spin tube to the upper surface of the coagulating liquid is termed the "stagnant layer."

Due to the large horizontal separation of filaments from the extreme ends of the spinneret and the relatively small diameter of the spin tube (the ratio may be 6:1 or larger), the outermost filaments enter the spin tube at a converging angle that may be 15° to 30° or more from the vertical. This convergence tends to drag the filaments along the walls of the tube.

Spin tube **8** is filled with liquid due to the pumping action of the filaments and gravity. The succeeding zones of larger cross section beyond the spin tube such as are afforded by the spin tube holder **20** and tube **11** are not filled with liquid and do not significantly increase the tension on the filaments beyond that obtained by passage through the spin tube. Tube **11** functions primarily as a support for the spin tube holder and should have a diameter such that there is no contact between its inner wall and the falling liquid or descending filaments. The length of tube **11** is not critical and can extend to change-of-direction means or even beyond with the provision of a side exit. Alternately, with a shallower vessel, this tube is not required and the yarn may exit from the spin tube into zones **22** and **24** (see Figure 2.3b) and thence into the atmosphere, or directly into the atmosphere.

The apparatus is arranged and the process conducted so that the filaments pass through the spin tube in less than 0.03 sec (e.g., 22.9 cm tube at 457 m/min) and preferably less than 0.01 sec (e.g., 9.1 cm tube at 549 m/min) in order to minimize the development of tension. Means **10** is located so that the filaments are in contact with liquid for at least 0.04 sec before running over the means (e.g., 30.5 cm at 457 m/min).

An embodiment of a spin tube with yarn deflection means **21** located at the top of the spin tube is shown in Figure 2.3b. It is believed that the yarn deflection means operates by physical displacement of the yarn and by providing a turbulent liquid layer at the wall that displaces the yarn. The circular shelf **21** of Figure 2.3b presents a sharp edge perpendicular to the preceeding spin tube to provide maximum liquid turbulence in liquid at the edge of the tube. The shape of the lower face of the shelf is less critical due to the expanded diameter of that zone. If a shelf is used between the ends of a spin tube it should present a near-perpendicular edge on both sides of the shelf in order to promote turbulent wall flow both up- and downstream of the shelf.

Preferably, the yarn deflection means is located at the entrance of the spin tube. The edges leading into and away from the deflection means must be sharp and near perpendicular to the tube. The effectiveness of the deflection means is significantly reduced as the angle to the tube approaches 135° or more because the liquid stream lines at such angles will pull the yarn into contact with the tube wall.

It may be desirable to provide a multiplicity of yarn deflection means along the tube. The yarn deflection means should have an inside diameter that is no larger than about 0.95 mm of the inside diameter of the spin tube and can be as small as 0.70 mm or less. Preferably, the inside diameter of the deflection means will be no larger than 0.9 mm (and no less than 0.75 mm) of the spin tube inside diameter.

The minimum size of the inside diameter of the deflection means and the spin tube is governed by the size of the yarn bundle and associated liquid and the mechanical problems of "stringing up" the process. The inside diameter should be from 3 to 15, preferably 5 to 10, times the diameter of the yarn bundle and associated liquid. For example, the diameter of a 1,000-filament, 1,500-denier yarn with liquid is about 1 mm.

The face of the deflection means located at the entrance of the spin tube, or midway, should preferably be relatively shallow, e.g., 0.15 to 0.3 mm thick, whereas deflection means at the bottom of the tube where the yarn is less sensitive can be much thicker, e.g., 1 to 5 mm.

The process is useful at spinning speeds of 300 to 800 m/min or higher, preferably 500 to 700 m/min.

Procedures used in the tests cited in the example were as follows: Fiber properties were measured at 24°C and 55% relative humidity on yarns that had been conditioned under the test conditions for a minimum of 14 hours. The nominal 1,500-denier yarns of the example were given about 0.8 turn twist per centimeter (i.e., 1.1 twist multiplier) and broken with a 25.4 cm gauge length at 50% strain per minute. Deniers were obtained by weighing a known length of yarn and corrected to a finish-free basis containing 4.5% moisture.

Example: Using an arrangement similar to that of Figure 2.3a, a spinning solution containing 18.8% by wt poly(p-phenylene terephthalamide) (PPD-%) having an inherent viscosity of 5.2 (H_2SO_4) is extruded at 85°C from a 4.8 cm diameter spinneret containing 1,000 0.064-mm diameter orifices arranged within a 4.2-cm diameter circle through a 9-mm air gap into 10°C aqueous coagulating liquid containing 3 to 4% by wt H_2SO_4. The stagnant layer is 19-mm thick. The filaments are pulled through a spin tube, then through a larger tube having an inside diameter of 24 mm, through the air and, after washing, neutralizing and drying, wound up at 608 m/min.

Part A – The above general procedure is followed using a straight spin tube known in the art, with a square entrance edge. The spin tube has an inside diameter of 9.5 mm and is 11.2 mm long.

Part B – The above general procedure is followed using a spin tube similar to the spin tube of Figure 2.3b. The spin tube has an inside diameter of 12.7 mm and is 7.6 mm long before a 2.5-mm thick square-edge circular shelf having an inside diameter of 9.5 mm located at the exit of the tube. The shelf provides a deflection of 1.6 mm (0.126 times the inside diameter of the spin tube).

Part C – The procedure of Part B is followed except the length of the spin tube is 15.2 mm before the shelf.

There is no statistically significant difference between the tenacities and breaking elongations of the yarns of Part B and Part C, for which tenacity equals 20.8 g/den and elongation equals 3.4%. By comparison, the yarn of Part A has tenacity equal to 18.8 g/den and elongation of 3.3%.

Reduced Salt Content in Arylamide Fibers

According to *G.B. Lewis and S.D. Moore; U.S. Patent 4,070,431; January 24, 1978; assigned to E.I. DuPont de Nemours and Company,* the salt content of neutralized aromatic polyamide fibers spun from an acid solution downwardly into a spin tube containing the fibers and a coagulating liquid is reduced when a gas is injected into the spin tube 0.25 to 20 cm downstream from the entrance of the spin tube.

In Figure 2.4a, on the following page, acid aromatic polyamide spinning dope is pumped through transfer line **1** to spinning block **2** then through spinneret **3** located above vessel **4** containing a liquid coagulating bath **5** supplied through pipe **7** and a layer of air **18** supplied by pipe **6**. The coagulating bath level and pressure of the air layer are controlled by pipe **22** and valve **23**.

A spin tube **8** immersed in the bath extends through the bottom of the vessel and connects with extraction and neutralization apparatus **10**, which is connected to further extraction and neutralization apparatus **11**. Extruded filaments **17** coming from the spinneret pass through the air layer and then through the coagulating liquid which is flowing from the vessel through the spin tube along with the filaments. The top of the spin tube may be fitted with a deflector **9** which aids in the circulation of the coagulating liquid in the vessel.

Air or another gas under pressure is inserted through pipe **21** into a manifold **24** (see Figure 2.4b) through peripheral openings **25** at an appropriate distance from the entrance of the spin tube. The filaments, coagulating liquid and air pass into extraction and neutralization apparatus **10** wherein entrained coagulating liquid may be removed by an air jet before treatment of the filaments with dilute caustic. The filaments may be further processed by passing them through a tube to further extraction and neutralization apparatus **11** where entrained dilute caustic wash may be removed by air jets before the filaments are wound on driven roll **12** and associated separator roll **13** and sprayed with very dilute caustic solution. The filaments then pass around heated rolls **14** and **15** for drying before being wound up on bobbin **16**.

Figure 2.4b shows the top of the preferred spin tube for use in the process along with the apparatus shown in Figure 2.4a. Other arrangements for the injection of a gas into the spin tube may also be used. Significant salt reduction has been obtained with tubes having initial tube lengths of between 0.25 and 20 cm before air injection.

Figure 2.4: Gas Injection Spinning System

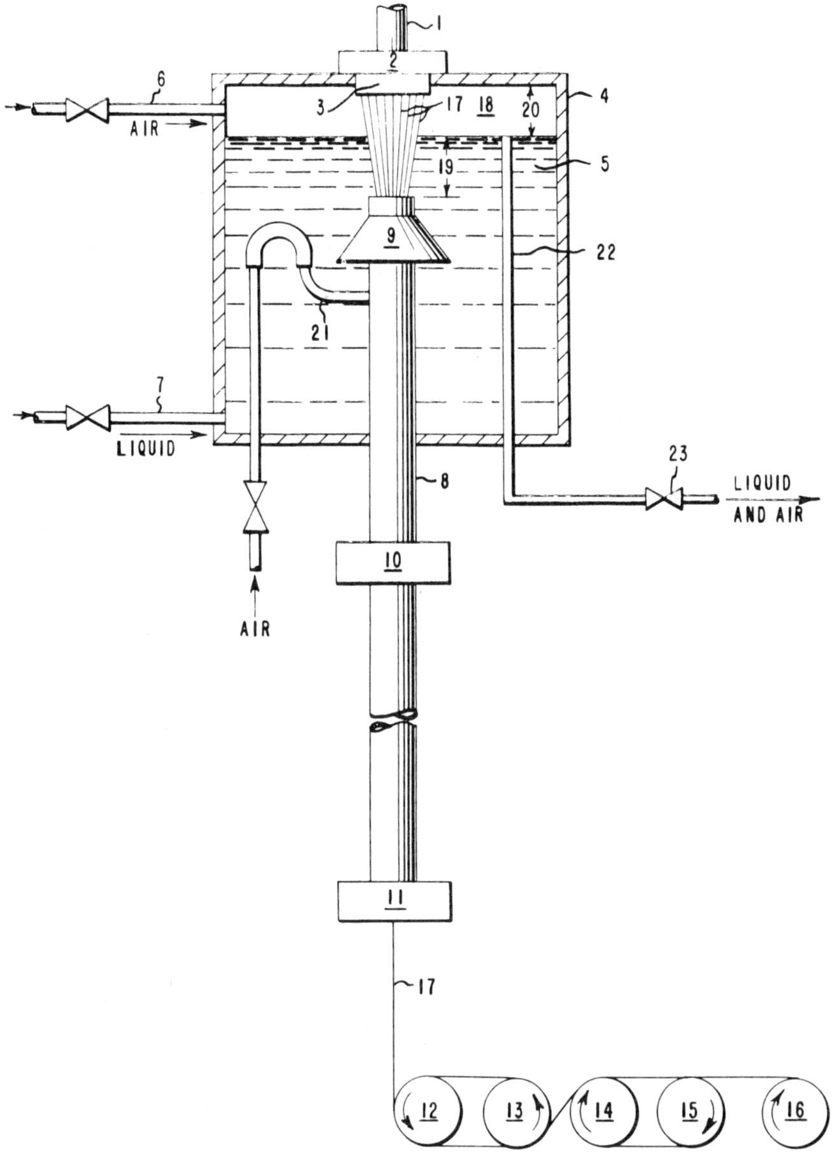

a.

(continued)

Figure 2.4: (continued)

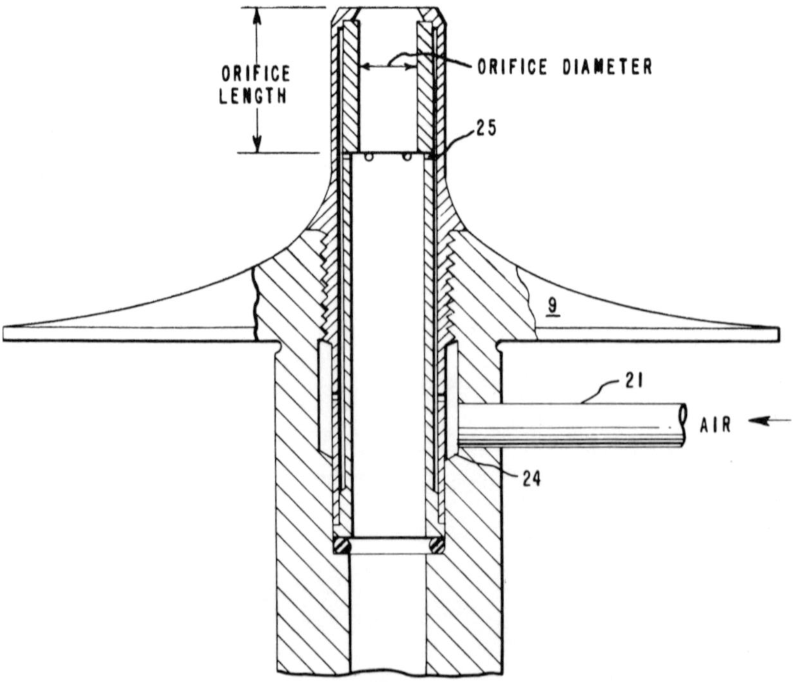

b.

(a) Spinning apparatus
(b) Top of spin tube **8** as shown in Figure 2.4a

Source: U.S. Patent 4,070,431

The salt (i.e., Na_2SO_4) content of the dried yarn is obtained by ashing a dry sample, dissolving the ashes in HCl, diluting volumetrically and measuring Na^+ in an atomic absorbing spectrophotometer. The weight percent Na_2SO_4 present in the yarn, as calculated from the total Na^+, is corrected for any Na^+ present in the initial polymer.

Fiber properties are measured at 24°C and 55% relative humidity on yarns that have been conditioned at 24°C and 55% relative humidity for a minimum of 14 hours. The nominal 1,500-denier yarns of the following examples are given about 0.8 turn twist/cm (i.e., 1.1 twist multiplier) and broken with a 25.4 cm gage length at 50% strain rate per minute. Deniers are obtained by weighing a known length of yarn and corrected to a finish-free basis containing 4.5% moisture.

In the examples, the following general procedure was used: A 19.3% by weight solution of poly(p-phenylene terephthalamide) having an inherent viscosity of 5.4 (H_2SO_4) in 100% sulfuric acid is extruded at 75°C from a spinneret containing 1,000 holes through a layer of air into the coagulating liquid (25°C water) using the apparatus of Figure 2.4a. The coagulated filaments are carried through the water in the spin tube for about 0.3 sec (2.7 m) before the liquid is removed by air jets. The bundle of filaments is then impinged with streams of a dilute (1%) aqueous solution of sodium hydroxide and the yarn advanced in contact with the sodium hydroxide solution in a tube for about 0.73 sec (6.7 m) before the liquid is removed by air jets. The yarn is then sprayed with a very dilute aqueous solution of sodium hydroxide (0.05% by wt) while passing from a driven feed roll to an idler roll with multiple wraps. The yarn (nominal 1,500 denier) is then passed over drying rolls into a package at 549 m/min.

Example 1: *Control* – The above procedure is followed using a conventional spin tube (inside diameter 8.6 mm) without gas with an air gap of 6 mm as measured vertically from the spinneret face to the upper level of the quench liquid (before the vortex) and a stagnant layer of 38 mm. Stagnant layer is the vertical distance from the top of the spin tube to the upper surface of the quench liquid, before the vortex. The coagulating liquid flows from the bath through the tube at a rate of 23 to 27 l/min.

Example 2: The above general procedure is followed using the same spinning solution with three different spin tubes of Figure 2.4b having an entrance inside diameter of 6.86 mm, an inside diameter of the following part of the tube of 7.11 mm and lengths of 2.5, 15, and 28 mm, respectively, before expansion to a continuing tube inside diameter of 8.64 mm immediately followed by the air injection holes.

Air from a 580 kPa (85 psi absolute) source and a valve and rotameter is delivered at the rate of 11.3 l/min (20°C and 101 kPa) through the air injection holes. A coagulating liquid flow of 13 l/min of water was used with an air gap of 6 mm, a stagnant layer of 25 mm, and a pressure of 120 kPa in the air gap. The air injected amounts to 0.87 v/v of coagulating liquid (water and air at 20°C and 101 kPa).

Results of the above examples are shown below.

	Example 1	Example 2
Tenacity, gpd	21.0	21.0
Elongation, %	3.8	4.0
Initial modulus, gpd	490	470
Salt content, %	2.0	1.6

Neutralization of Polyamide Spin Dope

J. Zimmerman; U.S. Patent 4,048,279; September 13, 1977; assigned to E.I. DuPont de Nemours and Company discloses an improved neutralization process for use in the high-speed spinning of polyamide spin dope containing an inorganic acid.

A spinning solution of 19.5% of poly(p-phenylene terephthalamide), intrinsic viscosity 5.8, in 100.1% sulfuric acid is extruded from a spinneret containing 1,000 holes through an 0.25-inch (6.4-mm) layer of air into water. The coagulated filaments are carried through the water for about 0.3 sec before the liquid is removed by air jets. The yarn of filaments (containing about 1% H_2SO_4 on a dry yarn basis) is then impinged with streams of dilute aqueous caustic (NaOH) at 50°C and the yarn advanced in contact with the caustic in a tube for about 0.8 sec before the liquid is removed by air jets. The yarn is then sprayed with water or very dilute aqueous caustic (50°C) for a final wash while passing from a driven feed roll to an idler roll with 12 wraps for a total residence time of about 1 sec. The yarn is then passed to drying rolls and to a package.

Example 1: In this comparative example the above procedure is followed with the dilute aqueous caustic concentration ranging from 0.6 to 0.7% and the flow rate of the water spray is varied. Properties of the dried 1,500-denier yarn collected at 600 yards per minute are given in the table below as items A through D.

Example 2: The above general procedure is followed using a dilute aqueous caustic solution of 0.6% and varying the flow rate of the final wash with very dilute caustic (0.02 to 0.05% NaOH). Properties of the dried 1,500-denier yarn collected at 500 yards per minute are given in the table as items E through H.

The acid/base analyses of dried yarns are obtained from other spins with about the same conditions.

The heat-aged strength retention (HASR) is determined as follows. Dry yarn is wound by hand around an aluminum plate. The plate is placed in an air-circulating oven at 240°C, heated for 3 hours at that temperature and cooled. The plate with yarn is conditioned for 2 hours at 75°F (23.9°C), 55% RH, and the yarn is removed from the plate. The heated yarn and similarly conditioned original yarn are each twisted to 3 turns per inch (per 2.5 cm) and broken with a 10-inch gauge on an Instron testing machine. HASR is calculated from the equation:

$$\frac{\text{Break strength heated yarn}}{\text{Break strength original yarn}} \times 100\%$$

It is noted that without the final wash (items A and E) a basic yarn is obtained with a low HASR value. High wash rates (items C and D) are found to yield an acidic yarn with low HASR values. It is seen that the use of water alone makes the process extremely sensitive to wash flow rate. The use of very dilute caustic gives an extended range of acceptable flow rates (items F and G) that affords an easier control of the process.

Similar yarn with an HASR of 82% is obtained using 0.4% caustic as the first neutralization followed by 0.03% caustic at a flow rate of 1.4 gallons per minute with drying and winding up at 600 yards per minute.

Item	Wash Liquid	Final Wash Rate (gal/min)	. . Dried Yarn Properties. . . HASR (%)	Acid or Base Content ($eq/10^6$ g)	
A	none	0	41	72	base
B	water	0.25	78	4.5	base
C	water	0.5	75	–	acid
D	water	1.5	69	–	acid
E	none	0	40	72	base
F	very dilute caustic	0.5	80	–	–
G	very dilute caustic	1.5	82	5.7	base
H	very dilute caustic	2.5	74	1.8	acid

Fibers from Anisotropic Dopes of Aromatic Polymers

A process is provided by *K. Nagasawa, M. Mitsuishi, H. Suzuki and T. Matsushita; U.S. Patent 4,016,236; April 5, 1977; assigned to Asahi Kasei Kogyo KK, Japan* for manufacturing fibers by a wet-spinning procedure from an anisotropic dope of a polymer having aromatic mononucleus or polynucleus divalent radicals in which the chain extending bonds from each aromatic nucleus are coaxial or parallel and oppositely directed, and which are connected by at least one link selected from –NHCO–, –NHCOCONH– and –CONHNHCO–. The filaments taken out from a coagulation bath are, without applying any substantial tension thereto, washed with water, dried and optionally heat treated, on a conveyor. The resulting fibers possess moderate initial modulus and elongation as well as high tenacity.

In Figure 2.5, shown on the following page, an anisotropic dope of the specified aromatic polymer is extruded into a coagulation bath through a spinneret **2**. The coagulated filaments **3a** are taken from the coagulation bath by take-up rollers **4** and made to fall onto a conveyor **6** by a cage roller **5**. The cage roller has a cagelike appearance and is composed of a plurality of rods which constitute the filament-guiding periphery.

The filaments are formed into an endless narrow fleece of piled loose filaments **3b** on the conveyor and are transferred onto another conveyor **7** while being turned over. This conveyor is moved continuously or intermittently by a suitable driving means (not shown) at a velocity substantially equal to that of conveyor **6**. The fleece of piled loose filaments travels together with the conveyor **7** through a washing apparatus **8**, a drying apparatus **9**, and then a heat-treating apparatus **10**. Then, the filaments **3c** are taken up from the conveyor **7** and are wound around a bobbin by a winding apparatus **11**.

When the piled loose filaments are liable to be disheveled and the individual filaments are liable to be entangled with each other by the washing water, hot air for drying or the fact that the filaments are electrostatically charged, a covering endless belt **12** is preferably provided. This covering belt sandwiches the filament fleece together with the conveyor to prevent the filaments from being disheveled. The covering belt travels at the same speed as the conveyor and imparts no tension to the filaments.

Figure 2.5: Spinning Apparatus with Conveyor for Heat Treatment

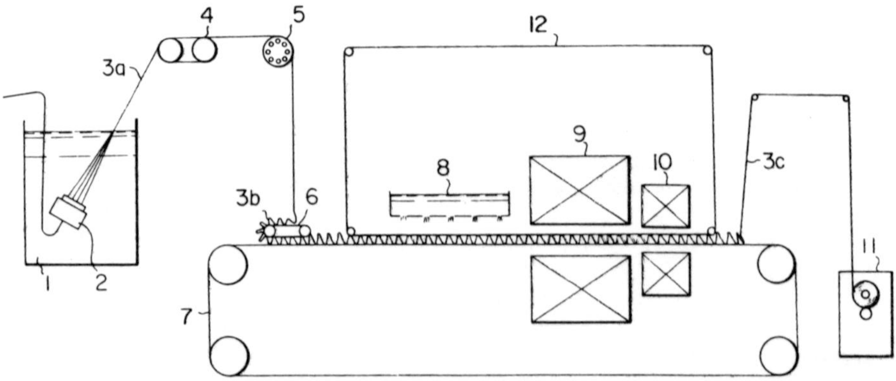

Source: U.S. Patent 4,016,236

Polymer Preparation: 378 g of p-phenylenediamine was dissolved in 10 liters of a mixture of 2 parts hexamethylphosphoramide and 1 part N-methylpyrrolidone, followed by cooling to 10°C. 710 g of powdered terephthaloyl chloride was added at once to the solution. The polymerization mixture gelled into the form of cheese within several minutes.

A semisolid mass so obtained was combined with approximately the same amount of water and pulverized using a Henschel mixer. The pulverized product was filtered, washed 3 times with warm water and then dried by hot air at 110°C.

830 g of light yellow polymer in the form of sawdust was obtained [poly(p-phenylene terephthalamide)]. The inherent viscosity was 5.9.

Example: Following the above procedure, a polymer was prepared wherein the preparation scale was enlarged with all other conditions remaining substantially the same. The polymer (referred to as polymer A) had an inherent viscosity of 6.1. Polymer A was stirred in 99.3% sulfuric acid at 70°C in vacuo for 2 hours, allowed to stand for 2 hours and then deaerated to prepare a dope of 18% concentration. The dope was extruded through a spinneret with 800 orifices, each having a diameter of 0.06 mm. The extrudate travelled 10 mm through the air and successively travelled through a coagulation bath of 30% aqueous sulfuric acid maintained at 3°C.

The filaments so formed were taken out at a rate of 100 m/min. Then the filaments were washed with water, dried and heat treated by using an apparatus as shown in Figure 2.5, where the speed of the filaments falling onto the conveyor was 100 m/min, the speed of the conveyor was 1 m/min, and the thickness, areal density and width of the filament fleece on the conveyor were 6.3 mm, 0.027 g/cm^2 and 50 mm, respectively. The drying was carried out by hot air at 120°C and the heat treatment by a heated plate at 220°C. A cover belt of a plain weave fabric of polytetrafluoroethylene fibers was used.

The heat-treated filaments had a fineness of 1,200 den, a tenacity of 21.5 g/den, an elongation of 6.4% and an initial modulus of 355 g/den. The dried filaments which were not subjected to the heat treatment had a fineness of 1,200 den, a tenacity of 21.3 g/den, an elongation of 6.8% and a 330 g/den initial modulus.

Both the heat-treated filaments and filaments not heat treated were heated at 200°C in an oven under a tension of 0.5 g/den. The resulting filaments exhibited tenacities of 21.0 g/den and 21.2 g/den, elongations of 6.2% and 3.3%, and initial moduli of 370 g/den and 650 g/den, respectively. These results substantiate that the filaments produced by the process exhibit tensile properties approximately similar to those which are not subjected to such after-heat treatment.

For comparison purposes, the abovementioned procedure for the production of filaments was repeated wherein the filaments taken out from the coagulation bath were washed in a water bath, dried in a tunnel drier and then heat treated on a heated plate at 220°C instead of using the apparatus as shown in Figure 2.5. In the water bath, water was forced to flow in a countercurrent manner to the travelling filaments, and in the tunnel drier, hot air of 150°C was forced to circulate. In these arrangements, delivery rollers were fitted at the inlet of the water bath, between the water bath and the drier, at the outlet of the drier, and downstream from the heated plate. These rollers were driven at controlled rates so that any tension, other than that generated by the countercurrent flow of water and filament guides, is given to the filaments to the smallest extent possible.

The heat-treated filaments had a fineness of 1,150 den, a tenacity of 17.0 g/den, an elongation of 2.5%, and an initial modulus of 680 g/den. The dried filaments which were not subjected to the heat treatment had a fineness of 1,180 den, a tenacity of 17.5 g/den, an elongation of 3.4% and a 555 g/den initial modulus. These filaments are poor in tenacity and elongation and possess high initial modulus as compared to the filaments obtained by the process.

Arylene Oxadiazole/Arylene N-Alkylhydrazide Copolymer Fibers

Fibers of arylene oxadiazole/arylene N-alkylhydrazide copolymer and of high tensile strength are prepared by *H.C. Bach; U.S. Patent 4,115,503; Sept. 19, 1978; assigned to Monsanto Company* by a unique solution spinning process, whereby an arylene oxadiazole/arylene N-alkyloxadiazolium hydrosulfate copolymer in sulfuric acid or oleum is extruded into an aqueous coagulation medium in which concurrently a fiber is formed and the latter-mentioned copolymer undergoes a chemical transformation to provide the first-mentioned copolymer.

Example: This example illustrates the preparation of fiber of p-phenylene oxadiazole/N-methyl terephthalhydrazide copolymer, wherein a 90 to 10 molar ratio of terephthalic acid to dimethyl terephthalate and a 1.25% molar excess of hydrazine sulfate were used in preparing the dope from which fiber was reaction spun.

984 g (515 ml) of fuming sulfuric acid (30% SO_3) was added with stirring at room temperature to a mixture of 58.18 g of terephthalic acid, 7.55 g of dimethyl terephthalate, and 51.26 g of hydrazine sulfate. The resulting reaction mixture was then heated to 120°C and maintained at this temperature for 6 hours. During this period the viscosity of the solution increased greatly and a total of 200 g of fuming sulfuric acid (30% SO_3) and 100 g of concentrated sulfuric acid were added to moderate the solution viscosity. The polymer in the resulting dope had an inherent viscosity of 3.8.

This dope was reaction spun to fiber by extruding it through a 6.5 mil (0.1651 mm) five-hole spinneret into a coagulation bath consisting of 8% aqueous sulfuric acid maintained at 23°C. The spinneret was heated to 120°C and kept a half-inch (12.7 mm) above the coagulation liquid. After passage under a guide in the bath, the fiber was passed over a water wash roll, a neutralizing roll rotating in an aqueous 4% solution of sodium carbonate, through a water bath at 95°C, over another wash roll, a steam-heated drying roll and then hot-stretched over a hot-shoe at 383°C and finally wound onto a bobbin at a speed of 138 ft/min (42.1 m/min). The fiber had a tenacity of 5.7 g/den, an elongation of 3.7% and initial modulus of 202 g/den.

Aromatic Oxadiazole Polymers and Copolymers

An economical process for continuously producing shaped articles of aromatic oxadiazole or aromatic oxadiazole/N-alkylhydrazide polymers from certain monomer solutions is provided by *J.C. Richardson; U.S. Patent 4,054,633; October 18, 1977; assigned to Monsanto Company*. The process involves continuously feeding monomer solution into a heated, pressurized continuous reactor to provide dope which is continuously extruded into an aqueous medium to provide a shaped article.

Example: This example illustrates the preparation of polymeric fiber consisting essentially of recurring units of the following formulas (1) and (2).

(1) 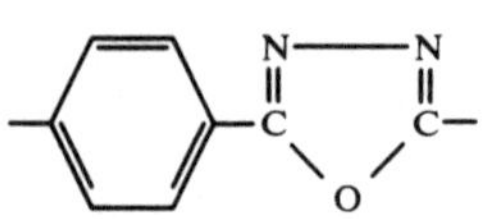

(2) 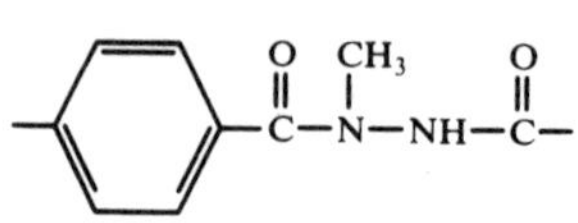

The recurring units (1) and (2) are used in a mol ratio of about 30 to 70, respectively, in this process.

A monomer solution consisting of about 15% by weight of DMT (dimethyl terephthalate) and $HySO_4$ (hydrazine sulfate) in a mol ratio of 1 to 1.03 in oleum (SO_3 utilization of 65%) was introduced at the rate of 19 g/min into a 150°C heated Type B tubular reactor by means of a positive displacement pump. A 150°C heated spinneret with twenty-five 6-mil (0.15-mm) holes was attached to the exit end of the reactor.

A dope, having a Brookfield viscosity of about 47,000 poises and a solids concentration of 7%, consisting of oleum and polymer having an inherent viscosity of 3.96 was formed in the reactor. This dope was spun into fiber by extruding the dope through the spinneret into a room temperature 8% aqueous sulfuric acid bath. The pressure in the reactor (140 psig, 9.5 atm) was sufficient to maintain the contents of the reactor in the liquid phase. The residence time of liquid in the reactor was 30 minutes.

The spinneret was positioned 1.27 cm above the water bath. After passage under a guide in the bath, the fiber was passed over a water wash roll, a neutralizing roll rotating in an aqueous 8% solution of sodium carbonate, over three additional water wash rolls, a steam-heated drying roll, drawn about 3.6 times over a hot-shoe at 370°C, and wound onto a bobbin at about 350 ft/min (106.68 m/min).

The fiber had the following tensile properties, as determined on an Instron Tester using a constant extension rate of 10% per minute with a gauge length of 10 centimeters: a tenacity of 5.15 g/den, an elongation of 4.03% and a modulus of 216 g/den.

VINYLS

Recovery and Recycle of Salt Solution in Vinyl Polymer Spinning

R.B. Hurley, F.B. Eastwood and J.A. Mann; U.S. Patent 4,029,725; June 14, 1977; assigned to Dow Badische Company provide an efficient, nonoxidative process for the purification of a recovered aqueous inorganic salt solution prior to the recycling thereof into the polymerization step in a continuous process for producing a filamentary vinyl polymer by the wet-spinning of a vinyl polymer solution.

Example: (All parts and percentages cited in the example are by weight.) A zinc chloride solution consisting of 58.2% zinc chloride and 41.8% water was used as a solution polymerization medium. To 85 parts of this medium was added 15 parts of a mixture consisting of 91% acrylonitrile, 7.5% methyl acrylate, 1.48% of a zinc salt of a vinyl sulfonic acid, and 0.02% of hydrogen peroxide.

Polymerization was effected in standard equipment, after which the vinyl polymer solution was spun through a standard spinneret into a coagulation bath consisting of 35% aqueous zinc chloride. The bath was maintained at a temperature of 30°C. The zinc chloride concentration of the bath was adjusted by a countercurrent flow of aqueous zinc chloride from a wash bath downstream of the coagulation bath. Overflow from the coagulation bath was concentrated by evaporation of water to 35% zinc chloride. The pH was then adjusted to approximately 5.5 (10% zinc chloride basis), and hydrogen peroxide, tartronic acid and activated carbon were added in that order in amounts sufficient to provide 25 ppm, 25 ppm and 0.2%, respectively. The activated carbon was Darco S-51.

The temperature of this mixture was controlled at 25°C and agitation was effected for 30 minutes with a mechanical stirrer. After filtration of the activated carbon, the filtrate was concentrated to 60% zinc chloride by evaporation of water and the pH was adjusted to 3.0 (10% zinc chloride basis). Tartronic acid was then added in an amount sufficient to provide about 25 ppm.

The so-treated solution was then utilized as a polymerization solvent without any deleterious effects on, or diminution of, polymerization activity.

Lithium Halides as Solvents for Polyhydroxymethylene

According to *W.S. Creasy and H.K. Reimschuessel; U.S. Patent 4,131,669; December 26, 1978; assigned to Allied Chemical Corporation*, certain solutions of lithium halides readily dissolve polyhydroxymethylene and copolymers of hydroxymethylene with up to about 30 mol % of ethylenically unsaturated comonomers at high concentrations suitable for solution spinning. Such solutions are desirable for commercial operation as well as preferred from an economical and toxicological point of view.

Example 1: An aqueous lithium chloride solution, saturated at room temperature (13 M) was prepared. 21.25 g was heated to 75°C and to this was added 3.7 g of polyhydroxymethylene (with hydrazine inherent viscosity of 1.3). After stirring for about 3 hours at 75° to 80°C, the polymer was found to have dissolved completely, giving a clear, light amber, viscous solution. On cooling to 25°C, the solution remained free flowing but highly viscous. Polymer recovered from this solution by precipitation in water had a hydrazine inherent viscosity of 1.20.

Example 2: To demonstrate the spinning of filaments of polyhydroxymethylene, the following experiment was conducted. To a 42% aqueous solution of LiCl was added at about 75°C polyhydroxymethylene having an inherent viscosity of 1.5 until a 20% polymer solution was obtained. This solution was filtered, degased, and transferred into a motor-driven syringe. Employing a needle with an open diameter of 0.006 inch, and a temperature of about 50°C, the solution was extruded first into an aqueous coagulation bath which was held at room temperature, then fed into a second water bath to complete extraction of any remaining salt. From this bath the formed filament was fed consecutively into take-up rolls which operated at different speeds resulting in a drawing of the filament to about four times its original length.

This operation yielded filaments with deniers between 30 and 65, tensile moduli between 120 and 220 g/den at ranges of 2 to 4 g/den and 6 to 20% elongation.

Example 3: When copolymers of vinyl carbonate with up to about 30 mol % of ethylenically unsaturated comonomers are hydrolyzed in accordance with the procedure of Example 1, then corresponding polyhydroxymethylene copolymers are obtained which are soluble in aqueous lithium chloride. Their solution in aqueous lithium chloride can be spun into filaments by extrusion into a spinning bath as abovementioned for polyhydroxymethylene.

Melt Spinning

Filaments of various polymeric materials such as polyolefins, polyamides, polyesters, etc., are produced by feeding the polymer, which is solid and in particulate form, to a screw extruder in which the polymer is melted and extruded through a spinneret. A spinneret is essentially a plate having a plurality of orifices therein. The orifices form the molten polymer into filaments which are then cooled or quenched by a cooling medium, generally conditioned air, and then further processed as known in the art.

ASSOCIATED APPARATUS

Spinneret Assembly Producing Plug Flow

The process developed by *R.D. Meyer; U.S. Patent 4,104,015; August 1, 1978; assigned to Phillips Petroleum Company* reduces the nonuniformity of filaments produced by melt spinning. Plug flow of a polymer through a spinneret assembly is approximated by employing an annular inlet means that initially distributes the polymer between the center and the outer wall of a spinneret assembly.

Referring to Figure 3.1, a spinneret assembly **10** comprises a housing **12** having side wall **14** and top **16.** The top has an inlet **18** positioned in the center thereof attached to pipe **20** connected to an extruder (not shown) for a source of molten polymer. A spinneret **22** is positioned on the opposite end of the housing from the inlet to form the outlet. The spinneret is secured to the housing by annular ring **24** which has apertures **26** for securing the spinneret assembly to a spinning block (not shown) as known in the art. The spinneret has a plurality of orifices **28.** Each orifice comprises an inlet portion **30,** a transition portion **32** and a land portion **34.**

Figure 3.1: Spinneret Assembly Producing Plug Flow

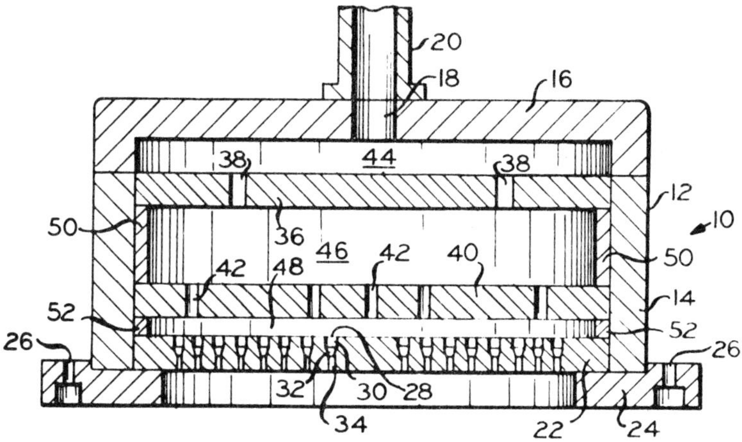

Source: U.S. Patent 4,104,015

Positioned just below inlet **18** is an annular inlet means illustrated in this embodiment in the shape of a circular plate **36**, which has a plurality of orifices **38**. These orifices are equally spaced, and the center of each orifice lies on a circle the center of which coincides with the center of the plate. In this particular embodiment the orifices are positioned 40° apart, and the centers of these orifices lie on a circle having a radius equal to 60% of the radius of plate **36**.

Immediately below the annular inlet means and above spinneret **22** lies a distributor means, also shown in this embodiment in the shape of a circular plate **40**. This plate is positioned parallel to plate **36** and the spinneret. This plate also has a plurality of orifices **42**. The centers of these orifices are positioned on three concentric circles. The concentric circles are equally spaced along the radius of plate **40**. The orifices **42**, located on any one circle are positioned 30° apart and these orifices positioned on adjacent circles are offset 15° so that these orifices are staggered.

In the operation of the spinneret assembly, a viscous fluid, such as a molten polymer, flows through pipe **20** and the inlet into an initial distribution zone **44**. The fluid then flows radially outward through orifices **38** of plate **36** as an annular zone of restriction and into a first distribution zone **46**. A portion of the polymer moves toward the center of this zone as well as toward **14** so that the fluid flows through orifices **42** of plate **40** as a second zone of restriction and into a second distribution zone **48**. Orifices **42** of plate **40** force the fluid uniformly from the first zone to the second zone.

The polymer is then formed into filaments as it passes through orifices **28** from the second distribution zone **48**. The polymer in the second distribution zone is moving substantially in "plug flow"; thus, the polymer flowing through orifices **28** is flowing through each orifice at substantially the same flow rate.

The position of orifices **38** of plate **36** must be positioned so as to generally form an annulus. Where orifices are used as the passages for the fluid as previously described, generally the centers of orifices **38** are positioned on a circle having a radius from about 25 to 80% the length of the radius of plate **40**; however, from the results of the example hereinafter described, it is believed that good results can be obtained employing a radius from about 35 to 65% the length of the radius of plate **40**.

Example: A spinneret assembly was constructed essentially as shown in Figure 3.1. The spinneret had 280 orifices 0.0177 inch (0.45 mm) in diameter. The spinneret plate was 0.750 inch (1.905 cm) thick and had a diameter of 5.780 inches (14.68 cm). The orifices in plate **36** were laid out such that the radius of the circle of orifices was 1.25 inches (3.175 cm), and the orifices in plate **40** were laid out such that the radius of the three circles of orifices were ¼, ½, and ¾ the radius of the plate, respectively, or 0.635 inch (1.83 cm), 1.445 inches (3.67 cm), and 2.1675 inches (5.505 cm), respectively.

Both plates **36** and **40** were the same diameter as the spinneret. The distance between plates **36** and **40** was 2 inches (5.08 cm), and the distance between plate **40** and the spinneret **22** was ¼ inch (0.635 cm). The orifices in both plates **36** and **40** were ⅛ inch (0.317 cm) in diameter.

In an effort to demonstrate the molten polymer flows through a spinneret assembly substantially in plug flow, the abovedescribed spinneret assembly was employed to spin polypropylene filaments. The polypropylene resin had a 4 melt flow and was unpigmented or natural in color. At time zero, while spinning unpigmented polypropylene, approximately 1 pound of red pellets of polypropylene was fed to the screw of the melt extruder.

After 3 minutes and 10 seconds the filaments extruded through the 3 outside rings of orifices of the spinneret changed to a medium red color, and the filaments from the orifices in the center of the spinneret were the natural color. After 4 minutes had passed, all the filaments were a dark red. After 6 minutes the color of all the filaments was turning to a lighter red, but all the filaments were the same color. The red color continued to lighten evenly until 13 minutes had elapsed at which time the filaments from the outside rings of orifices appear to be slightly lighter in color. After 17 minutes only the filaments from the inside rings of orifices appear to have a faint pinkish color.

This illustrates that the polymer was flowing through the area of the spinneret assembly designated as zone **48** essentially in plug flow and that the polymer was flowing through the orifices of the spinneret at substantially the same flow rate.

Manufacture of Crimpable Filaments

B. Boyes and A. Jones; U.S. Patent 4,038,357; July 26, 1977; assigned to Imperial Chemical Industries Inc., England provide a method for the manufacture of potentially crimpable, synthetic linear polymer filaments which comprises the steps of melt spinning a plurality of filaments, initially partially cooling the hot filaments evenly with a stream of gas, and asymmetrically cooling the filaments further by causing them to contact continuously renewed thin film of liquid.

Figure 3.2 comprises a perforated cylinder **1** having a porous shroud **2** mounted immediately below and sealed at its upper end by a spinneret **3**. The cylinder is mounted on a hollow support stem **4** which also serves as an air feed pipe to the cylinder. A cooling liquid applicator unit, generally designated **5**, surrounds the stem in the vicinity of the lower end of the cylinder.

The applicator unit **5** comprises a porous element **6** which defines a hollow chamber **7** with the stem. A supply pipe **8** for water passes interiorly of the stem and communicates with the chamber of the applicator unit.

Figure 3.2: Cross-Sectional Side Elevation of Apparatus for the Manufacture of Crimpable Filaments

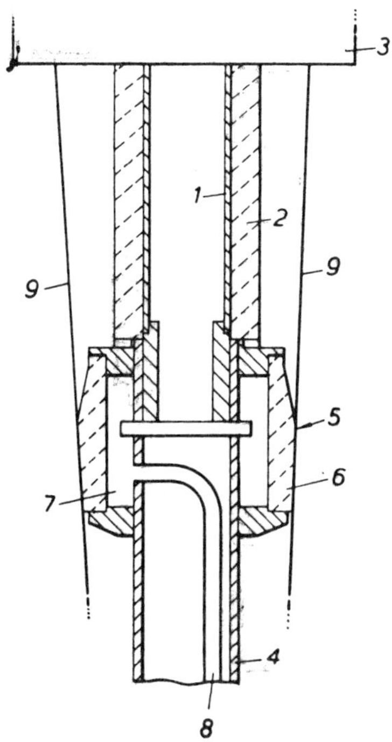

Source: U.S. Patent 4,038,357

In operation, air is supplied to the cylinder **1** via the stem **4** and passes from the cylinder and through the porous shroud **2** as a horizontally radially directed gas stream which uniformly cools the filaments **9** issuing from the spinneret **3**. The initially cooled filaments then contact a film of water on the surface of the porous element **6** which further cools them asymmetrically.

Example 1: Polyethylene terephthalate of intrinsic viscosity 0.675 was melt extruded through a spinneret with 424 holes each of 0.018 inch diameter arranged in a circle. The temperature of the wall of a filter sand pack immediately adjacent to the spinneret was maintained at 287°C. The spinning throughput of the polymer melt was 57 pounds per hour, and the spun filaments were wound up onto a bobbin at a linear speed of 2,350 fpm.

The hot filaments were initially evenly cooled by air at ambient temperature passing radially outwardly through a quench unit, comprising a 4 inch long perforated stainless steel cylinder **1** having a porous shroud **2** of sintered bronze at an air flow rate of 7 cfm. A cooling liquid applicator unit **5** was positioned immediately below the quench unit. Water was caused to flow from the inside of a porous carborundum element **6** to its outer surface at a flow rate of 250 cm^3/min. The spun filaments were creeled from bobbins to form a tow which was passed through a bath containing a textile finishing agent, drawn to a ratio of 3.5:1 in steam, and dried and cooled at substantially constant length. The tow was then laid down in a tensionless condition on the moving brattice of an oven and heat treated at 140°C for 2 minutes.

The percentage coefficient of variation of the spun filament diameter was between 5 and 8, the decitex per drawn filament was 4.4 (circa 60 kilotex for the final drawn tow), and the filament crimp was 50%. The runnability of the method was in no way critical; i.e., it was considered stable.

Example 2: Polyethylene terephthalate of intrinsic viscosity 0.675 was melt extruded through a spinneret with 252 holes each of 0.018 inch diameter arranged in a circle. The temperature of the wall of a filter sand pack immediately adjacent to the spinneret was maintained at 276°C. The spinning throughput of the polymer melt was 35 pounds per hour, and the spun filaments wound up onto bobbins at a linear speed of 3,000 fpm. A cooling liquid applicator was positioned at 5.5 inches below the spinneret. As in Example 1, water was caused to flow through a porous carborundum element at a rate of 250 cm^3/min. The spun filaments were then formed into a tow, drawn in steam, dried, cooled and heat treated in accordance with Example 1, except that a draw ratio of 3.0:1 was used.

The decitex per drawn filament was 4.4 (circa 60 kilotex for the final drawn tow), and the filament crimp was 50%. However, without the initial forced air cooling, the production rate was reduced (polymer throughput down from 57 pounds per hour to 35 pounds per hour), the percentage coefficient of variation of the spun filament diameter worsened to between 15 and 20.

The preceding conditions were repeated except that the melt polymer throughput was increased to 47 pounds per hour using a spinneret with 336 holes each having a diameter of 0.018 inch arranged on a circle. The cooling liquid applicator was positioned 7 inches below the spinneret and utilized a water flow rate of 220 cm^3/hr.

Again the drawn filament decitex was 4.4 and the filament crimp was 50%. However, the percentage coefficient of variation of spun filament diameter was in excess of 20 and, in terms of runnability, the method was too unstable for it to be considered commercial.

Multifilamentary Yarns of Uniform Quality

W.B. Fisher, D.W. Kim, O Wyatt, J.W. Showers, E.A. Swanson and G.C. Weedon; U.S. Patent 4,045,534; August 30, 1977; assigned to Allied Chemical Corporation describe an improved process and apparatus for production of synthetic multifilamentary yarns having uniform quality. The method comprises extruding a molten synthetic polymer downwardly through a spinneret, advancing the extruded filaments downwardly through a substantially stationary column of heated air, and substantially advancing the filaments downwardly through a quenching zone where they are in contact with cooling air introduced into the path of the filaments under controlled conditions of air velocity and direction of flow.

Example: A melt of polyethylene terephthalate having an intrinsic viscosity of about 0.92 was supplied at a rate of 60 pounds per hour, at a temperature of about 291°C, to the apparatus shown in Figure 3.3a. The molten polymer was fed by extruder **1** to spinning pump **2** which fed spinning block **3** containing a conventional spinning pot comprising a spinning filter and a spinneret; the spinning filter being disposed between the spinning pump and the spinneret.

The spinning filter consisted of a conventional sieve filter combination of 24 metal screen layers. The pressure drop through the spinning filter averaged 200 to 400 atmospheres. The spinning pot was enclosed in a controlled high temperature atmosphere so that loss of heat from the polymer was minimized. The melt enthalpy increase through the pump and sieve filter was sufficient to heat the melt at a point immediately above the spinneret to about 305° to 310°C, and the pressure at this point was about 50 atmospheres. The flow of polymer through the spinneret was maintained at a constant rate of 60 pounds per hour by spinning pump **2**.

The spinning pot spinneret was divided into two parts by means of an undrilled "stripe" wide enough to form a visible split between the multiple ends below the spinneret. The spinneret was preferably positioned with respect to the quenching chimney **6** so that the stripe was parallel to the side walls of the chimney and, therefore, to the flow of cooling air through the top portion of the chimney.

Figure 3.3: Spinneret and Quenching Chamber

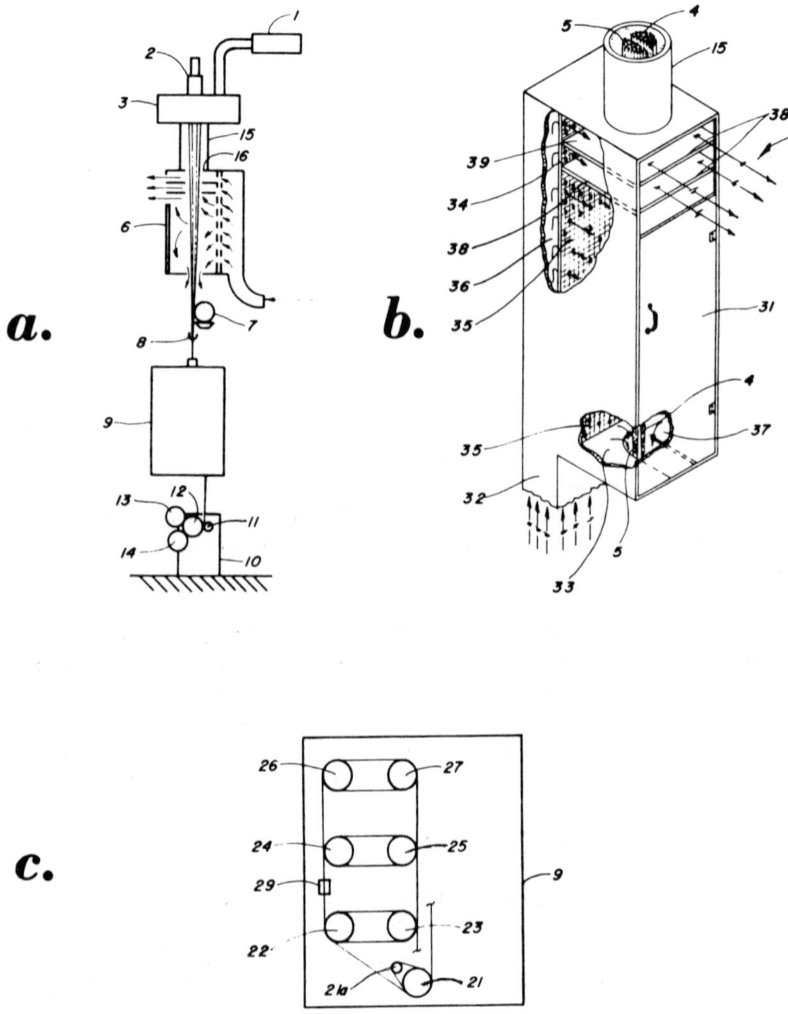

(a) Schematic side view

(b) Perspective view of the quenching chimney labeled **6** in Figure 3.3a

(c) Schematic of a two end embodiment of the draw panel labeled **9** in Figure 3.3a

Source: U.S. Patent 4,045,534

The spinneret plate had 384 holes (192 holes on each side of the undrilled stripe), each of 0.018 inch diameter, spaced so that the distance between the filaments formed was 0.28 to 0.40 inch immediately below the spinneret.

From the spinneret there was extruded two ends of multifilament, continuous filament yarn, and the two ends were passed downwardly into a substantially stationary column of air contained in a heated sleeve **15**, about 15 inches in height, disposed surrounding and immediately beneath the spinneret. The air temperature in the heated sleeve was maintained at about 300°C at the top of the sleeve, decreasing to about 115°C at the bottom. The temperature of the metal in the heated sleeve was about 330°C at the top and 220°C at the bottom of the sleeve.

The minimum distance between filaments at the bottom of the heated sleeve was about 0.24 inch. A heated sleeve baffle **16** was provided at the bottom of the heated sleeve forming an inwardly extending flange to minimize flow of cooling air into the heated sleeve.

Yarn leaving the heated sleeve was passed directly into the top of quenching chimney **6**, shown in detail in Figure 3.3b. The quenching chimney was an elongated chimney 70 inches in height, substantially rectangular in cross section and provided with imperforate top, rear and side walls. The front of the chimney was partially covered by an imperforate door **31** which terminated about 17.5 inches short of the top wall and presented an open passage for air discharging from the chimney.

The interior of the quenching chimney was partitioned by a perforated distribution plate **34** and distribution screen **35**, which formed the boundary between plenum chamber **36**, and quenching chamber **37**. Quenching air at about 25°C and 65% relative humidity was supplied to the plenum chamber from inlet duct **32**, at about 200 ft^3 of air per pound of filaments entering the quenching chamber **37**.

The two ends of multifilament, continuous filament yarn were advanced downwardly through the quenching chamber where they were in contact with the cooling air introduced into the path of the filaments. Four horizontal parallel baffles **38** approximately 3 inches apart vertically and extending down about 12 inches from the top quenching chamber were used to provide transverse contact of the cooling air with the filaments in the top portion of the quenching chamber. In the middle portion of the quenching chamber below the horizontal baffles, the cooling air was deflected upward by the imperforate door and flowed substantially countercurrent to the movement of the filaments, the air leaving the quenching chamber via the opening above the imperforate door. Part of the cooling air flowed out of the bottom of the quenching chamber with the filaments, and the air flow in the bottom zone of the quenching chamber was substantially cocurrent to the movement of the filaments.

This quenching system increased quenching efficiency significantly as evidenced by air temperature and air velocity profiles. The number of horizontal stack baffles **38** required to obtain improved undrawn yarn quality was preferably 2 to 5. The temperature of the cooled yarn at the bottom of quenching chamber **37** was about 50°C. In this example, where polyester was supplied at 60 pounds per hour, 4 parallel horizontal baffles **38** gave high quality yarn; the coefficient of variation of the evenness of the undrawn yarn was not above 10 over an extended period of operation.

The two ends of cooled, multifilament continuous filament yarn were advanced downwardly, preferably through a shaped baffle **33** at the bottom of quenching chamber **37**. The ends were separated by guides **8** located below a lubricating finish applicator **7**. Following lubrication, the ends passed through a guide separation to pretension roll **21** with its accompanying separator roll **21a** shown in Figure 3.3c.

The yarn was then passed over cold feed roll pair Godet rolls **22** and **23**, then through a draw point localizer which was a conventional steam jet, then to a draw roll pair of Godet rolls **24** and **25** operated at about 145°C and traveling at a speed 5.0 to 6.6 times faster than the feed roll, then to a relaxation pair of Godet rolls **26** and **27**, and optionally through an entangling apparatus such as a conventional air operated interlacing jet, and on to winder **10** as shown in Figure 3.3a. Typical yarn prepared at a draw ratio of 6 had the following properties: denier, 1,000; tenacity, g/den, 9.25; elongation, %, 13.5; and shrinkage, %, 9.5.

Filament Manufacturing Device of Small Height

K. Lehner; U.S. Patent 4,133,620; January 9, 1979; assigned to Didier Engineering GmbH, Germany provides an apparatus for the continuous manufacture of polymer filaments, particularly polyamide or polyester filaments, where the manufacture of such filaments may be carried out in a relatively small space and particularly in a limited vertical space, i.e., in only one floor or story of a building.

Referring to Figure 3.4, a conventional fusion extruder **1** is supplied with the necessary materials for the formation of the particular type of polymer filament involved from a supply container **2**. The fused material is fed to a spinning head **3** which has incorporated therein a plurality of spinning nozzles. From the spinning nozzles a plurality of threads **6** are extruded downwardly so that they pass vertically downwardly through a blowing shaft **4** arranged beneath the spinning head **3**. It is to be understood that elements **1**, **2** and **3** may be of any type or construction which are known and conventional in the art for the spinning formation of polymer, and particularly polyamide or polyester, filaments.

Adjacent to the bottom end or area of the blowing shaft **4** is arranged a drawing off roller **5** in the form of a guide roller which draws filaments **6** from the blowing shaft **4** and supplies such filaments over a guide roller **7** to a winding machine **8**.

A specific inventive feature of the process does involve the placement and arrangement of the guide roller and the winding machine with respect to the blowing shaft. Specifically, the winding machine is located laterally of the blowing shaft and not at a level below the blowing shaft. The necessity of the guide roller being arranged generally centrally above the winding machine is known.

Figure 3.4: Filament Manufacturing Device of Small Height

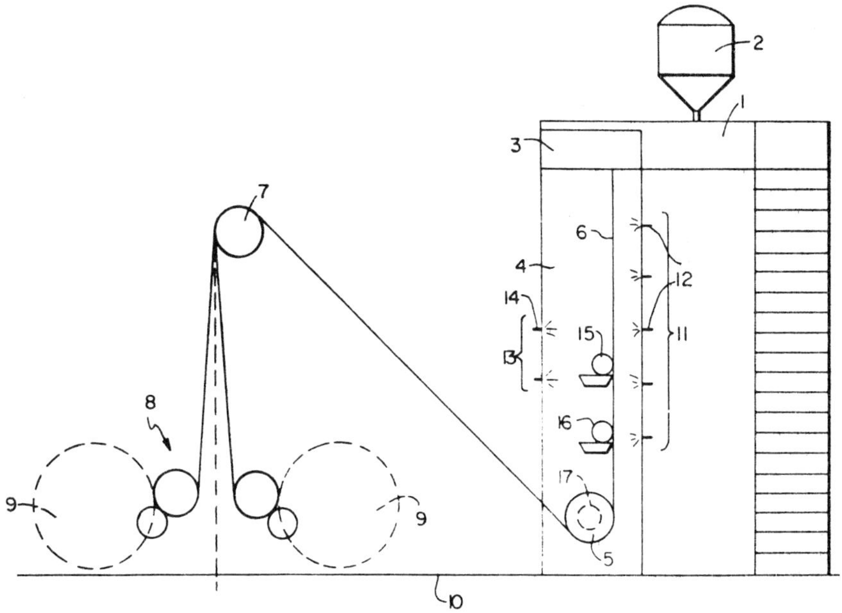

Source: U.S. Patent 4,133,620

However, in accordance with the process, the guide roller is also located at a level substantially above drawing off roller **5**. That is, the guide roller is arranged sufficiently above the level of the drawing off roller such that the winding machine may be positioned entirely laterally of and not at a level below the blowing shaft. Thereby, the entire arrangement shown in Figure 3.4 may be assembled to and mounted on a common frame, shown schematically at **10**, such that the entire filament manufacturing apparatus is extremely compact.

Filaments and Fibers Having Discontinuous Cavities

In a process developed by *E. Siggel, G. Wick, H. Linhart and E. Kessler; U.S. Patent 4,164,603; August 14, 1979; assigned to Akzona Incorporated* a filament of a thermoplastic synthetic polymer having a plurality of adjacent, separate discontinuous cavities is made by a process where a silicone oil and an inert gas or gas-forming substance are dispersed in a polymer melt, and the melt is extruded into a filament. The melt contains up to and including 1% by weight of the silicone oil, based on the weight of the melt, at the time it is extruded and up to and including 10% by weight of an inert gas.

Example: A single point extruder spinning machine is used to spin filaments according to the process from polycaprolactam chips of a solution viscosity of 2.72, measured as a 1 weight % solution in 90% formic acid at 25°C. The melt emerges from the extruder at a pressure of about 100 bars via the first pressure pump.

In the melt line to the mixer, including a shaft provided with radially aligned pins as mixing elements, the melt at a temperature of about 278°C is mixed via a piston melting pump with about 0.3 weight %, based on the weight of the melt, of a dimethyl siloxane (20 cp at 20°C) and nitrogen at about 117 bars supplied from a N_2 bottle at a pressure of 150 bars via a regulating valve and a VA capillary of a diameter of 0.7 mm. The mixer operates at a speed of about 180 rpm.

The second pressure pump runs at a slightly higher rpm than that of the first pressure pump and transports the largely homogenized mixture of melt, silicone oil and nitrogen contained in the melt to the metering pump located some 1.80 meters away. It is fed to the spinning head and extruded from a spinneret having 22 spinning orifices of 0.14 mm diameter. At a throughput of about 23 g/min, the filaments are wound at a rate of about 1,100 m/min. The spinning point operated over a period of several days virtually trouble-free with a scraping cycle of about 6 to 8 hours.

The filaments spun under these conditions were subsequently drawn to a ratio of 1:2.52 under conventional conditions. They exhibit a strength of 33 cN/tex and a water retention capacity of 25%. The effective denier is dtex 78 f 22. The density of the filaments is 1.0 g/cc corresponding to a cavity component of about 12%.

The cross section of the drawn filaments according to the process is essentially circular, has on the average some 12 to 18 cavities of a diameter of 1 to 2 μ, about 3 cavities of a diameter of 2 to 3 μ and occasionally (about 2) cavities of a diameter of 4 to 6 μ. Larger diameters were not observed. The surface of the filament is essentially smooth, i.e., there are practically no "burst" cavities. Cavities having a diameter of less than 1 μ have not been included in the count.

Spinning Pack Filter

A melt spinning process developed by *A.H. Johnson and I.P. Sanderson; U.S. Patent 4,088,431; May 9, 1978; assigned to Imperial Chemical Industries Ltd., England* utilizes a spinning pack which contains a filter comprising a shaped, rigid, porous structure comprising inert granular material, e.g., sand or alumina particles, bonded with a permanent binding agent, e.g. sodium silicate or a complex phosphate of aluminum. Synthetic thermoplastic polymers are melt spun, particularly polyesters.

Figure 3.5 illustrates a melt spinning assembly for the spinning of synthetic thermoplastic polymer.

As shown, a supply of molten polymer **1**, is forced under pressure via a metering pump **3** to a spinning pack **5** through a feed pipe **7**. This feed pipe communicates with the central cavity **9** and passages **11**. An extrusion disk **13** containing orifices **15** is provided through which fibers are extruded. A screen **17** may be positioned with filter elements **19** which are closed at their free ends **21**.

Figure 3.5: Spinning Pack Filter

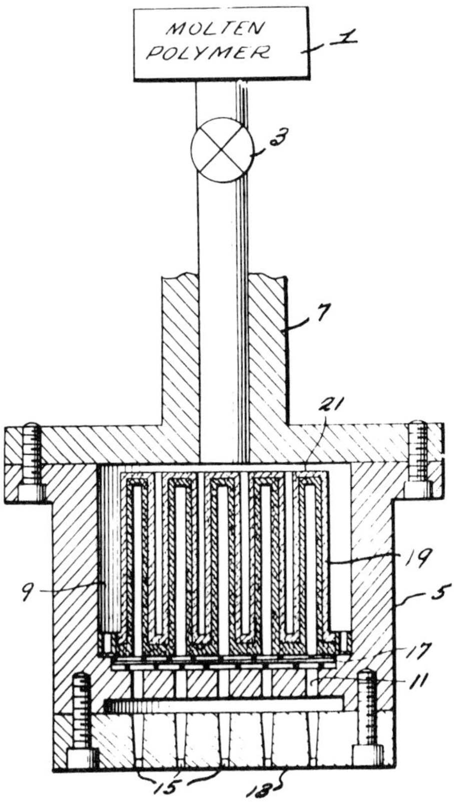

Source: U.S. Patent 4,088,431

Example: A solid complex aluminum phosphate binder having the empirical formula $AlPClH_{11}O_9$ was prepared by slowly adding 15.2 parts of anhydrous aluminum chloride to 40 parts of distilled water. The resultant solution was cooled to ambient temperature and 7.4 parts of an 88% by weight solution of orthophosphoric acid in water was added to it with stirring. The solution was then concentrated to about 20 parts. The resultant solution was allowed to stand, and the crystals which were eventually produced were filtered off, washed with ethanol, and dried in a vacuum desiccator. Analysis showed that the crystals corresponded to an empirical formula of $AlPClH_{11}O_9$.

A composition comprising 86 parts of 30 grade (U.S. Standard B74.18, 1968) alumina (Universal Abrasives Limited), 4 parts of high alumina ball clay, 0.8 part of magnesium oxide (Magnorite 100F, Norton Abrasives Limited), and 10 parts of the complex aluminum phosphate binder having the empirical formula $AlPClH_{11}O_9$ was thoroughly mixed in the dry state, 5 parts of water were then added slowly with thorough mixing.

The resultant composition was charged to a mold 6 inches in diameter and 1½ inches deep and profiled so as to produce a disc shaped molding having annular grooves in the upper and lower surfaces of the molding. The contents of the mold were subjected to a pressure of 10 tons, and the mold was heated for 10 min at a temperature of 50°C.

The resultant molding was removed from the mold and heated at 110°C for 2 hr. It was then heated up to 1200°C over a period of 4 hr, maintained at this temperature for 2 hr and cooled down over a period of 12 hr. The resultant strong, rigid filter was suitable for use in the filtration of molten polymers, especially in the filtration of molten polyethylene terephthalate as part of a fiber spinning process.

POLYESTERS

Highly Oriented Undrawn Yarn

A process is provided by *M. Mineo, K.-i. Oka, K. Nakagawa, H. Matsuda, S. Ohno and Y. Takagi; U.S. Patent 4,049,763; September 20, 1977; assigned to Toray Industries, Inc., Japan* for producing a highly oriented polyester undrawn yarn having a filament denier of about 2 to 6 comprising melt spinning a polyester from a spinneret and solidifying a melt spun yarn at a take up velocity from about 2,500 to 4,500 m/min. The spinning length defined as the length from the spinneret to the first take up roller is established within the range from (d + 2)/4 (meter) to (A/1.2)/(Faero/x) (meter). d is the filament denier of the undrawn yarn, A is the tension (gram) of the filament in an elongation range under a constant tension of a stress-strain curve, Faero/x is the rate of increase of yarn tension per meter subsequent to the yarn solidifying point by an air resistance which is calculated by $(Faero/x) = 0.229 \times 10^{-4} (D^{0.39})(V^{1.39})$ where

D is the diameter of a solidified filament (cm) and V is the velocity of the solidified filament (cm/sec).

Example 1: Polyethylene terephthalate having an intrinsic viscosity of 0.61 was melted at 285°C, extruded from a spinneret having 36 orifices each having a diameter of 0.3 mm at a rate of 40 g/min and taken up at a velocity of 3,000 m/min to obtain an undrawn yarn. Its particulars are shown in the table below.

Yarn denier	122
Filament denier	3.4
Filament diameter, cm	1.87×10^{-3}
Tension*, g/filament	1.7
Maximum spinning tension**, g/filament)	0.95

*In an elongation range under constant tension of the stress-strain curve.
**During spinning.

The spinning length at this time was 3.5 m and an oiling roller was disposed at a distance of 3.0 m from the spinneret. The upper limit of the spinning length sought from (A/1.2)/(Faero/x) was 5.1 m, and the lower limit of the spinning length sought from (d + 2)/4 was 1.4 m.

Using an Uster evenness tester, when the undrawn yarn was subjected to a half inert test at a yarn speed of 25 m/min, the yarn unevenness was 0.35%. When the interference fringe due to molecular orientation of this yarn was observed under a polarization microscope, no unevenness was recognized along the axial direction. Next, when the undrawn yarn was false-twisted while being drawn about 1.7 times, the number of broken filaments was not more than 2/10,000 m and from the resulting false-twisted yarn, very uniform knitted goods were obtained even after dyeing.

Example 2: Melt spinning was carried out under the same conditions as in Example 1 except that the spinning length was 6.5 m. The maximum spinning tension was 63 g or 1.8 g/filament.

When this undrawn yarn was false-twisted, while yarn was measured by an Uster evenness tester, it was 1.3%. When the undrawn yarn was observed under an interference microscope, an interference fringe was observed as undergoing an irregular change.

When this undrawn yarn was false-twisted while being drawn about 1.7 times, many filaments broke and occurrence of more than 10 broken filaments per 2,000 m was recognized often.

Multifunctional Chain Branches

According to *D.L. MacLean and R.T. Estes; U.S. Patent 4,113,704; Sept. 12, 1978; assigned to Monsanto Company*, significantly higher productivity of polyester feed yarn is achieved when the polymer from which the feed yarn is melt spun is chain branched with multifunctional chain branches having 1 to 15 μeq of reactive branching sites based on 1 g of the polymer. This productivity increase (over conventional production employing the same spinnerette in accordance with teachings of the prior art as applied generally to unbranched polymers) may be achieved specifically by spinning a feed yarn (of the same filament denier and draw ratio) at a higher throughput rate at a higher windup speed followed by draw-texturing (at the same draw ratio to the same drawn denier.)

It may also be achieved by spinning at a higher throughput rate employing the same windup speed, a feed yarn of higher denier and draw ratio followed by draw-texturing at the higher draw ratio of the same drawn denier or by any combination thereof.

Example 1: Polyethylene terephthalate is prepared in a laboratory scale continuous polymerization unit. The equipment consists of an apparatus to continuously feed a slurry of ingredients to a stirred reactor operated at 34.6 psi (178.93 cm Hg) and 270°C. After an average dwell time of 2 hours the mixture is metered to a 7 tray bubble cap column, passing downward at 280°C against a countercurrent flow of 7,500 cm/min of dry nitrogen. The low molecular weight polymer is metered to a horizontal tubular reactor stirred by a cage type agitator.

Average dwell time in this final reactor is approximately 2 hours; temperatures and/or pressures are adjusted to give a final polymer intrinsic viscosity of 0.65. Temperatures are commonly 275° to 290°C, pressures commonly 1 to 10 mm Hg. Water and glycol are removed continuously from all stages of the polymerization.

The slurry feed to this polymerization system consists principally of terephthalic acid and ethylene glycol in a ½ molar ratio. Minor components consisted of an antimony compound (antimony glycoloxide) at 290 ppm antimony (calculated on the weight of polymer theoretically produced) as a catalyst, titanium dioxide at 3,000 ppm as a delusterant, and triphenyl phosphite at 300 ppm as a stabilizer.

Feed rates are adjusted throughout to yield 5.0 lb/hr of a polymer of the specified viscosity. The molten polymer is extruded in a strand, quenched in water and cut into pellets. The pellets are dried under heat and vacuum to a moisture content of less than 0.005%.

The dried pellets are fed into a 1.5 inch extruder with extruder temperatures adjusted to feed a 270°C polymer melt into an attached melt spinning pump, filter, and spinnerette assembly. The molten polymer is extruded at 295°C through a spinnerette having 35 holes, each of 9.0 mil diameter. Polymer throughput is at a constant rate of 89.8 g/min. The molten threadline is cooled in a current of air at room temperature and wound up at speeds from 3,000 to 4,400 yd/min (2,743 to 4,012 m/min). The constant polymer throughput rate gives yarns that vary from 295 denier (8.43 dpf) at 3,000 yd/min (2,743 m/min) to 201 denier (5.74 dpf) at 4,400 yd/min (4,023 m/min).

The spun yarns are simultaneously drawn and false twist textured on a single position laboratory draw texturizer. The 40 inch (101.6 cm) primary heater and 16 inch (40.64 cm) secondary heater are at 200°C, false twist at 60 turns per inch (23.6 turns per cm) (of final yarn length) is inserted at a point between the heaters, and the forward speed of the yarn at windup is 200 yd/min (813 m/min). Draw ratios are adjusted to yield drawn, textured yarns of about 30% elongation at break with filament deniers of 3.4 to 4.8.

Example 2: This example illustrates the increase in spinning productivity obtained with branched polymers spun at high speeds.

Polyesters are prepared, spun and draw textured as in Example 1, but with the addition of a small amount of a polyfunctional chain branching agent. Pentaerythritol is added to the initial slurry in amounts from 100 ppm (based on theoretical polymer weight) to 750 ppm. With chain branching the residual elongation in yarns spun at high speed, above 3,000 yd/min (2,743 m/min), is surprisingly higher than the residual elongation of unbranched polyethylene terephthalate. The residual elongation of the spun yarn is defined as the elongation at the break point where the yarn is extended at the rate of 240% of its initial length/min. The increase in residual elongation over the unbranched control tends to increase as the spinning speed increases.

Two trends are apparent. First, productivity increases with increased chain branching, passes through an optimum level and declines. Second, productivity in the optimum chain branching range increases with increased spinning speed.

The optimum level of chain branching shown will not necessarily be applicable to polyesters differing in composition or processing from these examples. General experience with processing of polyesters would cause one skilled in the art to anticipate that the optimum level of chain branching for maximum productivity increase would be influenced by a number of factors including, but not necessarily restricted to, spinning temperature, intrinsic viscosity, quenching rate, denier per filament, and comonomers or unreactive additives introduced into the polymer.

The type of equipment has a significant effect on spun yarn and drawn yarn properties. Optimum levels of chain branching and the productivity increase obtained will vary with equipment. For this reason, comparisons of chain branched yarns with unmodified polyethylene terephthalate must be made on the same equipment to be meaningful.

Transfer System Between Melt Source and Spinning Position

In the production of polyester fibers in a melt-spinning process, the transfer system between the melt source and the spinning position has a special importance due to the thermal degradation of the polyester. *D. Pendlebury; U.S. Patent 4,072,663; February 7, 1978; assigned to Allied Chemical Corporation* provides an improved transfer process which ensures relatively low thermal degradation of the molten polymer. The process is characterized by employment of two or more indexes whereby transfer elements and polymer flows in the transfer system may be correlated for minimum degradation of the polymer.

Example: 41½ parts per hour of purified terephthalic acid, 23 parts per hour of ethylene glycol and 0.88 part per hour of diisopropylamine are continuously fed to a paddle wheel mixer where they are converted to a paste. The paste mixture is then pumped from the mixer by a feed pump to the inlet of a circulating pump. The paste mixture is pumped with 40 parts of recirculating mixture by the circulating pump through a multiple tube and shell heat exchanger where it is heated to 260° to 270°C.

After leaving the heat exchanger, the mixture enters an esterifying reactor maintained at 260° to 270°C by conventional heating means, and 90 to 110 psig pressure by means of an automatic vent valve. The reaction mixture leaving this reactor is split, with part returned to the inlet of the circulating pump where it is combined with fresh paste, and part reduced in pressure to near atmospheric in a series of three reactors maintained at 260° to 270°C. This part is then flowed for polycondensation to the first in a series of three polycondensation reactors operating at subatmospheric pressure.

The final polycondensation reactor in the series of reactors is an essentially horizontal totally enclosed cylindrical reactor having an essentially horizontal polyester flow, a pool of polyester in its lower portion, and driven wheels to create high surface area in the polyester to facilitate evaporation of volatiles from the polymer. The pool of polyester in the final reactor is maintained by a pump and any change in inventory of polyester in the reactor may be estimated by observing the change in level on a conventional level recorder. A preferred reactor is described in detail in U.S. Patent 3,976,431. Excellent results are also obtained with use of the reactor described in U.S. Patent 3,728,083.

The polyethylene terephthalate polyester polymer issuing from the final polycondensation reactor has an average intrinsic viscosity of 0.96 unit. This polymer is transferred to a melt spinning system in accordance with the method of this process. It is first fed at a temperature of 276°C to a conventional screw pump operating at about 1,920 lb/hr. The polymer melt exits this screw pump at a pressure of 350 psig and a temperature of 286°C and is fed through a first pipe zone maintained at 286°C to two gear pumps in parallel, operating at 13 rpm.

For minimum thermal degradation of the polymer, the pipes in the first pipe zone are sized so that the rate of flow of the polymer melt in each pipe is maintained in accordance with the limit formula $D_1 = K_1(Q_1)^{1/3}$ where Q_1 is the rate of flow of the polymer melt in the pipe in lb/hr, D_1 is the inside diameter of the pipe in inches, and K_1 is between 0.331 and 0.360.

The polymer melt exits the two gear pumps at a pressure of 2,150 psig and a temperature of about 292°C and is fed through a second pipe zone maintained at 292° to 294°C to 32 conventional metering pumps in parallel arrangement. For minimum thermal degradation of the polymer, the pipes in the second pipe zone are sized so that the rate of flow of the polymer melt in each pipe is maintained in accordance with the limit formula $D_2 = K_2(Q_2)^{1/3}$ where Q_2 is the rate of flow of the polymer melt in the pipe in lb/hr, D_2 is the inside diameter of the pipe in inches and K_2 is between 0.248 and 0.270.

The metering pumps operate at about 13 rpm. The polymer melt exits the 32 metering pumps at a pressure of about 3,000 psig and a temperature of 298°C and is fed directly to a spin pot comprising a spinning filter and a spinnerette, the spinning filter being disposed between the metering pump and the spinnerette. The spinning filter consists of a conventional sieve filter combination of 24 metal screen layers. The pressure drop through the spinning filter averages about 2,500 psig. The spin pot is enclosed in a controlled high temperature atmosphere so that the loss of heat from the polymer melt is minimized.

The increase in melt enthalpy through the metering pump and sieve filter is sufficient to heat the polymer melt at a point immediately above the spinnerette to about 305° to 310°C, and the pressure at this point is about 500 psig. The flow of polymer through the spinnerette is maintained at a constant rate of 60 lb/hr by the metering pumps.

The spinnerette is divided into two parts by means of an undrilled stripe wide enough to form a visible split between the multiple ends below the spinnerette. The spinnerette plate has 384 holes (192 holes on each side of the undrilled stripe), each 0.018 inch in diameter, spaced so that the distance between the filaments formed is 0.28 to 0.40 inch immediately below the spinnerette.

From the spinnerette there is extruded two ends of multifilament, continuous filament yarn, and the two ends are passed downwardly into a substantially stationary column of air contained in a heated sleeve about 15 inches in height, disposed surrounding and immediately beneath the spinnerette. The air temperature in the heated sleeve is maintained at about 300°C at the top of the sleeve, decreasing to about 115°C at the bottom. A heated sleeve baffle is provided at the bottom of the heated sleeve forming an inwardly extending flange to minimize flow of cooling air into the heated sleeve.

The two ends of multifilament yarn leaving the heated sleeve are passed directly into the top of a conventional quenching chimney, such as that described in U.S. Patent 3,936,253 where they are in contact with cooling air introduced into the path of the filaments. The undrawn yarn thus produced may be further treated by conventional procedures well known in this art.

In this example, the total residence time of the polymer melt in the transfer system and spin pot is 18 minutes, and the total carboxyl increase is 5 eq/10^6 g of polymer. Total decrease in intrinsic viscosity of the polymer is about 0.05 unit.

In further tests, it is shown that the preferred residence time of the polymer melt in the transfer system and spin pot should be less than 20 minutes. Thus, the spinning unit should be located as closely as possible to the final polycondensation reactor in order to minimize length of the transfer lines. The screw pump is preferably located at the base of the final polycondensation reactor, and the screw pump is designed for relatively low discharge pressure, e.g., 350 psig, to minimize temperature rise across the pump.

The gear pumps for transfer of the polymer should be located as closely as possible to the screw pump discharge, and the preferred pump is a rotary gear pump with low temperature rise characteristics. The number and size of the gear pumps should be selected so that normal operating speeds are 10 to 15 rpm. Preferred gear pump suction pressure is 100 to 250 psig. Preferred metering pump suction pressure is 200 to 1,000 psig.

The pipes in the transfer system are preferably cooled with a Dowtherm solution at a minimum liquid temperature of 265°C. If in-line static mixers are used in the transfer system, the mixers are preferably located before major flow divisions and immediately prior to the spinning system.

Enhanced Dyeability and Thermal Stability by High Speed Spinning

H.R.E. Frankfort and B.H. Knox; U.S. Patent 4,134,882; January 16, 1979; assigned to E.I. Du Pont De Nemours and Company have found that polyethylene terephthalate filaments spun using extremely high withdrawal speeds (e.g., over about 5,000 m/min) show excellent "dye at the boil capability," i.e., it is possible to dye such filaments at the boil within a reasonable period of time without the need for conventional carriers or chemical modifiers.

Prior commercial polyethylene terephthalate textile yarns having similar physical properties, e.g., tensile properties and boil-off shrinkage, have not shown this capability. When these dyeable yarns are textured, they may lose some of this capability, to an extent depending on the speed of withdrawal during spinning, but such textured yarns can be dyed with a reduced need for carriers. It has also been found that polyester filaments that have been spun at these extremely high withdrawal speeds have good thermal stability, i.e., relatively low shrinkage over a large temperature range. Prior commercial polyester textile filaments have not shown such stability.

It has also been found that most filaments spun at these extremely high withdrawal speeds are characterized by a high long period spacing (LPS) of over 300 A. These properties seem to be largely inherent in filaments spun at withdrawal speeds as described in U.S. Patent 2,604,667. Other useful characteristics have also been discovered in yarns produced at extremely high speeds, especially above 6,000 m/min.

Increasingly difficult problems with broken filaments have, however, been encountered as the withdrawal speed has been increased, to the extent that sometimes it has not even been possible to achieve continuity of winding at these extremely high speeds. Broken filaments and other yarn defects have also presented problems during subsequent textile operations, such as texturing, when using filaments spun at these extremely high withdrawal speeds.

It has been found that many of these problems during spinning at these extremely high speeds, or during subsequent textile operations on the resulting filaments, can be associated with a significant difference between the birefringence of the surface and the birefringence of the core of the filament, and that better filaments can, therefore, be obtained more reliably and consistently by controlling the spinning and cooling conditions so as to minimize such difference in the as-spun filament.

This difference is referred to herein as differential birefringence (Δ_{95-5}), being the difference in birefringence between points along the radius of the filament at the indicated 95 and 5 percentage distances from the axis, or more simply as "skin-core." The skin-core values generally increase with the spinning speed, i.e., the speed of withdrawal from the spinneret, which correlates approximately with the stress required to extend the as-spun yarn by 20% (δ_{20}). As the spinning speed increases from about 5,500 yd/min (about 5,000 m/min), it becomes increasingly more difficult to ensure that the skin-core value is low enough to reduce the likelihood of problems, such as broken filaments, to an acceptable level.

If the filaments are spun at about 5,500 yd/min (about 5,000 m/min), problems resulting from high skin-core values may be manifested only during subsequent textile operations, e.g., broken filaments during texturing, or breaks and other defects in e.g., woven fabrics. As the spinning speed increases, however, high skin-core values in the solidified filaments are more likely to cause continuity problems in the actual spinning process.

Problems with continuity in spinning or with yarn and fabric defects can also be caused by other factors, so that it is not a complete solution to such problems merely to arrange for the filaments to be spun with a low skin-core, and to ignore the effect of other factors. It has been found that the spinning of filaments having high skin-core values at these extremely high withdrawal speeds will generally cause such problems, despite care in controlling other factors.

There are provided, therefore, low shrinkage polyethylene terephthalate filaments having enhanced dyeability, a long period spacing (LPS) of more than 300 A and a differential birefringence (Δ_{95-5}) as herein defined according to the relationship $\Delta_{95-5} \leqslant 0.0055 + 0.0014\ \delta_{20}$ where δ_{20} is the stress measured in grams per denier (gpd) at 20% extension and is at least 1.6 gpd. When δ_{20} is between 3 and 4 gpd, however, Δ_{95-5} may be $\leqslant 0.0065\ \delta_{20} - 0.0100$. Δ_{95-5} is preferably less than 0.008.

When these filaments are annealed, the LPS decreases significantly. The filaments are, however, characterized by an amorphous birefringence (Δ_{am}) less than 0.07, and crystal size of at least ($1250\ \rho - 1670$) A where ρ is the density, which is preferably at least 1.37 g/cm^3, whether such filaments are as-spun or annealed. Preferrred features of these filaments are that the dyeability be such that the relative disperse dye rate (RDDR) be at least 0.050, the thermal stability be such that the dry heat shrinkage (DHS) be not more than 1% more than the boil-off shrinkage (BOS) and that the mean birefringence Δ_5 be between 0.09 and 0.14.

The BOS is measured as in U.S. Patent 3,772,872. A yarn having a low BOS of 2 to 2.5% can be package-dyed without the use of special packages or heat setting.

The DHS (160°C) is unusually low; generally being less than 1% more than the BOS, and is measured in the same manner as the BOS except that the sample is heated in air in a 160°C oven.

The dyeability of various yarns is compared herein by measuring their disperse dye rate (DDR) which is defined hereby as the initial slope of a plot of percent dye in filament by weight vs the square root of dyeing time which is a measure of a dye diffusion coefficient (if corrected for difference in surface to volume ratio). The values of the disperse dye rate are normalized to a "round filament" of 4.7 dpf having a density of 1.335 g/cm^3, i.e., of an "amorphous" 160-34 round filament yarn, as a RDDR, defined by the relation:

$$\text{RDDR} = \text{DDR (measured)} \times (\text{dpf}/4.7)(1.335/\rho)[100/(100 - \text{BOS})]^{1/2}$$

where ρ is the polymer density; dpf is the filament denier; and BOS is the yarn boil-off shrinkage. The RDDR value is approximately independent of the surface-to-volume ratio of the dyed filaments and reflects differences in filamentary structure affecting dye diffusion.

The improved dyeability of the filaments is partially attributed to their large crystals and low amorphous orientation. An increase in crystallinity and/or a decrease in crystal size will reduce potential dyeability. The above structural features appear characteristic of yarns spun at extremely high speeds, but to make a useful yarn with these desirable properties at these speeds, it is necessary to avoid forming a skin on the filaments. The absence of any significant skin is indicated by low $\Delta_{95\text{-}5}$ values.

Deep-Dyeing Textured Yarn Spun at High Speed

In the process developed by *J.E. Bromley, W.T. Mowe and F. Stutz; U.S. Patent 4,128,989; December 12, 1978; assigned to Monsanto Company*, polyester yarn is spun at a sufficiently high speed to produce substantial stress-induced crystallinity and low shrinkage. The resulting feed yarn is textured, yielding a textured yarn which dyes considerably deeper than conventional polyester yarns.

In the spun state (prior to texturing), the yarns dye progressively lighter as spinning speed increases. One would expect lighter dyeing textured yarns to be produced from lighter dyeing feed yarns. It is, therefore, entirely unexpected that the textured yarns produced from feed yarns spun at speeds above the noted transition speed range dye considerably deeper than textured yarns produced from feed yarns spun at speeds below the transition speed range. Depending on the texturing conditions, the dyeing conditions, dyestuff, type and amount of carrier (if any), the textured yarns according to the process dye as much as 50% deeper than textured yarns made from yarns spun at intermediate speeds such as 3,200 to 3,500 m/min.

The stress induced morphology in the spun state results in the morphology in the textured state, which is believed to be responsible for the observed deeper dyeing characteristics of the textured yarn.

Example: Linear polyethylene terephthalate polymer of normal textile molecular weight is extruded downwardly through 34 capillaries of a spinneret, each capillary having a diameter of 15 mils (0.38 mm) and a length of 30 mils (0.76 mm). The spinning temperature just above the spinneret is 290°C. The molten streams are quenched by horizontally directed air at room temperature and an air velocity of 14 m/min in a quench zone just below the spinneret.

The solidified filaments are converged into a filament bundle or yarn at a point 2 m below the spinneret and pass downwardly about a driven feed roll and its associated separator roll rotating with a peripheral speed of 6,000 yd/min (5,482 m/min), from which the yarn is fed to a winding mechanism where it is collected in an orderly fashion as a spun yarn having 180 denier, an elongation to break of 55%, and 4% shrinkage. The feed roll is located 5 m below the spinneret. The spun yarn has a crystallite dimension in the 103 direction of about 125 A, crystallite dimensions in directions lateral to the yarn axis of at least 60 A, and an average crystallite volume of 6.7×10^5 A^3.

The spun yarn is simultaneously draw textured at 340 m/min and a draw ratio of 1.14:1 using a friction false twist mechanism of the type disclosed in U.S. Patent 3,973,383, a primary heater temperature of 205°C, and a secondary or setting heater temperature of 200°C.

The resulting textured yarn is knitted into fabrics and dyed using various dyestuffs. The fabrics dye as much as 50% deeper than fabrics knit from textured partially oriented yarn (POY). X-ray examination of textured yarn according to the process reveals a crystallite skewness angle between 5° and 35°, typically between 20° and 30°, as compared with textured POY which exhibits skewness angles in excess of 38° to 40°. This observed lower skewness angle in yarns according to the process correlates with the observed deeper dyeing and is believed to define the internal structural parameter responsible for the deeper dyeing.

High Speed Production of Preoriented Yarn

E. Luzzatto; U.S. Patent 4,185,062; January 22, 1980; assigned to Snia Viscosa Societa Nazionale Industria Applicazioni Viscosa SpA, Italy discloses a process for the high speed production of preoriented yarns from synthetic linear polymers, in particular polyesters, comprising the steps of melt spinning the polymer by extrusion through a spinneret and, before winding up the extruded filament bundle, causing the bundle to pass through a zone where the ambient gas, in particular air, is caused to move in a direction parallel to and concurrent with that of the filament bundle progressing towards the wind up means.

With reference to Figure 3.6, numeral **10** diagrammatically indicates a spinneret from which the polymer, in particular the polyester, employed for the carrying out of the process, is extruded. The spinneret may have any desired number of orifices, correspondingly producing a filament bundle **11** having any desired number of filaments, and the orifices need not necessarily have diameters which differ from those employed in the known art. The flow rate of polymer melt through the spinneret orifices is easily calculable from the wind-up speed and the count of the yarn that it is desired to obtain and will have to be such as to furnish the quantity of polymer that is necessary to collect a yarn of the desired count at the wind-up speed employed.

Below the spinneret the yarn **11** comprises a zone **11A** in which it solidifies and in which it may be optionally subjected to a cooling action using cold gaseous streams, which are not illustrated because they have no direct relationship to the process.

Below zone **11A**, the yarn runs along the zone **11B** in a guide channel **12**. In the channel an air stream, or even a stream of a different gas, is caused to flow in the same stream direction as the yarn. The air may be at room temperature, or may be heated, according to the polymer employed and the characteristics desired from the yarn, and according to what appears suitable in each different case; the temperature is not a critical feature in carrying out the process.

The concurrent motion of the ambient gas may be determined in any desired way. Only for illustrative purposes it is assumed in the drawing that the ambient gas is fed in by a pump **14** through a pipe **15** which blends with an annular space **16** from which the air, through a plurality of holes **17**, radially enters into the accompanying tube **12**. A plate, schematically indicated at **22** closes the tube at the top, leaving an aperture **23** for the passage of the yarn; this aperture is narrow enough for the entraining effect of the yarn to prevent the air from leaking out of it.

Above the entrance of the tube finish may be applied by any suitable device, schematically indicated as a couple of rollers **18**, or an analogous device, indicated by **19**, may be located at the outlet of the tube, or finally both devices or any other known devices may be employed to lubricate the yarn. It is also acceptable that a finish may be applied inside the tube by means of devices easily conceivable by persons skilled in the art.

As it leaves the tube, the yarn is drawn by a drawing device generally indicated at **20** and proceeds to form a bobbin **21**. The wind-up speed is maintained uniform, as is normal in devices of this kind.

Figure 3.6: High Speed Production of Preoriented Yarn

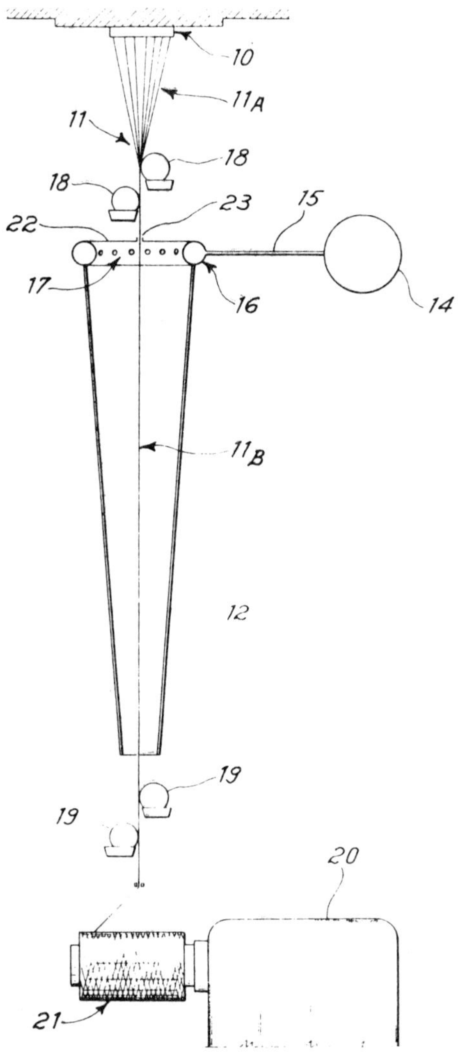

Source: U.S. Patent 4,185,062

Operating according to the prior art using polyethylene terephthalate with a relative viscosity in phenol/tetrachloroethane of about 22, if it is desired to obtain a preoriented yarn having an elongation at break of about 140%, it is necessary to operate at a wind-up speed of 3,000 to 3,100 m/min. If the wind-up speed is increased to about 4,000 m/min, the elongation at break decreases to

70 to 75%. Lower elongations at break do not permit an advantageous application of the concurrent drawing and texturization process.

If the air is fed to the tube **12** at such a flow rate that, taking into account the cross-section of the tube, a speed of flow of about 1,000 m/min in the same direction of the yarn is imparted to the air in the immediate vicinity of the yarn, it will be possible to operate at a wind-up speed of about 4,000 m/min and still have an elongation at break of about 140%. If an elongation at break of 70 to 75% is sufficient, with the same speed imparted to the air, it will be possible to increase the wind-up speed to about 5,000 m/min.

It will be necessary to specify that by "speed imparted to the air" is understood the speed, measured in the direction of the yarn axis, that would be reached by the air in the absence of any entraining by the yarn and while aperture **23** is closed, which speed is different from that reached by the air while the yarn is in motion, and it is possible to measure the speed before starting the spinning or by subsequently stopping the yarn.

When the tube **12** does not have a constant cross-section, the actual speed changes from cross-section to cross-section, and it is then necessary to refer to the maximum speed which is generally the one measured adjacent to the outlet cross-section of the yarn; that is, at the bottom. The air speed is always understood to be measured in the vicinity of the yarn, even though the outlet cross-section will be generally small enough to permit practical reference to the average speed.

Vinyl Copolymer to Reduce Pilling

K.-H. Magosch, R. Feinauer and J. Rüter; U.S. Patent 4,104,261; August 1, 1978; assigned to Chemische Werke Huls, AG, Germany disclose an improvement in the process of melt spinning ethylene terephthalate polyester to produce textile fibers with decreased pilling in the threads, fibers or fabrics made therefrom. The ethylene terephthalate is melt polymerized in the presence of a condensation catalyst to a final temperature of about 250° to 300°C and a final pressure of less than 10 mm mercury to form an anhydrous melt.

A vinyl copolymer having more than two carboxyl and/or hydroxyl groups is added to the melt polymerization at a concentration of about 0.005 to 2.0% referred to the dicarboxylic acid component. Melt spinning of the modified polyester is then carried out. The vinyl copolymer has the following general formula:

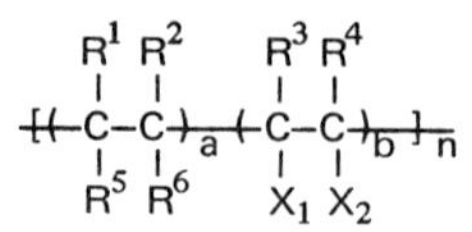

where

n is an integer from 10 to 10,000;

a is zero or an integer from 1 to 100;

b is an integer from 1 to 100;

R^1 thru R^6 are hydrogen, an alkyl group with 1 to 3 carbon atoms or an oxalkyl group $-OR^7$;

R^7 is an alkyl group with 1 to 6 carbon atoms or an aryl group with 6 to 10 carbon atoms;

X_1 is hydrogen, an alkyl group with 1 to 3 carbon atoms, or a carboxyl group, an esterified carboxyl group; and

X_2 is a carboxyl group, a hydroxyl group, an esterified carboxyl group or an esterified hydroxyl group.

The pilling behavior of the yarns from the polyesters of the process is suitably determined by the wire kink rupture coefficient [(DKZ) are obtained with the wire kinking abrasion instrument of K.H. Gruenewald, *Chemie-fasern* 12, 853 (1962)].

Example: The following ingredients are placed into an ester interchange reactor holding 100 liters and provided with a stirrer, a double jacket and a fractioning column: 38.8 kg of dimethyl terephthalate; 0.145 kg of a methanol esterified ethylene-maleic acid (1:1) copolymer having a molecular weight of about 20,000; 24.8 kg ethylene glycol; and 0.0388 kg calcium acetate.

The mixture is heated and methanol starts distilling off at about 150°C. After methanol dissociation has ceased at 200°C, which generally will be the case no later than after 3 hours, the ester interchange product is fed into the polycondensation reactor of the same capacity.

0.0388 kg triphenyl phosphate and 0.0116 kg of antimony trioxide are added. The temperature is continuously raised to 280°C while the pressure is gradually lowered to about 0.1 torr (torr = 1 mm Hg). Polycondensation is stopped when power absorption while stirring the reactor contents corresponds to a melt viscosity of about 500 to 800 poises at 285°C.

The polyester melt is extruded in the form of thin strands through the bottom of the reactor where the strands are cooled in cold water. They are then cut into small cylinders of about 5 mm length and 2 to 3 mm diameter.

The reduced specific viscosity (RSV) values of the colorless polyesters (measured in phenol/tetrachloroethane 60/40 at 25°C) amounts to 0.48 dl/g; the melting point (determined with the monoscope of Bock, Bad Oeynhausen) is 257°C. The melt viscosity (determined with a high pressure capillary viscosimeter of Zwick) is 540 poises at 285°C.

The material is extruded through a spinneret with 8 holes of 0.25 mm diameter at a spinning temperature of 307°C to obtain yarns which are wound at a speed of 640 m/min and which are then stretched at a ratio of 1:3.88.

The thread properties are as follows: titer, 44/8 dtex; tear resistance, 2.1 p/dtex; and elongation, 39%. Subsequently the DKZ is determined from 48 individual capillaries; the average value from these individual determinations is obtained as 270 lifts.

Yarns prepared in similar manner but without the addition of the methanol esterified ethylene-maleic acid (1:1) copolymer, and of the following properties: RSV, 0.60 dl/g; melting point, 258°C; titer, 44/8 dtex; and tear resistance, 35.7% show a DKZ of 2,250 lifts.

Polyethylene terephthalate with RSV, 0.48 dl/g and melting point of 256°C which does not contain the ethylene-maleic acid (1:1) copolymer, may not be spun under the same conditions because of the low viscosity of the melt.

Anti-Pilling Filaments with High Tenacity and Low Knot Tenacity

H. Sugiyama and H. Yasuda; U.S. Patent 4,071,502; January 31, 1978; assigned to Toyo Boseki Kabushiki Kaisha, Japan produce a polyester with reduced pilling by using an undrawn filament having a high birefringence, which may be produced by melt spinning at a high velocity, for the manufacture of a drawn filament. By the use of such filament, it has been made possible to effect drawing evenly under a high tension and to produce a uniformly drawn filament having a high tenacity. It is possible to obtain a good anti-pilling polyester fiber from such drawn filament of high tenacity by suppressing the knot tenacity sufficiently.

The fiber has an intrinsic viscosity of 0.32 to 0.48 and has the following physical properties: tenacity, 3.5 to 5.0 g/den; knot tenacity, not more than 3.0 g/den; and ratio of tenacity and knot tenacity, not less than 1:5.

The anti-pilling property was determined as follows. A specimen was applied to an ICI pilling tester which was operated at 60 rpm. After 5 hours the surface condition of the specimen (knitted fabric) was evaluated macroscopically in comparison with the standard photographs for judgment ranked into 1 to 5 grades. The value as indicated is the average value of 4 specimens (JIS L 1076).

Spinnability was determined by counting the frequency of yarn breakages per hour on 400 spinning spindles. The spinning spindle rotation was 14,000 rpm.

Example: Polyethylene terephthalate of 0.38 intrinsic viscosity was subjected to melt spinning through a spinning nozzle having 300 outlets at a spinning temperature of 270°C and a take-up velocity of 2,500 m/min to obtain an undrawn tow having an intrinsic viscosity of IVf equals 0.37 and a birefringence of Δn equals 0.025. Two devices provided with double heating plates (i.e., upper and lower plates) between two rollers which have different rotation speeds were equipped, and the undrawn tow which elapsed more than 7 days after melt spinning was passed through them to give two-stage drawing.

The drawing conditions were 3.0 times at 180°C for the first stage and 1.2 times at 180°C for the second stage. The breakage of the tow at the time of drawing was nil, the monofilament denier of the drawn tow was 1.5 den and the birefringence was Δn equals 0.197.

The drawn tow was crimped by a stuffer box type crimper and subjected to dry heat treatment at 160°C for 1 minute, after which it was cut into 38 mm lengths to prepare the spinning staple.

The fiber quality was tested to show tenacity (DT) equals 3.9 g/den, knot tenacity (KT) equals 2.4 g/den and DT/KT equals 1.62. It displayed a favorable spinnability and an anti-pilling property of grade 5.

Low Carboxyl Polyester Fibers Using Alkali Metal Salt as Catalyst

G.L. Korver; U.S. Patent 4,071,504; January 31, 1978; assigned to The Goodyear Tire & Rubber Company describes an improved process for preparing textile and industrial fibers by melt extruding a polyester resin from melt spinning apparatus to form filaments while adding to and reacting with the polyester resin while in the melt spinning apparatus a monofunctional epoxide. The improvement comprises incorporating in the polyester resin prior to extrusion thereof at least one catalytic compound consisting of an alkali metal salt in an amount from about 0.01 to 10.0 pbw (calculation based on the alkali metal) per 1,000 pbw of the polyester.

Example: A sample of polyethylene terephthalate was prepared by the esterification of a mixture of terephthalic acid and ethylene glycol and subsequent polycondensation of the esterification product employing conventional techniques. To the initial reaction mixture of terephthalic acid and ethylene glycol was also added 0.026 pbw of sodium (as sodium acetate) and 0.025 pbw of lithium (as lithium acetate) per 1,000 pbw of the polyester resin. The melt polymerized resin containing the sodium and lithium salts was determined to have an intrinsic viscosity (I.V.) of about 0.6.

This resin was diced into cubes and solid state polymerized by tumbling the resin cubes in a blender-dryer at a temperature of 240°C and a pressure of 0.5 mm of mercury pressure. This final solid state product was divided into equal portions (Samples A and B) and dried to a moisture level of about 0.004%. Samples A and B were then extruded into fibers using a 1 inch NRM extruder. In the case of Sample B 0.36% by weight of phenyl glycidyl ether (PGE) was added. The results are shown below.

Samples	I.V.	Carboxyl Content
Polyester resin	1.05	10 eq/10^6 g
Sample A	0.80	17 eq/10^6 g
Sample B	0.78	7 eq/10^6 g

From the preceding it is readily apparent that an alkali metal salt catalyzes the reaction between the carboxyl ends in polyester and an epoxide compound. It is also apparent that this catalyzed reaction can be carried out during the spinning of the resins into fibers, thus avoiding the economical drawbacks of prior art solutions.

Antistatic Polyether-Polyester Block Copolymer

T. Endo, M. Takeuchi, T. Hanada, K. Nakagawa, H. Komatsu and I. Nakamura; U.S. Patent 4,035,441; July 12, 1977; assigned to Toray Industries, Inc., Japan have prepared a polyester filament in which a polyether-polyester block copolymer is dispersed as fine striae along the filament axis, at least one of which is substantially endless. The process for preparing the same includes a melt spinning apparatus with static mixing elements interposed between separate molten polymer feed passageways and spinneret holes.

There is shown in Figure 3.7 an assembled spinning apparatus consisting of pack block **11**, spinneret **1**, mixing plate **4**, pressure supporting plate **6**, spacer **8** and annular bolt **14** together with filters **3**, **7**, **31** and gaskets **2**, **5**, **9**, **13**, to which apparatus (sometimes referred to herein as a "pack") is connected tip **15** of a polyether-polyester block copolymer (polymer A) supply system via gasket **16**, making it possible to carry out melt spinning with instant mixing.

The polyester (polymer B) constituting the main feed, passes through conduit **17** and reaches space **19** via annular space **18**, passing through filter layer **12**, thereafter, rising in annular space **22** via conduit **20** and space **21**, meeting in meeting space or chamber **28** with polymer A separately supplied via conduit **25**, filter layer **26** and small hole **27**. Thereafter, polymers A and B pass through the opening of static mixing part **23** and are immediately spun from spinneret orifices in spinneret **1**.

Example: In a pack structure as shown in Figure 3.7, the static mixing part consisted of nine static mixing elements fixed in a pipe having an inner diameter of 6.3 mm. Each of the mixing elements was formed by twisting a rectangular plate 180°. The elements were arranged so that elements of right-handed and left-handed curvature alternated with one another, the angle between contacting edges of adjacent elements being about 90°, all as disclosed in U.S. Patent 3,286,992.

In this example polymer B was obtained by melting at 290°C polyethylene terephthalate having an intrinsic viscosity of 0.66. This was supplied to conduit **17** at a rate of 60 g/min. Morandum particles having an average particle diameter of 60 mesh formed filter layer **12**, the height of which was about 20 mm. Polymer B passed through this layer and thereafter was introduced to meeting space or chamber **28** via the spaces or passageways **21** and **22**.

To form polymer A, 102 parts of polyethylene glycol (molecular weight 20,000) and 0.15 part of calcium acetate as a catalyst for an ester interchange reaction was added to 69.05 parts of dimethyl terephthalate and 43.90 parts of ethylene glycol. An ester interchange reaction was carried out while distilling off methanol at a temperature of 140° to 220°C. Thereafter, 0.07 part of trimethyl phosphate and 0.17 part of Irganox 1010 as an antioxidant, and 17 parts of sodium dodecylbenzenesulfonate were added to the reaction product, and the mixture was polymerized at a temperature of 255°C under a highly reduced pressure of about 0.2 mm Hg for 4 hours to obtain an opaque polymer A having a good color tone and a relative viscosity of 2.8 measured in a 1% solution of o-chlorophenol at 25°C. The specific resistance of the polymer A was 2×10^6 Ω-cm.

Figure 3.7: Melt Spinning Apparatus Incorporating Mixing System

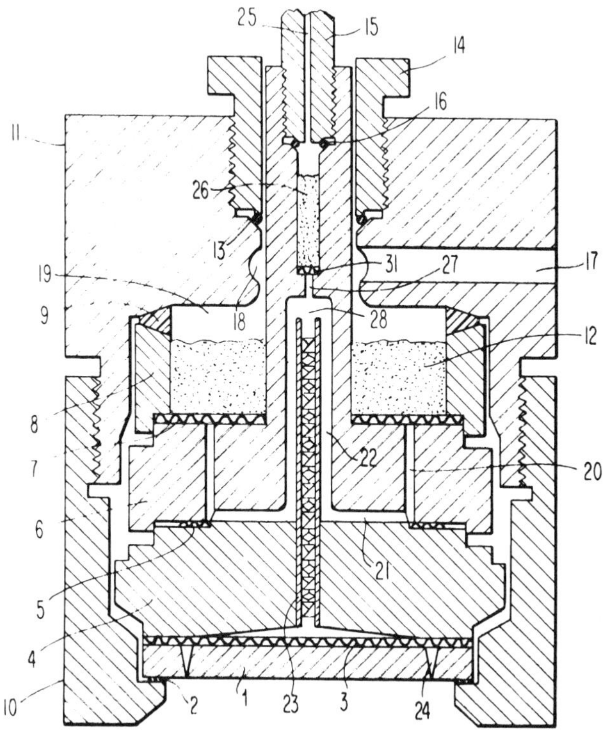

Source: U.S. Patent 4,035,441

For solid polymers, specific resistance is based on the resistance between both ends after the polymer is made into a specimen in a state of gut or rod, and a voltage of 500 V is impressed; for liquid polymers, it is a value determined from the resistance between 2 silver electrodes inserted in a specimen (impressed voltage again 500 V) measured in an atmosphere at 20°C and at a relative humidity of 40% for more than 200 hours until the obtained value of resistance is considered to substantially converge to a constant value, while keeping the temperature at 20°C during the period. The specific resistance of the polymer is a value calculated from the resistance (R) in accordance with the following equation:

$$\rho = \frac{R \times S}{L} = \frac{R \times W}{L^2 \times d}$$

ρ is specific resistance (Ω-cm) of the specimen;

R is measured resistance (Ω);

W is weight of the specimen (g);

S is cross-sectional area (cm^2) of the specimen;

d is distance between electrodes (cm); and

L is length (cm) of the specimen.

The polymer A, melted and kept at 275°C was discharged into the space **28** via the conduit **25**, filter layer **26**, and small hole **27** of the apparatus shown in Figure 3.7. The rate of feed of polymer A was controlled so the filament product was 0.48% by weight. After passing through static mixing elements, with an average residence time therein of 2.6 sec, the polymer A-polymer B mixture was then spun within about 12 sec thereafter.

A spinneret having 48 spinneret orifices, each having a diameter of 0.3 mm was used, and the spun filament was taken up at a rate of 1,000 m/min by a conventional method to obtain an undrawn yarn, which was then drawn by a conventional method to obtain a 150 den drawn yarn having an elongation of about 30%. The number of the endless stria was about eleven, and the average maximum diameter was about 0.6 μ.

The polyester drawn yarn had a specific resistance of 3.5 x 10^8 Ω-cm, and it had very good antistatic properties.

Next, the 150 den/48 filament yarn thus obtained was false twisted using a false twist machine designated CS-9 (Earnest Scragg Co.) at a spindle revolution rate of 180,000 rpm, a set temperature of 240°C, number of false twist of 2,500 T/m and a feed rate of 2%.

When the crimp resiliency (CR %) of the obtained textured yarn was sought, it was 41.1%.

The CR value of a comparable textured yarn obtained by false-twisting a 150 den/48 filament polyester yarn without polymer A dispersed therein, was 44%. (Filaments with a CR value less than 40% have insufficient elasticity for application in certain fields.)

The crimp resiliency was estimated by a value obtained by the method of H.A.T.R.A. (which is the same as JIS L 1077) after heat-treating the textured yarn at 90°C for 30 minutes.

Process for Textured Yarn

J.E. Bromley and J.-P. Yu; U.S. Patent 4,176,150; November 27, 1979; assigned to Monsanto Company produce a textured melt spun filament having alternate S-twisted and Z-twisted helical sections connected by twist reversal regions; the filament having a cross-section comprising a spiral where the outer portion of the spiral lies at the inside of the helical sections.

Figure 3.8a shows an S-twisted section of a filament **20** according to the process where the filament cross-section is in the form of a spiral having an outer portion **22** and an inner portion **24**. As shown in Figure 3.8a, the outer portion **24** of the spiral lies at the inside of the coils of the helically coiled region while the remainder of the spiral lies at the outside of the coils of the helically coiled region.

Figure 3.8b illustrates a spinneret orifice which may be used for spinning the Figure 3.8a filament. Orifice **26** is formed in spinneret plate **28**, extending in a spiral from an inner end **30** to the outer end **32**. Preferably, the spiral extends over more than 360°, as illustrated. If the clearance between the inner end and the nearest intermediate portion of slot **26** is sufficiently small, the molten stream issuing therefrom will bridge the gap between the inner end of the spiral cross-sectioned stream and the nearest intermediate portion of the stream cross-section, forming a filament with a spiral cross-section closed at its inner end as illustrated in Figure 3.8a.

Figure 3.8: Production of Textured Yarn

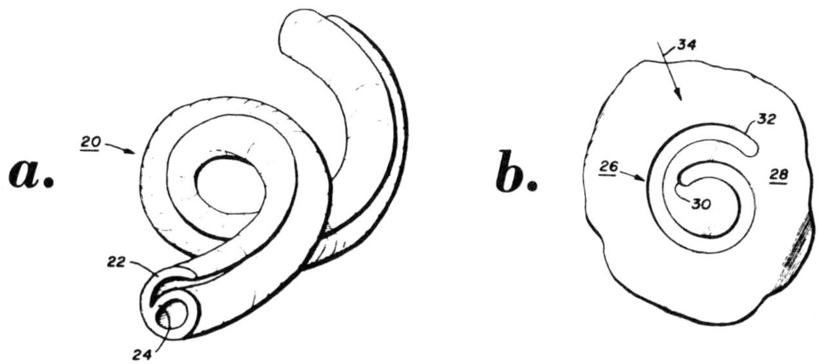

(a) Portion of extruded filament
(b) Bottom plan view of spinneret

Source: U.S. Patent 4,176,150

On the other hand, if the noted clearance is slightly larger, the bridging will not occur, and the resulting filament will have a spiral cross-section open at its inner end. Selection of the proper clearance to provide either a closed inner end or an open inner end while using particular spinning and quenching conditions can readily be made by one skilled in the art.

Generally speaking, the filament having a cross-section comprising a spiral closed at its inner end will have a more powerful crimp than one having a cross-section comprising a spiral open at its inner end. The latter will, however, have substantially increased moisture transport and moisture holding capacity as compared to the former, which is itself superior to ordinary round filaments.

Example: An orifice similar to that in Figure 3.8b is used, the slot being 0.1 mm wide and 4 mm long along its spiral length. Polyethylene terephthalate polymer of normal textile molecular weight is extruded at a temperature of 290°C through the orifice and is solidifed by transversely directed quenching air into a filament which is wound at 3,000 m/min. The polymer extrusion rate is selected such that the filament has a denier of 8.5. The quenching air has a temperature of 18°C and 68% relative humidity, and is directed horizontally at the molten stream in a direction parallel to arrow **34** in Figure 3.8b, the quenching zone being 1.5 m long.

The quenching air has an average velocity of 20 m/min and impinges on the relatively thin finlike outer portion of the spiral cross-section while the remainder of the molten stream is shielded from the quenching air by the outer portion.

The resulting filament has latent crimp and an elongation of 85%. Upon being hot drawn at a temperature of 100°C to an elongation of 10 to 30%, e.g., 20%, the yarn develops more than about 12% crimp with alternating S and Z helical sections, the finlike portion (the outer portion of the helix which was exposed to quenching air) forming the inside of the helical crimp and the remainder of the filament cross-section forming the outside of the helical crimp.

C-Shaped Filaments

H.G. Jackson; U.S. Patent 4,176,211; November 27, 1979; assigned to Phillips Petroleum Company has developed a method for producing polyester filament having a cross-section of C-shape by melt spinning polyester through a spinneret so that each filament is made through an array of holes having a center hole with eight holes equally spaced around the center hole with the size and spacing of the holes within a specified relationship.

Referring to Figure 3.9a, the alignment of the holes in an array on the face of the spinneret is shown which is described below as used in the example. The figure shows eight circular holes **1** of substantially equal diameter arrayed in even alignment around the ninth central hole **3** with the center of the surrounding holes removed two diameters from the center of the central hole.

Figure 3.9: Production of C-Shaped Filaments

a.

b.

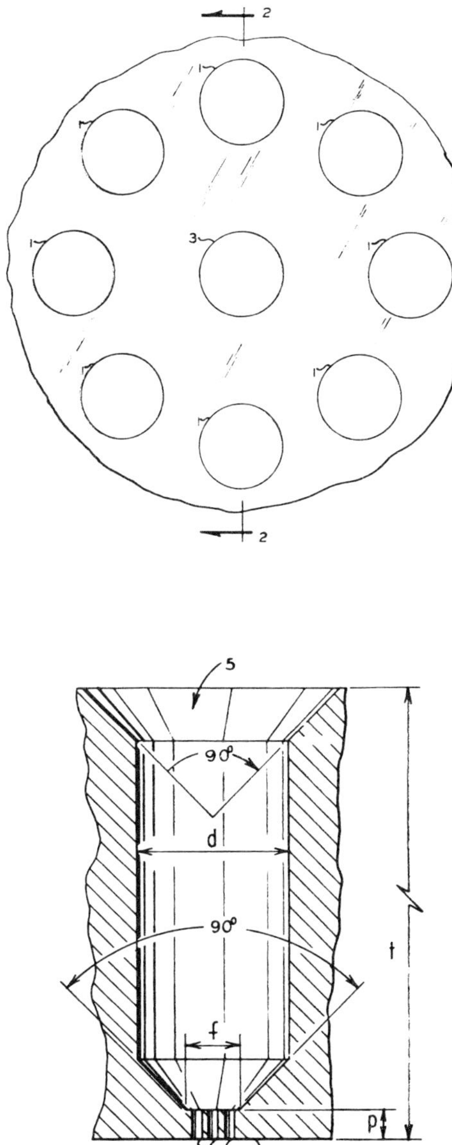

(a) Array of holes on spinneret face
(b) Side view through **2-2** of (a)

Source: U.S. Patent 4,176,211

In Figure 3.9b is seen a cross-section of a spinneret plate showing the bore **5** into which molten polymer is passed to be pressured through an array of spinneret holes **7** at the face **9** of the spinneret plate. The conditions for producing molten polyester and the conditions required for expressing this material through a spinneret are well known in the art.

As shown in Figure 3.9b, for each array of spinneret holes a bore is passed almost through the thickness **t** of the spinneret plate with counterbore at the top and the bottom of the bore to provide a flat surface of a desired thickness **p** at the bottom of the bore through which the spinneret array is drilled. In a most preferred embodiment a bore diameter **d** of 3 mm is drilled through a plate of thickness **t** of 50 mm so that with a 90° chamber at the bottom of the bore a circular, flat surface having a face diameter **f** of 1.25 mm is provided at a plate thickness **p** of 0.8 mm from the outside face of the plate. The top of the bore is countersunk with a 90° chamber.

Upon passing through the spinneret hole array, the polyester from the nine individual holes coalesces into a single filament that has a cross-section that can be best described as roughly C-shaped. The surface of the filaments does not form as a continuously smooth line but rather is irregular in outline.

When yarns of the type used for double knit fabrics are melt spun, they are conventionally made as a partially oriented yarn (POY). The yarn leaves the spinneret in a vertical direction and travels for a sufficient distance to be quenched by crosscurrent air and then taken up at speeds from about 1,800 to 3,200 m/min. The yarn is taken up on a spool in a large package which may weigh as much as 11 to 12 pounds. At this point, the yarn does not have bulk and is termed a flat yarn, which means a yarn that has not been textured. In a second step the yarn is either sequentially drawn and textured by false twist texturing or can be simultaneously draw textured. After texturing, the yarn exhibits bulk and can be used to make double knit fabrics.

Example: Polyester was melt spun through a spinneret comprising arrays of nine round holes, one center hole and eight equally spaced holes of diameter of 0.20 mm with the radius of the center of each of the outside holes from the center of the central hole a distance of 0.40 mm. The spinneret consisting of 34 such clusters was therefore capable of making 34 fibers of the coalesced polymer.

The filaments were contacted with crosscurrent quench air at 78°F (25.5°C) at a velocity of 80 to 100 cfm when 3.812 inches (9.68 cm) from the spinneret face. In operation, each cluster produced a fiber that coalesced into a C-shaped filament while being quenched by traveling through air and wound on winder at 1,800 m/min. The yarn produced at a total denier of 415. This partially oriented yarn was draw textured to produce a yarn 150/34 (150 denier with 34 filaments) which was suitable for knitting. The yarn was conventionally knitted to produce a fabric which exhibited good nonglitter characteristics.

NYLONS

Polycaproamide Reacted with Cyclic Tetracarboxylic Acid Dianhydride

It has been suggested that a high strength polycaproamide fiber having excess number of carboxyl end groups over amino end groups can be produced by melt spinning a polymer prepared by polymerizing ε-caprolactam and reacting the polymer with a dibasic carboxylic acid containing at least six carbon atoms. However, with incorporation of this dicarboxylic acid reactant in continuous operation, serious problems have been encountered in melt spinning of the polymer due to the frequent occurrence of "nubs" or enlarged places in the extruded polyamide filament.

R.A. Lofquist and J.C. Haylock; U.S. Patents 4,061,621; December 6, 1977; and 4,075,271; February 21, 1978; both assigned to Allied Chemical Corporation have now found that the occurrence of the nubs in the fiber can be greatly reduced by utilizing certain cyclic tetracarboxylic acid dianhydrides and reacting the dianhydride at the end of the polymerization. Moreover, dyeing properties of the fiber may be more readily controlled.

Example 1: Over a period of about 3 weeks about 100 parts per hour of ε-caprolactam was continuously fed to a stirred reactor operating at a temperature of 265°C together with steam at a pressure of about 50 psig to form a prepolymer melt. A trace amount of copper compound soluble in the reaction mixture was incorporated in the reaction mixture as heat stabilizer.

Polymerization was continuously accomplished in a series of three stirred reactors operated at conditions described in U.S. Patent 3,109,835, i.e., the water and part of the unreacted lactam was continuously removed from the prepolymer melt by smoothly stirring the reaction melt at about atmospheric pressure at a temperature of about 265°C while sweeping the surface of the smoothly stirred reaction mixture with dry nitrogen gas at a rate of about 3 to 4 unit volumes of gas measured at standard temperature and pressure (STP) per hour per unit volume of the reaction mixture for approximately 4 hours. The final reactor may be operated at reduced pressure if desired.

The resulting polymer was continuously extruded into a warm water bath and chopped into pellets $\frac{1}{10} \times \frac{1}{10}$" in size. The pellets were then hot-water washed at 100°C to reduce the content of hot-water-soluble constituents to about 1 to 2%, and the pellets were dried to less than 0.1% moisture. The polymer was a white solid having a relative viscosity of about 55 to 60 as determined by concentration of 11 g of polymer in 100 ml of 90% formic acid at 25°C (ASTM-D-789-62T).

The total primary amino group plus carboxyl group analysis of the polymer was 105 to 115 meq/kg of polymer. The polycaproamide pellets were melted at 260° to 285°C and reacted for 15 minutes with about 0.2 mol of pyromellitic dianhydride per 100 mols of lactam in the polymer, then melt extruded under a pressure of about 1,500 psig through a 70 orifice spinnerette to produce 3,900 denier fiber. The fiber was then collected and drawn about 3 times its extruded length to produce a 1,300 denier yarn. For convenience this yarn hereinafter will be called Yarn A. The polymer of Yarn A had a total primary amino group plus carboxyl group analysis of 107 meq/kg of polymer and a primary amino group analysis of 30 meq/kg of polymer.

In order to provide for comparative testing of polycaproamide made with a dicarboxylic acid additive, Yarn B having about the same total primary amino group plus carboxyl group analysis and primary amino group analysis per kilogram of polymer as Yarn A was prepared by the general procedure of this example except that sebacic acid at a concentration of 0.2 mol % of the lactam in the polymerization reaction mixture was fed to the prepolymer reactor with the ϵ-caprolactam. Yarns A and B both showed a good tensile strength of about 9.5 g/den. Yarns A and B were also tested for the number of nubs per pound as shown in Example 2.

Example 2: This example outlines the method used for locating, identifying and calculating the nubs per pound in Yarn A and Yarn B as prepared in Example 1. In this method a nub is defined as an enlarged place in a filament which is no more than several filament diameters in length. This method may be used for either monofilament or multifilament yarns. However, it is not applicable to most types of crimped yarn.

In accordance with the test, the 1,300 denier yarn is drawn directly from the package by means of an air aspirator and is passed through an opening of known width, specifically, 0.006" in width. Such an opening is conveniently provided by use of a ceramic cleaner gap, which is well known in the art. The presence of a nub is detected when it stops the yarn passage through the opening. The filaments are separated and the cause of the yarn stopping identified as a nub or as the twisted end of a broken filament. For representative results, about 10 pounds of yarn is passed through the gap and the number of nubs counted. Yarns A and B had 4.2 and 14.4 nubs per pound, respectively.

Example 3: In further tests, a polycaprolactam polymer having a relative viscosity of 50.6, a primary amine group analysis of 50 meq/kg of polymer and a carboxyl group analysis of 50 meq/kg of polymer was melted at 260°C and agitated with 0.088 mol of 3,3',4,4'-benzophenone tetracarboxylic dianhydride per 100 mols of lactam in the polymer. After 1, 2, 5 and 10 minutes of agitation, samples of the polymer were withdrawn from the agitated melt, and the samples were analyzed for relative viscosity and primary amine groups. The results are shown in the following table.

Agitation Time (min)	Relative Viscosity	Primary Amine Groups (meq/kg of polymer)
1	72.9	16.3
2	71.2	16.9
5	71.1	17.8
10	69.2	18.2

These results show that the reaction between the polymer and cyclic tetracarboxylic dianhydride is relatively fast, i.e., a reaction time of 1 minute was ample. In other tests, reaction times up to 30 minutes were found suitable.

Polypyrrolidone with Alkylamines for Improved Extrudability

According to *J.D. Neafsey; U.S. Patent 4,076,696; February 28, 1978; assigned to Chevron Research Company* C_{2-30} alkylamines admixed in small quantities with polypyrrolidone substantially improve the extrudability of the polymer.

The amines are normally added by applying a solution of the additive to the polypyrrolidone pellets before they are dried and extruded.

The chief advantage conferred by the process is that the extrusion may be performed at a melt temperature of less than about 277°C, and preferably even at melt temperatures less than 270°C, to produce a yarn having a tensile factor of at least about 18. The melt extrusion of 100% normally solid polypyrrolidone cannot be performed at such low temperatures to yield a yarn of high tensile strength (i.e., of tensile factor about 18 or greater). The tensile factor is defined as the tenacity in grams per denier times the square root of the percent elongation at break.

The lowering of the extrusion melt temperature and the throughput rate, with retention of tensile strength, is regarded as a significant demonstration of the superiority of the method over the extrusion of pure polypyrrolidone.

Spinning trials demonstrating these factors were carried out on polypyrrolidone of MW 60-70 x 10^3, extruded at the rate of 5.5 lb/hr through a 15/1 extruder. Results are shown below.

Amine, wt %	Yarn MW x 10^{-3}	Tensile Factor $T\sqrt{E}$	Melt Temperature (°C)	Throughput Rate (lb/hr)
None	37	20	281	5.5
Hexamethylene-diamine, 0.3	41	19.2	269	3.8
Octadecadiamine, 0.5	42	19	262	3.7

Nylon 66 Spinning Process

In the process developed by *M.M. McNamara, W.T. Mowe and D.E. Walker; U.S. Patent 4,123,492; October 31, 1978; assigned to Monsanto Company* polyamide yarn is melt spun at high speed and drawn almost immediately (between 0.002 and 0.25 sec) after solidification. A turbine driven feed roll replaces the conventional feed and separator roll. The process displays unusually low drawing tension and exceptionally uniform yarn.

As illustrated in Figure 3.10, molten yarn 66 polymer is metered and extruded from a nonillustrated conventional block through spinneret **22** into quench zone **24** as a plurality of molten streams. The streams are cooled and solidified in this zone by a flow of transversely moving air into filaments which constitute yarn **26**. This yarn passes in a partial wrap around feed roll **28** into the draw zone, then around optional intermediate roll **30** prior to entering insulated chamber **32**. Driven heated draw roll **34** and its associated or paired skewed separator roll **36** are mounted within the chamber for drawing and forwarding the yarn, which passes in several separated wraps around rolls **34** and **36** prior to leaving the chamber. The yarn next passes in a partial wrap around roll **38** and then downwardly to schematically illustrated yarn winding apparatus **40**.

In this embodiment, spin finish is applied by slowly rotating conventional finish roll **42**, whose lower surface is immersed in liquid finish carried in trough **44**. A conventional gauze finish skirt **43** transfers the finish from the roll to the yarn, the skirt being anchored at **45**.

Examples: Table 1 discloses several exemplary processes for operating the Figure 3.10 apparatus so as to produce the yarns. The polymer contains 2% TiO_2 by weight and is selected so that the resulting yarn will have a relative viscosity of about 48 to 50. For all items, quenching air is supplied at a temperature of 20°C and a relative humidity of 98%. The average velocity of the quenching air is 25.389 m/min (83.3 fpm), and the height of quench zone **24** is 116.84 cm (46 inches).

Figure 3.10: Nylon 66 Spinning Process

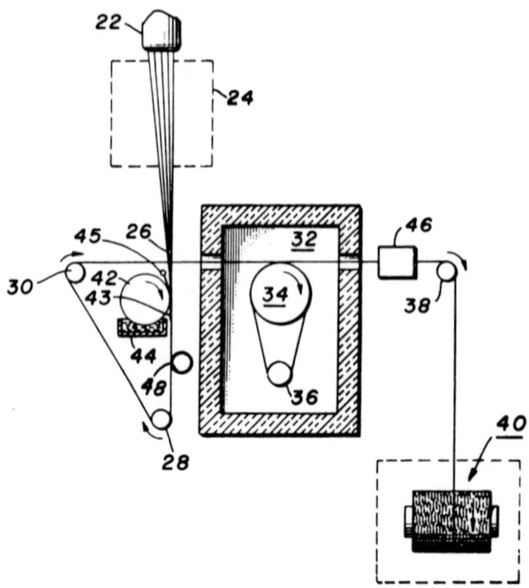

Source: U.S. Patent 4,123,492

The reported tensions are as follows: t_1 is measured down stream of roll **38**, t_2 is measured between device **46** and roll **38**, t_3 is measured as the yarn leaves chamber **32**, t_4 is measured between roll **30** and chamber **32**, t_5 is measured between rolls **28** and **30**, t_6 is measured between roll **28** and roll **48**, and t_7 is measured just above roll **42**. A Rothschile Tensionmeter Model R1092 is used for measuring all tensions.

Table 1

Process Conditions	Sample A	Sample B	Sample C
Feed roll surface temperature, °C	30	120	185
Roll 28 speed, m/min	3,306	3,318	3,061
Roll 28 turbine air press, kg/m^2 gauge	-21,093	-21,093	-21,093
Feed roll 34, m/min	4,573	4,573	4,573
Winding speed, m/min	4,396	4,517	4,504
Underdrive, %	3.8	1.2	1.5
Roll **30** turbine air press, psig	0	0	0
Roll **38** speed, m/min	4,445	4,475	5,020
Tensions, g			
t_1	8.5	8.0	7.5
t_2	12.0	11.0	16.0
t_3	7.0	8.0	8.8
t_4	55	50	62
t_5	54	48	50
t_6	42	44	38
t_7	28	30	33

Table 2 shows the physical properties of the yarns produced by the processes disclosed above, Samples A, B and C, and compares these properties with those of commercially available yarns having the same nominal denier and the same number of filaments. The data reported are the average of at least five bobbins for all items. Sample D is a commercially available nylon 66 premium quality yarn produced by a single stage draw coupled process. Sample E is a commercially available nylon 66 yarn produced by the split process. Sample D is a relaxed yarn, that is, heat treated under appropriate tension so as to reduce the shrinkage. Sample E was not heat treated and is not a relaxed yarn, as evidenced by the high shrinkage. All items are flat (untextured) yarns.

Table 2

Properties	Sample A	Sample B	Sample C	Sample D	Sample E
Drawn denier, D_0	70	68	68	69	71
Tenacity, g/den	4.4	4.5	4.8	5.3	5.0
Elongation, %	52	46	44	39	31
Shrinkage, S, %	6.2	6.6	6.0	5.0	10
Uster unevenness, % U	0.57	0.63	0.52	0.67	1.5
Initial modulus, M_i, g/den	21	25	24	25	28
10% modulus, M_t, g/den	11	13	16	30	31
Final modulus, M_f, g/den	8.5	9.0	8.9	7.4	5.2
Modulus ratio, R	1.3	1.5	1.8	4.2	6.1
Stress index, α	23	21	18	-6	-2

Magnesium Oxide Incorporated into Polycaprolactam

T. Kato, A. Ishitani and K. Kobori; U.S. Patent 4,142,031; February 27, 1979; assigned to Toray Industries, Inc., Japan provide a polycaprolactam composition comprising a compound which can easily be handled and is not poisonous and which can prevent occurrence of bending in a yarn during melt spinning and also prevent subsequent enhancement of unevenness in the yarn and increase of yarn breakages without having bad influence on the resulting polycaprolactam yarn.

The foregoing object can be attained by a polycaprolactam composition comprising polycaprolactam and, incorporated therein, 0.05 to 2.0% by weight of magnesium oxide in which the half-width of the x-ray diffraction profile at (200) as determined according to a specific method described hereinafter is at least 0.3.

In order to attain the intended effect of the process, namely, the effect of preventing contamination of the edge of a spinneret nozzle, it is indispensable that not the morphological size that can be discriminated by a microscope but the crystallite size that can be discriminated by the x-ray diffractometer should be smaller than a certain value.

An x-ray diffractometer manufactured by Rigaku Denki KK is used. It includes an x-ray generator Model D-8C, a goniometer Model 2155D, a scintillation counter as the detector and a recorder Model 5055. The applied x-ray is a Cu-Kϵ ray passed through a nickel filter. The voltage and current of the generator are adjusted to 40 kV and 20 to 40 mA, respectively. The slit is characterized by a divergence angle of 1°, a receiving length of 0.15 mm and a scattering angle of 1°. The scanning speed is adjusted to ¼°/min.

Example: According to a customary method, ϵ-caprolactam containing 15% by weight of water was polymerized to obtain pellets. The pellets were treated in hot water maintained at 98°C to remove unreacted ϵ-caprolactam and oligomers. The so obtained polycaprolactam contained 1.0% by weight of oligomers. The relative viscosity was 2.61 as measured at 25°C in 98 wt % H_2SO_4, and the terminal amino group concentration was 4.6×10^{-5} mol/g of the polymer. Then magnesium oxide powders differing in the crystallite size determined according to the x-ray diffractometer described in the text of the specification, were separately incorporated into the above polycaprolactam pellets. In each case, the amount incorporated of the magnesium oxide powder was 0.09% by weight.

The resulting pellets were fed to an extruder and molten at 270°C, and the molten polycaprolactam composition was extruded in the form of a filament at a rate of 21.5 g/min through a spinneret having 24 nozzles of a Y-shaped cross-section and wound up at a rate of 1,000 m/min. Melt spinning was continued under these conditions for 10 days, and contamination of the edge of the spinnerette nozzle and occurrence of yarn breakage were examined.

Polymethylphenylsiloxane (manufactured by Shinetsu Kagaku Kogyo KK) was coated on the spinneret surface in advance, and every time yarn breakage took place the spinneret surface was cleaned by a wire brush and polymethylphenylsiloxane was coated again.

When magnesium oxide powders (samples C, D, E and F) having a half-width larger than 0.3 at (200) and a small crystallite size were incorporated in polycaprolactam, even in an amount as small as 0.09% by wt, a high effect of preventing accumulation of tar and other carbonized materials around the spinneret nozzle edge could be attained and the frequency of occurrence of yarn breakage was reduced to 1/8 of that of control sample A which was free of magnesium oxide. Magnesium powder (sample B) having a half-width smaller than 0.3 at (200) and a large crystallite size did not exert the above effect when incorporated in polycaprolactam, and the frequency of occurrence of yarn breakages was not substantially different from that of sample A.

Trilobal Filaments

In a process developed by *C.S. Shah; U.S. Patent 4,091,065; May 23, 1978; assigned to E.I. Du Pont de Nemours and Company,* the ability to maintain a constant differential between modification ratios of trilobal filaments under cospinning conditions is provided by spinning a filament of lower modification ratio (MR) through a spinneret orifice consisting of three radially intersecting tapered slots and a filament of a higher modification ratio through a spinneret orifice configured as three radially intersecting reverse tapered slots.

The orifices configured as three radially reverse tapered slots provide a high modification ratio with low sensitivity to normal spinning process fluctuations which in combination with orifices having tapered slots for filaments of lower modification ratio facilitate maintenance of a more constant differential in modification ratios between the filaments.

In Figure 3.11a, symmetrical capillary **20** consists of three radially intersecting slots **22** whose imaginary center lines **23** intersect at center point **24.** Each slot has the same length **25** measured between the center point and flat tip **28** which is perpendicular to the center line. Each slot is tapered such that the base width **26** is greater than the width of the tip to define a taper angle **B** between imaginary extensions **29** of the sides of the slot. Angle **C** between the adjacent slots is shown equal in each instance (120°).

In Figure 3.11b, symmetrical capillary **30** consists of three radially intersecting slots **32** whose imaginary center lines **33** intersect at center point **34**. Each slot has the same length **35** measured between the center point and flat tip **37** which is perpendicular to the center line. Each slot is reverse tapered such that the base width **36** is less than width **38** of the tip to define a taper angle **D** between imaginary extensions **39** of the sides of the slot. Angle **E** between the adjacent slots is shown equal in each instance (120°).

In the following example, filaments are extruded from a supply of polyhexamethyleneadipamide containing 0.02% by weight TiO_2 delusterant as very fine dispersed particles. A screw melter converts the flake polymer to polymer melt.

Figure 3.11: Spinneret Capillaries for Spinning Trilobal Filaments

a.

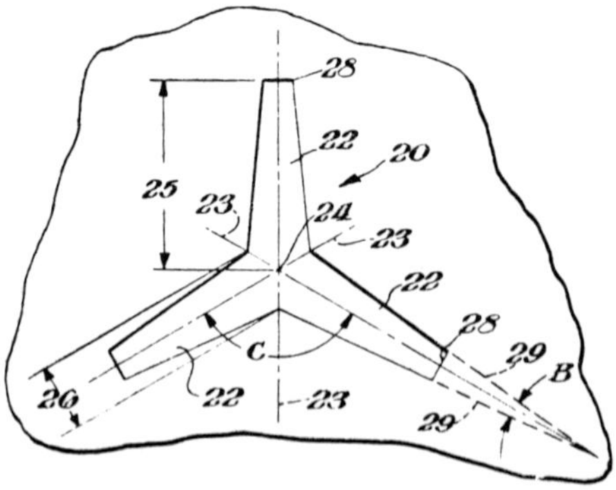

b.

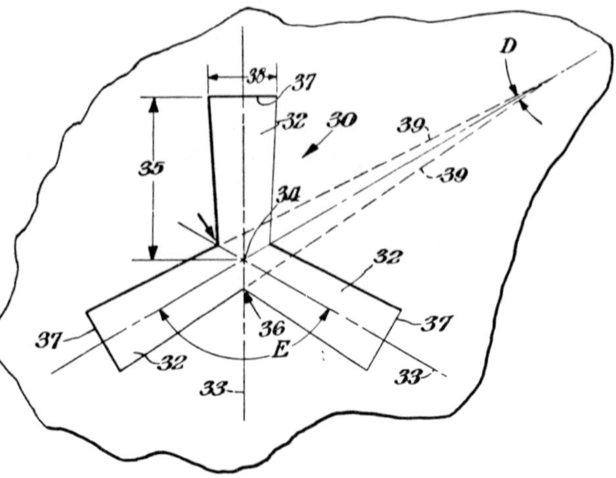

(a) Magnified transverse cross-section of a spinneret capillary comprised of three radially intersecting tapered slots

(b) Magnified transverse cross-section of a spinneret capillary comprised of three radially intersecting reverse tapered slots

Source: U.S. Patent 4,091,065

Relative viscosity (RV) of the melt is varied as desired by controlling temperature and relative humidity of recirculating inert gas in a conditioner through which flake passes before being melted. Nominal RV of the extruded polymer is about 66, but as specified, RV is varied over a range of 60 to 72 to test the effect of

RV on MR. Unless otherwise specified, extrusion temperature of the melt is 290°±2°C.

Filaments in each example are produced at a single position fitted with a spinnerette plate having 332 extrusion capillaries arranged in 8 parallel rows in staggered array such that each odd numbered row has 42 and each even numbered row 41 capillaries. All capillaries in odd numbered rows are identical with a given trilobal cross-section, and all capillaries in even numbered rows are identical with a different trilobal cross-section. Exact cross-sections are specified hereinafter.

The polymer melt is spun to filaments at the rate of 110 lb/hr (49.9 kg/hr), and the filaments are quenched in a chimney using crossflow air at 45°±3°F (7.2°±7°C) and quench air flow rates from 290 to 380 ft^3/min (8.21 to 10.76 m^3/min), as subsequently specified. The quenched filament bundle is then collected as a tow which, in a separate operation, is drawn at a draw ratio of 3.75 times and crimped conventionally in a stuffer box crimper. All filaments so prepared are nominally of 18 dpf (20 dtex).

Example: This example shows cospinning two species differing in MR by at least 0.3 MR units and having a high MR in excess of 1.9 and a low MR less than 1.9. The odd numbered rows of the spinneret plate (producing the low MR component) have tapered capillaries (Figure 3.11a) with slot length **25** of 14.0 mils (0.36 mm), base width **26** of 7.0 mils (0.18 mm), width of flat tip **28** of 4.3 mils (0.11 mm), taper angle **B** of 12.8°, and symmetrical slot angle **C** of 120°. Capillary length is 4.0 mils (0.10 mm).

The even numbered rows (producing the high MR component) have reverse tapered capillaries as shown in Figure 3.11b. Slot length **35** is 18.3 mils (0.46 mm), base width **36** is 5.7 mils (0.14 mm), flat tip width **38** is 7.6 mils (0.19 mm), reverse taper angle **D** is 6.5°, and symmetrical slot angle **E** is 120°. Capillary length is 8.0 mils (0.20 mm).

Changes in modification ratios obtained with changes in quench air flow and RV range from 0.05 to 0.10 for low MR and 0.20 to 0.25 for high MR.

Comparison of the MR changes for the high MR component with those of the low MR component reveals that the reverse tapered high MR capillary of Figure 3.11b is only slightly more sensitive to process variables than is the tapered capillary of Figure 3.11a. The ranges of RV and quench air-flow investigated in the example are broader than any variations normally anticipated in a given commercial production process. Thus, using the tapered capillary of Figure 3.11a for a low MR component cospun with a high MR component utilizing the reverse tapered capillary of Figure 3.11b yields a MR differential which is constant within the normal accuracy of detection of shifts in MR which affect product quality.

In a related patent by *C.S. Shah; U.S. Patent 4,001,369; January 4, 1977; assigned to E.I. Du Pont de Nemours and Company* improved process control for cospinning synthetic trilobal filaments having different modification ratios is provided by spinning filaments of one modification ratio through spinneret orifices consisting of three radially intersecting slots where each slot has parallel sides and spinning filaments of another modification ratio through spinneret orifices consisting of three radially intersecting slots which taper with increasing distance from the center of the orifice.

The modification ratio of filaments spun from the orifices having tapered slots is less sensitive to changes in normal spinning process conditions than is the modification ratio of filaments spun from the orifices of slots having parallel sides. This differential response of filament modification ratio to spinning conditions provides better process control of the modification ratio differential between the filaments.

High Speed Spinning of Polyamides

G. Koschinek and D. Wandel; U.S. Patent 4,181,697; January 1, 1980; assigned to Zimmer AG, Germany describe a method for draw off of polyamide threads at very high speeds, i.e. 1,800 to 5,000 m/min, while maintaining high quality threads and producing stable packages. The method comprises heat treating the threads, after an initial air blast cooling, by passing them through a spinning duct the walls of which are heated to a preselected temperature below 140°C and above T_{min} which is calculated from the following equation:

$$T_{min} = \frac{5{,}500 - v}{33} + T_a(°C)$$

where v is draw off speed in m/min, and T_a is a temperature value of zero for spinning ducts over 1,500 mm in length, or 6°C for each 100 mm reduction in duct length under 1,500 mm. The heat treated threads are subjected to a finishing and wetting operation and then drawn off at the high speed.

Example 1: Polyamide 66 of the relative viscosity (n_{rel}) of 2.4 was melted at 295°C and spun in a quantity of 34 g/min through a spinneret plate with 24 holes, each hole having a diameter of 0.25 mm.

The emerging threads were cooled to a temperature of 70°C in a blow duct of a length of 1,750 mm by means of a transverse flowing air stream of the temperature of 20°C and the velocity of 0.8 m/sec. The thread temperature was measured by a commercially available convection measuring probe known as Transmet.

Subsequently the threads passed through a vertical spinning duct 2,000 mm in length, having its walls heated to a temperature of 70°C. The thread temperature below the spinning duct was about 70°C. The threads were then treated with finish, moistened and wound up at a speed of 3,500 m/min. The titer of the spun thread was 100/24 dtex (yarn weight in grams per 10,000 meters in length). Further quality data characterizing the thread structure are compiled in the table. The wound thread produced a stable and uniform package, and the package hardness was very uniform. A thread end subjected immediately after the spinning and take up operation to a length measurement showed no thread length deviation with time. During the winding operation the thread length remained unchanged which resulted in an excellent yarn evenness.

Threads produced in such a manner were then reprocessed in a simultaneous draw-texturing process applying a temperature of 200°C and a draw ratio of 1:1.27. The quality characteristics of the yarn produced in such a manner are compiled in the table. Knitted stockings made from this yarn are characterized by a uniform and streak-free dyeability.

Example 2: Polyamide 66 threads are spun and wound up under the same conditions as those described in Example 1 except that the temperature of the heated spinning duct was set to 55°C. Such treatment resulted in a collapsing of the package on the tube so that a yarn package could not be produced. The measurement of the change in thread length directly after the spinning operation revealed a thread lengthening of about 1% within a few minutes so that an unstable package was the unavoidable consequence.

Thread Properties

Spun thread	
Titer, dtex	100/24
Breaking load, g	340
Elongation at break, %	80
Birefringence	43×10^{-3}
Draw textured yarn	
Titer, dtex	81/24
Tenacity, g/dtex	4.2
Elongation at break, %	35
Draw ratio	1:1.27

ACRYLICS

Acrylonitrile/Styrene/Isobutylene Copolymer Needing No After-Stretch

M.H. Richmond; U.S. Patent 4,107,252; August 15, 1978; assigned to Polysar Limited, Canada provides a process for the production of a synthetic filament which comprises spinning by extruding molten polymer which contains no solvent or plasticizer through a spinneret, in which the polymer is a solid copolymer of acrylonitrile, styrene and isobutylene. The filament so produced has a high tenacity without the need for being after-stretched.

Example: *Polymer Preparation* – A polymer suitable for forming into filaments was prepared by an emulsion polymerization process. To a reactor equipped with inlet means, agitating means and temperature regulating means was added 200 pbw of water, 2 pbw of the disodium salt of a straight chain ethoxylated alcohol half ester of sulfosuccinic acid, 2 pbw of sodium monoalkylphenoxybenzenedisulfonate, 0.2 pbw of sodium bisulfite and 15 pbw of isobutylene. The temperature of the contents of the reactor was raised to 122°F, and the agitation means was put into operation. In a separate container was prepared a mixture of 70 pbw of acrylonitrile, 15 pbw of styrene and 0.5 pbw of tert-dodecyl mercaptan.

Also in a separate vessel was prepared a 0.5 wt % solution of potassium persulfate in water. Sufficient potassium persulfate solution was added to the reactor to correspond to 0.5 pbw of potassium persulfate. Essentially simultaneously, the addition was started of the acrylonitrile/styrene/mercaptan mixture, the rate of addition being uniform and such that the addition was complete after a total of 5 hours.

At 4 hours from the start of the addition of the acrylonitrile/styrene/mercaptan mixture, an increment of potassium persulfate solution, corresponding to 0.25 pbw of potassium persulfate was added to the reactor. The polymerization was stopped at 7 hours after the start of the addition of the polymerizable monomers, the conversion of total monomers to polymer being 95%. An aqueous slurry of an alkylated aryl phosphite was added to the latex in an amount equivalent to 1 pbw/100 pbw of polymer.

The polymer was recovered by adding the latex to a 1% aqueous solution of calcium chloride at a temperature of 190°F and separating off the coagulated polymer which was then water washed and dried in a forced air dryer at a temperature of 180° to 190°F. The so dried polymer was then further dried by being put through a vacuum extruder, and the extruded product was pelletized. The polymer was found to contain, by weight, 69% of acrylonitrile, 16% of styrene and 15% of isobutylene. It had a melt index of 0.6 g/10 min (2.16 kg load and 445°F temperature).

Filament Preparation – The pelletized polymer was dried in a vacuum oven for 1 hour at 80°C and then fed to an extruder. The experimental filament spinning system contained an extruder having a diameter of 1 inch and a length:diameter ratio of 25:1 with a three-zone temperature control system on the barrel. The output from the extruder was fed through a valve adapter to a spinning pack containing a 100 mesh screen and into a spinneret having 104 holes of 0.02 inch diameter.

The filament from the spinneret was cooled by a stream of room temperature air and was passed to a wind up bobbin. The extruder was operated at a screw speed of 18 rpm and at a temperature of 360°F on all three barrel sections, and the spinning pack was maintained at 485°F, the pressure in the spinning pack being 2,700 psi. The cooling air was maintained at a pressure of 10 psi, and the wind up bobbin was run at a speed such that the rate of wind up was 48 fpm. Draw-down of the filament from the spinneret was 40 to 1. No solvent or plasticizer was added to the polymer being fed to the extruder.

Filament Properties – The filament so produced had an average denier of 59 and an average diameter of 126 μ. The average tensile strength of the straight filament was 1.9 g/den, and the elongation to break was 27%. Knotted filament had an average tensile strength of 1.5 g/den and an elongation of 16%, showing about 80% retention of the strength of the knot.

The filaments produced in this example were found to show little evidence of sticking to an iron when subjected to conventional ironing at medium temperature settings.

This example shows that a filament having good strength properties without having undergone after-stretching, is readily produced from the polymer herein described.

Extrusion of a Single Phase Melt of Polyacrylonitrile and Water

In a process developed by *H. Porosoff; U.S. Patent 4,163,770; August 7, 1979; assigned to American Cyanamid Company,* a single phase fusion melt of an acrylonitrile polymer and water is extruded into a steam pressurized solidification zone maintained under conditions such that the continuous filament resulting retains sufficient plasticity so that stretching can be effected at a stretch ratio of at least about 25 relative to the linear speed of the single phase fusion melt through the spinnerette.

In Figure 3.12, there is shown generally an extruder **11** provided with a spinnerette **12** at its outlet and a pressurized solidification chamber **13** positioned to receive extrudate issuing from the spinnerette. As illustrated, the extruder is shown as a piston extruder where cylinder **15** is provided with a closely fitted piston **16** movable by means, not shown, to force the contents of the cylinder through the spinnerette directly into the pressurized solidification chamber. With the cylinder, single phase fusion melt **17** is heated to the proper temperature by heating means, not shown, such as steam jackets or electrical heaters in the walls of the cylinder.

The cylinder is also provided with a thermometer **18** and a pressure gauge **19** for monitoring the temperature and pressure within the extruder during melt spinning. While the extruder is shown as a piston extruder, other types of extruders, such as screw extruders, gear pumps, etc., as are known for melt spinning other organic polymers, may be used.

At the outlet of the extruder, a spinnerette is mounted. The spinnerette may be provided with circular or noncircular orifices for spinning filaments or fibers. This process contemplates one or more filaments. A filament issuing from the spinnerette, here shown as filaments **21**, goes directly into the pressurized solidification chamber from which it is drawn, under tension, by rapidly rotating godet or thread-advancing rolls **22** which produce the extremely high stretches of this process. The pressurized solidification chamber is provided with an inlet **24** through which steam under pressure and at elevated temperature can be admitted, an outlet **25** from which water can be withdrawn as necessary, and a thermometer **26** and a pressure gauge **27** for monitoring the temperature and pressure within chamber **13.** Chamber **13** is also provided at its outlet with a pressure seal **28,** illustrated herein as a long, thin slot only slightly larger than the diameter of the bundle of filaments **21** passing therethrough.

Figure 3.12: Spinning Apparatus for Polyacrylonitrile/Water Melt

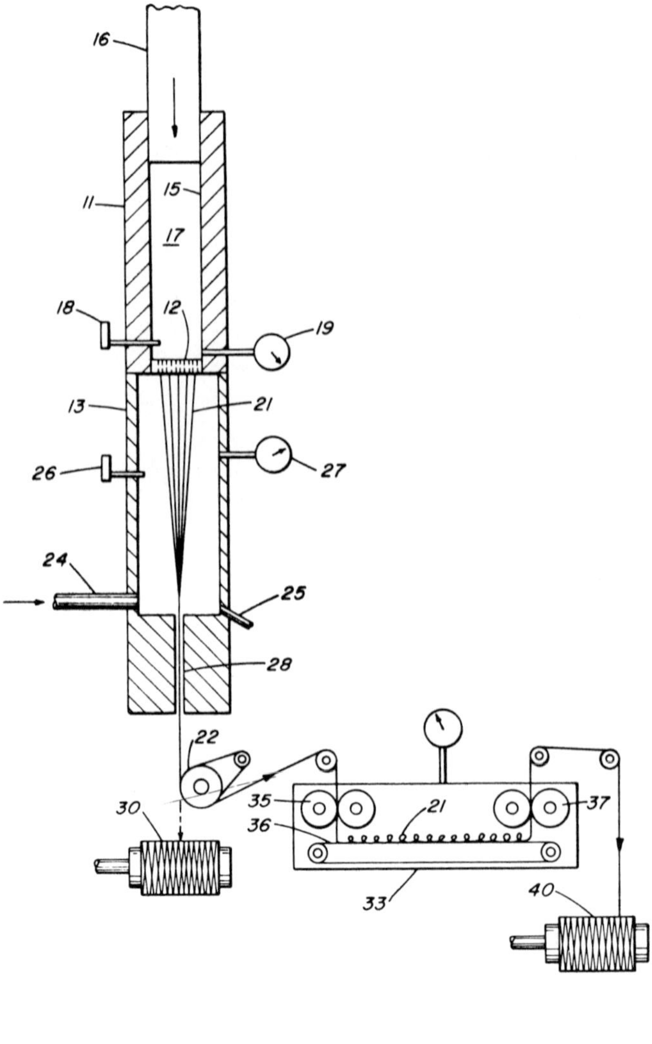

Source: U.S. Patent 4,163,770

From godet or thread-advancing rolls **22**, which may optionally be located within pressurized chamber **13** or outside thereof as illustrated, filaments **21** may be wound up on yarn package **30** by a suitable winder, not shown, or preferably, filaments **21** may be relaxed in steam chamber **33** where steam under superatmospheric pressure and at elevated temperature is permitted to contact the filaments. In the steam chamber, the filaments are fed through inlet pressure seal, not shown, by inlet rolls **35** onto a conveyor belt **36** where they are conveyed through the steam chamber to exit rolls **37** which feed the relaxed filaments through exit pressure seal, not shown, out of the steam chamber to be wound onto yarn package **40** by a suitable winder, not shown.

Example: The phase diagram for an acrylonitrile polymer-water system where the acrylonitrile polymer was a copolymer of 89.3 wt % acrylonitrile and 10.7 wt % methyl methacrylate and had a kinetic molecular weight (Mk) of approximately 58,000 was determined. The resulting phase diagram shows that a single phase fusion melt region of temperature and composition exists which can usefully be melt spun into shaped articles. Having determined the phase diagram for mixtures of this polymer and water, the following melt-spinning was conducted using apparatus substantially as schematically illustrated in Figure 3.12.

To 15 g of dry polymer was added 3.3 g of water; thus providing an acrylonitrile polymer-water mixture containing 18% water. The mixture was sealed in a jar and mixed on a roll mill for 30 minutes to insure complete blending. The blended mixture was then placed in a piston extruder **11** equipped with a spinnerette **12** having a single orifice of 16 mil diameter and 128 mil orifice length. The orifice opening was temporarily sealed to prevent any premature loss of moisture during start up. The extruder and the spinnerette were heated to 154°C under sufficient pressure to prevent water vaporization. The spinnerette orifice was unsealed and 800 psi of force was applied by piston **16** in order to extrude a filament at a rate of 0.446 m/min.

Pressurizable solidification chamber **13** had previously been sealed and saturated steam had been introduced through inlet **24** until a pressure of 38 psig, corresponding to a temperature of 140°C, prevailed. This temperature was 140°C; that is, 14°C below the fusion melt temperature in the extruder and about 10°C below the minimum fusion melt melting temperature (T_m) of 150°C for this polymer. Under these conditions, the maximum take-up speed achieved was 38 m/min for a draw down stretch ratio of 85 (8,500% stretch).

A quantity of fiber so produced, having a denier per filament of 15, was collected and relaxed in saturated steam under pressure at 127°C in a free-to-shrink condition in an autoclave. The denier increased to 19.5, indicating that about 23% shrinkage was achieved during relaxation. Physical properties of this relaxed fiber were: straight tenacity, 3.5 g/den; straight elongation, 43.0%; loop tenacity, 1.98 g/den; loop elongation, 19%; and initial modulus, 58.0 g/den. These fiber properties and the fine filament denier indicates a textile fiber eminently suitable for use as a carpet fiber.

OTHER POLYMERS

Polyethylene Oxide Monofilament

Polyethylene oxide monofilaments together with their preparation and use are disclosed by *R.J. Hartigan, Jr.; U.S. Patents 4,097,652; June 27, 1978; and 4,119,693; October 10, 1978; both assigned to Albany International Corp.* The monofilaments are characterized in part by their properties which are useful for weaving or sewing and by their solubility in a wide variety of solvents including water at room temperatures. The monofilaments are useful in weaving as temporary binder yarns and as basting threads.

Referring now to Figure 3.13, a schematic drawing of a preferred process embodiment, there is seen an extruder **10** providing melt spinning of a monofilament **20** into a quench bath **22** of a saturated aqueous solution of potassium carbonate. The quench tank **21** is disposed in a cooling vessel **24** through which cooling water **25** is circulated via inlet **26** and outlet **28**. The temperature of the cooling water is selected to balance the thermal input of the molten extrudate received by the quench bath. A stirrer **30** powered by motor **32** provides circulation of the quench bath. The distance between spinnerette **18** and the quench bath is preferably less than 6 inches so that the molten homopolymer has minimal exposure to the air.

The fixed-by-quench monofilament **20** is picked up by a grooved guide roller **34** in the bottom of the quench tank and carried under slight tension to guide roller **36** outside of the quench. The monofilament is then taken up by a first godet roller **38**. A slight cold drawing may take place between the roller and the godet roller due to the slight tension on the monofilament. The monofilament is then drawn in a single draw through a dry air draw oven **42** capable of heating the traveling monofilament to a temperature of between about 90° and 130°F. The draw is made by second godet roller **40**. The speed ratio of godet **38** to godet **40** is within 5.0:1.0 to 5.5:1.0, preferably 5:1. This provides a stretch of about 500 to 550% to orient the molecules of the monofilament.

The drawn monofilament is then allowed to cool to ambient temperature, i.e., room temperatures, and taken up on spool **44**. The cooled and spooled monofilament should be immediately packaged under a hermetic seal to prevent degradation by exposure to moisture.

The test procedures used to characterize the monofilaments are as follows: percent of elongation, ASTM Test Method D-638-58T; tensile strength, ASTM Test Method D-638-58T; tenacity, the average tensile strength in pounds as measured by ASTM Test Method D-638-58T converted to grams and divided by the average yarn weight per unit length in denier (g/9,000 m); and MIT flex, based on ASTM Folding Endurance Test Method D2176-63T, but modified for single yarns and monofilaments by welding $\frac{1}{32}$ inch carbide rods to the tops of 0.06 to 0.07 inch gapped jaws.

Figure 3.13: Spinning of Polyethylene Oxide

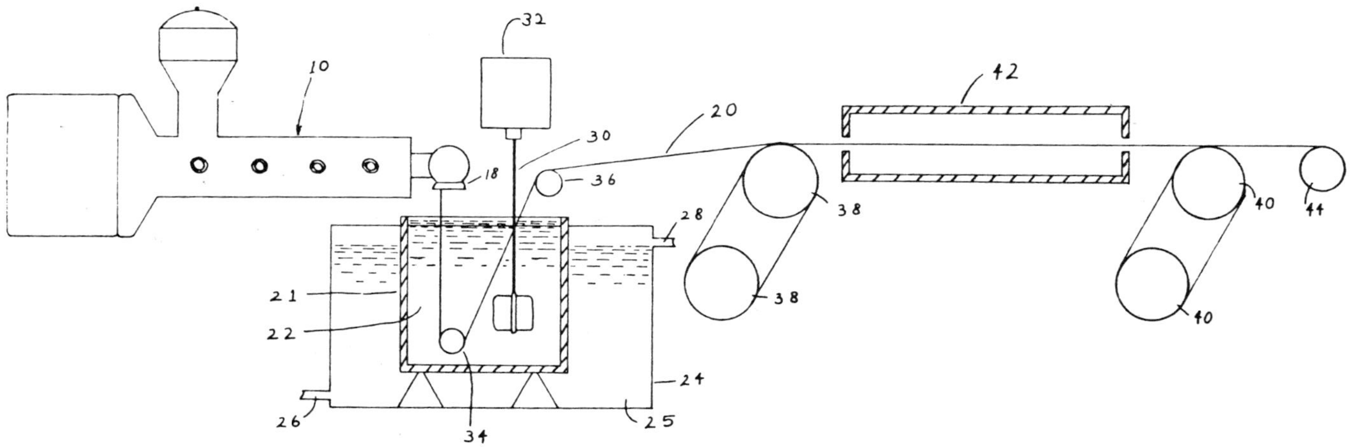

Source: U.S. Patent 4,119,693

Example: The apparatus employed in carrying out the process described in this example is a conventional single stage screw mixer extruder with a 1" (internal diameter), nonvented barrel. The length to diameter ratio is 20:1, and there are three heating zones including one at the extrusion head. The extrusion head is a 90° monofilament head without a spin pump. The spinneret has 10 monofilament holes, each hole having a diameter of 0.40 inch. No screen pack is employed.

In carrying out the continuous process, the hopper and feed section of the extruder is charged with tablets (11/32 inch diameter x 5/32 inch thickness) of polyethylene oxide homopolymer characterized by an approximate average molecular weight of 600,000, a melting point of 150°±4°F, a specific gravity of 1.21, a heat of fusion of 33 cal/g and an aqueous viscosity of 4,100 to 8,000 cp (5% aqueous solution, 25°C, Model RVF Brookfield Viscometer, Spindle No. 2 at 2 rpm).

Operating at an extruder screw speed of 50 rpm, under a head pressure of approximately 4,000 psi, the charge is melted and moved to the initial heating zone maintained at 270°F. The melted charge progresses through the second heat zone maintained at 390°F and onto the head which is maintained at 395°F. The melt is spun downward into a quench bath (positioned about 1 inch below the die) of a saturated aqueous solution of potassium carbonate maintained at a temperature of between 70° and 80°F and taken up by a first godet roller at a speed of about 25 fpm. The resulting monofilament is drawn through a draw heater operating at a temperature of approximately 110°F by a second godet roller operating at a speed of 125 fpm, thereby giving a draw ratio of 5:1.

The monofilament produced is smooth and uniform. The product is immediately protected from humidity by packaging under hermetic conditions. The properties of the monofilaments obtained at each screw speed are:

Average diameter, mils	10.3
Weight, g/100 yd	86.4
Denier	551
Tensile, lb/end	3.4
Tenacity, g/den	2.8
% Elongation at 1 g/den	35.3
Break	56.6
MIT flex (cyclic; 500 g wt)	134
Cold water solubility, 70°F	*

*Disintegrates in 30 seconds; complete solution in <10 minutes.

Nylon Modified Phenolic Resin Fibers

According to *J. Economy, F.J. Frechette and L.C. Wohrer; U.S. Patent 4,110,277; August 29, 1978; assigned to The Carborundum Company* phenolic resins used for melt spinning fibers are modified by the incorporation therein of 5 to 30% of an aliphatic nylon. The melt spinning characteristics of the phenolic resins are improved, and the tenacity of the melt spun fibers is increased.

Example 1 (Control): A novolac is prepared conventionally by condensing formaldehyde with a slight molar excess of phenol in the presence of a catalytic amount of oxalic acid. After purification to remove any particulate impurities and residual phenol, the resin has a number average molecular weight of about 700. The resin is fiberized, i.e., formed into fibers, by melt spinning, 38 filaments being simultaneously drawn from a melt of 127°C through a spinnerette having 38 circular orifices 0.25 mm in diameter and 0.5 mm long, at a rate of 615 m/min. The melt temperature employed is the optimum temperature with respect to minimum filament breakage.

As the filaments are drawn and after they have solidified, they pass through a gathering device which assembles them into a 38 strand continuous multifilament yarn, and then through a traversing mechanism which distributes the yarn evenly over the surface of a revolving take up spool. The melt spinning is continued until approximately 30 g of fibers have been collected on the spool and, during this period, the average frequency of filament breakage at and just below the orifices is observed to be 10 breaks/min; i.e., one of the 38 filaments breaks on the average of every 6 seconds. The resulting uncured phenolic resin fibers have an average tenacity of 0.3 g/den.

Curing of the phenolic resin is effected as follows. The fibers on the spool are immersed in 2 liters of an aqueous solution containing 18% paraformaldehyde as a source of methylene groups and 18% HCl as a catalyst, at room temperature (about 25°C). The solution is heated to 40°C over a 2 hour period, then to the boiling point (103°C) over a 1.5 hour period, and the temperature is held at the boiling point for 2 hours, whereupon the fibers are removed, washed with water, and dried in air at about 60°C. The resulting infusible cured phenolic resin fibers have an average tenacity of 1.7 g/den and an average break elongation of 41%.

Example 2: A series of comparative melt spinning runs is carried out employing the same phenolic resin as used in Example 1, but having an aliphatic nylon (6,10; 11; 6; 6,6; 6,12; 12; and a mixture of nylon 6, nylon 6,6, and nylon 6,10) incorporated therein. In each run, the nylon is incorporated in the phenolic resin by heating the latter to 180°C and dissolving the nylon in the resulting melt with the aid of occasional stirring. In each instance melt spinning is carried out as in Example 1, with the same apparatus and at the same drawing speed.

However, since the nylon modifier increases the viscosity of the phenolic resin melt at any given temperature, the optimum melt temperature for spinning with minimum filament breakage is somewhat higher than in the case of the unmodified phenolic resin and each run is carried out at the optimum melt temperature. The average diameter of the filaments in the yarns produced is from 13 to 17 μ.

Aliphatic nylons in amounts of 5% or more are shown to be extremely effective modifiers for phenolic resins to reduce the incidence of filament breakage during melt spinning. The average tenacity of the uncured phenolic resin fibers increases with increasing proportions of nylon. While 5% nylon is sufficient to effect a marked decrease in the incidence of filament breakage, the preferred amount of about 10% is still more effective. However, 20% nylon is not significantly more effective than 10% nylon in reducing filament breakage, and the larger amount is less preferred, tending to be more detrimental to the flame resistance of the fibers after curing.

Mixtures of two or more different aliphatic nylons may be employed, if desired, but no advantage has been found in doing so. Certain types of aliphatic nylons (6,6; 6,12; and the aforementioned mixture) may tend to have a somewhat adverse effect upon one or more mechanical properties of the cured phenolic resin fibers. This is probably due to the susceptibility of certain nylons to hydrolysis under the acid-curing conditions employed in the examples; it is not thought to be a factor in heat-curing resole fibers or in curing novolac fibers under nonacidic conditions.

Nonetheless, novolacs are generally much preferred to resoles for purposes of the process since it has been found that they lend themselves more conveniently to fiberization and to subsequent curing, the latter step preferably being carried out under acidic conditions. When employing a novolac for melt spinning and curing under acidic conditions, it is usually preferred to employ an aliphatic nylon which does not have any significant adverse effect upon the mechanical properties of the resulting cured phenolic resin fibers, nylon 6,10 being especially preferred in this regard.

However, depending upon the intended use for the cured phenolic resin fibers, the improvement in melt spinning characteristics of phenolic resins effected by the aliphatic nylons may be a factor of much greater significance than the mechanical properties of the cured phenolic resin fibers.

NONWOVEN WEBS

Reinforced Matting

H. Stapp; U.S. Patent 4,012,249; March 15, 1977; assigned to Akzona Incorporated provides a reinforced matting comprising interlooped, amorphous, continuous synthetic filaments which are self-adherent and applied to a flat, latticed structure without any need to use binders, impregnating agents or similar bonding materials. The process has the object of creating such a matting in a single operation, relying essentially on the self-adherent properties of the melt spun thermoplastic filaments to achieve a special mat construction having a strength sufficient to handle high loads without tearing or rupturing.

A specially designed spinning head together with other required apparatus is shown in Figure 3.14a which will also serve to explain a preferred embodiment of the process which essentially incorporates methods and similar apparatus as taught in U.S. Patent 3,691,004. In order to avoid undue repetition, the process and apparatus are explained herein to the extent it is necessary to modify the apparatus used in U.S. Patent 3,691,004.

Referring now to Figure 3.14a, the spinning head **7** is shown schematically in cross-section as having a rectangular nozzle plate **8** divided into two compartments **9** for the thermoplastic polymer melt which is maintained under a pressure as supplied from a metering pump and conventional extruder (not shown). The spinning head has a central feed slot **10** extending therethrough in place of one row of spinning bores or nozzles. Conventional spinning nozzles **11** form three outer rows of filaments while the spinning nozzles **12a** and **12b** are inclined on either side and directly adjacent to the feed slot so as to spin the filaments **16** at an angle, α, to the usual vertical spinning direction of the nozzles **11** from which the filaments **15** and **19** fall freely down to the water bath surface **17**.

Figure 3.14: Apparatus for Production of Reinforced Matting

a.

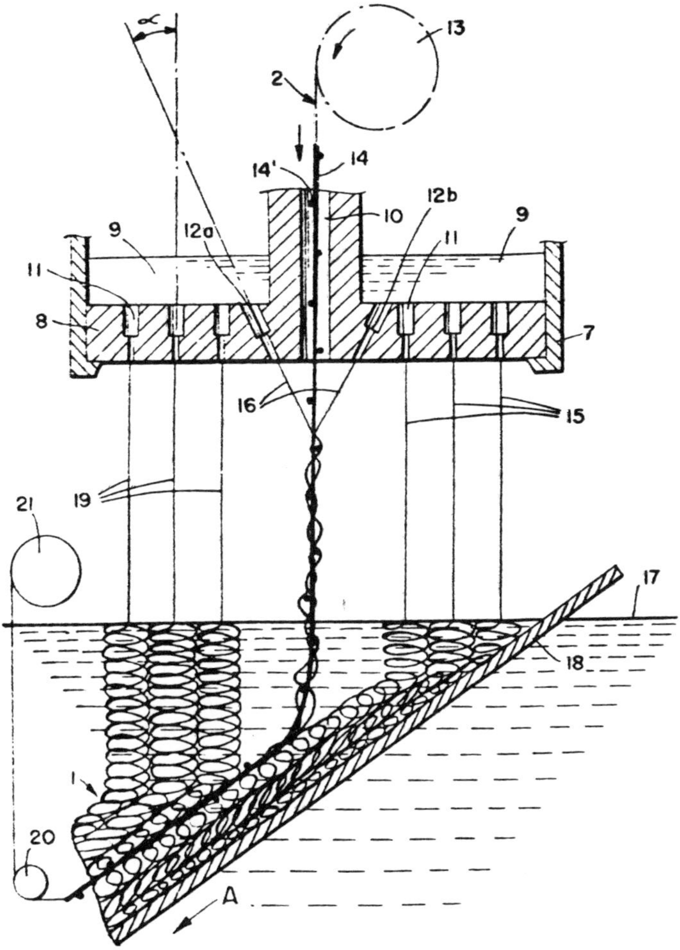

(continued)

Figure 3.14: (continued)

b.

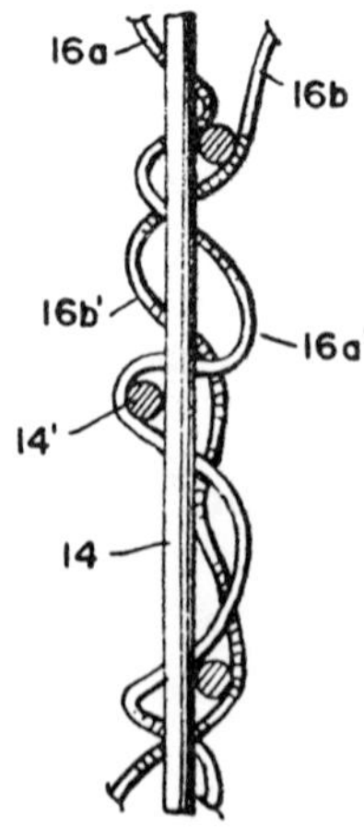

Source: U.S. Patent 4,012,249

They fall without interference from the reinforcing member **14** being fed continuously from the supply reel **13.** It is also preferable to maintain the filaments **15** and **19** free of contact of the inwardly angled freshly spun filaments **16** as these are applied to the reinforcing member in the free fall zone extending from the bottom of the nozzle plate **8** down to the bath surface **17.**

Each row of spinning nozzles **11** and **12** can provide from about 20 to 200 spinning openings by way of example, preferably of the same size and substantially equally spaced in each row and with a uniform spacing between adjacent rows. In general, the interval between the nozzle openings can range from about 3 to 8 mm, depending upon the diameter of the filaments and the desired density of the matting. If desired, the spacing of the rows and size of filaments may also be varied over the base of the nozzle plate.

The angle, α, can vary between about 10° and 70°, but is preferably between about 15° and 35°. In this example, the angle is about 25°.

The distance of the nozzle openings **12a** and **12b** from the feed slot opening **10** should be kept as small as possible, consistent with a frequent penetration of the latticed reinforcing member **14** from both sides. Sufficient pressure can be exerted to create a jet effect which will span the gap between the openings and the slot; the angle, α, also being set to ensure that the jet has enough momentum to pass through the mesh openings of member **14** in a random manner from each side of the downwardly conducted reinforcing structure.

Figure 3.14b provides an enlarged view of a segment of this reinforcing structure between the point of first contact of the two rows of filaments, **16a**, **16b** and the bath surface. The warp filaments, strands or threads **14** are maintained in a substantially vertical path with the horizontal weft threads **14'** alternating on either side or in a plain weave pattern. There is a frequent penetration of the freshly spun filaments to form loops **16a'** and **16b'** projecting partly through the reinforcing member and contacting the filaments applied from the opposite direction frequently enough to entangle or envelop the weft threads **14'**.

This reinforcing structure by itself is unique as a special means for subsequently adding another thermoplastic filamentary layer on one or both sides, especially if the penetrating and enveloping filaments **16** are composed of a relatively low melting point thermoplastic material, e.g., copolymers of polyethylene terephthalate of lower melting point than the homopolymer. In this case, these filaments **16** can act as a bonding agent as in more conventional multilayer fleeces, but they are preferably combined immediately in a single continuous operation with one or more additional filamentary layers.

The outer three rows of filaments **15** and **19** begin to loop and spread laterally in helical to sinuous fashion just as they enter the bath surface and immediately overlap for self-adherence near the bath surface in a bonding zone of the bath, preferably extending at least down to the point where all the filamentary layers are joined together by self-adherence; e.g., where the looped filaments **19** are collected on the upper surface of the reinforcing member **14** carrying looped filaments **16**.

The guide plate **18** serves to lay over the bottom row of loops of the right hand row of spun filaments so as to be substantially parallel to the horizontal or longitudinal plane of the matting, this bending or reorientation of the bottom loops yielding a relatively flat base structure of higher density. For example, the second row of loops from the bottom could be closer to 45° while the third row reaches almost 90°; i.e., with loops almost perpendicular to the plate. The upper three rows of looped filaments may then also approach this 90° angle.

Other variations in the position of these sinuously to helically shaped loops can also be achieved as noted in detail in U.S. Patent 3,691,004. In fact, one can generally obtain the same type of matting as in this prior patent except for the newly incorporated reinforcing structure of continuous mesh band or web **14** through which filaments **16** are interlooped.

While the distance from the nozzle plate **8** to the bath surface may range between about 2 and 30 cm, it has been found that good results with amorphous poly-ε-caprolactam filaments are usually obtained at free fall distances of about 4 to 20 cm for the outer sets of filaments **15** and **19**. As the entire matting **1** is formed, it is drawn off in the direction of **A** around guide roller **20** and onto a take up roll or winder **21**. The speed at which the matting is drawn off through the bath is adjusted so as to avoid pulling out the loops of filaments **15** and **19** while still tacky or deformable in the bonding zone or along the plate, which may be internally heated if desired.

Example: The following example was carried out with a spinning head similar to that illustrated in Figure 3.14a. The inclined guide plate had been omitted.

The essential part of the spinning head is a spinneret whose length is 402 mm, its width being 182 mm and its height being 42 mm. 634 openings having a diameter of 0.250 mm are arranged at equally spaced intervals of 6 mm, each of these openings possessing a counter bore hole having a diameter of about 3.0 mm. The length of the spinning openings is about 0.400 mm, that of the counter bore holes about 18 mm. The center feed slot has a length of 259 mm and a width of 3.5 mm. Spinning nozzles are inclined on both sides of and directly adjacent to the feed slot, the angle, α, being 19°. The measurements of these spinning nozzles are the same as those of the other spinning openings.

A poly-ϵ-caprolactam melt is spun through the openings and spinning nozzles at a temperature of about 270°C, the delivery rate being 1.080 g/min.

A metallic screen consisting of warp and weft wires composed of steel and having a diameter of 0.45 mm (the measurements of the mesh openings being 25 x 25 mm) is fed by two rolls from the supply reel through the central slot with a feeding rate of 2 m/min.

The freshly spun filaments and the metallic screen are deposited onto a water bath whose temperature is kept at 45°C. The gap between spinneret and bath surface is about 15 cm.

The so produced reinforced matting is vertically forwarded through the water bath and then drawn off to a take up roll outside from the water bath. It has a height of 40 mm, a width of 280 mm, a weight of 2,000 g/m^2 and a strength of 25 kp/25 cm length measured in a direction perpendicular to its running direction.

Webs of Continuous Thermoplastic Filaments

J. Barbey; U.S. Patent 4,017,580; April 12, 1977; assigned to Rhone-Poulenc-Textile, France discloses a process for the manufacture of nonwoven webs of continuous filaments by extrusion, stretching, deflection and deposition of a bundle of filaments on a receiving table and executing a reciprocating movement in a direction transverse to the direction of advance of the receiving table.

The continuous filaments are preferably filaments of synthetic polymers such as, for example, polyamides, polyesters, polyolefins and their copolymers and mixtures of polymers. The filaments can also have a heterogeneous structure such as a core sheath or side by side structure.

If reference is made to Figure 3.15a, one notes the spinneret block **1**, the bundle of extruded filaments **2**, the stretching nozzle **3**, the fixed part **4** of the tube, the sliding part **5** of the tube, the predeflection by blow jet **6**, the deflector **7**, the receiving table **8** and the drive carriage **9** for the reciprocating movement.

In Figure 3.15b, one sees the drive carriage **9** as well as the device which provides the reciprocating movement, comprising chains **10**, a motor **11**, a frame **12** suspended from an external support element which is not shown, and lateral shock absorbers **13**.

In Figure 3.15c, showing the movable tube **5** sliding in the fixed part **4** of the tube, one sees the bearings **14** for holding the fixed tube, and the collar **15** for coupling the movable tube to the carriage **9**.

Figure 3.15: Unit for Manufacturing Nonwoven Web

a.

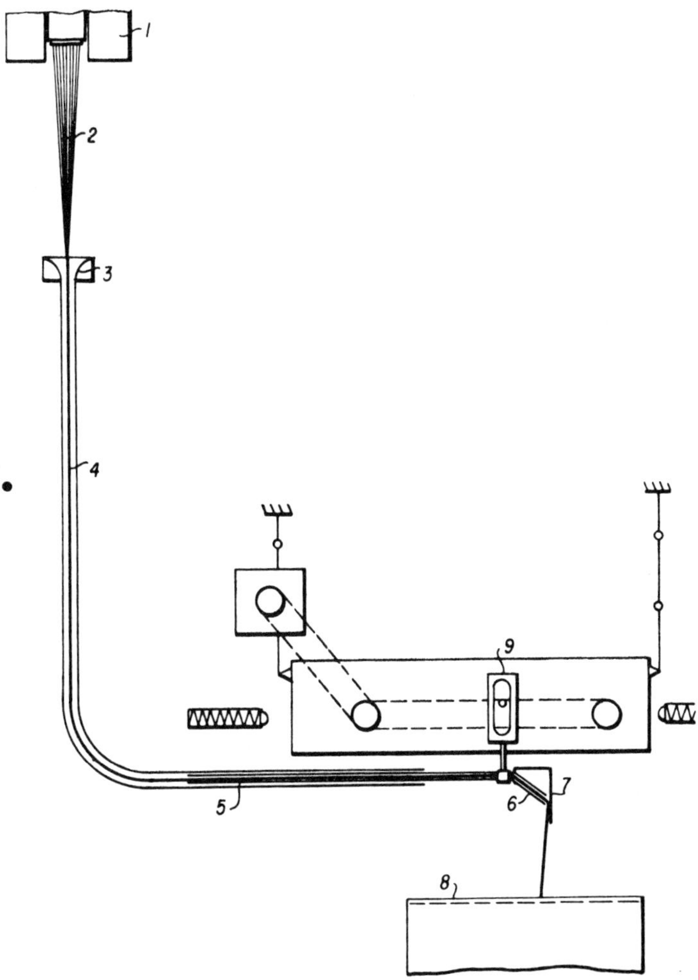

(continued)

Figure 3.15: (continued)

b.

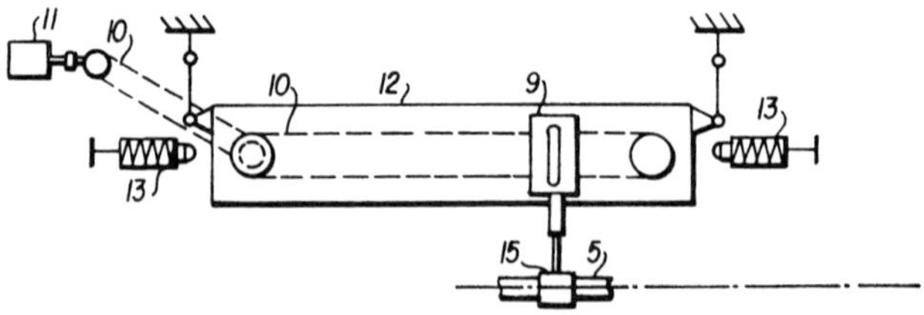

c.

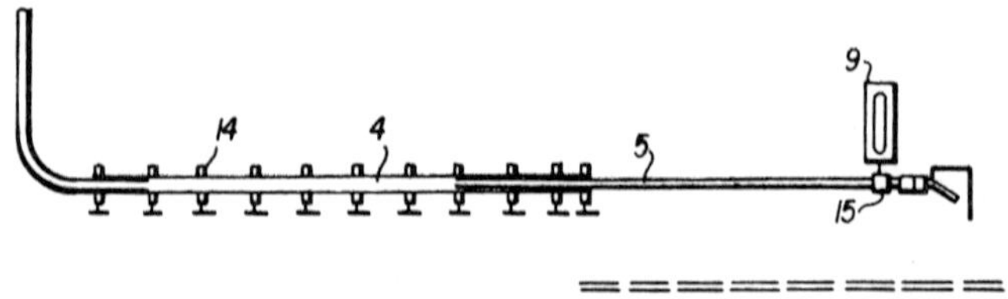

(a) Partially schematic view
(b) Reciprocating means
(c) Sliding movement of movable tube over fixed tube

Source: U.S. Patent 4,017,580

Referring to the figures, the filaments **2** coming from the head of the spinnerette **1** pass through a stretching nozzle **3** such as that forming the subject of French Patent 1,582,147 and U.S. Patent 3,576,284. Upon issuing from this nozzle, the stretched filaments travelling at the stretching speed, enter the fixed part **4** of the tube which has an elbow, the angle of which is immaterial.

Still travelling at the same speed, they then enter the movable part **5** of the tube and, thereafter, the blow jet predeflector **6**, such as that described in French Patent 2,128,216. Upon issuing from this predeflector, the filaments encounter the deflector **7**, the lower part of which is optionally subjected to a vibrating movement such as is described in French Patent 2,166,281 and U.S. Patent 3,853,651. The control system of the carriage causes the reciprocating movement of the latter and the carriage, being firmly fixed to the movable part of the tube, causes this to execute a complete cycle equivalent to one to-and-fro movement while the belt advances by an amount equal to the width of the voile or veil.

In this process, a single source of traction, namely, the stretching nozzle, draws the filaments and imparts to them the requisite speed to transport them as far as the point at which they are deposited on the table.

Continuous Production of Tubular Modular Filter Elements

A process is provided by *D.B. Pall; U.S. Patent 4,116,738; September 26, 1978; assigned to Pall Corporation* for the continuous production of modular tubular filter elements using nonwoven webs in cylindrical or sheet form spun from thermoplastic fibers, spinning the fibers continuously from a melt onto a rotating internal tubular core for the filter element, serving as a mandrel, and winding them up on the core to form a wound tube.

The apparatus shown in Figure 3.16 includes a spinning die **1** whose face **2** has an array of orifices **3** (see Figure 3.16 e and f). The spinning die is supplied with molten thermoplastic polymeric material, such as polypropylene from a reservoir **4** which is fed by the inlet line **5** carrying molten thermoplastic polymeric material from an extruder or other source of supply (not shown) and is further supplied with compressed air or steam under pressure sufficient to project the resin from the orifices of the die, thereby forming a plurality of molten fibers **10**.

The fibers traverse a short air space **11**, in the course of which they are attenuated and solidify, and are collected on a rotating array of interconnected tubular cores **12**, interconnected and rotatably interlocked with spacers **13**, the entire array being self-supporting. The array is held between two sets of three canted rollers **20, 21**, which are driven by an electric motor (not shown) in unison at the same speed, and rotate the array of the cores and the spacers at a relatively slow speed, approximately 3 to 5 fps in this instance. The fibers are randomly oriented and heterogeneously intertwined as they are wound up on the core, forming a generally wound layer of nonwoven fibrous material **15**, serving as the porous filtering sheet of the resulting filter element.

The spinning die is parallel to the cores so that the orifices at the lower end **7** are the same distance from the cores as the orifices at the upper end **8**. Consequently, the portion at **16** of the nonwoven fibrous mat formed of fibers laid down from the orifices at the end **7** has the same density as the portion at **17** of the mat formed of fibers laid down from the orifices at the end **8**.

The cylinder **15** as it is being rotated and formed continuously is also drawn continuously to the right by the canted rollers. By the time an element of the cylinder has reached the point **17**, it has reached its final thickness.

The core unit lengths **12** are preformed in selected lengths and have a plurality of circumferential surface ridges **9a** spacing the fibrous layer **15** from the slots **9b** defined therebetween and the openings **9c** at the base of the slots **9b** for flow of liquid to or from the open interior **28** of the cores through the cores to the exterior thereof. Each core length has an interdigitated end, interconnecting with one end of the spacers. The other end of the spacers is interdigitatingly locked to the next adjacent core length. This locks the core sections together so that they rotate in unison and can also be drawn continuously to the right. Each core section may, however, be withdrawn axially from the spacer as will now be seen.

Figure 3.16: Continuous Production of Tubular Modular Filter Elements

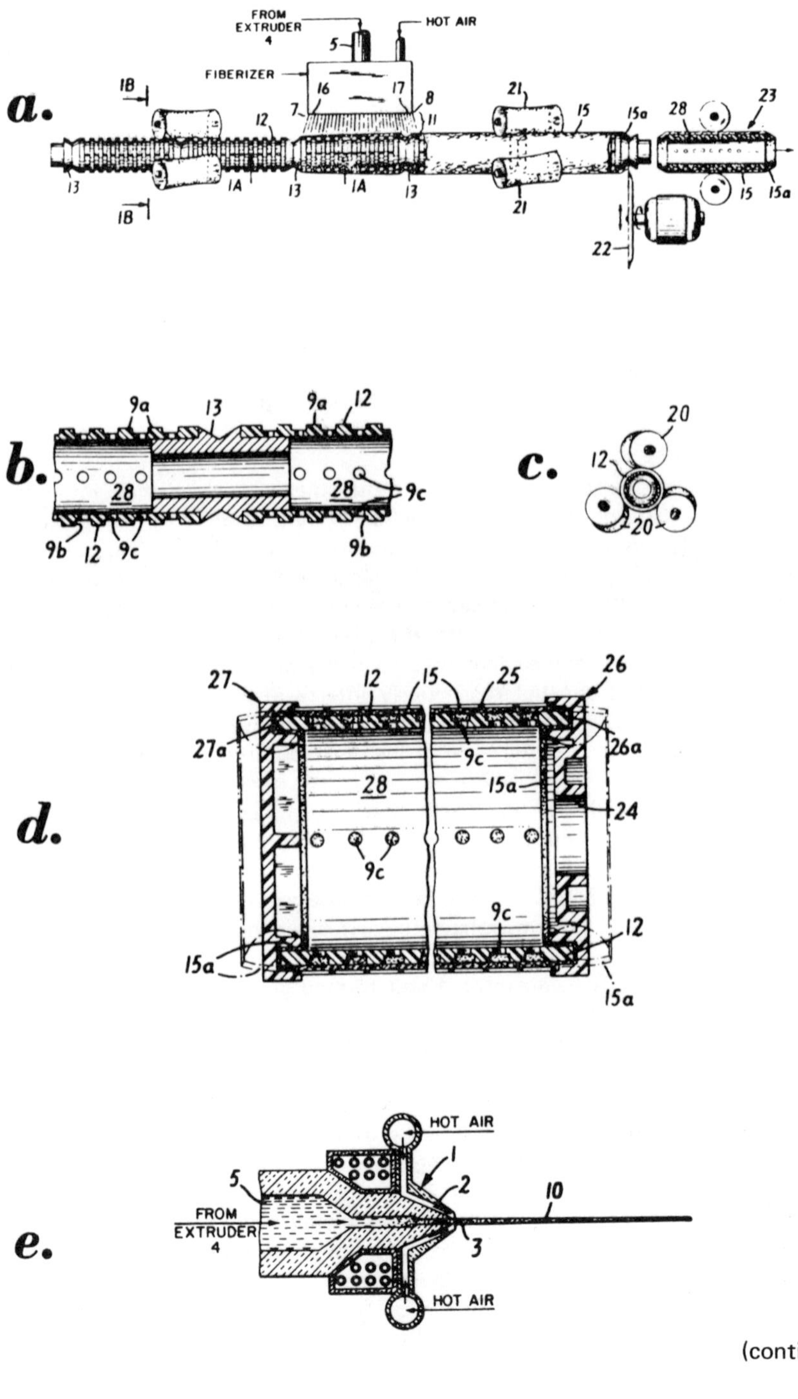

(continued)

Figure 3.16: (continued)

f.

(a) Apparatus for producing cylindrical filter elements upon a rotating array of interlocked modular preformed tubular core lengths, cutting through the fibrous layer to form filter element sections and applying end caps to each end of each length

(b) Detailed view on an enlarged scale of a two core and spacer assembly of portion **1A** in Figure 3.16a

(c) Cross-sectional view taken along the line **1B** to **1B** of Figure 3.16a

(d) Detailed view of the filter cylinder prepared in the process shown in Figure 3.16a with end caps applied to cap off the ends of the cylinder

(e) Longitudinal section of one orifice of the spinning die of Figure 3.16a having an array of orifices embraced by nozzles for directing a gaseous blast for attenuating and disrupting the molten filaments spun through the orifices

(f) The spinning die of Figure 3.16e in perspective

Source: U.S. Patent 4,116,738

The filter cylinder is now ready to be separated and approaches the cutoff wheel **22**, where the fibrous layer **15** is cut through at about midpoint of the spacers **13** leaving a lap **15a** extending beyond the core **12** at each end, making it possible to pull off the filter length **23** by withdrawing the core portion from the next following spacer **13**.

The resulting cylinder lengths **23** can be used as filter cartridges by enclosing the filter element in an external sheath **25**, and then applying end caps **26, 27**.

The end caps (shown in detail in Figure 3.16d) are provided with narrow circumferential slots **26a, 27a** having parallel sides and a width such that when assembled to the end of the core section and the fibrous web is folded inward, the web is compressed to about ¼ to ¾ of its full thickness. One end cap **26** has central aperture **24**; the other end cap **27** is closed.

The layer **15** when cut through is longer than the core **12**, and the lap **15a** is folded over the end of the core and tucked into the open center **28** of the core. The end caps **26, 27** are then applied so that slots **26a, 27a** receive the folded portion of lap **15a**, compressing the layer to three-fourths or less of its normal thickness there and making a tight seal without the need for a bonding agent (although one can be used, if desired). The porosity of the compressed portion is less than that elsewhere in layer **15**, which is adequate for sealing purposes at the end cap. The filter cartridge is then complete.

Bonded, Low Density Matting

A. Rasen, R. Vollbrecht and K. Schenesse; U.S. Patent 4,177,312; December 4, 1979; assigned to Akzona Inc., describe an improved coherently bonded, low density matting article composed of a primary matting layer in which synthetic fiber-forming thermoplastic macrofilaments are irregularly looped and intermingled in a highly porous or open peak and valley three dimensional sheet structure, preferably having peaks, humps or projections in a wafflelike pattern and a secondary footing layer or "sole" applied and fused or bonded by its surface filaments to some of the filaments in the peak portions on one face of the primary matting layer.

A special process and apparatus are provided to produce this matting article wherein transverse strength is increased while maintaining a low apparent density and further improving its filling characteristics for use as a soil retention matting.

The manufacture of the complete matting article is explained with the aid of Figure 3.17 which illustrates one technique for applying the footing layer or sole **S** to the primary matting **M**. The formation of the primary matting layer **M** is accomplished by a number of rows of filaments **5**, e.g., seven rows, from a first spinnerette **1** vertically downwardly onto the moving support element, i.e., roll or drum rim **3** with its projections **4**.

Figure 3.17: Apparatus for Manufacture of Layered Matting

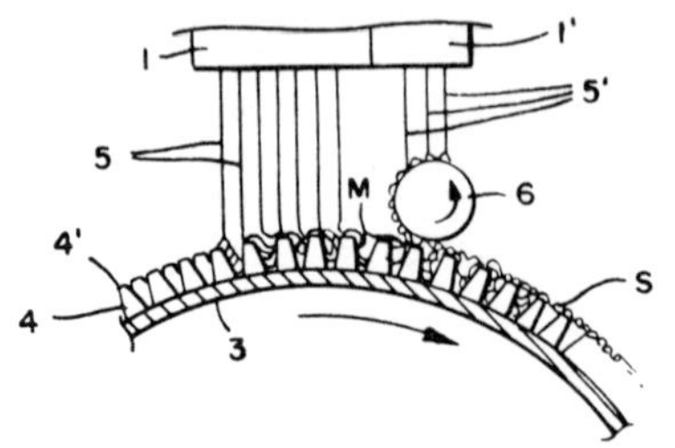

Source: U.S. Patent 4,177,312

Between another spinnerette **1'**, which can be part of spinnerette **1** or else can be mounted adjacent thereto as a second spinnerette, and the depositing element, an essentially smooth roller **6** is rotatably mounted in such a manner that the spun filaments **5'** of spinnerette **1'** are deposited as a thin, flat, nonwoven web or sheet on the roller surface **S** moving counter to the rotational movement of the much larger support drum. The nonwoven sheet **S** is lightly pressed at the same time onto the freshly formed matting **M** while the filaments of both matting and nonwoven sheet are still tacky and readily adherent to each other.

During cooling of both of these layers or sheet structures, the filaments of nonwoven sheet **S** fuse or bond to the filaments of matting **M** at surface contact points. The roller **6** should be quite small in diameter since it does need to handle more than a few rows of filaments **5'** and, at the same time, is preferably arranged in the gap space between the spinnerettes **1** or **1'** and the larger moving drum support. When the gap space between the drum and spinnerette **1** is very small, the second spinnerette **1'** and the roller can be removed farther to one side. However, it is essential that the filaments **5** and **5'** remain tacky and adherent while being pressed together along the contacting surface portions of the nonwoven sheet **S** and the primary matting **M**.

Example: Polycaprolactam is spun as a large number of macrofilaments from 7 rows of spinning orifices of a spinnerette onto a rotating drum provided with truncated pyramids to form a primary matting layer with a waffle-like surface structure. On an essentially smooth and even roller, friction driven by the drum as in Figure 3.17, another group of macrofilaments is spun from 3, 4, 5 or 6 rows of spinning orifices to form an essentially flat nonwoven web or sheet which is surface bonded to the structured primary matting sheet as the two sheets are squeezed slightly between the drum and roller.

Spinning and draw off speeds are adjusted so that the matting with its applied sole or footing layer according to the process has a total weight per unit area (surface weight) of 400, 500 or 600 g/m^2.

Each matting article as a finished product has a total thickness of 15 mm. Weight in g/m^2, ranged from 400 to 600. Surface weight per unit thickness varied from 26,000 to 40,000 g/m^3. Transverse strength per unit surface weight was 3.95 to 6.51 Nm/g.

Wet Lay Process

In a wet lay process for the production of nonwoven webs, short cut staple fibers, either natural, synthetic or a blend of same are dispersed in an aqueous medium to produce a fiber slurry. A chemical binder is added to the slurry or to the formed web to promote bonding of the fibers into a unified structure. The fiber slurry is fed to a papermaking machine, such as a Fourdrinier machine, where it is positioned on a porous support. Suction is applied beneath the porous support and removes the majority of the aqueous medium from the slurry, leaving a damp web of fibers across the forming area. The newly formed web is then further processed to yield a final nonwoven web where the fibers are bound to adjacent fibers to define a unitary structure.

A nonwoven web produced by the wet lay process can only be as good as the initial fiber dispersion. The fiber dispersion determines uniformity across the width of the web, the presence or the absence of voids in the web, later bondability of the fibers into a unitary structure, and the presence or absence of globs of fiber in the web. It is, therefore, quite important that a proper fiber dispersion be achieved to produce a good fiber slurry and, thereafter, that the fiber slurry be properly processed to realize a quality nonwoven web.

A composition for short cut, synthetic polymeric staple fibers that are used in a wet lay application for the production of nonwovens is disclosed by *R.L. Hawkins; U.S. Patent 4,179,543; December 18, 1979; assigned to Hoechst Fibers Industries, Division of American Hoechst Corporation* along with a fiber having the finish thereon and the process for dispersing the fiber in an aqueous medium.

Synthetic polymeric filaments are cut into staple lengths, generally from about ¼ to 3 inches in length, and are dispersed in an aqueous medium in conjunction with a composition that includes as an essential ingredient an ethoxylated primary emulsifier that contains at least 5 mols of ethylene oxide and exhibits a surface tension of at least 30 dynes/cm in a 0.10 weight percent aqueous solution at 25°C plus or minus 2°C.

Optionally, a lubricant may be added to the finish composition to improve processability of the fiber during manufacture. The lubricant should be compatible with the primary emulsifier. A secondary emulsifier may also be added to emulsify the lubricant and may additionally assist in dispersing the fiber. The composition may be added to the fiber during the manufacture of same or may be added to the aqueous medium. A preferred composition for addition to the fiber at the time of manufacture includes 50 parts of an ethoxylated castor oil, 25 parts of an ethoxylated sorbitol hexoleate and 25 parts of n-octyl, n-decyl adipate. The composition should have a low foaming propensity and not adversely affect the bondability of the fibers.

Examples 1 through 16: A 1.5 denier per filament (dpf) polyester fiber was treated with various concentrations of formulation with a finish of the composition: 50 parts POE (16) castor oil; 25 parts n-octyl, n-decyl adipate; and 25 parts POE (50) sorbitol hexoleate. A 10 g fiber sample was placed in a container with the various concentration finish formulations. After the fibers had wet out, the fibers were removed from the container and placed in a hose leg. The hose leg having the wet fibers therein was centrifuged for 1 minute, and thereafter the fiber was allowed to dry overnight.

The following day methanol extractions were conducted on the various fiber samples to determine the amount of finish pickup. These data are reported below. Though some inconsistency appears to be present, the reported data do appear to represent a relationship between concentration of oil phase in the emulsion, and fiber finish add on, and dispersion quality.

Fiber Pickup of Finish Compositions

Example Number	Finish Concentration (%)	Finish on Fiber (wt %)	Fiber Dispersion
1	0.008	0.13	fair
2	0.02	0.12	fair
3	0.04	0.15	fair
4	0.06	0.17	fair
5	0.08	0.16	good
6	0.2	0.23	good
7	0.4	0.34	good
8	0.6	0.37	good
9	0.8	0.46	good
10	1.0	0.22	good
11	2.0	0.45	good
12	3.0	0.55	good
13	4.0	0.70	good
14	5.0	0.89	good
15	6.0	1.02	good
16	7.0	0.72	good

Plant trials were conducted where 1.5 dpf polyester filament was extruded, quenched and treated with a 3.0 weight percent finish formulation of the composition. Thereafter, the fiber was processed as described above and cut into ¼ and ¾ inch staple fiber lengths. The staple fibers with the finish thereon were then placed in an aqueous medium in a mixing tank for a wet lay nonwoven process, and the nonwoven web was produced according to the general process steps set forth above. Little foaming was evident throughout the wet lay process and highly suitable nonwoven webs were produced.

COATINGS AND FINISHES

Fiber Finishes

G. Koerner, G. Schmidt, J. Langner and H.-J. Patzke; U.S. Patent 4,064,057; December 20, 1977; assigned to Th. Goldschmidt AG, Germany disclose melt spin finish compositions composed of 1 to 99% by weight of compounds having the general formula:

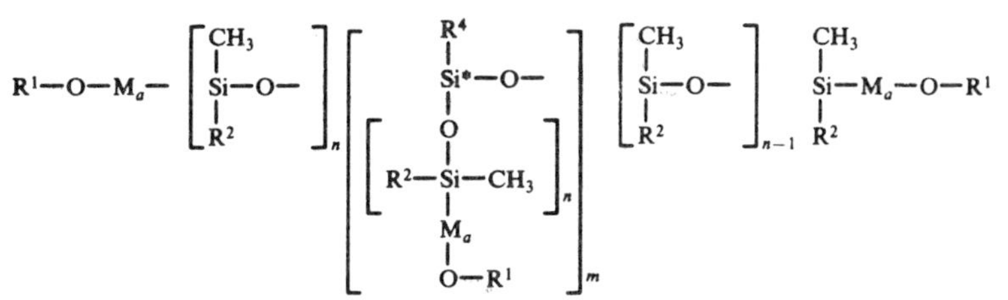

where

R^1 is a mono-, di- or trialkylphenyl residue, in which the sum of the carbon atoms per phenyl residue, bound in the form of alkyl residues, is at least 6 and at most 12, and/or the trimethylsilyl residue;

R^2 is an alkyl residue with 1 to 16 carbon atoms, where the carbon chain may be interrupted by the groupings

$$-O-\overset{\overset{O}{\|}}{C}- \qquad \text{or} \qquad -\overset{\overset{O}{\|}}{C}-O-$$

and/or the M_a-O-R^3 residue; where

R^3 is a mono-, di- or trialkylphenyl residue, where the sum of the carbon atoms per phenyl residue, bound in the form of alkyl residues, is at least 6 and at most 12, and/or is

$$-\underset{\underset{CH_3}{|}}{CH}-C_6H_5$$

R^4 is an alkyl residue with 1 to 16 carbon atoms and/or

$$-O-\left[\begin{matrix}CH_3\\|\\Si-O-\\|\\R^2\end{matrix}\right]_n M_a-O-R^1$$

M is an alkyl residue with 1 to 3 carbon atoms,

n is 2.5 to 15;

m is 0 to 5; and

a is 0 or 1

and the average molecule contains 5 to 30 Si atoms of which at most 20 mol percent are Si* atoms, and 0.5 to 10 mono-, di- or trialkylphenyl residues, in addition to 99 to 1% by weight of known finish components.

The finish compositions possess not only exceptional thermal stability but also confer excellent frictional properties to the melt spun fibers.

Example: 112.2 g (0.51 equivalent) of nonylphenol are treated with 996 ml toluene. Subsequently, 200 ml toluene are distilled off while stirring in order to azeotropically dry the nonylphenol. The contents of the flask are brought to a temperature of 50°C and treated with 286.3 g (0.50 equivalent) of a chlorosiloxanyl sulfate of the idealized average formula:

$$Cl-\left[\begin{matrix}CH_3\\|\\Si-O\\|\\CH_3\end{matrix}\right]_{13.7}-\begin{matrix}CH_3\\|\\Si-Cl\\|\\CH_3\end{matrix}$$

where a part of the Cl is replaced by ($SO_4/2$). Subsequently, the mixture is stirred for 30 minutes at 50°C. Thereupon ammonia is passed in at the same temperature until the mixture becomes alkaline. Stirring is then continued for a further 30 minutes under a mild NH_3 cover, and the temperature is subsequently lowered to below 30°C.

This is followed by an intermediate filtration through a filter press covered with a filter layer of cellulose fibers. The clear filtrate is returned to the flask which is then evacuated. While stirring vigorously all the distillate is taken off until the contents of the flask reach a temperature of 100°C. When the distillation ceases, stirring is continued for a further ½ hour at a flask content temperature of 100°C under vacuum (20 mm Hg). The contents of the flask are then cooled to approximately 30°C, and the vacuum is released. The reaction mixture is filtered again through a filter layer of cellulose fibers. The clear product so produced has a viscosity (at 20°C) of 44.8 cp.

Stabilized Silicone Oil Coating for Melt Spinning Nozzles

W. Eberius and H. Lorenz; U.S. Patent 4,034,034; July 5, 1977; assigned to Akzona Incorporated provide a process for melt spinning a synthetic linear fiber-forming polymer into filaments by extrusion through a nozzle treated on its discharge face with a thin layer of silicone oil stabilized with a cerium compound consisting of the reaction product of a cerium compound with a polymethylsiloxane bearing reactive hydrogen atoms.

Example: In a 1 liter Erlenmeyer flask, 6.51 g of cerium nitrate and 4.5 g of acetylacetone are dissolved in 300 ml of distilled water. A 2% ammonium solution is added in drops, shaking vigorously to a pH of 8. This results in a light yellow precipitate of cerium acetylacetone monohydrate which is suctioned off by a Buchner funnel and rinsed with 300 ml of distilled water. The precipitate is dried under high vacuum without heating until a dustlike product is obtained.

4.10 g of the cerium acetylacetone monohydrate are refluxed in 2.5 liters of distilled benzene for 5 to 10 minutes while air is introduced into the solution at a rate of 5 l/min. To the resulting solution, there is added 100 ml benzene containing 6.25 g of a siloxane having a reactive hydrogen atom, and the mixture is refluxed for 1 hour while introducing 5 l/min of air. Then 5 kg of the silicone oil to be stabilized are added, and the benzene is distilled off under constant introduction of 15 l/min of air. The internal temperature of the mixture is increased to 275°C, the amount of air is increased to 80 to 100 l/min, and heating is continued for another 4 hours at 275°C internal temperature.

The silicone oil thus stabilized is completely transparent. The nozzles of melt spinning apparatus are covered with a thin film of this oil by spraying and are then used to produce profiled nylon 6 filaments with a size of 17 dtex. With a time interval of 8 hours in the routine scraping procedure, only 0.83 additional interim scrapings are needed per 12 spinning positions per day. If a typical unstabilized commercial silicone is used for coating the nozzles, 5.36 interim scrapings are needed under otherwise identical conditions.

Extrusion of Filaments, Films and Shaped Articles

OLEFINS AND VINYLS

Controlled Degradation

Extruders have been used conventionally for many years to process all types of polymeric materials and especially polyolefins. Generally, the polymer is melted and worked to some extent in the extruder and conveyed to a particular molding means so that it is in the proper state to be handled by that specific means.

In a process developed by *R.A. Steinkamp and T.J. Grail; U.S. Patent 4,001,172; Jan. 4, 1977; assigned to Exxon Research & Engineering Co.,* modified polymers, particularly polyolefins having improved flow and in some instances improved adhesion properties over that of a polymeric, e.g., polyolefin, base stock used as a starting material, are produced by a controlled reaction often involving degradation in an extruder, in which an initiator is injected under conditions of either maximum distribution or intensive mixing wherein appreciable rheological, i.e., molecular weight distribution, changes in the base polymer occur. In some embodiments monomers are also grafted to the base stock, during the degradation process. In such instances, grafted polymers with high melt flow properties and other useful properties are obtained.

Various extruder designs may be employed to carry out the graft copolymerization process. However, the common characteristic of each extruder design is that thorough, instantaneous mixing of the initiator and acrylic acid with the polymer, i.e., polypropylene, occurs.

The extremely high degree of mixing which characterizes the process is evidenced by appreciable degradation of the polymer. Evidence for the appreciable degradation of the polyolefin is shown by the substantial increase in the melt flow rate or melt index of the copolymer over the base resin. Evidence for a narrowing of the molecular weight distribution is seen from the fact that the die swell of the graft copolymer is lower than the die swell of the polyolefin base stock used in making the copolymer. It is to be emphasized that a change in molecular weight distribution leads to many useful properties of the resulting polymers.

Die Swell: Some high molecular weight polymers such as polyolefins when forced through a capillary die of a relatively short length produce an extrudate of a larger diameter than the diameter of the capillary.

This property of polymers has been characterized as die swell which is expressed numerically as the ratio of the diameter of the extrudate to the diameter of the capillary (by some the ratio to the first power and by others to the second power). The term "die swell" as used herein is defined as follows:

$$\text{die swell} = \left(\frac{D_e}{D_o}\right)^2$$

where: D_e is the extrudate diameter and D_o is the capillary diameter.

The numerical value of die swell is also dependent on the geometry of the rheometer used to force the polymer through the capillary. In obtaining the numerical values set forth herein, and in the table which follows, a rheometer having a rheometer barrel of $\frac{3}{8}$ inch i.d. (inside diameter) was used wherein the barrel was heated to a temperature controlled to $\pm 2°F$ of the recorded temperature and the polymer was forced through a capillary having a 0.03018 inch i.d. and which was 1.006 inches long. The capillary had a 90° entry angle.

The measurements were made by forcing the polymer through the capillary by a plunger operating at a constant speed or a constant shear rate (γ°) ranging from 13.5 reciprocal seconds to 338.3 reciprocal seconds (sec^{-1}). The polymer was forced through the capillary into ambient air at room temperature (70°-80°F).

The measurement of die swell is frequently used as a gross measure of molecular weight distribution in polyolefins; high die swell resins possess broader molecular weight distribution than polymers having lower die swells.

Example: Compositions were prepared by introducing polypropylene having a melt flow rate (MFR) of 0.4 into an extruder. Into an injection line connected to a side port was introduced a mixture of acrylic acid and a peroxide (VulCup R) in the ratios as set forth in the table below. For composition A, the acrylic acid and peroxide were added with essentially no back pressure on the injection line which was located over the decompression portion of the screw.

For composition B, in contrast, the acrylic acid and peroxide were added to a standard extruder under conditions where the reaction zone was filled with polymer and considerable melt pressure was evident against the injection line.

Copolymerization Conditions	A	B
Feed zone, °F	520-490	520-490
Reactor zone, °F	305-400	300-400
Metering zone, °F	450-400	450-400
Output, lb/hr	92	80
Screw speed, rpm	160	160
Peroxide:acrylic acid, g:1,000 g	15	15
Acrylic acid injection rate, wt %	3.26	3.84
Total acrylic acid in product, wt %	2.66	3.18
Conversion of acrylic acid, %	81	83
Melt pressure against injection line, psi	0	150

The compositions were evaluated. The thorough dispersion and mixing of the initiator with the resulting appreciable degradation of the polypropylene for composition A is clearly apparent in the lower die swell and higher melt flow rate data obtained for these to the base polymer.

In contrast, the polypropylene acrylic acid graft copolymers of composition B had a die swell either the same or higher than the polypropylene base stock and suffered very little molecular weight loss. This clearly indicates no appreciable degradation of the composition B during the grafting of the acrylic acid.

Furthermore, the increase in MFR of the graft copolymers in composition A over the MFR of the base resin shows that appreciable degradation occurred in contrast to the slight increase in MFR of the graft copolymers of composition B over the MFR of the base resin.

Continuous Preparation of Fibrous Polymer Crystals

According to *C.E. Meihuizen, A.J. Pennings and A. Zwijnenburg; U.S. Patent 4,137,394; Jan. 30, 1979; assigned to Stamicarbon, BV, Netherlands* filament-like polymer crystal fibers are prepared from a solution of a crystallizable polymer, such as polyethylene or polypropylene, in a vessel containing a spinning rotor, preferably having a slightly roughened surface. The fiber thus formed is taken up and removed at a rate equal to the crystal growth rate. The longitudinal crystal growth rate is of sufficient speed for commercial application while at the same time yielding fibers of outstanding mechanical properties.

Example: A high density polyethylene was dissolved in p-xylene to form a 0.5% by weight solution. The polyethylene, which is a commercial product (Hostalen GUR) had the following characteristics: the I.V. in decalin at 135°C was 15 dl/g; the molecular weight (number average) $\overline{M}n$ was 10×10^4 determined osmometrically; and the molecular weight (weight average) $\overline{M}w$ was 1.5×10^6, determined by light scattering in α-chloronaphthalene at 135°C.

The various polyethylene solutions prepared were stabilized with 0.5% by weight of the antioxidant Ionel DEPC which is di-tertiary-butyl-para-cresol, and all tests were conducted under pure nitrogen. Fibrous polyethylene crystals were chosen as seed crystals. These crystals were obtained from an 0.1% p-xylene solution of the polyethylene polymer as described above. The crystals were 40 mm long and had a cross section of 0.25 x 0.10 mm.

The device shown in Figure 4.1 for conducting the following tests included a cylindrical vessel **1** shut off at the top by stopper **2**, a rotor **3** bearing-mounted in Teflon at **4** and **5**, which was rotated at the indicated speed via shaft **6**. A thin Teflon tube **7** was attached at the outside of the vessel, in contact with the inside and mounted slightly tangentially. The seed crystal could be introduced through the aperture **8**. In the apparatus as described the external diameter of the rotor was 112 mm and the internal diameter of the vessel 135 mm leaving about a 22 mm space **11** to be filled by the polymer solution. The filament **9** was wound on a reel **10**. The space **11** was filled with a polymer solution, which could be supplied through an aperture **12**; the tube **7** was filled with solvent to externally clean the filament of adhering solution. The entire device was submerged in a thermostatic bath which kept the temperature constant to within ±0.01°C.

First, two comparative tests were conducted: (1) a test wherein only the tip of the growing crystal was close to the rotor; and another series of the tests (2) in which 20 cm of the growing crystal was contacting the rotor.

Figure 4.1: Continuous Preparation of Fibrous Polymer Crystals

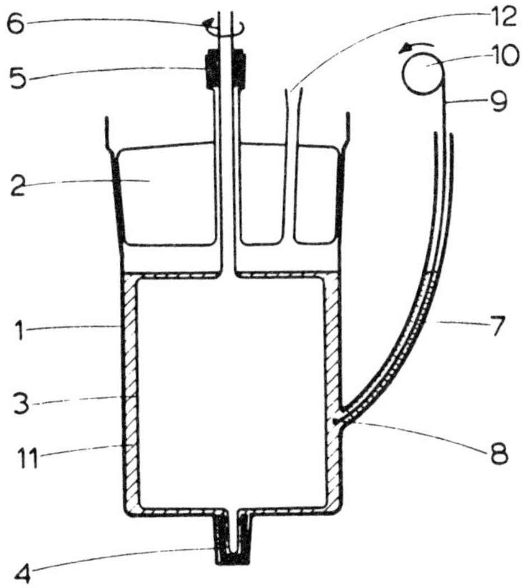

Source: U.S. Patent 4,137,394

(1) In an 0.5 polyethylene solution the longitudinal growth as a direct function of the reeling speed at 103°C and a rotor speed of 20 rpm, indicating rpm, was only 0.8 cm/min.

(2) Under the conditions of temperature and rotational speed the longitudinal growth or reeling speed could be increased to 20 cm/min at only 2 rpm of the rotor.

Under the conditions of the comparative tests, part 2, and within the rotational speed range from 0.8 to 4 rpm, the growth rate can be varied between 8 cm/min and 31 cm/min. The mass of the fiber could be increased from 27 x 10^{-5} mg/cm.

The influence of the character of the rotating contracting the growing crystal appears from the following table, the tests being conducted at 2 rpm at 103°C and 20 cm/min.

Fiber Mass, mg/cm	Growth Rate (reeling speed), cm/min	Surface Character of Rotor
15	20	Smooth (Teflon)
40	31	Sandblasted glass
59	31	Silanized, sandblasted glass

Contrary to original expectations, it was found that the tensile strength of the fiber so produced in fact increases with the reeling speed. For instance, starting from a solution of 0.5% polyethylene in xylene at 110°C the tensile strength is: 200 kg/mm^2 at a reeling speed of 20 cm/min and 300 kg/mm^2 when increasing the reeling speed to 80 cm/min.

Polyoxymethylene Filaments

B.A. Egorov, A.S. Grzhimalovsky, A.V. Judin, K.E. Fishman and L.D. Vernaya; U.S. Patent 4,060,582; Nov. 29, 1977 provide a method of manufacturing polymethylene filaments, wherein polyoxymethylene having the molecular weight from 30,000 to 100,000 and containing stabilizing additives in a quantity from 0.1 to 3.0% of the weight of the polyoxymethylene is subjected to thermal treatment at a temperature from 100° to 150°C and residual pressure from 1 to 100 mm mercury to attain constant weight. The thus thermally treated polyoxymethylene is melted at a temperature from 170° to 230°C, whereafter the melt is forced through the orifices of an extrusion nozzle. The jets of the melt, leaving the orifices of the extrusion nozzle, are cooled at a temperature from 70° to 169°C. After the cooling the obtained filaments are drawn at a temperature from 120° to 165°C to a length exceeding from 7 to 14 times the initial length.

The method offers a simple technology of producing filaments as strong as 100 g/tex. To produce low-shrinkage fiber (i.e., with a shrinkage rate from 0 to 5% at 150°C) the drawn filaments are thermally treated at a temperature exceeding that of the drawing by 2° to 50°C, the filaments being maintained under tension. Alternatively, the drawn filaments may be first tensioned and then thermally treated, as indicated hereinabove.

Example: Granulated copolymer of trioxane and 1,3-dioxolane (the weight content of the latter being 4%) with as additives 2,2'-methylenebis(4-methyl-6-tert-butylphenol) in a quantity of 0.5% by weight as the antioxidant and formaldehyde dicyandiamide in a quantity of 0.5% by weight as the acceptor, with characteristic viscosity in dimethylformamide equalling 0.56 is thermally treated at 100°C and residual pressure of 5 mm mercury, melted in an extrusion machine in an air atmosphere at 190°C, and the melt is forced by a metering pump through an extrusion nozzle with a single orifice 1.2 mm in diameter. The rate of feed of the melt is 5 g/min. The jet of the melt is cooled in a 0.5 m long tube. The temperature of the air cooling the jet of the melt is 90°C. The filament molding rate is 250 m/min.

The molded filament is drawn to a 10:1 draft at a speed of 125 m/min over a 250 mm long "iron" at 150°C. The obtained filament has 72 g/tex tensile strength and elongation at rupture equalling 11.5%. The shrinkage of the filament is 18%, with 80% of the initial strength remaining after the shrinkage.

After the drawing operation the filament is thermally treated in air under a tension of 2 kg/mm^2 at 155°C for 30 minutes. After this treatment the shrinkage value is 4%, with 95% of the initial strength remaining after the shrinkage.

Sheets from a Mixture of Vegetable Fiber and Polyolefin

M. Dereppe and A. Deryckere; U.S. Patent 4,144,304; March 13, 1979; assigned to Solvay & Cie, Belgium disclose a process for the manufacture of sheets having a high resistance to water absorption by hot calendering of a mixture produced from 25 to 75% by weight of vegetable fibers and from 75 to 25% by weight of molten polyolefin. Such a mixture does not stick to the calender rolls during the sheeting step.

Example: To carry out this example, the apparatus shown schematically in Figure 4.2 was used, which comprises a plunger-type internal mixer **1**, a conveyor belt **2** which transfers the charges **3** delivered by the internal mixer to the rolls of an external mixer **4**, a calendering installation with four heated rolls arranged in an L, **5**, a conveyor **6** which transfers the strip **7** taken from a roll **8** of the external mixer into the first nip of the calendering installation, and a cooling train **9**.

5.1 kg of a composition comprising, in parts by weight, 500 parts of polypropylene, 500 parts of wood flour (mean size 0.4 mm), 0.5 part of calcium stearate, 0.5 part of 2,6-di-tert-butyl-4-methylphenol and 0.5 part of glycerol monostearate are introduced in successive charges into the internal mixer **1** of Figure 4.2.

Figure 4.2: Calendering Apparatus

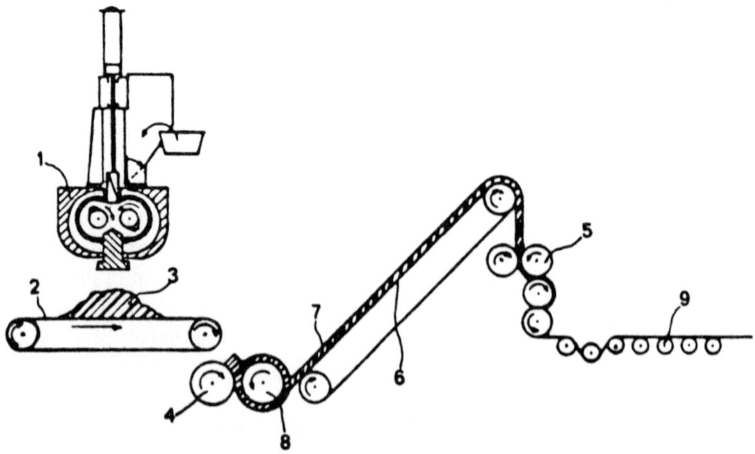

Source: U.S. Patent 4,144,304

The internal mixer (of the Werner and Pfleiderer GK 5 type) is preheated to 170°C, and the mixing cycle is 90 seconds; the work is from 40 to 50 kg-m. The charges delivered by the internal mixer are received on the rolls of a Berstorff external mixer heated to 175°C. On roll 8, a 5 cm wide strip is taken off and is transferred into, and spread in, the first nip of the calendering installation. The successive rolls of the calendering installation are respectively kept at 185°, 182°, 183° and 172°C. The sheet issuing from the last calendering roll has a uniform thickness of 1.5 mm and its surface appearance after cooling and solidification on the cooling train 9 is excellent.

A sample of the sheet is subjected to a flexural strength test and it is found that the strength amounts to 380 kg/cm^2. A sample of size 100 x 100 mm is cut from the sheet and is subjected to the following test.

After having coated the cut edges of the sample by means of paraffin, the sample is weighed and is then immersed in water, kept at 23°C for 24 hours. After this period of time, the sample is withdrawn, blotted by means of a filter paper, taking care not to remove the paraffin, and again weighed. The difference in the weight before and after immersion, relative to the weight before immersion, gives the percent absorption of water. It is found that this percentage amounts to 4% for the sample examined.

A fresh flexural strength test is carried out on the sample which has been subjected to the water absorption test and it is found that the strength is still 380 kg/cm^2. By way of comparison, a sheet was produced with the utmost care from the same composition by the extrusion technique, employing a sheet die.

During the water absorption test, it is found that the percentage amounts to 8%. Furthermore, it is found that the flexural strength of the sheet after immersion is no more than 75% of the flexural strength of the sample not subjected to the immersion test.

Reinforced Thin Wall Plastic Bag

W.J. Clayton, R.H. Olson and D.F. Kutniewski; U.S. Patent 4,076,121; Feb. 28, 1978; assigned to Mobil Oil Corporation describe a plastic bag such as trash can liners, grass and leaf bags, sandwich bags and the like which is reinforced by placing spaced ribs therein, integral with the wall structure, the ribs having a rib-to-rib spacing in the order of ⅛ to 2 cm, the ribs being peaked and merging smoothly from a projecting peak to the thickness of the walls between the ribs, the peaks extending about 1.5 to 10 times the wall thickness of the film which is preferably, in the order of about 0.4 to 2.0 mils.

Upon extrusion, the extrusion die is formed with small notches with rounded corners so the ribs are extruded while the material is extruded; the notches provide additional material in the region of the ribs which merges smoothly with the material being extruded to form the film. Such rib-reinforced bag structures may be produced by forming a continuous roll of such bags, the individual bags in such a roll being readily separated along predetermined lines of weakness.

Fibrous Products of Improved Fineness

In a process developed by *O.-B. Rasmussen; U.S. Patent 4,125,581; Nov. 14, 1978,* fibrous products, including single fibers, yarn and fibrous webs, of improved fineness are produced from extruded sheet material extruding in the form of a sheet-like, multilayer composite stream a plurality of extrudable materials, at least one of which is a synthetic thermoplastic fiber-forming polymer and another of which is removable by mechanical, chemical, or solvent action or a combination thereof.

The layers extend generally parallel to the faces of the stream, with each such layer containing either the fiber-forming polymer or the removable extrudable material with the different layers being interspersed in the composite stream. This composite stream is solidified into a multilayer sheet, then the resultant solid sheet is split to convert the same into a coarse fibrous product, and the thus-produced fibrous product is treated by mechanical, chemical, or solvent action or a combination thereof to at least partially remove the removable extrudable material and produce a fibrous product of increased fineness.

The apparatus shown in Figure 4.3a comprises a row of slots, three for the reinforcement material and four for the filling material, above which is a collecting chamber consisting of parts **5** and **6** which narrow down to an extrusion slot. The two parts of the collecting chamber may be rotated together relative to the row of slots so that the drag exerted by the bottom of the collecting chamber on the lamellae **1** and **2** as they are extruded from slots **5** and **6** causes

Figure 4.3: Multilayer Extrusion

a.

b.

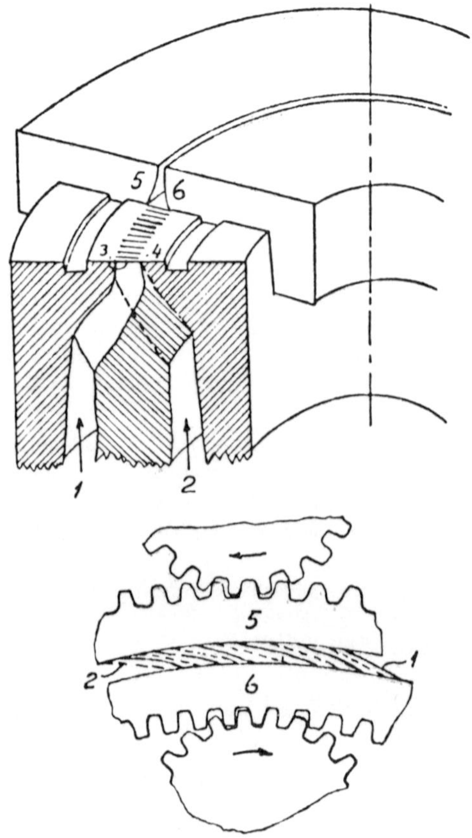

(a) Perspective view, partly in section, of a ring die comprising a circular row of extrusion slots, and a collecting chamber ending with a slot for haul-off of the sheet.
(b) Diagrammatic view from above of an apparatus as in Figure 4.3a, showing the drive and the emerging S-structure.

Source: U.S. Patent 4,125,581

the lamellae to be laid substantially flat along the row. Simultaneously, however, the lamellae are forced upwards by fresh polymeric material being extruded through the slots and as they are forced upwards their sides drag against the sides of the collecting chamber, and in particular against the neck (not shown), and a U-form results.

However, as shown in Figure 4.3b, the two parts **5** and **6** of the collecting chamber may also be moved in opposite directions to exert a symmetrical drag on the two sides of the lamellae, whereby the S-form is produced. For this purpose, the reduction of thickness within the collecting chamber ought to take place less rapidly, and it is even possible to carry out the process without the reduction.

Example: The S-method was used in producing a sheet with high contents of an inorganic filler for use as substitute of book-print paper. First polymeric material: High density polyethylene of melt index 0.2 (ASTM condition E). Second extrudable material: 50% talc powder + 50% low density polyethylene of melt index 70 (ASTM condition E). Proportion between reinforcement material and filling material about 1:1. The sheet material should be biaxially drawn at about 110°C at a ratio of 1:2 in both directions. Thickness before drawing about 0.1 mm.

Resin Tubes with Thick-Walled and Thin-Walled Portions

In a process by *S. Murai, T. Kita and K. Mochizuki; U.S. Patent 4,036,930; July 19, 1977; assigned to Sekisui Kagaku Kogyo KK, Japan* resin tubes having thick-walled portions and thin-walled portions are formed by providing a specially configured orifice on an extruder and varying the rate at which the extruded resin tube is pulled from the extruder. The orifice is shaped to form a thin tubular opening and a thick tubular opening communicating with one another, the thick tubular opening being adjacent and downstream of the thin opening. When the extruded resin tube is pulled at a high speed the thin tubular opening is filled with resin that passes through but does not fill the thick tubular opening. At the slow pulling speed the resin fills at least the open-end portion of the thick tubular opening thereby forming a thick-walled portion of the resin tube.

All resins having a thermoplastic property can be used in the manufacture. The preferred resins are rigid vinyl chloride resins.

Figure 4.4 is a partially cutaway sectional view schematically illustrating the process. In Figure 4.4, there is shown an orifice, which comprises an outer die **1** and an inner die **2**, the inner die being fixedly positioned within the outer die to form a tubular opening therebetween. The inner die **2** is cut at **21** on the resin discharge side of the orifice to form two tubular openings. These two openings consist of a small-width tubular opening **A** and a large-width tubular opening **B**. The small-width opening **A** is positioned in the inner part of the orifice while the large-width opening **B** is positioned at the discharge portion of the orifice.

The outer die **1** is separated from an outside diameter control device **3** by a washer **8**. The outside diameter control device **3** is cylindrical and has an inside diameter which is approximately equal to the inside diameter of outer die **1**. In this way, a resin tube emerging from the orifice is passed into the device **3** without having the outer surface of the resin tube brought into contact with the atmosphere. The device **3** includes a cavity **31**, through which cooling water is circulated to cool the resin tube. The device **3** also has annular grooves **4** formed in an internal surface thereof and in communication with a pressure re-

Figure 4.4: Configured Orifice for Extrusion of Resin Tubes

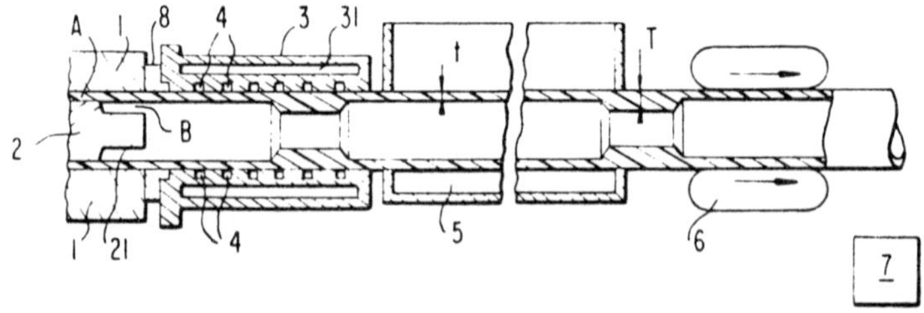

Source: U.S. Patent 4,036,930

ducing pipe through which air is discharged. This results in a partial vacuum causing the resin tube to come into close contact with the internal surface of the device **3**. In this manner, the external surface of the resin tube may be controlled by the device **3**.

The resin tube which emerges from device **3** enters a water vessel **5**, in which cooling water is stored, so that the external surface thereof is cooled by the cooling water. The resin tube is cooled to such a degree that it maintains its shape when subsequently pulled by means of a take-up or pulling machine **6**. The take-up machine **6** comprises, for example, a pair of endless belts or rollers, which pull the resin tube by holding it between the rotating belts or rollers. The rotation of the belts or rollers may suitably be adjusted by an instrument housed in a control box **7**.

The rotation of the take-up machine **6** may be controlled to have at least two stages. That is, the machine **6** is, in one case, rotated in a direction as indicated by the arrow at a high speed V_1 and in the other case, rotated in the same direction at low speed V_2. The high speed V_1 is within a range such that the small-width tubular opening **A** is filled with resin within the orifice but the large-width tubular opening is not filled.

In this manner, the resin discharged through the orifice is exclusively controlled by the small-width opening **A** to thereby form a resin tube having a thin wall (**t**). On the other hand, the low speed V_2 is within a range such that the large-width tubular opening **B** is filled with resin within the orifice or at least the open-end portion of the tubular opening **B** is filled therewith. In this manner, the resin discharged through the orifice is exclusively controlled by the large-width opening **B** to thereby form a resin tube having a thick wall (**T**). Thus, the rotational speed of take-up machine **6** is controlled at least in a manner of high and low stages to alternately repeat operation of high-speed take-up and low-speed take-up, whereby the resulting resin tube has alternately formed thin and thick wall portions.

Reduced Buildup of Resinous Materials at Orifice of Extrusion Die

E.D. Cooper; U.S. Patent 4,056,597; Nov. 1, 1977; assigned to Phillips Petroleum Company provides a process which reduces the buildup of resinous material around the orifice of an extrusion die.

Referring to Figure 4.5, there is shown a die body **12** having bolt holes **14** and **16** for mounting the die to an extruder. Near the center of the die is orifice **18** having an inlet portion **20**, a cylindrical portion **22** and an outlet portion **24**. Inlet portion **20** converges in the direction of flow of the resinous material and is attached to one end of cylindrical portion **22**. The cylindrical portion also is frequently referred to as the land.

As is known in the art the diameter of the cylindrical portion or land is determined by the desired diameter of the strands of resinous material and the length of the cylindrical portion is a function of the melt properties of the resinous material used and the overall extruder and die design. Outlet portion **24** diverges in the direction of flow of the resinous material and is attached to the other end of the cylindrical portion.

Figure 4.5: Extrusion Die Designed to Reduce Resin Buildup

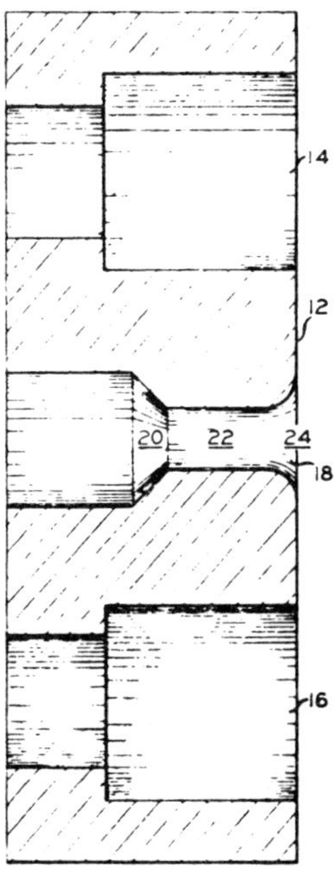

Source: U.S. Patent 4,056,597

Generally inlet portion **20** is in the shape of a converging truncated cone; however, other converging shapes can be used. Frequently the inlet portion **20** is chamfered at an angle in the range of about 40° to about 60°. The use of a converging inlet portion is important in the prevention of stagnation of resinous material inside the extruder near the entrance of the orifice of the die and to reduce the pressure drop across the inlet portion. Resinous material subjected to high temperatures in extrusion processes due to stagnation in the extruder have substantially altered properties as compared to the resinous material which flows evenly through the extruder. Since various amounts of the stagnated material occasionally pass through the orifice, stagnated material causes impurities and nonuniformity in the extruded resinous material.

The use of a divergent outlet portion is important to reduce the problem of accumulation or buildup of resinous material at the outlet of the orifice. A buildup or accumulation of resinous material at the outlet of the orifice causes oxidizing and crosslinking of the material which results in problems similar to those caused by stagnated material previously discussed.

As known in the art, outlet portion **24** is normally a continuation of cylindrical portion **22** and is generally a square-edged opening. As the resinous material in the form of a strand passes out of the orifice it expands. This expansion of the resinous material causes a portion of the material to contact the outer surface of the die, and frequently a small part of the resinous material sticks to the surface of the die and oxidizes, etc. Surprisingly, it has been found that accumulation of the resinous material is sharply reduced by constructing the outlet portion of the die so that the outlet portion is diverging in the direction of flow of the resinous material.

The resinous materials suitable for use in accordance with the process vary widely. Some of the useful materials include polyethylene, polypropylene, polystyrene and styrene-butadiene copolymers.

Melting Frozen Plugs of Polymer

An extrusion pelleting process and apparatus is disclosed by *D.W. Cuff; U.S. Patent 4,112,039; Sept. 5, 1978; assigned to Leesona Corporation* wherein the outermost portion of an extrusion orifice is surrounded by a heat reservoir maintained at a temperature significantly higher than the softening temperature range of the polymer being extruded. The high-temperature heat reservoir constitutes a heat source from which individual extrusion orifices may be heated above the polymer's softening range to assist in melting frozen plugs of polymer which may block the orifice from time to time.

Turning to Figure 4.6, an extrusion die **10** is depicted. Such die **10** is generally of ring-like configuration and includes a plurality of extrusion orifices **12** adapted for operative engagement generally with an extruder as is known in the art. The die includes an initial zone for feeding a thermoplastic resinous material such as polypropylene, polyethylene, or the like in a heated fluid condition and

Figure 4.6: Extrusion Die with Heat Reservoir

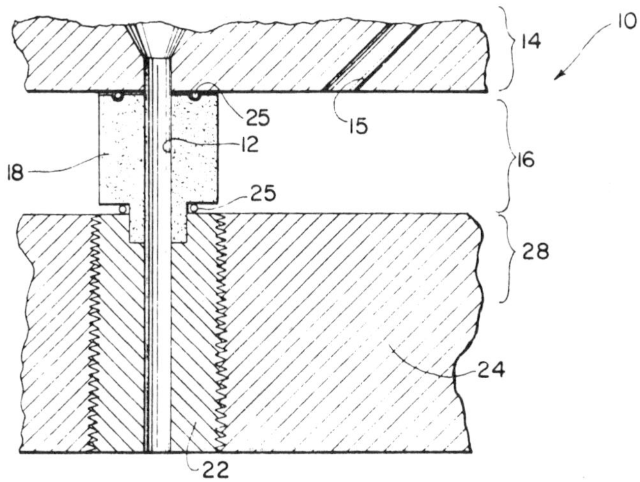

Source: U.S. Patent 4,112,039

forcing such material through the plurality of such extrusion orifices **12**. Such zone is bracketed and indicated by the reference numeral **14**. Thereafter, the material passes through an intermediate zone **16** wherein coolant under pressure, i.e., water is introduced as through opening **15** into direct contact with the outer surface of the resinous material filling the extrusion orifice **12** as through body **18** preferably of open porous metal construction as depicted. As soon as the coolant comes in contact with the material, by reason of an accompanying pressure reduction, a part of the coolant is immediately vaporized and accordingly removes an extensive amount of heat from the material contained therein.

With such systems one or more of the extrusion orifices **12** by reason of the strong cooling action provided by the depressurized coolant and particularly by reason of the induced ebullition (heat of vaporization effect) upon the aforementioned surface of such material and for other reasons may become blocked or plugged with chilled solidified masses of material so that strand emergence from such extrusion orifice or orifices is temporarily blocked.

The porous member **18** may be secured to the die **10** by means of an insert **22** threadably or otherwise positioned within an outer section **24** of the die body. Appropriate seals **25** may be provided therebetween as is known. The portion of the die downstream from the die body portion **24**, provides a final extrusion zone **28** downstream of the aforementioned intermediate zone in which the material is contained in the form of rod or strand-like configurations having an outer solidified surface during its passage through the die. Upon emergence from the die face, a cutter is positioned to cut the rod or strand-like material into discrete separate pellets.

Example: Polypropylene having a melt flow index of 5 and with the midpoint of its workable melting temperature range about 450°F was extruded at a rate of approximately 2,500 lb/hr through a two-stage 400 hp extruder coupled to a die face cutter. The outer platen, i.e., the heat reservoir was maintained at temperatures varying from 500° to 600°F. Water coolant was introduced under high pressure via open cell sintered metal inserts surrounding each orifice at a rate wherein the rods of polymer emerging from the die face exhibited an outer hardened cooled polymer shell which when cut produced individual pellets without evidence of smearing.

ACRYLICS

Melt Shaping Acrylonitrile Polymers

In a process developed by *K. Odawara and S. Takeuchi; U.S. Patent 4,108,818; August 22, 1978 assigned to Japan Exlan Company Limited, Japan,* an acrylonitrile polymer composition obtained by uniformly mixing an acrylonitrile polymer, water and acrylonitrile is heated at the autogenous pressure or above and the resulting melt is shaped into a desired form. The shaping processability is improved thereby.

Example: An acrylonitrile copolymer consisting of 90% acrylonitrile and 10% methyl acrylate, water and acrylonitrile were mixed in the mixing ratios shown in the table below. Each composition was put into a pressure-resisting cell through which the interior can be observed, and then the cell was hermetically sealed. Thereafter, the cell was heated and the melting behavior of the acrylonitrile polymer composition was observed. After the melt was cooled, it was removed from the cell to judge the coloring of the shaped article (the thermal stability of the melt). The melting points of the acrylonitrile polymer compositions and the coloring of the shaped articles are shown in the table.

As comparative examples, there were prepared two kinds of acrylonitrile polymer compositions, one obtained from 60 parts of the abovementioned acrylonitrile copolymer and 40 parts of water, and the other obtained from 80 parts of the acrylonitrile copolymer and 20 parts of water; and two kinds of acrylonitrile polymer compositions, one obtained from 60 parts of the abovementioned acrylonitrile copolymer and 40 parts of acrylonitrile, and the other obtained from 80 parts of the acrylonitrile copolymer and 20 parts of acrylonitrile, and their melting behavior was observed using the abovementioned cell.

Acrylonitrile Copolymer (%)	Acrylonitrile (%)	Water (%)	Melting Points (°C)	Coloring of the Shaped Articles*
60	32	8	137	O
60	24	16	125	O
60	16	24	116	O
60	8	32	113	O

(continued)

Acrylonitrile Copolymer (%)	Acrylonitrile (%)	Water (%)	Melting Points (°C)	Coloring of the Shaped Articles*
60	4	36	127	O
80	16	4	148	O
80	12	8	143	O
80	8	12	143	O
80	4	16	141	O
80	2	18	143	O
60**	40	–	152	X
80**	20	–	153	X
60**	–	40	>180	X
80**	–	20	>180	X

*X, colored; O, colorless.
**Comparative

From the results shown, it is easily understood that, in the case of the acrylonitrile polymer compositions containing in mixture therewith water and acrylonitrile, their melting points were markedly lowered as compared with those containing only water or acrylonitrile, and moreover the shaped articles were colorless.

The acrylonitrile polymer composition obtained by uniformly mixing 60 parts of the acrylonitrile copolymer, 8 parts of water and 32 parts of acrylonitrile was fed to a Plunger-type flow tester melt-spinning apparatus (type 301; produced by Shimadzu Seisakusho Ltd.), and heated under the autogenous pressure to melt. The melt maintained at 140°C was melt-spun through spinning orifices, each 0.5 mm in diameter, with the draft ratio maintained at 2 times under an extrusion pressure of 45 kg/cm^2. No substantial filament breaking occurred at the time of extrusion, which allowed continuation of stable spinning. The physical properties of the finally obtained acrylic synthetic fiber were by no means inferior to those of ordinary acrylic synthetic fibers. In addition, the fiber represented excellent whiteness.

Shaped Articles from Multiblock Copolymers

V. Jungr, A. Stoy, V. Stoy and J. Zima; U.S. Patent 4,053,442; Oct. 11, 1977; assigned to Ceskoslovenska akademie ved, Czechoslovakia describe a method for manufacturing shaped articles from multiblock copolymers of acrylonitrile with acrylamide and/or acrylic acid, using hot pressure molding, preferably in swelled condition.

Example: 15% solution of acrylonitrile in a 65% nitric acid was polymerized by adding 0.1% of diisopropyl percarbonate and 0.1% of urea, related to the solution as a whole. After 48 hours' standing at 15°C a clear highly viscous solution was obtained which was then kept for 200 hours at 8°C. The solution, heated to room temperature, was extruded through a 10 mm opening into water and the coagulating stream was drawn off at a rate of 15 cm/min. After thorough washing in water the strand, about 8.5 mm in diameter, was cut to 1-mm-thick cylinders. The blanks thus obtained were put into a concave hemispherical mold heated to 95°C and pressed by means of a spherical die.

The tightly closed mold, kept under pressure by means of a spring, was cooled to ambient temperature and opened. The material of the mold was polypropylene, that of the die was steel. The molding in the shape of a contact lens was washed in water, kept for 8 hours in physiologic saline and sterilized therein by adding 0.08% of hydrogen peroxide, sealing the lens in a polyethylene foil and heating it for 10 min at 100°C steam. The lens had about 60% water at swelling equilibrium, tensile strength of 70 kg/cm^2, elastic elongation about 600%. The permeability for oxygen was approximately 50% higher than that of sparingly cross-linked ethylene glycol methacrylate gel, utilized for manufacturing contact lenses, and the lens could be worn for a considerably longer time.

Semipermeable Membrane

According to *K. Kamada and S. Minami; U.S. Patent 4,025,439; May 24, 1977; assigned to Mitsubishi Rayon Co., Ltd., Japan,* dry semipermeable membranes of polyacrylonitrile or copolymers containing more than 65 wt % of acrylonitrile are prepared by dissolving the polymer in a solvent in a solids concentration of 15-30% by weight; uniformly coating the solution onto a surface in a desired shape under an atmosphere whose relative humidity is 60-85% and dipping the coated surface into a nonsolvent medium so as to remove the solvent which forms the membrane; heat-treating the membrane under moist conditions at 60°-90°C; and drying the membrane at a temperature less than the temperature of the heat treatment under conditions such that the coefficient of contraction is less than 1%, whereby a dried semipermeable membrane having an open-celled structure with a pore size of a diameter less than 0.5 μ and a porosity of 0.40-0.7 is obtained.

Example: A copolymer of 93 wt % acrylonitrile and 7 wt % vinyl acetate was dissolved in dimethylacetamide to form a 23 wt % solution. The solution was coated onto a glass plate by an applicator and immediately thereafter it was dipped into a bath for 10 minutes. The bath was a mixture of dimethylacetamide and water in a ratio of 20:80 and kept at 40°C. The resulting semipermeable membrane was washed with water and was heat-treated in hot water at 75°C for 10 minutes while being maintained at a constant length.

The resulting heat-treated membrane and a non-heat-treated membrane as a reference were dried at room temperature for 5 days. The dried semipermeable membranes were dipped again in water, and the characteristics of the semipermeable membranes were measured. In the reverse osmosis tests, a 0.5% solution of potassium ferricyanide was treated under a pressure of 40 atm, and the amount of water which permeated through the membrane was measured and the concentration of potassium ferricyanide in the permeated water was measured by electroconductivity and the rejection rate (%) of the solute was calculated. The porosity of the membrane was calculated by measuring the weight of the membrane before and after drying.

The porosity of the heat-treated membrane was 0.515 and the structure of the membrane was not changed after the drying step.

Accordingly, the reverse osmosis test could be achieved by using the heat-treated membrane, and the water permeation velocity was 18 gal/ft^2-day and the rejection rate of the solute was 95%. The characteristics were the same with those of the membrane before drying. On the other hand, the porosity of the non-heat-treated membrane was 0.670 before drying. However, the membrane was deformed by the drying step and the tests could not be performed.

Comparative Example: A commercial cellulose-acetate reverse osmosis membrane was dried at room temperature for 5 days. The dried membrane was contracted and deformed. When the dried membrane was dipped in water, the membrane did not recover and the reverse osmosis test could not be performed. The porosity of the semipermeable membrane was 0.696 before drying.

POLYESTERS

Processing Polyester to Minimize Acetaldehyde Formation

In a process developed by *S.A. Jabarin and E.A. Lewis; U.S. Patent 4,142,040; Feb. 27, 1979; assigned to Owens-Illinois, Inc.,* gaseous oxygen is excluded during thermal processing of saturated polyesters in the molten state and, as a result, acetaldehyde formation is minimized.

Figure 4.7 illustrates an important embodiment of the process. Particulate molding grade polyester, such as poly(ethylene terephthalate) having an inherent viscosity of 0.72, is delivered to screw extruder **1**, having the usual feeding, transition and metering zones, through hopper **2** while screw **4** is turning to advance polymer toward the opposite end of the extruder. An inert gas, such as nitrogen, is introduced through one or more conduits **3** into the bottom of the hopper or through one or more conduits **3a** into the feeding zone (or both).

Figure 4.7: Screw Extruder for Plasticizing, Mixing, and Delivering Polyester Resins to Molding Means

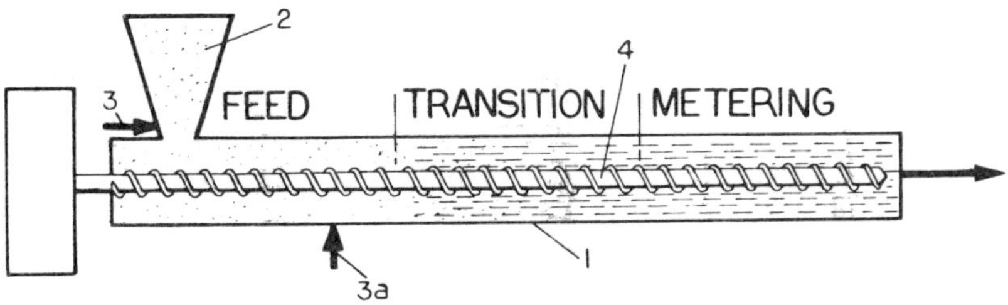

Source: U.S. Patent 4,142,040

The inert gas flushes essentially all air from the polyester as it advances through the initial part of the feeding zone. Polymer is delivered from the exit end of the extruder into molding means, which can be, for instance, an annular die outfitted on the exit end of the extruder to deliver a tubular shape or which can be an injection mold, not shown, to form a desired shaped plastic product. The temperature maintained in the extruder after the polymer becomes molten in the latter part of the feeding zone is for this polymer about 275°-285°C. Acetaldehyde formation is minimized because of the absence of oxygen. Of course, before the initial start-up of the extruder, the entire extruder is preferably flushed with inert gas such as dry nitrogen.

Poly(Ethylene Terephthalate) Fibrillated Tape

F.R. Lowder and J.D. Gibbon; U.S. Patent 4,036,003; July 19, 1977; assigned to Celanese Corporation and Fiber Industries, Inc. produce low denier, high tenacity poly(ethylene terephthalate) fibrillated tape yarn by extruding a poly(ethylene terephthalate)/polypropylene polymer mixture of about 0.1 to 20% polypropylene through a modified slit die wherein the depth of the die slit is greater at the ends than at the center, melt drawing the extrudate at a ratio from about 26:1 to 75:1, rapidly quenching the extrudate to maintain a uniform polypropylene dispersion and thereafter drawing the tape in two stages and fibrillating preferably in accordance with U.S. Patent 3,707,837.

Example: A mixture of 0.67 I.V. poly(ethylene terephthalate) and 3% polypropylene (melt flow index 15) are extruded through a slit die 1.0 inch in length and 0.005 inch wide. The depth of the slit is 0.040 inch at the center and increases in depth at the edges as a result of curving the entry port with a radius of 0.188 inch. The die slit has 80 serrations spaced approximately inversely along the slit to the spacing of ribs which would result on a tape extruded through an evenly serrated die under the same processing conditions. The serrations are perpendicular to the main slit, 0.015 inch long and 0.005 inch wide.

The extruded tape is quenched in water with the water level being positioned about 0.3 inch from the face of the die. The extruded tape is fed to a feed roll at a rate of 300 fpm and has a denier of 1,600. This tape is then drawn in saturated steam to about 4 times its original length. Following this drawing step a second stretching operation is carried out over a hot plate at 200°C giving an overall stretch ratio of 5.35:1. The tape is then relaxed by passing the drawn tape over hot rolls at 215°C and then onto cold rolls running 6% slower. The relaxed tape moving at 1,500 fpm is then fibrillated by passing through adjacent pairs of false twisting air jets and wound up on a winder which provides constant tension to the fibrillating tape. The yarn produced has the following properties:

Denier	320
Tenacity, gpd	4.85
Elongation, %	16
Initial modulus, gpd	75
Hot air shrinkage at 350°F, %	3.5

This yarn when twised to 8 tpi (turns per inch) had the appearance of a spun staple yarn with a lustrous but not sparkly appearance.

CELLULOSE AND STARCH

Process for Making Shaped Articles of Cellulose

H. Rodier; U.S. Patent 4,129,640; Dec. 12, 1978; assigned to Rhone Poulenc Textile, France discloses a process for producing shaped articles of cellulose, wherein the process includes steps of contacting a shaped solution of cellulose in dimethyl sulfoxide (DMSO) and formaldehyde, containing at least 6% by weight of cellulose based on the volume of DMSO, and containing formaldehyde in a weight ratio of formaldehyde/cellulose of about 0.2 to about 0.6, with a coagulating solution of DMSO and water at a temperature no greater than 10°C. The coagulant bath generally contains 30 to 50% by weight of DMSO, and the solutions are readily coagulated at coagulation speeds appropriate for commercial practice, and generally the coagulation occurs in less than 20 sec.

The products produced by this process are shaped articles such as filaments, fibers, films, sheets and membranes, and especially filaments which are suitable for use in the textile field.

Preferably the cellulosic solutions which are used in the process have an Np value less than 250, more preferably less than 200, and most preferably less than 100, while the solutions have a I_c value greater than 70 and preferably greater than 80.

The Np, or number of fibrous particles per cm^3 of cellulose solution is determined by examination under a polarizing microscope. Counting is effected on the solution introduced into a rectangular microscope cell, using a magnification of 50X. Particles are counted if their diameter is greater than or equal to 25 microns. This test principally counts the undissolved fibers which form highly birefringent particles which are only slightly swollen, and of which the internal structure can be discerned in polarizing light, as well as certain more or less extensively swollen gels which are more or less birefringent and of which the internal structure can still be distinguished. The solutions are classified, depending upon the number of particles observed, as follows:

Np of 0 to 100 – excellent
Np of 100 to 200 – good
Np of 200 to 400 – good to satisfactory
Np greater than 400 – mediocre.

In addition to the particles counted by the Np method described above, the cellulose solutions can also contain very extensively swollen particles or gels which virtually have no detectable internal structure, have a refractive index very close to that of the surrounding medium, and are therefore quite difficult to detect by the Np test.

The extent of these gels can be assessed by measuring the clogging index value. This clogging index is determined by filtering the solution under a constant pressure of 2 bars through a filtering medium composed of three superposed assemblies, each composed of a stainless steel gauze of 34 microns mesh size and a stainless steel gauze of 0.610 mm mesh size.

The time of passage t_1 corresponding to the passage of a volume of 120 cm^3 of solution is determined and then the volume V, which passes through the filtering medium during a time of passage which is 4 x t_1, is measured. Clogging index is calculated, using the following equation: $I_c = 100\ V/480$.

The solutions are classified, depending upon the clogging index value I_c obtained, as follows:

I_c of 90 to 100 – excellent
I_c of 80 to 90 – very good
I_c of 60 to 80 – satisfactory
I_c less than 60 – mediocre.

Example: 90 grams of sulfate pulp, viscose grade, having a degree of polymerization of 450 and containing 5.5% of moisture, were dried to a moisture content of less than 1% and then added to 1,000 cm^3 of DMSO containing 600 ppm of water and 106 grams of 96% strength paraformaldehyde, corresponding to a paraformaldehyde/cellulose weight ratio of 1.20. The temperature of the mixture, initially at ambient temperature, was raised to 130°C in 1 hour with slow stirring, and was maintained at this temperature for 3 hours with continued stirring. The dissolution of the cellulose was complete, as observed under a polarizing microscope.

The temperature of the resulting solution was reduced to 120°C and then a stream of dry nitrogen at 120°C was bubbled through the solution until formaldehyde had been removed to the point that the formaldehyde/cellulose weight ratio was 0.27. DMSO was added to the solution to compensate for DMSO losses during the nitrogen bubbling step, and the solution, which was free of gels, had a viscosity of 1,100 poises at 20°C, a Np value of 80 and a I_c value of 84. The solution was extruded through a spinneret containing 200 orifices each of a diameter of 0.06 mm into a coagulant bath which consisted of a mixture of water and 30% by weight of DMSO. The coagulant bath was maintained at a temperature of +2°C. During the extrusion step the cellulose solution was maintained at a temperature of 80°C to prevent spinneret head blockage.

The extruded filaments traveled a path of 100 cm in length in the coagulant bath, and were then withdrawn by a positively controlled mechanical device on which the filaments executed several turns and were washed countercurrently by water at a temperature of +2°C.

The filaments then underwent two successive drawing operations in air, with the first stage drawing to an extent of 44% and the second stage drawing to an extent of 30%. The peripheral speed of the last pair of drawing rollers was 25 meters per minute. The filaments were then oiled before being collected on a bobbin and dried. These filaments exhibited the following physical characteristics:

Gauge: 3 dtex
Tenacity, condition state: 20.8 g/tex
Elongation, condition state: 13%
Tenacity, wet: 12 g/tex
Elongation, wet: 14%.

Comparative Example: The above example was repeated, except the formaldehyde/cellulose ratio reduction step, which was achieved by bubbling nitrogen through the cellulose solution, was eliminated. The cellulose solution having a formaldehyde/cellulose weight ratio of 1.20 was extruded into the same coagulant bath, under the same extrusion and coagulation conditions as above. A gel which was devoid of cohesion and impossible to extract from the bath was obtained at the spinneret outlets.

Medicament Carriers in the Form of Film

Pharmaceutical composition unit dosage forms adapted for enteral or topical administration are described by *P. Fuchs and J. Hilmann; U.S. Patent 4,136,145; Jan. 23, 1979; assigned to Schering AG, Germany* comprising a safe and effective amount of a pharmaceutically active medicament compound dissolved or uniformly dispersed in a flexible, water-soluble film carrier therefor. Admixtures of medicament and carrier are drawn into a film, which can be cut to desired unit dosage content.

Example: *Preparation for 1,000 Units* – 0.25 gram of d-norgestrel, 0.05 gram of ethinyl estradiol and 0.84 gram of polyoxyethylene-polyoxypropylene copolymer are dissolved in 95.00 grams of ethyl alcohol wih stirring. Into the resultant solution there is homogeneously introduced a powdered mixture of 16.93 grams of hydroxypropylcellulose and 16.93 grams of cellulose. The resultant suspension is drawn on a suitable film drawing apparatus to a sheet having a thickness of 500 μm, and is then dried.

One unit corresponds to an area of about 3 cm^2. Appearance of the film: white, paper-like. The dry film has a thickness of about 170 μm.

Highly Absorbent Filaments and Webs

According to *W.C. Sigl and F.O. Lassen; U.S. Patent 4,049,764; Sept. 20, 1977; assigned to Kimberly-Clark Corporation,* in the formation of highly absorbent and rapidly wicking filaments and web structures by extrusion and solvent drying of extrudates composed of highly swollen hydrophilic chemically modified cellulose fibers, improved results are obtained in the form of a softer and better wicking product by treatment of the extrudate prior to extrusion.

The extrudate is a mass containing the swollen chemically modified cellulose fibers and a water content of up to 98% at the time of extrusion. By adding to the extrudate, prior to extrusion, an amount of from about 2 to 50% of the solvent to be used in drying in the form of a mixture containing solvent but which may contain water or other nonsolvent and the solvent to be used in drying, a much better performing extrudate is produced which results in better filament formation, improved bonding between filaments, lower water content in the extrudate, faster drying of the extruded product, and a softer, better wicking product.

Example 1: An extrudate of swollen phosphorylated cellulose was prepared in the following manner. 2,383 g (bone-dry) of Northern spruce pulpboards were soaked for 5 minutes at 50°C in a solution of 47% urea, 23.5% orthophosphoric acid and 29.5% water by weight. Following this soaking the pulp was nipped using a commercial manually powered washing machine wringer set at its highest nip pressure. The pressed pulp was then resoaked in the above solution for one-half minute and then renipped. Weighing of the pressed pulp showed it to have retained about 180% of its own weight of the solution. The pulp was then cured for 1 hour at 160°-170°C in a forced-draft oven.

Following a water wash, the fibers were treated for one-half hour with 3.7% HCl (w/w) at 55°-65°C. Following another water wash, the fibers were soaked in 5.0% Na_2CO_3 (w/w) for 1 hour. The fibers were removed from the base solution by centrifugation using a Sharples Model P-660 continuous centrifuge operating at a bowl speed of 5,700 rpm, a conveyor differential of 16 rpm, a dam setting of No. 4, and an effluent flow rate of 10 l/min. The collected fibers were washed by slurrying them in water and re-collected by centrifugation similar to above at an effluent flow rate of 8.0 l/min.

Following reslurrying the collected fibers in water, the pH was adjusted to 7.2 using dilute HCl solution. Refining of the fibers was accomplished by the mechanical action of centrifuging the pulp from water slurries of about 2% (w/w) consistency 5 additional times. The centrifuge was operated at settings similar to above with the following respective effluent flow rates: 10 l/min, 7.2 l/min, 10 l/min, 8 l/min and 4.5 l/min. The final centrifugation yielded a stiff extrudate of phosphorylated pulp fibers containing 93.5% (w/w) H_2O.

Thirty milliliters of the above extrudate was extruded through an 18 gauge (i.d. ≈0.032 inch) syringe needle into a web-like structure 6 inches by 7 inches, solvent-dried in reagent-grade acetone, and the remaining solvent evaporated off. The average dried-down weight of the web was 0.41 g/in^2. The resulting web was stiff and harsh-feeling with very weak contact bonding.

Example 2: A less viscous extrudate was formed by mechanical mixing of additional water into the extrudate produced in Example 1 using a Hobart M-50 dough mixer. Water was added in the ratio of 150 ml of water for each 375 g of Example 1 extrudate. The resulting extrudate contained 95.7% water (w/w).

Thirty milliliters of this extrudate was extruded and solvent-dried in a manner similar to that of Example 1 to yield a web of an average weight of 0.28 g/in^2.

The resulting web was more flexible and less harsh than that of Example 1, had similar retention values as determined for Example 1, but had significantly faster wicking times than those of the web of Example 1. Unlike the web of Example 1, the resulting web showed significantly stronger bonding of both the contact and fused types.

OTHER MATERIALS

Anionic Polymerization of Lactams in an Extruder

In the process described by *M. Biensan and P. Potin; U.S. Patent 4,067,861; Jan. 10, 1978; assigned to Ato Chimie, France,* anionic polymerization of lactams is carried out in an extruder into which at least one lactam, at least one catalyst and at least one activating agent are introduced and wherein the mixture of the substances is polymerized prior to being extruded through an extrusion die. The output rate of the polymerized substance is controlled by transfer means, such as a metering pump, arranged between the outlet of the body of the extruder and the extrusion die.

As shown in Figure 4.8, the device comprises an extruder or mixer **1**, the body **2** of which is represented in the drawing; the extruder comprises a screw **3**, and a metering pump **6** adapted to feed an extrusion die **7** is mounted onto the outlet **4** of extruder body **2** by means of a connecting piece **5**.

The extruder is fed in a conventional manner with a mixture containing a lactam, a catalyst and an activator. This mixture is heated to its polymerization temperature, and the resulting polymerized substance reaching the outlet of the extruder body is taken up by pump **6** and extruded through the die in the form of a rod or profiled section **8**.

Figure 4.8: Extruder with Controlled Output Rate

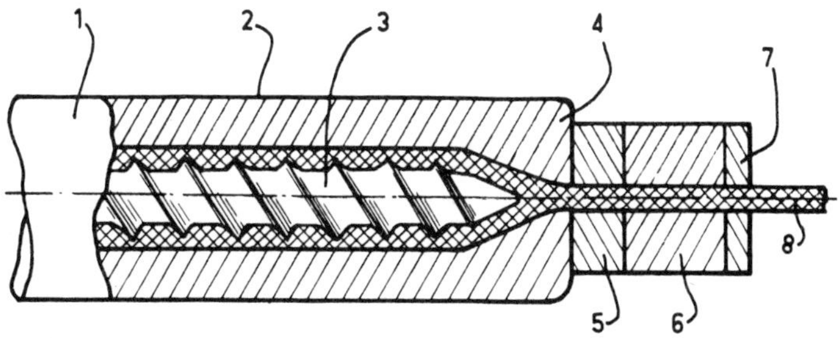

Source: U.S. Patent 4,067,861

Example: When carrying out a test using the device illustrated in Figure 4.8, the mixture to be polymerized comprised lactam-12, 0.65 mol % sodium hydride and 0.65 mol % activator. The pump was operated so as to produce a normal output of about 35 kg/hr, and the polymerization temperature was about 250°C. A pump having an output of 20 ml per revolution and speed controlled by means of a "motor-variator" group was used.

The test was continued for 24 hours with successive loads of 200 kg; substantially only the first ½ hour of operation produced a substance which was not in conformity with the official standards. During the remainder of the test the adjustment of the pump was not modified; the pressure upstream of the pump was at least 50 kg/cm^2.

Thus, the average inherent viscosity being 1.45 dl/g, with a tolerance on the order of 0.02 during the test using a pump, the viscosity varied from 1.40 to 1.55 dl/g during a test without a pump. Also, the rate of polydodecalactam monomer (as determined by sublimation at 230°C under vacuum of 1 torr) during the test using a pump was 0.28 with a tolerance on the order of 0.08, as compared to a value varying between 0.1 and 0.7 obtained during the test without using a pump.

The flowability of the polydodecalactam during the test using a pump was 2,150 with a variation on the order of 275, as compared to the flowability of 870 to 2,800 obtained in a test carried out without using a pump. The flowability of a substance is defined by the weight of flowing substance per unit of time—particularly the number of mg/min—which flows under the effect of a pressure of 2 kg/cm^2 through a die having a diameter of 3 mm, at a temperature of 240°C.

During the test using a pump, the variations of the output rate were particularly small; the average value of the output was 34.7 kg/hr with a tolerance on the order of 0.8% (relative value), whereas the corresponding relative tolerance value obtained in a test without using a pump was about 2.5%.

It has been noted that the quality of the polymer obtained in the installation according to the process was satisfactory practically from the moment the operation of the installation was started when granules were produced, and that the quality was satisfactory after about ½ hour from the start of the operation, when profiled sections were produced.

It must be noted particularly that the regular output provides for satisfactory reproducibility, in regard to the properties of the products obtained.

The anionic polymerization of lactams in an extruder allows on the one hand for obtaining stability of the polymer flow at the die inlet and consequently an excellent homogeneity of the extruded product, and allows on the other hand, if desired, for increasing the pressure upstream of the extrusion die, or of intensifying the agitation of the reaction mixture by conveniently controlling the velocity of the screw or screws of the mixing extruder without increasing the output rates of the extruded product, whereby the homogeneity of the product and consequently its mechanical properties are further improved.

Polyglycolic Acid Sutures

In a process developed by *A. Glick, and L.D. Chirgwin, Jr.; U.S. Patent 4,008,303; Feb. 15, 1977; assigned to American Cyanamid Company,* polyglycolic acid surgical elements, particularly sutures, are colored green to contrast with tissue, blood, and surroundings by extruding while melted the polyglycolic acid and concurrently in an optically homogeneous dispersion, from about 0.03 to 0.5% by weight of 1,4-bis(p-toluidino)anthraquinone (D&C Green No. 6).

Example 1: Into a suitable reaction vessel there is charged 400 parts of a commercial glycolic acid which is then heated from room temperature to about 200°C over a period of about 4 hours. When the pot temperature has reached 185°C, the pressure of the system is reduced from atmospheric pressure to 15 mm of Hg causing the water of condensation and/or esterification to distill off. The residue is allowed to cool and is pulverized into about 280 parts of a powder which is then added in small increments to a suitable pyrolysis chamber maintained at a temperature of about 250°-285°C at a pressure of less than 15 mm of Hg.

The distillate which weighs about 238 parts is dissolved in a minimum amount of hot ethyl acetate, and after decolorizing and purifying with active carbon, the distillate is recrystallized from the above solution to provide 160 parts of product having a melting point of about 82.5°-84.0°C. The infrared spectrum confirms that the product is substantially identical to a substantially pure glycolide composition.

Into a heavy walled glass tube having a bore about 3⁄10 inch and sealed at one end is charged about 3 parts of substantially pure glycolide composition produced as above, 0.04 part of a 0.1% ether solution of $SnCl_2 \cdot 2H_2O$ (about 0.0013% based on the weight of the substantially pure glycolide composition), 0.0166 part of lauryl alcohol (0.346 mol percent based on the mols of the substantially pure glycolide composition), and a magnetic steel ball 5⁄32 inch in diameter.

The tube is evacuated and purged with argon. The tube is evacuated again to a vacuum of less than 1 mm of Hg and the top is sealed. The reaction tube is placed in a vertical position in a closed glass chamber throughout which dimethyl phthalate is refluxed at 222°C. The boiling point of the dimethyl phthalate is controlled by decreasing the pressure of the system. At periodic intervals after melting, the viscosity of the reaction mixture is measured by raising the steel ball by means of a magnet and measuring the rate of the fall of the ball in sec/in. Ninety minutes after the melt is first achieved, the ball drop time is 550 sec/in or about 7,200 poises, and after 120 min, the ball drop time is 580 sec/in or about 7,600 poises.

The thus-formed polymer, while maintaining protection from ambient moisture, is cooled, ground to a coarse mesh and is then ready for melt spinning.

Example 2: 100 parts of polymer prepared as above is ground with 1 part of 1,4-bis(p-toluidino)anthraquinone, the polymer melted, stirred until homogeneous and again cooled and ground. This dried polymer containing 1% of 1,4-bis(p-toluidino)anthraquinone is mixed with 10 times its quantity of uncolored polyglycolic acid, placed in a melt spin extruder, filtered, and forced through spinning apertures under high pressure, in accordance with conventional melt spinning techniques.

When finished and braided into a surgical suture, an excellent green colored suture is obtained.

Controlling Uniformity in Tire Tread Stock

J.L. Bayonnet; U.S. Patent 4,087,499; May 2, 1978; assigned to The Goodyear Tire & Rubber Company improves uniformity in weight per unit length of continuously extruded tire tread stock measuring a dimension thereof such as thickness in the tread stock between the extruder die face and the conveyor and controlling the speed of the conveyor in response to deviations from an arbitrary norm of the dimension. A laser beam measuring device measures the thickness variations in the tread stock before it reaches the conveyor and while the tread stock remains dimensionally unstable following its extrusion.

In Figure 4.9, an extruder **10** provides forming means to shape the tread profile desired in the extruded tire tread stock issuing from the extruder. Upon issuing from the die, the tire tread stock moves first downwardly and is then turned about a stabilizing roll **12** when it travels horizontally in an unsupported span to conveyor **16.** From the conveyor the tread stock moves continuously to further processing, in accordance with usual known practice.

Figure 4.9: Extruder with Laser Beam Measuring Device

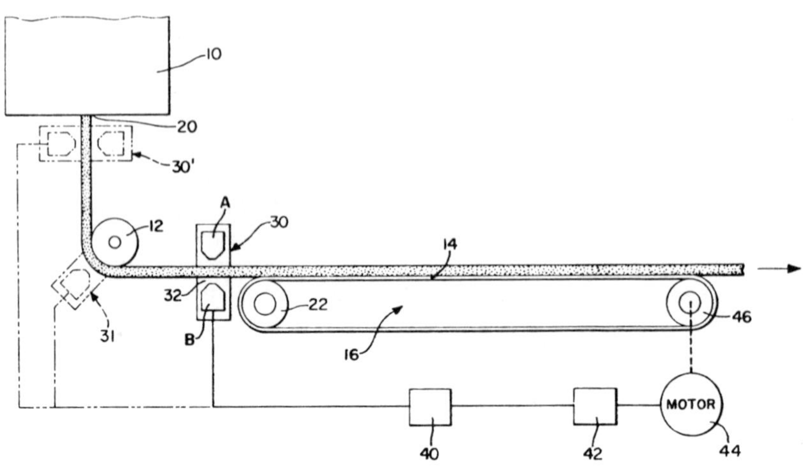

Source: U.S. Patent 4,087,499

In the process herein described, measurement of the tread stock is made continuously, upstream from the conveyor, and deviations from an empirically determined dimension are used to adjust the speed of the conveyor to effect correction. This procedure is referred to herein as a "feed forward" system.

Between the die face **20** of the extruder **10** and the first roll **22** of the conveyor, the tread stock is unsupported and generally in only light contact with the roll **12** which is free-wheeling and rotated only by the movement of the tread stock. In this unsupported reach of the tread stock, as well known, the dimensions of the tread stock, namely the weight, the thickness, and the density, are continuously changing at any particular location therealong as the tread stock travels through the reach. As the stock exits from the die, the cross-sectional area of the tread stock tends to enlarge or swell, a phenomenon commonly known as die swell. Additionally, the action of the screw of the extruder and the flow characteristics of the compound traveling to and through the die yield additional irregularities in the amount of die swell.

It is found convenient to measure a representative thickness of the stock by measuring means comprising a thickness measuring device **30** by which a continuous scan of thickness of the stock, that is, a distance between oppositely facing surfaces of the stock is made. A measurement useful for control of the process leading to improved uniformity of weight per unit length in the tread stock produced is thereby provided.

The thickness measuring device can be any suitable commercial measuring means capable of continuous measurement of thickness and, preferably, having a resolution of 0.025 mm or better. In the preferred practice, the device employed is a model EPG laser gauge obtained from the Autec Corporation.

The laser dimension gauge is a differential measuring system having two heads **A,B** which are mounted in opposing relation, with the tread stock therebetween. A suitable frame **32** supports the gauge rigidly, upstream of the conveyor roll **22** where the unsupported reach of tread stock extending from the die face **20** to the conveyor passes between the two heads.

The two heads of the Autec gauge each direct a laser beam against a surface of the tread stock. Each beam is then reflected along two paths and by a system of mirrors is directed to an image converter as two light points the distance between which represents the actual location of the surface. The face of the converter is periodically swept to generate a pair of pulses produced by the light points on the converter face and which control a gate. The gate allows high frequency clock pulses to flow from a clock oscillator into a converter only during the time period of the sweep between the light points. The number of clock pulses in the counter is thus representative of the actual location of the surface. The respective actual location of the upper and lower surfaces of the tread so measured are compared with the fixed known distance between the two heads to determine the actual thickness of the tread stock.

The device **30** provides, in addition to a direct visual readout of thickness, a signal representative of the continuous variation and thickness being measured. This signal is conducted to a conventional filtering amplifying circuit **40** from which the amplified signal is communicated to a speed control means **42** controlling a direct current motor **44** which drives the conveyor drive roll **46** and the conveyor belt. The speed of the conveyor belt is thus made directly responsive to the thickness of the tread as measured by the device **30** in the unsupported reach of the tread stock upstream from the conveyor.

The correction of the weight per unit length and thickness of the tread stock is effected by adjusting the speed of the conveyor in response to a feed forward signal from the device **30**. This is contrary to and a useful improvement over the prior art, in which the corrections were made by feedback from measurements of a weigh-scale measuring the weight of treads cut to length from the tread stock.

Filaments from an Unconfined Source of Molten Material

R.E. Maringer; U.S. Patent 4,124,664; Nov. 7, 1978; assigned to Battelle Development Corporation describes methods and apparatus for forming filamentary material directly from a pendant drop of molten material, and provides for a significant improvement in the quench rate. A typical method of making filamentary material comprises the steps of:

(a) heating a solid material, typically, a metal or metal alloy, so as to form a pendant molten drop of the material, which is at a temperature within 25% of its equilibrium melting point in °K, the molten material having a surface tension of 10 to 2,500 dynes/cm and a viscosity of 0.001 to 1.0 poise at the temperature,

(b) rotating a hollow cylindrical member about its longitudinal axis at a peripheral speed in excess of 3 fps, the member having a circumferential, heat-extracting lip projecting from its inner periphery, and the lip having a narrow peripheral edge of arcuate cross section; and

(c) bringing the surface of the pendant molten material drop and the peripheral edge of the lip into contact so as to withdraw a solidifying filament from the drop while maintaining the form and stability of the drop.

One preferred embodiment is illustrated in Figure 4.10. Cylinder **30**, having its lower end enclosed, is shown rotating about its longitudinal axis, with the cylinder being driven by shaft **5** connected as shown. Shaft **5** may be driven by any conventional driving means (not shown).

Figure 4.10: Partial Cross Section of Apparatus for Making Filamentary Material

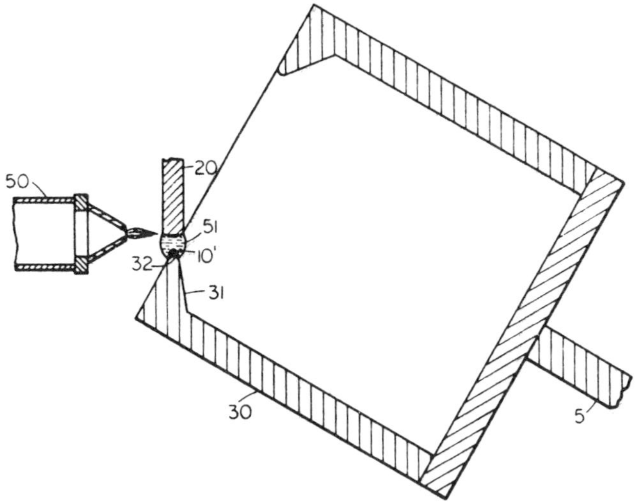

Source: U.S. Patent 4,124,664

A lip **31** projects inwardly from the inner periphery of the cylinder at its open, upper end. In this embodiment the lip is roughly V-shaped in cross section, with the apical region of the V-shape comprising a narrow peripheral edge **32** of an arcuate cross section. For the production of continuous filamentary material according to this embodiment it is preferred that the lip extend only a relatively short distance (on the order of a few tenths of an inch) from the inner periphery of the cylinder.

The member **20**, which constitutes the material supply for this embodiment of the process, may be of any convenient shape, such as, for example, rod **20**. As shown, the end portion of member **20** may be melted by any conventional means, such as for example, oxyacetylene torch **50**. The local heating of the end portion of the member **20** creates a molten zone **51**.

The process is initiated by bringing the rotating edge **32** and the pendant drop **51** into contact. Surprisingly the surface tension of the material in the drop **51** is sufficient to maintain stability even with the edge **32** entering and inducing a shear flow within the drop.

When the edge **32** passes through the pendant drop **51**, a portion **10'** of the molten material solidifies on the edge **32**. Further rotation of the cylinder **30** draws this (solidified) filamentary portion **10'** out of the pendant drop **51**. The filament **10'** tends to adhere tightly to the edge **32** for a short distance (several centimeters or less), and then it spontaneously ruptures this bond (apparently due to thermal shrinkage) and finally becomes a collectable solid filamentary material.

Example: A drop of molten tin was formed on the end of a solid bar of tin by locally heating the end with a propane torch. The drop was manually brought into contact with the peripheral edge of a V-shaped, heat-extracting lip of a rotating hollow aluminum cylinder similar to that shown in Figure 4.10.

The axis of rotation of the cylinder was at an angle of approximately 25° from the horizontal, and the diameter of the circle circumscribed by the peripheral edge of the lip was approximately 4.7 inches. The angle described by the two sides of the V-shaped lip was about 45° and the radius of curvature of the peripheral edge of the lip was about 0.002 inch. The cylinder was rotated so as to yield a linear velocity at the peripheral edge of the lip of approximately 5 feet per second. Tin filament was formed which spontaneously released from the forming edge.

PROCESSES AND PRODUCTS USING A VARIETY OF POLYMERS

Self-Gripping Devices

Self-gripping devices are disclosed by *G.C. Brumlik; U.S. Patent 4,001,366; Jan. 4, 1977; assigned to Ingrip Fasteners, Inc.* and include a base having one or more integral trains of generally upright gripping elements which are adapted to penetrate and become lodged in a receiving material and especially receiving material containing fibrils to form a self-gripping connection. The gripping elements in each train have a common origin in a continuous rib integral with the base.

The self-gripping device can be formed from materials including metals, metal alloys, plastics, ceramics, glass, cements, fibrous materials and the like. Suitable metals include aluminum and aluminum alloys, magnesium and its alloys, copper, copper alloys, such as beryllium copper, iron, carbon and stainless steels and the like. Suitable plastics include thermosetting phenolic compositions, melamines, epoxy resins and the like and thermoplastics such as polyethylene, polypropylene, polystyrenes, polycarbonates, polysulfones, nylons, fluorinated polymers and the like. Fibrous materials include paper, glass fiber, ceramic fiber, metal fiber and organic fiber, reinforced plastics and webs and sheeting made therefrom.

Suitable apparatus for carrying out the process by extrusion is shown in Figure 4.11 wherein an extruder **50** is provided with an elongated slot-type die **52** having an opening **54** which corresponds to the cross-sectional profile of the integral structure to be extruded. Extrudable material is melted and conveyed in the extruder **50** and forced out of the die **52** in a hot plastic state. The extruded shape as shown in Figure 4.11 includes a base beneath substantially parallel integral ribs **22**. It should be noted that the ribs **22** can also be wavy or zig-zag or vary in thickness using two or more part dies and reciprocating the one or more parts thereof transverse to the direction of extrusion. Such dies are known in the art.

Figure 4.11: Apparatus for Extrusion of Self-Gripping Device

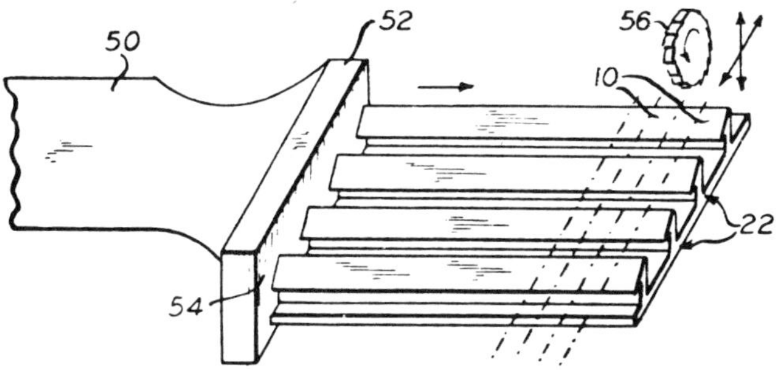

Source: U.S. Patent 4,001,366

Fiber-Expanded Low Density Reinforced Materials

In a process developed by *E.A. Cole, F.N.Cogswell and J. Huxtable; U.S. Patent 4,178,411; Dec. 11, 1979; assigned to Imperial Chemical Industries Limited, England,* a very low density material is produced by extruding a composition comprising a settable fluid containing long resilient fibers through a narrow die under conditions which cause the extrudate to expand to form an open fibrous structure as it leaves the die, and then causing the fluid to set. This is applicable to both thermoplastic and thermosetting materials. Various coating layers may be applied to provide rigid lightweight panels and shaped articles for various uses including furniture and building applications.

The settable fluid may be a molten thermoplastic material which is extruded in its molten state and then set by cooling until it freezes. Examples of such material include the thermoplastic polymeric materials, such as for example polyolefins, polyamides, polyvinyl or vinylidene compositions, acrylic polymers such as polymethyl methacrylate, and thermoplastic polyesters. Thermosetting materials, such as epoxy or polyester resins may also be used as alternative fluids.

The apparatus used is shown in Figure 4.12. In Figure 4.12 a ram extruder denoted generally by **10** consists of a barrel **11** and a piston **12** slidable therein. The extruder is positioned for extrusion in a vertically downward direction and is provided with a detachable die **13**. In the apparatus shown the die **13** has zero die length.

In operation the settable material to be extruded, in this case glass-filled polypropylene, is charged to the barrel by removing the piston. Heating is applied by means of heaters **14** in a jacket **15** surrounding the barrel **11** until the polypropylene is molten. The piston **12** is then advanced to force the mixture of polypropylene and randomly dispersed glass fibers **16** (shown much enlarged) through the die **13**. As the mixture approaches the die the glass fibers become distorted in shape as indicated by **17**. On passing through the die the restraining forces on the fibers are removed and the fibers immediately tend to revert to their original shape and orientation under their own resilience. This relaxation of stress results in the extrudate **18** expanding to a diameter considerably greater than the die bore to provide a low density extrudate **1** consisting of a mass of randomly oriented fibers designated as **16** in an open expanded polypropylene matrix.

Figure 4.12: Ram Extruder

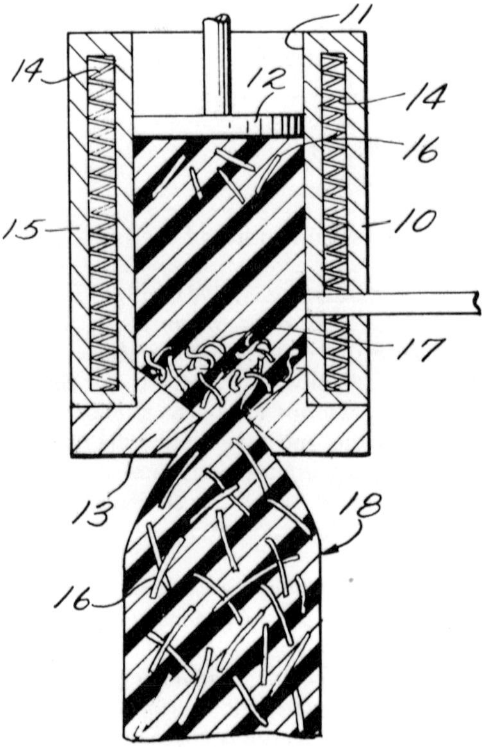

Source: U.S. Patent 4,178,411

Prevention of Deposits on Nozzle Orifices

G. Wilhelm, A. Jäschke, H.-J. Simon, J. Schmidt and S.S. Sandhu; U.S. Patent 4,124,666; Nov. 7, 1978; assigned to Bayer AG, Germany describe a method of keeping nozzle bodies or breaker plates clean during extrusion of polymer which can be shaped from the melt which comprises preventing the formation of deposits near the nozzle orifices by gassing these orifices at the point of extrusion of the melt, the gas consisting entirely or partly of vapors of substances which have a depolymerizing action on the deposits formed from the melt at the edge of the nozzle orifice.

The polymer melts to which the process is applicable are mainly polymer condensation and addition products which are capable of forming strands, fibers or sheets, such as polyamides, polycarbonates, polyesters and polyurethanes and their copolymers. The depolymerizing action on the abovementioned deposits formed on the edges of the nozzle orifices of the melt when such polymer melts are extruded is obtained from compounds which are reactive in the gaseous or liquid form.

In this respect, the process differs fundamentally from those in which a protective gas is used to produce an "inert" atmosphere at the die orifice. Among the most important substances which have a depolymerizing action in the vapor or liquid form is water. Other substances having a similar effect include for example alcohols and amines.

Figure 4.13 illustrates the application of the process to the extrusion of melts from nozzle orifices to form strands. Concentric inserts **2** are rigidly fixed to the die body **1**. The melt extruded as strands from apertures **3** is washed on all sides with vapor which is emitted from the vapor chamber **4** above the external cone of the insert **2** through annular gaps **5**. The vapor chamber **4**, which has a supply inlet **6**, is closed off from the atmosphere by a plate **7**.

Figure 4.13: Nozzle Plate Allowing Gassing with Steam

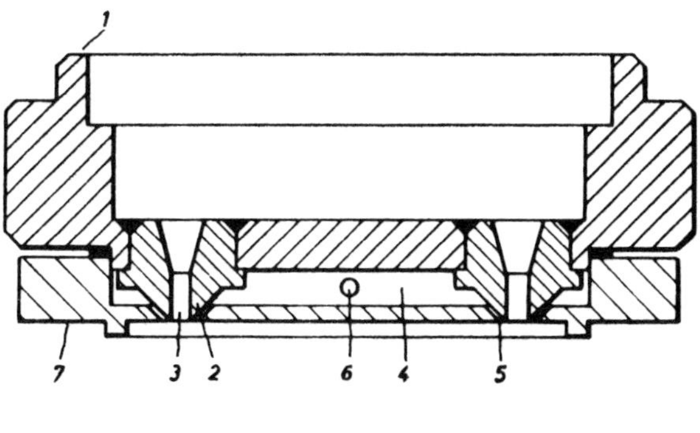

Source: U.S. Patent 4,124,666

Example: A polyamide-6 melt was extruded downwards through vertical, cylindrical nozzle orifices 4 mm in diameter at a temperature of 250°C and at an extrusion rate of 15 kg/hr and drawn off at such a rate that the strand of melt had a diameter of about 2 mm when cooled. At the arbitarily chosen beginning of the period of observation, melt droplets were already forming on 18% of the nozzle orifices under observation. The increase in the number of droplet-forming nozzle orifices over a period of observation of 16 days is found to be about 8%.

Nozzle plates constructed as illustrated in Figure 4.13 to provide the possibility of gassing with steam were then used, and the polymer melt mentioned above was extruded under the conditions already described. During the initial period of the experiment, however, extrusion was again carried out without the supply of steam.

The lips of the nozzle orifices were still free from drops of melt when the nozzles had been in operation for 6 days. It was only when, on the 18th day of operation of the nozzle, 10% of the nozzle orifices showed droplets, that the ends of the nozzles were gassed with superheated steam at 250°C and a pressure of 1 bar emitted from the annular gap 5 shown in Figure 4.13.

Spinning and Extrusion of Hollow Fibers

Hollow fibers are an outgrowth of textile spinning of synthetic organic polymers in order to produce high-bulk, low-density fabrics. Hollow fibers are manufactured from textile grade resins and may be spun by wet spinning (spinning from a polymer solution into a liquid coagulant); dry spinning (spinning from a solution of a polymer in a volatile solvent into an evaporative column); or by melt spinning. The tubular cross section is formed by extruding the molten polymer or polymer solution through an annular dye or spinneret.

ACRYLICS, VINYLS AND OLEFINS

Acrylic Ion-Exchange Hollow Fibers

An ion-exchange hollow fiber is prepared by *A. Rembaum, S.-P.S. Yen and E. Klein; U.S. Patent 4,045,352; August 30, 1977; assigned to California Institute of Technology* by introducing into the wall of the fiber polymerizable liquid monomers, and polymerizing the monomers therein to form solid, insoluble, crosslinked, ion-exchange resin particles which embed in the wall of the fiber.

Excess particles blocking the central passage or bore of the fiber are removed by forcing liquid through the fiber. The fiber has a high ion-exchange capacity, a practical wall permeability and good mechanical strength even with very thin wall dimensions. Experimental investigation of bundles of ion-exchange hollow fibers attached to a header assembly have shown the fiber to be very efficient in removing counterions from solution.

Textile grade acrylic fibers, suitably polymers of acrylonitrile, have been found to be compatible with the polyquaternary polymer reactants. The polyacrylonitrile may be homopolymers or copolymers containing up to 20% by weight of comonomers such as ethyl acrylate or vinyl acetate.

The acrylonitrile polymers are wet spun from a 15 to 25% solids solution in a highly polar solvent such as dimethyl acetamide, dimethyl formamide or 40% $Ca(SCN)_2$ through an annular spinneret into a liquid coagulant such as water. The pore size and distribution can be controlled by selection of solvent.

The polyquaternary, water-insoluble, crosslinked materials are prepared by reacting a vinyl pyridine with a dihalo organic compound of the formula:

$$X{-}R_1{-}X$$

where X is halo, preferably bromo, chloro or iodo and R_1 is a divalent organic radical such as alkylene, alkenylene, alkynylene, arylene, alkarylene or aralkylene. R_1 may also be alkylthioalkylene or alkyloxyalkylene of the formula:

$$(CH_2)_x(Z)_y(CH_2)_z$$

where Z is oxygen or sulfur and x, y and z are integers from 1 to 100. R_1 may also be of prepolymer or polymeric length of up to 6,000 molecular weight such as bromo-terminated polybutadiene, but, preferably has a carbon content of from 1 to 20 carbons to provide an increased charge center density per unit volume and weight of the polymeric product. R_1 may be substituted with other groups that do not interfere with the polymerization reaction or properties of the polymer product such as hydroxyl, alkyl, aryl, nitro, cyano or similar groups.

Example: A 20% solids solution of fiber grade polyacrylonitrile in dimethyl acetamide was wet spun through an annular spinneret and coagulated in a bath of water to form lengths of hollow fibers having an i.d. of 0.0250 cm and a wall thickness of 0.0040 cm.

The diffusivities of certain solutes through the wall of the fiber are: creatinine, 100 min/cm; glucose, 160 min/cm.

Polyvinyl Alcohol Ion-Exchange Fibers

According to *M. Uzumaki, M. Takashio and T. Utsumi; U.S. Patent 4,125,486; November 14, 1978; assigned to Nitivy Co., Ltd., Japan,* a partially polyenized polyvinyl alcohol fiber is obtained by heating and dehydrating polyvinyl alcohol type fibers at temperatures of about 80° to 350°C so that the dehydration weight loss falls within the range of about 5 to 40%. An ion-exchange radical is introduced into the partially polyenized polyvinyl alcohol fiber to obtain fibers with an ion-exchange capacity. The ion-exchange fiber thus obtained exhibits an exchange capacity comparable to ion-exchange resins and has practical properties such as good chemical resistance and flexibility.

Example: A commercially available polyvinyl alcohol (PVA) with the average degree of polymerization of 1,200 and the degree of saponification of 99.9% was heated and dissolved in water to obtain an aqueous 30% solution, and 5% by weight of an ammonium salt of polyphosphoric acid based on the PVA was added. The PVA solution obtained was extruded in hot air at 150°C through a spinneret and filaments were formed by dry spinning.

When the filaments obtained were stretched seven times in hot air at 200°C, a white fiber having a strength of 5 g/den and an elongation of 10% was obtained. The white fiber was heated under tension in air at 320°C for three hours and thus, a partially polyenized PVA fiber with a weight loss of 30% was obtained having a strength of 2,200 kg/cm^2 and elongation of 10%. The partially polyenized PVA fiber is not attacked by 60% sulfuric acid at 60°C and has good chemical and heat resistances.

Next, when the above fiber was treated in a 90% sulfuric acid bath at 100°C for 30 minutes, a fiber with a cation-exchange capacity having a sulfur content of 4% was obtained. This treatment had little influence on the elongation and tenacity of fiber. Ion-exchange capacity was 1.2 meq/g.

Separation Membrane Fibers from Polyamide, Polyvinyl Alcohol and Sulfoxide Mixtures

Hollow membrane fibers are formed by *E.F. Steigelmann, R.D. Hughes and J. Gabor; U.S. Patent 4,039,499; August 2, 1977; assigned to Standard Oil Company* from a mixture of N-alkoxyalkyl polyamide, polyvinyl alcohol, di(lower alkyl) sulfoxide and water. The fibers are useful for separating chemicals, e.g., aliphatically-unsaturated hydrocarbons, from mixtures containing them.

Example: In this example extrusion of the polymer mixture was conducted under a nitrogen pressure between 200 and 1,000 psi on a feed tank and with the extruder having a heated head. Hollow fibers of the polymer blend were formed by extrusion through a die having an opening in its center. During extrusion, air or nitrogen was blown through the center of the fiber by passage through a hypodermic needle extending into the opening in the middle of the die.

After extrusion the fibers were stretched under their own weight by allowing the fibers to drop below the extruder head. The stretched fibers were crosslinked by immersion in a 3% p-toluene sulfonic acid in 10% aqueous sodium sulfate bath for 60 minutes at 55°C. The fibers were then washed repeatedly with water to remove the salt from them and allowed to dry.

Fibers were made while using two different extruder heads. One head (Head 1) had a hole 0.067 inch in diameter with a 0.031 inch o.d. needle in the center of the hole, while the other head (Head 2) had a hole 0.040 inch in diameter having a 0.020 inch o.d. needle in the center of the hole. During extrusion the polymer compositions are forced through the annular space between the opening in the extruder head and the needle, and air is forced through the needle to keep the extruded hollow fiber from collapsing.

A number of different fibers were extruded using various head temperatures, driving pressures, air flow rates and fiber falling distances for drawing. The compositions employed were formed by mixing the named ingredients at ambient temperature and then raising the temperature of the mixture to approximately 260°F to effect melting. The mixture is heated at this temperature for at least one hour while undergoing vigorous stirring before extrusion. The compositions contained the following:

Component	. .Composition. . 1	2
Polyvinyl alcohol (0 to 0.5% acetate), grams	80	40
Nylon, Belding BCI-819, an N-methoxymethyl 6:6 nylon, grams	120	160
Dimethyl sulfoxide, ml	200	200
H_2O, ml	20	10

The polyvinyl alcohol employed was Borden's high molecular weight (0 to 0.5% acetate) grade, and was determined to have a number average molecular weight of about 12,360 by gel permeation chromatography. A 4% aqueous solution of the polymer was determined to have a viscosity of 60 centipoises at 20°C by the capillary tube technique, which may indicate a MW in the 200,000 range.

Hollow fibers up to fifty feet long were made and were free of holes in the fiber wall and free of plugs in the fiber bore.

Fibers for Reverse Osmosis, Ultrafiltration and Gas Separation

In a process developed by *E. Klein, J.K. Smith and F.C. Morton; U.S. Patent 4,051,300; September 27, 1977; assigned to Gulf South Research Institute,* hollow synthetic fibers useful in reverse osmosis, ultrafiltration, gas separation and the like, having high hydraulic permeability to the solvent and capable of operation at pressures from 600 psi and up, are prepared by forming a solution of a fiber-forming polymer in a suitable solvent, adding to the solution a second polymer soluble in the solvent but with limited compatibility with the first polymer when their total concentration increases on coagulation.

The resulting solution is extruded through an orifice equipped for coaxial extrusion so that coagulating fluid within tube flow results, and precipitated with a liquid which is miscible with the solvent for the fiber-forming material, which is a nonsolvent for the first polymer and a solvent for the second.

The extruded solution is contacted with the precipitating liquid either coaxially through the extrusion device or by passing the extrudate through the precipitating liquid. The resulting hollow fiber is finally washed free of residual solvents and nonsolvents, and air dried. The coagulating fluid and the precipitating liquid may be the same material. The coagulating fluid may be air or another gas.

The specific fibers described here are prepared from aromatic polysulfone polymers, and aromatic polyamides, but the process is not restricted to these materials. Other polymers having an initial modulus in excess of 2,000 psi, and a glass transition temperature at least 20°C higher than the projected use temperature of at least about 15°C can also be employed.

Example 1: A polysulfone hollow fiber was produced by extruding a solution containing 25% polysulfone and 15% polyvinylpyrrolidone in dimethyl acetamide (DMA) coaxially with a 50:50 DMA:H_2O mix. The fiber had a 455 micron outer diameter and a 160 micron wall. The hydraulic permeability of the fiber was 17 x 10^{-5} cm/sec-atm.

Example 2: The solution prepared as in Example 1 above was extruded at a rate of 2.8 cc/min through a die having a center tube located in it through which a mixture of water:DMA in the ratio 50:50 was forced coaxially with the extruding polymer solution. Within a traverse of 10 cm the fluid had formed a hollow fiber with a final o.d. of 340 microns and an i.d. of 160 microns. The fiber was wound up on an advancing reel at a rate of 20 m/min. After washing and air drying, the fiber exhibited a hydraulic permeability of 400 x 10^{-5} g/cm^2/sec-atm.

Example 3: The polymer solution prepared as in Example 2, but containing 24% polysulfone and 5% polyvinylpyrrolidone, was extruded through the coaxial extrusion jet, but only dry air was used as the coaxial fluid. The hollow extrudate was collected in water containing a small percentage of DMA. The hydraulic permeability of the fiber after drying was 2 x 10^{-5}, and the tensile strength was 8 kg/cm^2. The outer diameter was 510 microns and the inner diameter was 255 microns.

Microporous Polyethylene Fibers for Dialysis or Reverse Osmosis

Normally hydrophobic, polyethylene hollow fibers having contiguous microporosity are prepared by *R.D. Mahoney and S.E. Schneider; U.S. Patents 4,020,230; April 26, 1977 and 4,115,492; September 19, 1978; both assigned to The Dow Chemical Company* by extruding a heated solution of a polyethylene and an ester from a hollow fiber spinneret, simultaneously cooling and drawing the forming fibers to a ratio of up to 40 to 1, drawing the gelled fibers to an overall ratio of from about 1.5/1 to 800/1 and then removing the ester. The maximum pore radius ranges up to about 50 A, and the fibers have gas permeabilities approaching 10^{-2} cc (STP) per cm^2 per second per cm Hg transmembrane pressure differential. A significant increase in permeability over polyethylene hollow fibers made according to prior art is achieved.

Typical Procedure: *Stage 1* – A solution of the polyethylene and the ester is formed in a stirred and heated vessel, solidified by cooling and cut into pieces. The particulated mixture is placed in a heated tube, melted and ram-extruded through a hollow fiber spinneret, air being pumped through the center of the spinneret to keep the forming fiber continuously hollow. The nascent fiber is passed through an 8-inch air cooling zone and then, at room temperature, through a quench bath consisting of two vessels of isopropanol in series, the path length in each vessel being 40 cm. Upon exiting from the quench bath, the fiber is taken up at a preselected rate (depending on the purge rate at the spinneret and on the initial draw ratio desired) on a rotating spool.

Stage 2 – The hollow fiber is unspooled, and drawn at a preselected temperature to a preselected overall draw ratio. The drawing is done manually, as by immersing the fiber portion to be drawn in hot water, waiting about 10 seconds, drawing over another 10 second interval to the desired length, waiting about 10 seconds and then removing the drawn fiber from the water. The same sequence of steps can be carried out in a continuous, mechanical mode of operation by methods readily apparent to those skilled in the art of fiber manufacture.

Stage 3 – The hollow fiber is leached in a preselected solvent for a period of from ¼ hour to 24 hours.

Example: Hollow fibers were spun at 200°C from a mixture of butoxyethyl oleate (BEO) (47 wt %) and Dow high density polyethylene resin 85965 (53 wt %; melt index 0.85; density 0.965, tensile strength 4,200 psi, flexural modulus 14,500 psi). The relative rates of takeup and extrusion were such that a draw ratio of 14.3/1 resulted. The nascent fibers were quenched with air only.

The dimensions and mechanical properties (averaged for three replicate specimens each) of the quenched fibers were as follows: i.d. 380 μ; o.d. 480 μ; wall thickness 50 μ; tensile strength 1,171 psi; tensile yield stress 648 psi; ultimate elongation 886%; tensile yield strain 7.5%; and elastic modulus 8,600 psi. Some of these fibers were immersed in hot (83°C) water for 10 seconds, drawn over a period of about 2 seconds to a wall thickness of 23 microns (post-spin draw ratio 3.4/1), allowed to soak in the hot water for about 10 seconds more and then placed on paper to cool and dry.

Post-spin drawing, and then leaching, is essential to the attainment of permeabilities above about 1.7 x 10^{-5} cc per cm^2 per second per cm Hg transmembrane pressure.

Treating Polyethylene with SO_3 to Produce Cation Exchange Fiber

T. Sano, A. Kobayashi and H. Inoue; U.S. Patent 4,113,666; September 12, 1978; assigned to Sumitomo Chemical Company, Limited, Japan provide a process for producing strongly acidic cation exchange fiber, by allowing fibrous polyethylene to react with gaseous sulfur trioxide.

Example 1: Fibrous polyethylene with three-dimensional network structure obtained by flush spinning from high density polyethylene (Sumikathene Hard, Sumitomo Chemical Co.; melt index 6.2) was cut and beaten into pulp form. In a 50-liter rotary tank, there was charged 1 kg of the resulting pulp-like high density polyethylene. A stream of gaseous sulfur trioxide diluted with nitrogen to about 50% by volume was introduced at a rate of 1 liter per minute into the tank while it was rotated. In about 10 hours, 1 kg of sulfur trioxide was introduced. Thereafter, the unreacted sulfur trioxide remaining in the reaction system was purged by use of nitrogen. The sulfonated polyethylene pulp was discharged from the tank into 30 liters of water, collected by filtration, and washed with water. The first washing showed an acid concentration of about 0.1 N and was easily treated.

For comparison, the same high density polyethylene pulp as used above was immersed in chlorosulfonic acid and allowed to react at 60°C for 3 hours. The sulfonated polyethylene was collected by filtration and poured into ice water. The quantity of chlorosulfonic acid adhered to the sulfonated polyethylene after customary filtration was nearly the same as that of polyethylene. The washing showed an acid concentration of as high as about 3 N, rendering the treatment of waste liquor very difficult. The loss in acid was large enough to make the process uneconomical. The sulfonated polyethylene pulp showed an ion exchange capacity of 1.5 meq/g.

Example 2: A glass column, 50 mm in diameter, was filled with 50 grams of the same high density polyethylene in pulp form as used in Example 1. A nitrogen stream containing about 25% sulfur trioxide was passed at a rate of 5 l/min through the polyethylene bed at room temperature. After about 8 hours, nitrogen was passed through the column to expel the unreacted sulfur trioxide. The sulfonated polyethylene, which had been turned black, was poured into 1 liter of water. The washing showed an acid concentration of 0.08 N. The sulfonated polyethylene pulp, after having been washed and dried, showed an ion exchange capacity of 1.0 meq/g.

Hollow Polyolefin Fibers Polymerized on Organic Fibers

H.W. Dougherty; U.S. Patent 4,012,342; March 15, 1977; assigned to Exxon Research and Engineering Company discloses a process comprising polymerizing olefins in the presence of organic fibers and then removing the fibers to obtain hollow polyolefin resins.

Example: A glass vessel which had previously been dried in an oven was charged with 10.0 grams of paper fibers which had the appearance of cotton and a density of 0.020 g/cc. The paper fibers had been dried in a vacuum oven under an absolute pressure of about 20 mm of mercury and at a temperature of 60°C for 24 hours. Normal heptane (250 ml) containing 5.0 ml of 1 molar solution titanium tetrachloride in n-heptane was added to the dried paper fiber in the reaction vessel. The resulting mixture was stirred; and, after about 5 minutes, 10 ml of a 1 molar solution of triethyl aluminum in normal heptane, diluted with 50 ml of dry normal heptane, was added to the mixture over a period of 10 minutes. It was noted that upon the addition of the alkyl aluminum solution, the mixture turned dark brown, indicating that the titanium tetrachloride was reduced by the aluminum compound. The temperature during the reduction step was maintained at about 50°C by means of an external heat source. After all of the reducing compound was added to the mixture, 1 liter of dry normal heptane was added to further dilute the reaction mixture, and thereafter heat was applied to bring the temperature to 50°C and pure propylene was introduced into the liquid reaction mixture. The propylene was passed into the reaction mixture at the rate of 3,600 cc/min during the first 5 minutes of the polymerization reaction. No exit gas was recovered during this period, indicating total absorption of the propylene; and, furthermore, cooling was required in order to maintain the reaction temperature at about 50°C.

Thereafter, during the next hour, the propylene feed rate was reduced to 1,500 cc/min, and it was necessary to apply heat during the latter part of the aforementioned period in order to maintain the reaction temperature. From the quantity of exit gas recovered, it appeared that the polymerization was substantially complete after the first 15 minutes.

The reaction mixture was cooled and 45 cc of isopropyl alcohol was added to quench the reaction. If desired, a small amount (i.e., about 1 to 10 weight percent) of a conventional chelating agent, such as 2,4-pentanedione, 2,3-butanedione, diphenyl diketone or other diketone, may be added with the alcohol to reduce the ash content of the coated fibers to about 0.005%. Following the addition of the isopropyl alcohol, the reaction mixture was poured into an additional amount of isopropanol (2 liters) and stirred. The solid portion of the reaction mixture was separated from the liquid by means of filtration, and the coated fibers recovered were dried in a vacuum oven at about 20 mm of mercury absolute pressure and at about 60°C. The paper fibers coated with propylene weighed 14.6 grams. The product appeared to be homogeneous in that one could not identify paper fibers with the naked eye.

The fibers prepared in the aforementioned manner can be pressed in a mold at about 30° to 260°C, e.g., 150°C, to produce a film which is resistant to many substances, including water.

Alternatively, a sheet of paper may be used in place of the paper fibers in the above example and an encapsulated article is obtained.

In lieu of the cellulosic fibers used above, hay, straw, flax and hemp may also be utilized, as well as fibers of plastic polymers having recurring amide groups in their molecules.

Polypropylene Hollow Filaments for Reverse Osmosis

The porous polypropylene hollow filaments described by *K. Kamada, S. Minami and K. Yoshida; U.S. Patent 4,055,696; October 25, 1977; assigned to Mitsubishi Rayon Co., Ltd., Japan* are characterized in that the thickness of the surrounding wall portion is less than 60 μ, preferably 40 μ, a number of fine pores communicating with each other are present in the surrounding wall portion, and the distribution curve of radii of the fine pores has at least one maximum in the range of 200 to 1200 A. The method of making the filaments is characterized by melt spinning polypropylene by a nozzle for production of hollow filaments at a spinning temperature of 210° to 270°C and a draft of 180 to 600, then subjecting the resultant filaments to a first heat treatment at a temperature of not higher than 160°C, thereafter stretching them by 30 to 200% at a temperature lower than 110°C and then subjecting them to a second heat treatment at a temperature not lower than the temperature of the first heat treatment and not higher than 175°C.

Example 1: Polypropylene (having $[\eta]$ = 1.40 in tetralin at 135°C) was spun from a nozzle of bridge type having a diameter of 4 mm at a spinning temperature of 230°C, at a takeup velocity of 610 m/min and at a draft of 445. Thus obtained unstretched hollow filaments were wound on a metal reel and subjected to the first heat treatment at 140°C for 30 minutes in a hot air box. Then, they were stretched at various stretching ratios of 50, 100, 200, 300 and 400% of the original length at room temperature and were subjected to the second heat treatment at 145°C for 30 minutes in a previously provided hot air drier while maintaining the filaments at the stretched state.

Surface area of the thus obtained porous hollow filaments was measured by the gas adsorption method. Furthermore, distribution of the radius of the pores of the hollow filaments different in stretching ratio was obtained by Mercury Pressure Porosimeter. The gas permeability of the hollow filaments was measured by pressurizing the hollow filaments from outside with N_2 gas of one atm gauge and measuring the flow amount of the gas permeating from the hollow portion by gas meter.

From these results, it is shown that the distribution curves of the pore radius of the hollow filaments of stretching ratio 50 to 200% have at least one maximum point between 200 and 1200 A and that the hollow filaments of this stretching ratio have much higher gas permeability than those of the higher stretching ratios.

Example 2: The porous hollow filaments obtained in Example 1 were surface treated with a surfactant to make then hydrophilic. The outside of these hollow filaments were allowed to contact with water containing 100 ppm of oil under the pressure of one atm. The flow amount of water permeating from the hollow part of the hollow filaments was measured. The oil content in the water which permeated from the filaments was determined based on the percent transmission of light. The results are shown below. The filaments stretched by 50 to 200% had excellent oil removability and water permeability in treatment of oil-containing wastewater.

Stretching Ratio of Hollow Filaments Used (%)	Flux of Permeating Water (gfd)	Oil Content in Permeating Water (ppm)
50	2.5	<1
100	2.2	<1
200	1.1	<1
300	~ 0	–
400	~ 0	–

In the table "gfd" is an abbreviation of "gallon/(feet)^2-day" which means the amount (in gallons) of liquid which permeates the film area of 1 ft^2 per day and the film area was calculated from the outer surface of the hollow filaments.

Comparative Example: The unstretched hollow filaments obtained in the same manner as in Example 1, except that a takeup velocity of 200 m/min and a draft of 90 were employed, were subjected to the same first heat treatment as in Example 1 and then were stretched by 100% at room temperature and subjected to the second heat treatment at 145°C.

The distribution curve of the pore radius of thus obtained porous polypropylene hollow filaments had a maximum value at 1100 A and the filaments had an outer surface area of 15 m^2/g. The stretched hollow filaments had an outer diameter of 230 μ, an inner diameter of 90 μ and a thickness of the surrounding wall of 70 μ. The N_2 gas permeability of the porous hollow filaments was examined in the same manner as in Example 1. The gas permeability K was less than 10^{-6} and substantially no permeation occurred.

CELLULOSICS

Regenerated Hollow Cellulosic Fiber

Hollow cellulosic fibers are produced by *A.F. Turbak, R.B. Hammer, N.A. Portnoy and A.C. West; U.S. Patent 4,086,418; April 25, 1978; assigned to International Telephone and Telegraph Corporation* by nitrosating and dissolving a cellulosic material in an aprotic solvent selected from the group consisting of a dialkylacylamide and dimethylsulfoxide and regenerating the dissolved cellulose by contacting the cellulose solution with a regenerant. The regenerant is n-octanol where the solvent is a dialkylacylamide and n-hexanol where the solvent is dimethylsulfoxide. The fibers are hollow as regenerated and require no special spinneret designs nor special spinning conditions.

Example: A prehydrolyzed kraft pulp having a DP of 1,050 was alkaline aged to a DP level of 450, neutralized, washed, dried, then fluffed and defibered by means of a rotary cutter. A cellulose solution containing 8 parts of cellulose, 25 parts of N_2O_4 and 67 parts of DMF was prepared by charging 160 grams of the alkali aged prehydrolyzed kraft pulp (DP 450) and 1,340 grams of dimethylformamide (DMF) into a 2-liter four neck resin reaction flask equipped with a stainless steel mechanical stirrer, thermometer, and a 250 ml equalizing pressure addition funnel.

The resulting slurry was stirred and 500 grams of liquid nitrogen tetroxide (N_2O_4) was added dropwise over about a 60 minute time period. The solution was deaerated and vertically spun using a glass 300 hole (0.0025 inch) spinneret. The fiber tow was passed through a primary bath of octanol at 15°C, then through a secondary bath of water at 22°C. The fiber tow was cut from the takeup godet, cut into staple fiber lengths, neutralized with bicarbonate, washed with water, finished with a 0.3% lubricating finish, air dried and carded. Godet stretch was 100%. Godet stretch is the difference between the circumferential speed of the initial and final godet, divided by the speed of the initial godet, times 100. The fiber produced was hollow and segment frequency was 25 nodules per inch. The fibers had tenacities of 2.25 g/den (cond.) and 0.96 g/den (wet), elongations of 8.65% (cond.) and 15.95% (wet) for a 1.21 denier filament of 0.38 wet modulus.

Fibers were also prepared from a large number of coagulants outside the scope of the process, including other monohydric alcohols. Specifically, methanol, ethanol, isopropanol, cyclohexanol and water coagulants all produced solid fibers prepared from dopes in which dimethylformamide was the solvent. Similarly, water, isopropanol, methanol and octanol coagulants produced either solid fibers or fibers of inadequate physical properties from dopes in which dimethylsulfoxide was the solvent. No coagulant other than octanol produced a hollow fiber from dialkylacylamide solvents and no coagulant other than hexanol produced a hollow fiber from a dimethylsulfoxide solvent.

Dry-Fiber Cellulose Acetate Hollow Fiber Membranes

A dry-wet process for forming interiorly asymmetric hollow fiber cellulose acetate membranes is disclosed by *R.E. Kesting; U.S. Patent 4,035,459; July 12, 1977; assigned to Chemical Systems, Inc.*

The process comprises extruding a hollow fiber from a spinning solution into a gas phase whereby the outer skin and structure of the fiber is formed by the dry phase inversion process while injecting into the interior of the fiber a core liquid which causes the interior wall of the hollow fiber membrane to gel according to the wet phase inversion process, thus giving an asymmetric dry-wet phase inversion produced membrane.

Example: A spinning solution of 20 parts cellulose acetate (Eastman E-394-60), 60 parts of methyl formate and 64 parts of isopropyl alcohol was spun using ethanol as the core liquid to a 0.025 inch inside diameter, 0.030 inch outside diameter, hollow fiber membrane and the internal skin was hydrolyzed by circulating a 0.4 molar sodium hydroxide solution through the fibers for 7.5 minutes, then flushing the fibers with water. All parts are by weight.

The process yielded a hollow fiber membrane which had an ultrafiltration rate of 3.3 cm/hr mm Hg x 10^{-4} with a vitamin B_{12} dialysis permeability of 26.5 cm/min x 10^{-4} and a creatinine dialysis permeability of 170 cm/min x 10^{-4}.

Hollow Fiber with Selective Gas Permeability

Asymmetric hollow fibers useful for effecting separations are prepared by *K. Arisaka, K. Watanabe and K. Sasazima; U.S. Patent 4,127,625; November 28, 1978; assigned to Daicel Ltd., Japan* by spinning a solution of a fiber-forming cellulose derivative through the outer tube of a double tube-type spinneret into an aqueous coagulating bath to form a hollow fiber and simultaneously injecting an aqueous salt solution through the inner tube of the spinneret to contact the interior of the hollow fiber. The hollow fibers are dipped in warm water, then the free water is removed and the fibers are frozen and dried under vacuum.

Referring to Figure 5.1 which illustrates a part of the process for preparing the hollow fiber, a cellulose acetate dope formed by dissolving cellulose acetate in a solvent and removing gas therefrom is fed to an outer tube **4** of a spinneret **3** from a tank **1** by a gear pump **2** and is spun into a coagulating bath **14**. An aqueous liquid is fed into the central portion of the fiber from a tank **5** through an inner tube **7** of the spinneret by a metering pump **6** to form the core of the hollow fiber. In this step, a gas can be used instead of or together with the aqueous liquid. The spun and coagulated fiber **8** is wound on a bobbin **13** after passing around a guide **9**.

When a prescribed length of the fiber is wound on the bobbin, the fiber in the wet state is transferred to the next step of the warm water dipping treatment. Solvent-free water maintained at a prescribed temperature is supplied to the coagulating bath tank **10** from a supply coagulating water inlet **11**, while a part of the coagulating liquid is discharged by a siphon **12** so that an increase of the solvent concentration is prevented.

Figure 5.1: Diagrammatic Representation of Part of the Process for Forming Hollow Fibers

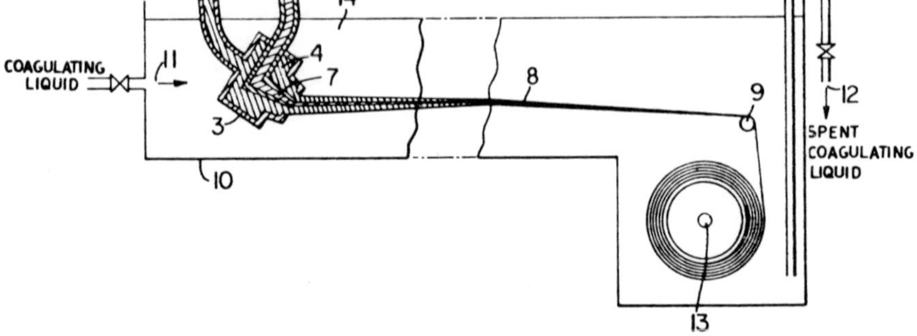

Source: U.S. Patent 4,127,625

Especially good results are obtained with respect to the gas permeation speed when the spinneret is submerged in the aqueous coagulating bath, namely, the extruded fiber is not exposed to air, and the value of the gas permeation speed in this case is more than several hundred times greater than the values attainable according to the conventional methods.

The gas separation capacity of the fibers is improved by using an aqueous solution of various water-soluble salts, such as sodium chloride, instead of water, as the injection medium fed to the central opening of the hollow fibers during spinning of the fibers. By this procedure a compact layer also is formed on the interior surface of the hollow fiber whereby the hollow fiber has compact layers including capillary vessels on both the outer and the inner surfaces of the hollow fiber.

Examples 1 through 4 and Comparative Example: A mixture comprising 23% by weight of cellulose acetate (RO-CA5430, Daicel, degree of acetylation 54%, viscosity 30 seconds), 52% by weight of acetone and 25% by weight of 1,4-dioxane was thoroughly blended to form a solution, and the solution was filtered, defoamed and injected into an outer tube of a spinneret of the double tube structure by means of a gear pump. Simultaneously, an aqueous solution containing sodium chloride at a concentration of 1 to 20% was fed to the inner tube of the spinneret by a metering pump. Both the solutions were extruded into a coagulating liquid (water) maintained at 3°C and the extrudate was wound at a rate of 12 m/min in the water.

Then, the extrudate was allowed to stand in the coagulating liquid (water) for at least 30 minutes, during which time one end of the extruded fiber was hung down from the vessel to permit the liquid in the hollow core to flow out by capillary action, whereby to replace the aqueous solution of sodium chloride in the hollow core by pure water. Then, the fiber was dipped in water maintained at 80°C for 15 minutes and then free water was removed from the outside and inside of the fiber.

Then, the fiber was immersed in gasoline maintained at -50°C to rapidly freeze the fiber and the fiber was then placed in a vacuum drier maintained at -40°C and 0.01 mm Hg for 24 hours to effect vacuum drying. The thus-obtained dry hollow fiber membrane was placed in a pressure vessel and a gaseous mixture comprised of 99 volume percent of nitrogen and 1 volume percent of krypton was flowed into the vessel under a pressure of 5 kg/cm^2. In this state, the permeated gas was sampled from the hollow core of the hollow fiber at prescribed time intervals and the volume ratio K_1 of nitrogen/krypton was determined by gas chromatogram.

The volume ratio K_0 of nitrogen/krypton in the starting gas was similarly determined and the degree R of separation of krypton from nitrogen was calculated from the following equation:

$$R = \frac{K_1}{K_0}$$

The flow rate of the permeated gas from the hollow portion was measured when the sample gas was collected. The results obtained are shown below.

Examples	Amount of NaCl in Injection Water (wt %)	Fiber Diameter (mm) Outer Diameter	Inner Diameter	N_2 Permeation Speed*	Degree R of Separation of Krypton
Comparative	0	0.85	0.65	2.40×10^{-4}	0.89
1	1	0.86	0.65	1.32×10^{-4}	0.88
2	3	0.84	0.63	0.61×10^{-4}	0.87
3	5	0.85	0.65	0.15×10^{-4}	0.74
4	20	0.84	0.62	0.019×10^{-4}	0.75

*Unit of permeation speed = cc/cm^2 sec cm Hg.

When the sodium chloride concentration was higher than 20%, reduction of the permeation speed was extreme and the resulting fibers had no practical utility.

High Strength, High Crimp, Hollow Rayon Fibers

According to *E. Costa, Jr.; U.S. Patent 4,130,689; December 19, 1978; and E. Costa, Jr. and M.P. Godsay; U.S. Patent 4,182,735; January 8, 1980; both assigned to International Paper Company* high strength hollow rayon fibers or filaments which will recover their hollow configuration after being immersed in water and subsequently dried are provided by an in-line process whereby a viscose solution containing a blowing agent is extruded into an aqueous acid coagulating bath of relatively high acid concentration.

The conditions of the process result in hollow filaments that are substantially irreversible since they remain hollow and do not collapse even after repeated washing and drying cycles. The hollow filaments also possess high strength nearly equivalent to that of high wet modulus rayon fibers. Modifiers may be employed in achieving this effect, but are not necessary.

The hollow filaments possess high crimp, such as about 25 to 30 crimps per inch, which will permit ease in carding and blending with other fibers.

The processes comprise first the spinning of a viscose solution containing cellulose in an amount of from about 6 to 9.5% (preferably 6.5 to 7.5%) of the weight of viscose; alkali-metal hydroxide, such as sodium hydroxide in the amount of from about 6 to 8% (preferably about 6.5 to 7.5%, optimally 7%) of the weight of viscose; as a blowing agent, from 3 to 5% (preferably 3.5 to 4.5%, optimally 4%) of alkali-metal carbonate, such as sodium or potassium carbonate or sodium or potassium bicarbonate based on weight of cellulose. Optionally, polyoxyethylene adduct of phenol, such as containing about 5 to 25 mols of ethylene oxide per mol of phenol, preferably 15-dendro phenol which has 15 mols of ethylene oxide per mol of phenol, in the amount of from about 0.25 to 0.75% (preferably 0.2 to 0.5%, optimally 0.3%) of the weight of cellulose is also employed. The viscose solution having a viscosity of from between about 90 and 140 poises (preferably 110 to 130 poises, optimally 120), ripened to a salt index from about 6 to 12 milliliters of sodium chloride is desirable, preferably 8 to 10, optimally 9. The viscose solution will desirably contain about 35 to 45% by weight of carbon disulfide, preferably about 36 to 40%, optimally 37%.

The resulting viscose solution is extruded through a spinneret which comprises capillaries, preferably each having a diameter of from about 30 to 75 microns (preferably 50 microns), into a first coagulating or aqueous acid bath. This bath comprises from about 150 to 220 grams per liter of sodium sulfate as a coagulating agent, preferably about 180 to 210 (optimally 200) grams per liter of sodium sulfate, and from 30 to 80 grams per liter of zinc sulfate, preferably about 40 to 60 (optimally 50) grams per liter of zinc sulfate, and from about 120 to 160 grams per liter of H_2SO_4, preferably about 130 to 145 (optimally 135) grams per liter. The coagulating bath shall have a temperature of at least about 25°C. No advantage is obtained by exceeding a temperature of 100°C. A preferred temperature of the bath is about 25° to 65°C or 25° to 45°C.

The coagulated fibers from the first coagulating or acid bath are then stretched from about 40 to 180%, preferably 90 to 100%, either in air or optionally within an aqueous stretch bath. A stretch bath, when employed, comprises from about 5 to 30 grams per liter of H_2SO_4, (preferably about 10 to 20 grams per liter) and about 2 to 20 grams, preferably 5 to 15 grams, per liter of zinc sulfate. No advantage is obtained by exceeding about 30 grams per liter of zinc sulfate. The preferred concentration of zinc sulfate is about 9 grams per liter. A stretch bath, when employed, is held at a temperature of from about 80° to 100°C, preferably 95° to 100°C.

In those cases where a subsequent formaldehyde treatment is to be employed, the viscose solution may be extruded into a first coagulating or aqueous acid bath of somewhat broader concentration ranges. Such bath may comprise from about 100 to 400 grams per liter of sodium sulfate, preferably about 180 grams per liter, from 0 to 100 grams per liter of zinc sulfate, preferably about 50 grams per liter, and from about 100 to 200 grams per liter of H_2SO_4, preferably about 130 grams per liter. The coagulating bath shall have a temperature of at least about 30°C. No advantage is obtained by exceeding a temperature of 100°C. A preferred temperature of the bath is about 45°C.

When formaldehyde treatment is employed, the coagulated fibers from the first coagulating or acid bath are then optionally passed into an aqueous stretch bath. A stretch bath, when employed, comprises from about 5 to 30 grams per liter of H_2SO_4 (preferably about 12 grams per liter) and at least about 2 grams per liter of zinc sulfate. No advantage is obtained by exceeding about 30 grams per liter of zinc sulfate. The preferred concentration of zinc sulfate is about 9 grams per liter. A stretch bath, when employed, is held at a temperature of from about 80° to 100°C. The preferred temperature is about 98°C.

Desirably, after coagulating the filaments, they are stretched up to 185% of their original length; preferably to from 40 to 130% of their original length.

In the aldehyde or formaldehyde embodiment of the process, after the cellulose has been regenerated and the fibers have been stretched, whether in air or in a stretch bath, as described above, they are then passed to an aldehydic treatment bath containing an aldehyde modifying agent. The aldehyde shall comprise at least about 5 grams per liter and no advantage is obtained by exceeding about 60 grams per liter. The preferred range is between about 15 and 45 grams of aldehyde per liter with excellent results being obtained at a concentration of about 40 grams per liter of aldehyde. The preferred aldehyde is formaldehyde and any available form of formaldehyde may be employed, such as aqueous solutions of urea-formaldehyde which provide formaldehyde by release in acid solution.

The fibers produced not only have permanent hollowness, but also exhibit high strength and sufficient crimp to permit ease in carding. The table below is a comparative study of the physical properties of this product as compared to regular rayon, high wet modulus rayon and cotton by means of the Instron data in a single fiber test.

Product	Regular Rayon Staple	High Wet Modulus Rayon	High Strength Hollow Rayon of this Process	Cotton
Conditioned Test*				
Tenacity, g/den	2.1	3.2	3.2	3.3
Elongation, %	10.7	9.5	11.0	9.0
Initial Modulus, %	60	90	80	50
Wet Test				
Tenacity, g/den	1.0	1.6	2.1	3.9
Elongation, %	26	18	17	10.0
Modulus at 5% extension, g/den	3.5	7.3	7.0	10.1
Crimps/inch	8	7	12	20

*Conditioned environment, 70°C, 65% relative humidity (yarn exposed for a minimum of 16 hours).

It is apparent from the above table that the fibers of this process have properties similar to commercial high wet modulus rayon and are approaching that of cotton.

The table below shows data taken on fabrics made from the high strength hollow fibers, the rayon fibers and the cotton fibers. All fibers were converted to 28's/1 yarn and woven into an identical fabric construction using both a plain and a 2 x 2 twill weave.

Fiber	. . . Cotton . . .		.Hollow Rayon.		. Bright Rayon .	
Weave	Plain	2 x 2 Twill	Plain	2 x 2 Twill	Plain	2 x 2 Twill
Open area, %	10.5	2.8	6.0	1.9	7.0	2.6
Dye absorption, ppm	3.1	3.5	2.8	3.3	1.9	2.6
Air permeability, $ft^3/min/ft^2$	103	76	114	74	237	125
Opacity	70.7	75.7	73.8	79.5	66.2	75.6
Brightness, %	75.5	77.5	65.1	67.9	70.1	69.6

It is apparent from these foregoing data that the fabrics made from the hollow rayon fibers exhibit bulk coverage and a dye absorption equivalent to that of cotton, and an opacity that is duller than either rayon or bleached cotton. Thus, this fiber also has the covering power and dye uptake equivalent to cotton.

Hollow-Fiber Devices for Diagnosis and Treatment of Oral Diseases

J.M. Goodson; U.S. Patent 4,175,326; November 27, 1979; assigned to Forsyth Dental Infirmary for Children provides capillary hollow fibers containing therein therapeutic and indicator agents and a method of employing such fibers, which method comprises placing the agent-filled, hollow capillary fibers into the oral cavity, and particularly about, adjacent or in contact with the teeth, at or about those areas where treatment of a dental disease is desired, or where indication of the presence of an oral or a dental disease is suspected, whereby such agent permeates from such capillary-hollow-fiber treatment into the localized area to treat the disease or to serve as an indication of dental disease in the area.

Example: A length of cellulose-acetate hollow fiber, with a stated molecular weight cutoff of 200 daltons, was cemented into the lumen of a 26-gauge hypodermic needle. A saturated solution of tetracycline hydrochloride in water was pumped into the hollow fiber by means of a peristaltic pump with silicone-rubber-pump tubing. The needle was attached by a standard Luer fitting to Teflon tubing which fit the inner bore of the silicone-pump tubing. After filling the fiber with the suspension, the distal end was occluded and the pumping maintained. The solvent (water), being smaller than a tetracycline molecule, more readily permeated the fiber wall.

After a time, the fiber became completely filled with tetracycline. The average content of fibers so constructed was 224 micrograms per linear centimeter, 75% of the maximum theoretical content of 300 micrograms per centimeter.

The release of tetracycline from a test segment of this fiber into distilled water was measured spectrophotometrically and was found to persist for 3 to 6 hours, with the highest concentration being delivered after 30 minutes.

Tetracycline-filled fibers thus constructed were tied around the necks of teeth of a patient afflicted wth periodontal disease. Prior microbiological diagnosis revealed a preponderance of large spirochetes. A comparison of spirochete counts before and after placement of the fiber revealed that the spirochetes had been eliminated completely by a single application of tetracycline-filled fibers. Concomitantly signs of inflammation in the treated area disappeared. Thus a beneficial therapeutic result followed the placement of tetracycline-filled fibers about the necks of teeth. Other antibiotic drugs effective against *Streptococcus* and other microorganisms also may be used.

Glass Fibers

APPARATUS AND METHODS

Fibers produced from a heat-softenable material, for example, glass fibers, can be classified roughly into two types. One type is what is called short fibers which have a diameter of several microns to several tens of microns and a length of several millimeters to several hundred millimeters. Prior suggestions for the manufacture of these short fibers include, for example: (1) a method utilizing a centrifugal force (a centrifugation method or a rotary method); (2) a method which comprises producing solid fibers from a viscous mass of a heat-softenable material, heating the fibers to render them viscous and at the same time blowing the fibers by a high speed gas stream to draw them (a blow method or flame method); and (3) a method which comprises directly rendering the heat-softenable material viscous, and blowing it by a high speed gas stream (a slit method or vertical blowing method).

The second type of fibers is called filaments or long fibers for use in reinforcing plastics which have a diameter of several microns to several tens of microns and a markedly long length. A drum wind-up method has been largely used for producing these long fibers.

A toration method has been developed which can be applied to the production of both short fibers and long fibers. It is well known that a heat-softenable material can be made into fine particles or fibers by rendering it liquid or viscous, and placing it in a high speed gas stream.

High Speed Gas Streams in the Formation of Glass Fibers

N. Ohsato, K. Tanaka and E. Mizushima; U.S. Patent 4,135,903; January 23, 1979; assigned to Nippon Sheet Glass Co., Ltd., Japan provide a method and an apparatus for producing short fibers or long fibers from a heat-softenable material. Basically, the method comprises: (1) heating a heat-softenable material to form a viscous melt and flowing it continuously; and (2) jetting out a specified high-speed gas stream to the melt that has flowed out.

Example 1: Using an apparatus of the type shown in Figures 6.1a and 6.1b, the following experiment was conducted. The results are shown in the table below.

Figure 6.1: Apparatus with High Speed Gas Streams

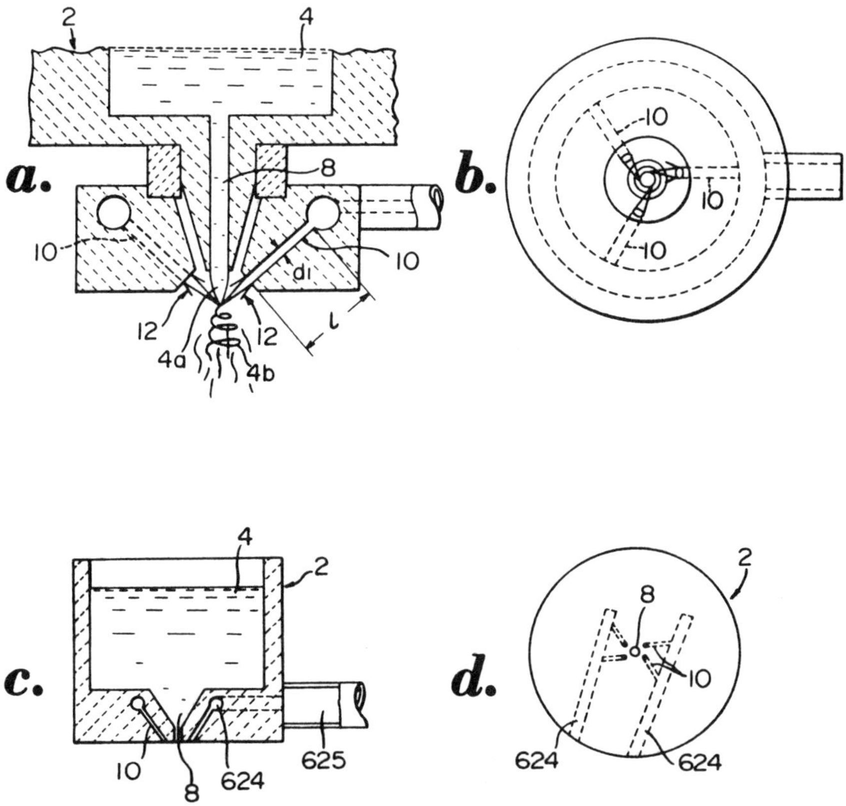

(a) Sectional view of one embodiment
(b) Bottom view of Figure 6.1a
(c) Sectional view of a second embodiment
(d) Bottom view of Figure 6.1c

Source: U.S. Patent 4,135,903

In Figures 6.1a and 6.1b, the molten glass was flowed out from crucible **2** through flow-out nozzle **8**. It was formed into a conical shape **4a** by high speed air stream **12** jetted out from air jet nozzle **10**, and left in the form of a fibrous melt **4b**. The gas stream **12** most closely approached the central axis of the cone **4a** at a position about 2 mm below the tip of the flow-out nozzle **8**. The air jet nozzle **10** was of a circular shape with a sectional diameter of 1.0 mm with an angle of 45° with respect to the horizontal surface. Three nozzles **10** formed an angle of 120° to each other when determined by their projected images onto the horizontal surface. The outlet of the jet nozzle **10** did not contact the outlet of the flow-out nozzle **8**, but was situated as close thereto as possible. The test was performed by adjusting the depth of the tank so that the amount of glass flow was 1 kg per hour. Air was at a pressure of 10 kg/cm^2 gauge, and the glass had a viscosity of about 50 poises at the outlet of the flow-out nozzle **8**.

Effective Hole Diameter of Flow-Out Nozzle (mm)	Fiber Diameter (μ)	Proportions of Unfiberized Masses (%)
3	30	50
2.5	10	30
2	5	10
1.5	4	7
1	3	2
0.5	<2	<1

When the effective hole diameter of the melt flow-out nozzle was 0.5 mm, it was impossible to adjust the amount of glass flow to 1 kg/hour.

It can be seen from the table that when the effective hole diameter of the flow-out nozzle is 3 mm, the proportion of unfiberized masses reaches 50%, but when the effective hole diameter decreases to 2.5 mm or below, the proportion of the fiberized masses decreases, and the fiber diameter also decreases.

Example 2: In an integral glass fiber-forming unit **2** shown in Figures 6.1c and 6.1d, a molten glass **4** having a viscosity of 50 poises and a temperature of 1100°C was stored, and in the lower part of the unit **2**, a molten glass flowing nozzle **8** having an inverted frustoconical shape at the upper part with the diameter of the thinnest part being 1.5 mm was perforated. Around the flow-out nozzle **8** were provided four gas stream jet nozzles **10** having a diameter of 0.6 mm. Air at high temperatures and pressures was conducted by nozzles **10** through conduits **625** and gas feed hole **624**, and was jetted out to form gas streams.

Each two of the nozzles **10** constituted a pair, and formed an angle of 60° to each other at the horizontal surface. Each of the nozzles was symmetrical with respect to the central axial line of the fiber-forming unit. Using this apparatus, air at 1000°C and 8 kg/cm^2 gauge was jetted out to form glass fibers having an average fiber diameter of 8 microns.

Forming Fibers by Toration

The term "toration" was coined by *M. Levecque, J.A. Battigelli and D. Plantard; U.S. Patent 4,015,963; April 5, 1977; assigned to Saint-Gobain Industries, France.* The terms "toration," a noun, and "torate," a verb, as used herein, constitute contractions of the words "tornado" and "attenuation" and are used to denote the action on an attenuable material in the zone of interaction created when a blast, or first jet, is penetrated by a transversely oriented second jet, the latter being completely encompassed or enveloped by the former. Clearly defined and forceful currents, including two counterrotating tornadoes, or whirls, or vortices, having substantial angular velocity, are generated by the interaction of the two jets, and these currents, by frictional licking or sweeping against the surface of the attenuable material, exert forces tending to induce the attenuable material to flow toward and into the zone of interaction; the material then is brought more and more under the influence of the tornadoes and forms itself into an elongated cone from the tip of which a fine fiber is drawn by the combined flow of the jets.

It is surprising that, although located in a whirling zone with very high velocities, the glass cone nevertheless is stable and that its cross section reduces progressively from its base at the plane of emission to its tip, from which a single fiber starts. It is also very surprising that this fiber, although, at least part of the time, has a nearly helicoidal movement at an increasing amplitude and velocity, continously emanates from the cone in a continuous attenuation process.

Referring to Figure 6.2, a blast, or primary jet source, is indicated at the left, being arranged to deliver the blast along a surface, in this case the under-surface, of a plate or wall **10**. The source of a secondary jet is also indicated, being arranged to deliver through the plate **10** so as to penetrate into the blast. The attenuable material, such as glass, is also delivered through the plate **10**, and in the embodiment of Figure 6.2, the point of delivery of the glass into the blast is just downstream of the point of delivery of the jet into the blast. An appropriate fiber collection means is indicated at the right.

Figure 6.2: Toration Process

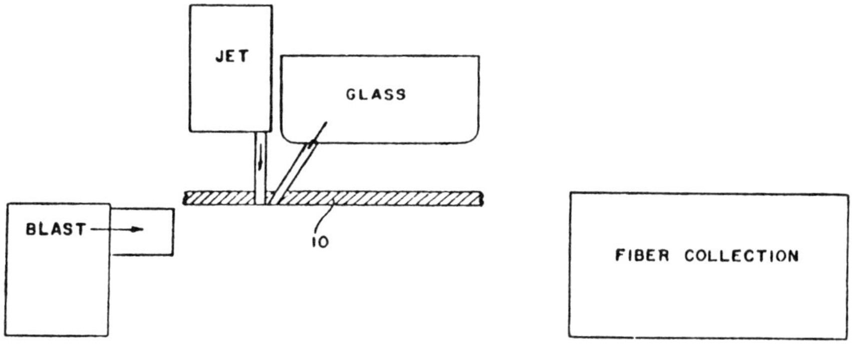

Source: U.S. Patent 4,015,963

In a related patent, *M. Levecque, J.A. Battigelli and D. Plantard; U.S. Patent 4,015,964; April 5, 1977; assigned to Saint-Gobain Industries, France* show in Figure 6.3 an embodiment in which aluminum oxide fiber type of insulation (for instance, 60% Al_2O_3) is applied to both sides of the crucible **200,** as indicated at **207**, thereby reducing heat loss from the crucible and serving also to protect the jet tubes **203** from the high temperature of the crucible. In order to withstand the temperature of the molten glass, the crucible, such as indicated at **200,** is ordinarily formed of platinum and the insulation described just above aids in making possible the construction of other components, such as the jet tube **203** of less expensive metals, for instance stainless steel.

Figure 6.3: Toration Apparatus

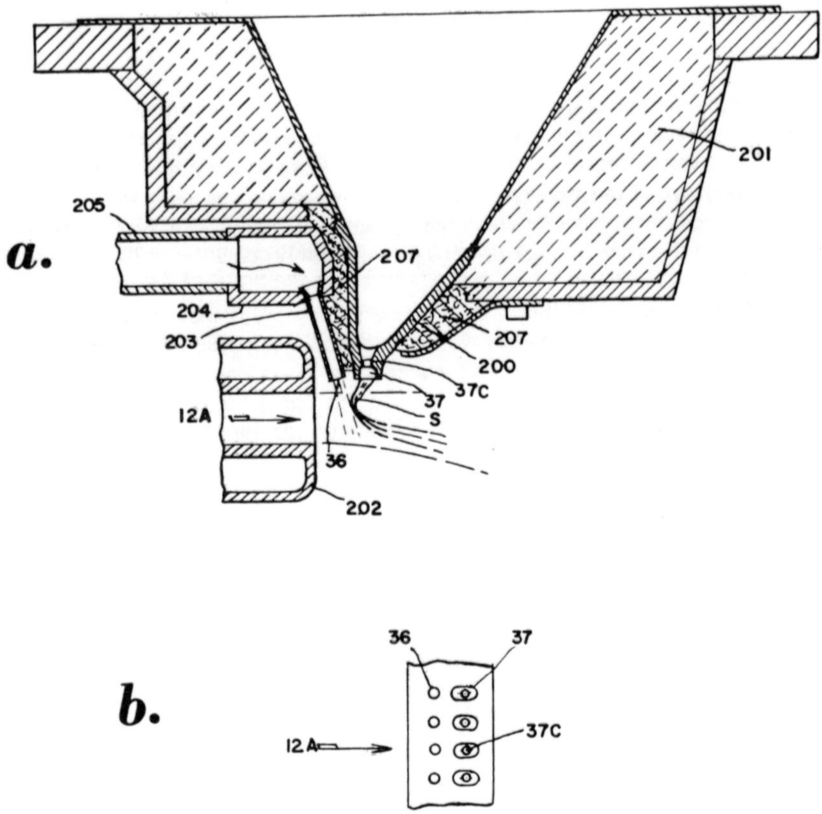

(a) Vertical sectional view of fiberizing center
(b) Glass admission orifices of Figure 6.3a

Source: U.S. Patent 4,015,964

With provision for insulation such as shown at **207** between the crucible and the jet and jet supply components, it is also possible to maintain a larger temperature differential between the glass on the one hand and the gas fed through the jet on the other. In this way even where the secondary jet is of relatively low temperature, the insulation **207** makes possible maintaining a relatively high glass temperature without excessive heat loss. This facilitates operating at a high production rate.

Attention is called to the configuration of the glass admission passage as shown in Figure 6.3. It will be seen that the glass passage in the bottom of the crucible has a central calibrating or metering opening **37c** and that the passage is enlarged both above and below the calibrating opening. The enlargement above the calibrating opening is funnel shaped thereby assisting glass flow through the calibrating opening into the enlargement or storage pocket from which the glass stream is delivered. The lower enlargement serves in effect as a small reservoir from which the streamlet **S** is delivered.

The perimeter of the delivery orifice **37** may be from two to three times that of the calibrating opening **37c**, and the entire perimeter of this delivery pocket is wetted by the glass, in consequence of which better stability of the glass bulb or cone drawn from the orifice is provided, especially at high glass temperatures. Because of the delivery pocket, the base of the glass bulb or cone which is formed is of greater size and in turn results in a bulb of greater length, thereby providing greater spacing from the point of glass admission to the point of drawing the fiber from the bulb. This increase in the downward spacing of the point where the fiber is first formed reduces tendency for fibers to stick and accumulate on parts of the equipment in the vicinity of the fiberizing center.

Although the pocket which terminates in the glass delivery orifice may be circular, it is advantageous to employ a pocket enlargement of ovoid form having its major axis arranged in the upstream and downstream direction with respect to the flow of the blast **12A** as is shown in Figure 6.3b.

With a glass admission means of the kind described with reference to Figure 6.3, the metering orifice indicated at **37c** constitutes the actual or controlling glass admission orifice.

In the embodiment of Figure 6.3, secondary jet pressures from about 2 to 10 bars may be employed with an orifice diameter of about 2 mm, and with an orifice of about 1 mm the pressure may be from 2 to 25 bars.

The capability of using high pressures and high kinetic energy for the secondary jet tends to increase the flow of the glass delivered from the admission orifices and this improves the uniformity of the flow or drawing of the glass and improves the uniformity of the glass bulbs which are formed, which in turn improves the uniformity of the fibers being produced.

With an arrangement as described above in connection with Figure 6.3, glass of the following formulation was used.

Formulation	% by Wt
SiO_2	63.00
Fe_2O_3	0.30
Al_2O_3	2.95
CaO	7.35
MgO	3.10
Na_2O	14.10
K_2O	0.80
B_2O_3	5.90
BaO	2.50

The operation was carried out as indicated by the following table of parameters and results:

Glass output	35–40 kg/day/glass orifice
Micronaire	4.0/5 grams
Fiber diameter	5 millimeters
Blast 12A	
Temperature	1550°C
Pressure	0.2 bar
Velocity	450 m/s
Secondary jet	
Temperature	600°C
Pressure	7 bars
Velocity	560 m/s

$$\text{Ratio of Kinetic Energies } \frac{\text{Secondary Jet}}{\text{Blast 12A}} = 6$$

In another related patent, *M. Levecque, J.A. Battigelli and D. Plantard; U.S. Patent 4,102,662; July 25, 1978; assigned to Saint-Gobain Industries, France* show somewhat schematically in Figure 6.4 a typical overall arrangement of equipment.

Toward the left there is shown in outline at **15'** a portion of a burner or blast producing structure having an associated nozzle **16'** with a discharge aperture **17'** of substantial width so as to deliver a blast **18'** with which a plurality of fiberizing centers may be associated. A supply line for a gaseous fluid under pressure is indicated at **19'** and this supply line is connected to jet manifold boxes **20'** which cooperate in supplying the jet fluid to and through jet orifices, one of which appears at **21'**.

A bushing **22'** associated with a forehearth or other appropriate glass supply means indicated at **23'** is provided with glass orifice means indicated at **24'**, and the stream of glass is delivered into the flow of the jet and is carried downwardly to the zone of interaction in the blast **18'**. Fiberization occurs in the jet and also in the blast, and as the blast delivers the fibers toward the right as viewed in Figure 6.4, a fiber blanket indicated at **25'** is laid down upon a perforated traveling conveyor or belt **26'**, having a suction box **27'** below the upper run of the conveyor, the box **27'** connecting with a suction fan diagrammatically indicated at **28'** to assist in laying down the desired fiber blanket on the perforated conveyor **26'**.

Figure 6.4: Arrangement of Equipment in Toration Process

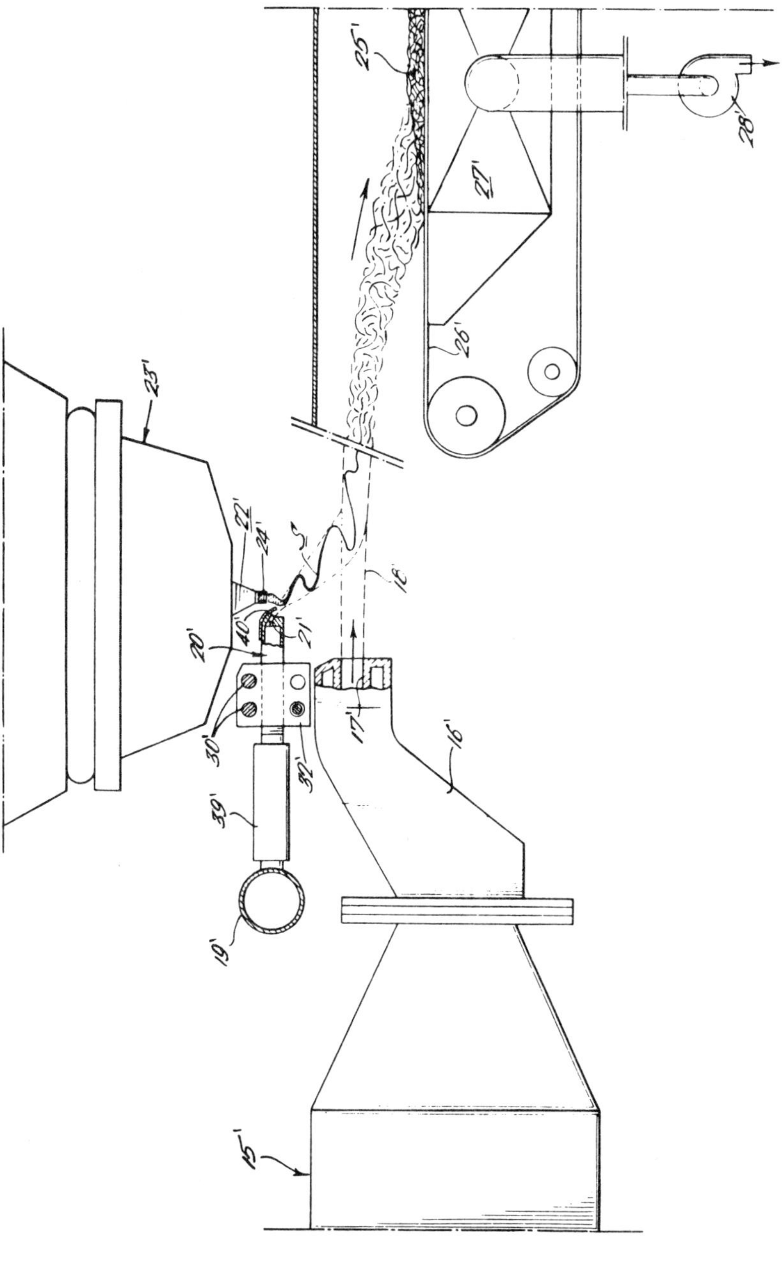

Source: U.S. Patent 4,102,662

Example: In equipment of the kind illustrated in Figure 6.4 and having 70 fiberizing centers, a glass of the following composition was attenuated.

	% by Wt
SiO_2	63.00
Fe_2O_3	0.30
Al_2O_3	2.95
CaO	7.35
MgO	3.10
Na_2O	14.10
K_2O	0.80
B_2O_3	5.90
BaO	2.50

The bushing temperature was about 1500°C and the jet and blast temperatures were respectively about 20° and 1500°C. The ratio of the kinetic energy of the jet to the blast was about 10 to 1. The pull rate was 55 kg/hole per 24 hours. The fiber diameter after both stages of attenuation averaged about 6 microns.

Gas Blast Attenuation in Toration Process

In the toration process, *M. Levecque, J.A. Battigelli and D. Plantard; U.S. Patent 4,146,378; March 27, 1979* propose that the jet which penetrates transversely into the blast have a temperature lower than that of the blast and preferably close to ambient temperature to thereby avoid erosion of the structural element positioned adjacent to the jet path.

The lower temperature and velocity contemplated still provides the required kinetic energy ratio between the jet and blast, i.e., a jet having kinetic energy higher than that of the blast so that the jet will penetrate the blast and create a zone of interaction. The reason why this desired kinetic energy ratio is still present with the lower velocity of the jet is because of the higher density of the jet fluid at the lower temperature. The density, of course, increases with decrease of temperature and since the kinetic energy is determined not by the velocity alone but also by the density of the jet fluid, a jet may readily be provided having a higher kinetic energy per unit of volume than the blast, even at velocities lower than the velocity of the blast.

By employing the lower jet temperature (for instance, a temperature approximating ambient or room temperature) a number of advantages are attained. In the first place when utilizing such temperature for the jet, the erosion and warpage problems of the jet guiding or deflecting elements are extensively reduced or even eliminated. In addition it becomes feasible to employ a commonly available source of compressed air as the source of fluid for the jet. The lower temperature also makes it practical to use commonly available materials, such as stainless steel both for the jet generating device and the guiding or deflecting elements, rather than more sophisticated and expensive materials, such as platinum alloys or ceramics, which are needed where very high temperature jets are used.

Still further, the use of the lower temperatures for the jet facilitates introducing the fiber being formed into a relatively cool environment as soon as the fiber is attenuated, this consideration being of importance in the toration technique.

It is also to be kept in mind that by employing a jet fluid at a temperature approximating ambient or room temperature, consumption of energy to heat the jet is eliminated.

In connection with the above it is pointed out that a jet of adequate velocity is readily obtained when using a source of compressed air as the source of the jet fluid. This is in distinct contrast to the employment of relatively high jet temperatures, with which it is technically more difficult to obtain high velocity, and because of the lower density of the fluid at high temperature it is therefore much more difficult to attain sufficient velocity to achieve the kinetic energy ratio required to effect penetration of the jet into the blast.

Suppression of Pollution in the Toration Process

M. Levecque and J.A. Battigelli; U.S. Patent 4,052,183; October 4, 1977; assigned to Saint-Gobain Industries, France provide for effective and extensive suppression of pollution originating from glass fiber blanket manufacturing plants, especially plants utilizing toration for fiber production as herein disclosed.

Considering the fiberization equipment shown in Figure 6.5, reference is first made to the blast or principal jet generators **154, 156** and **158** and secondary jet generators **148, 150** and **152**. Still further the molten glass streams are admitted from orifices in the crucibles **142, 144** and **146** which are provided transversely of equipment. Thus when viewed as in Figure 6.5, each blast generator, for instance the generator **154,** represents only one of a series which are aligned with each other. All of the fiberizing centers established by the groups of generators deliver the attenuated fibers into a common hollow guide **168, 170** or **172.** These guides in effect form channels for directing the fibers from the several groups of fiberization stations downwardly in an inclined path from the zone of fiberization toward the perforated fiber collecting conveyor **180.** Guides of this type are desirably tapered in plan form.

As will be understood, the gases from the blast generators and the jet generators in each group flow with the fibers into the upper or inlet ends of the channels formed in the hollow guides **168, 170** and **172**, and each of these fiber and gas streams is indicated in Figure 6.5 by the reference numeral **12.**

As seen in Figure 6.5, the fiberization equipment is arranged in spaced relation to the fiber collecting conveyor **180** and is located in a fiber receiving chamber **100** formed by various wall elements and preferably being substantially closed. The fiber collecting conveyor **180** defines at least in large part one of the walls of the collecting chamber and this conveyor serves to carry the formed blanket to the left out of the chamber below the lefthand end wall. As will be understood, appropriate openings are provided for the introduction of the fuel and air needed

Figure 6.5: Equipment for Fiberization and Fiber Collection

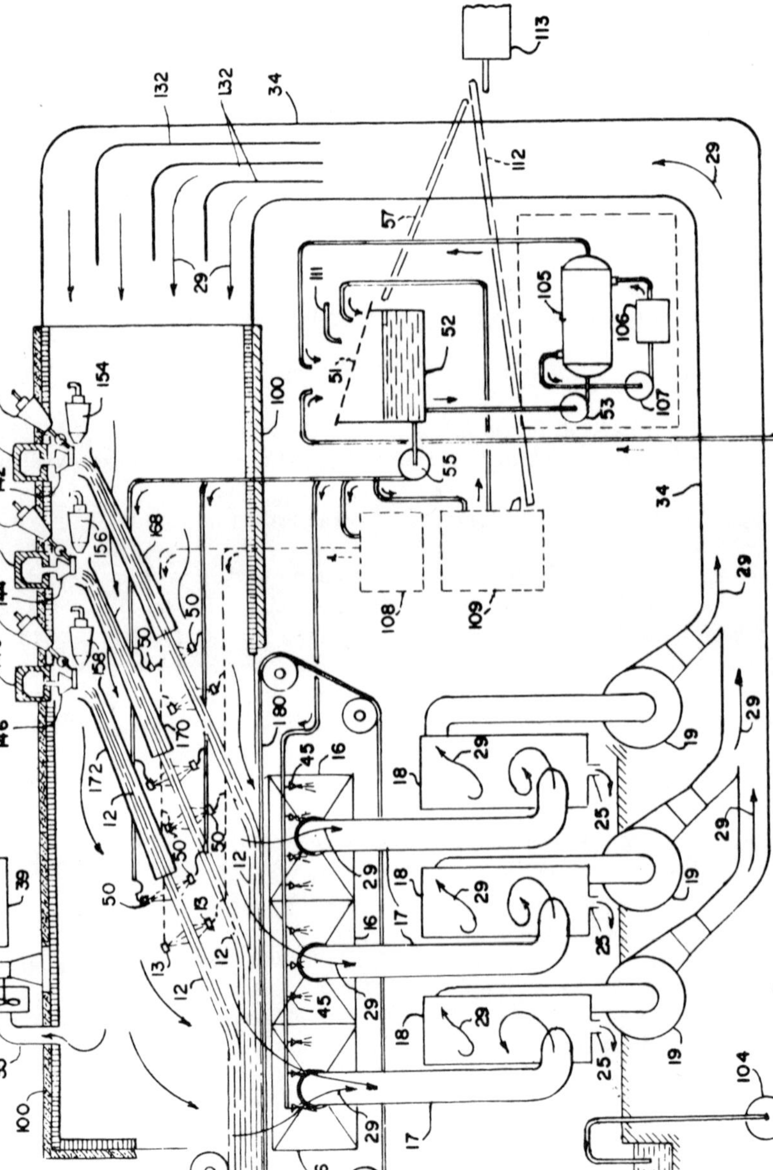

Source: U.S. Patent 4,052,183

for the blast and jet generators, and in addition there are openings appropriate to the accommodation of the molten glass forehearth branches and crucibles to provide for admission of the molten glass into the fiberization equipment.

To provide for collection of the fibers on the perforated fiber collecting conveyor suction chambers **16** are positioned beneath the fiber collecting run of the conveyor, the chambers being open at the top and having ducts **17** connected therewith and respectively communicating with cyclone separators **18.** Each cyclone separator has an off-take connection extended to a blower or exhaust fan **19** which discharges the exhausted gases into the duct **34**, constituting a recirculating duct which is connected with one end of the fiber receiving chamber **100.** Baffles **132** in the region where the duct **34** is connected with the chamber **100** serve to uniformly distribute the recirculating gases in the fiber receiving chamber.

For the purpose of cooling the fibers as they are delivered from the guides **168, 170** and **172**, water sprayers **50** are provided, preferably both above and below the fiber and gas stream **12** delivered from each of the guides. Downstream of the water sprayers, additional spray nozzles **13** are provided and these additional nozzles serve to spray onto the fibers a liquid binder resin material, preferably a binder which is adapted to harden or cure upon subsequent heating of the formed blanket, for instance in an oven through which the blanket is delivered after being discharged from the left of Figure 6.5.

Because of the spraying of the fibers with the water and with the liquid binder resin material, the gases which are withdrawn through the suction chambers **16** carry with them substantial quantities of moisture and also resin constitutents. It is contemplated that such constituents and also small fiber pieces which may be carried through the collecting conveyor by the gases being exhausted, should be removed from the gases before recirculation thereof back into the fiber receiving chamber. This separation is effected in the embodiment illustrated by means of the cyclone separators **18.** The separation is enhanced and assisted by the scrubbing action of additional water spraying nozzles **45** which are arranged within the suction chambers **16.**

The general flow pattern of the gases in the recirculating system shown in Figure 6.5 is indicated at various places by the arrows **29.** In the fiber receiving chamber the gas flow is not only established by the exhaust fans **19**, but also tends to be augmented by the action of the blast and jets at the fiberizing centers. Because the upper ends of the guides **168, 170** and **172** are open in the zones of the fiberizing centers, portions of the recirculating gases are induced to enter the upper ends of the guides and other portions join the gas and fiber streams **12** beyond the discharge ends of the guides.

The liquid constituents which are separated in the cyclone separators **18**, together with various constituents entrained thereby are discharged from the lower ends of the separators through the discharge openings **25** and collect in the sump **103**, and in this way various of the liquid and solid constitutents which are picked up or entrained by the gas flow stream are separated and isolated, so that they are not returned into the fiber receiving chamber with the recirculating gases. For further suppression of pollution, these separated liquids are specially treated as will be described hereinafter, but it is first noted that although virtually all of the gases which are drawn from the receiving chamber through the fiber collecting conveyor are recirculated to the fiber receiving chamber, some gases are withdrawn from the receiving chamber through the duct **35** under the influence of the fan **44**. This fraction of the circulating gases represents a quantity on the order of 5 to 10% of the total amount flowing through the perforated collecting conveyor and represents approximately that percentage of additional gases which are constantly being introduced under the action of the blast and jet generators employed at the fiberization stations.

The gases withdrawn under the action of the fan **44** are delivered into and through a burner device **39** in which the temperature is elevated to a point preferably above 600°C, after which the treated gases may be expelled to the atmosphere substantially without objectionable pollution. All of the remaining 90 to 95% of the gases are recirculated, and thus do not pollute the atmosphere.

In addition to recirculation of the gases, the system in Figure 6.5 also contemplates treatment of an recirculation of water delivered from the cyclone separators **18**. For this purpose, a pump **104** transfers the water from the sump **103** into the tank **52**, a screen or filter diagrammatically indicated at **51** being interposed in order to separate solids before the water is delivered into the tank. The water is recirculated from the tank **52** by a pump **53** through the heat exchanger **105** for the purpose of cooling the water, the cooled water being returned to the tank **52** through the screen or filter device **51** providing for separation of solids.

The cooler **105** is adapted to be cooled by a heat transfer medium recirculated by means of a pump **107** through the cooling unit **106**. The details of construction of these parts may take a variety of forms and need not be considered herein as they form no part of this process per se.

Water is also withdrawn from the tank **52** by means of pump **55** and is delivered thereby, preferably under suitable adjustable controls not illustrated, to the water supply nozzles **50** and **45**. The pump **55** may also deliver water to the station **108** for preparation of the fiber binder or sizing which may take any suitable form and from which the binder material is delivered to the spray nozzles **13**.

Still further, some of the water is preferably delivered to the station **109** for treatment to separate resin constituents carried in solution. This is preferably affected by subjecting the water to an increased pressure and to an increased temperature, after which it is cooled. This treatment results in insolubilizing resin constituents, which may then readily be separated, for instance by means of a centrifuge. The purified water is then returned to the tank **52** for reuse. The solid constituents separated at the station **109** and also by the filter **51** associated with the tank **52** are transferred by means of suitable conveyors **112** and **57** to the waste treatment station **113** which may consist of a heater or burner in which the solid waste materials are brought to a temperature of the order of 600° to 700°C so as to burn the resin binder constituents present and also to sinter any fibers present. The latter may if desired be reintroduced into the fiber circulation system, i.e., into the constituents from which the molten glass is prepared for fiberization.

Makeup water may be introduced into the system by means of the supply connection **111** delivering into the tank **52.**

Transporting Glass Fibers with Reduced Tension from Air Drag

D. Kleist; U.S. Patent 4,058,385; November 15, 1977; assigned to Owens-Corning Fiberglas Corporation provides a method and apparatus for reducing tension due to air drag in continuous filaments moved at high linear speeds by establishing a contoured air passage surrounding the filaments and moving air in the direction of the movement of the filaments at a velocity substantially equal to the velocity of the moving filaments. The method and apparatus are described in relation to a glass fiber process wherein the filaments are attenuated at high linear speeds.

Figure 6.6 shows continuous glass filament forming apparatus. The apparatus is shown in operation during package build. The embodiment uses a rotary filament pulling and winding device to form and collect continuous glass filaments at a constant filament forming speed. The air control means or duct is below a source of molten streams from which the glass filaments are drawn and above the filament gathering device and collecting winder. The air control means surrounds the filaments shielding them from extraneous air currents and producing complementary air flow in the direction of filament movement. Hence, the resistance to flow normally caused by the air is minimized. The results are reduced stresses in the filaments and a reduction in the amount of energy required by the winding device to pull the filaments.

Figure 6.6 illustrates a container or feeder **10** that holds a body of molten glass. The feeder **10** can receive a continuing supply of molten glass by several known ways. For example, a forehearth can supply molten glass to the feeder **10** from a furnace heating batch materials to molten glass. Also, a melter associated with the feeder **10** can supply molten glass to the feeder by reducing glass marbles to a heat-softened condition. At the ends of the feeder are terminals **12** that connect to a source of electrical energy to heat the feeder **10** by conventional resistance heating. Such heating keeps the molten glass in the feeder **10** at proper fiber-forming temperatures and viscosities. The feeder **10** has a bottom **14** with orifices or passageways for delivering streams **16** of molten glass from the feeder **10**. As shown depending orificed projections or tubular members **18** define the openings in the bottom **14**.

The feeder **10** is normally made of platinum or an alloy of platinum. Located adjacent the orificed projections **18** are metal fins **15** which are connected in heat-conducting relationship to a water cooled manifold **17**. The fins divide the total number of projections **18** into smaller groups and are effective in cooling the hot molten streams **16** and filaments **20**.

Figure 6.6: Fiber-Processing Apparatus with Moving Air to Reduce Drag

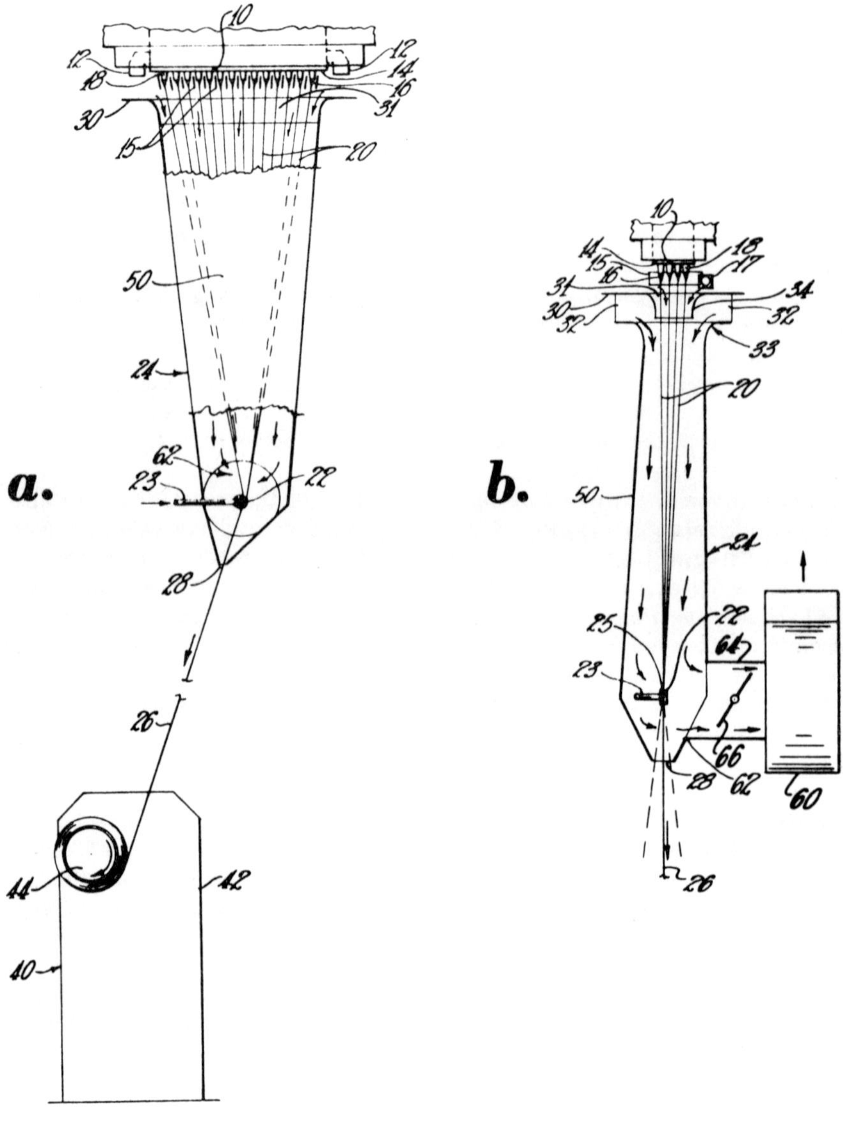

(a) Front elevational view, partially in section
(b) Side elevational view, partially in section

Source: U.S. Patent 4,058,385

The molten glass streams **16** are attenuated downwardly into individual continuous glass filaments **20**. Immediately after the glass filaments are formed they enter the air control channel or duct **24** below the feeder **10**. The exhaust fan **60** exhausts air through exit passage **62** thus inducing air flow into the duct **24** through its upper opening **32**. The fan **60** is regulated to induce an air flow speed matched to the linear speed of the filaments. In this way the air drag on the individual filaments is minimized.

A sizing applicator and gathering shoe **22** supported within the duct **24** applies a liquid sizing or other protective coating material to the swiftly traveling continuous glass filaments and combines the filaments into a bundle or strand **26**. The applicator may be any suitable type of applicator known to the art. However, the applicator preferably presents a minimum of interference with air flow through the duct **24**. As shown in Figure 6.6, the applicator gathering shoe **22** is hollow and connected to a supply of liquid coating material by means of a tube **23**.

From the applicator **22**, the combined filaments or strand **26** proceeds to a winder **40** which applies the necessary tension force for attenuating the molten glass streams **16** supplied by the feeder **10** into glass filaments. The collet **44** is rotatably mounted on a housing **42** located below the gathering shoe **22**. A motor (not shown) within the housing **42** rotates the pulling wheel at a constant high angular speed sufficiently fast to withdraw the continuous glass filaments **20** from the streams **16** at linear speeds to 12,000 feet per minute and faster.

The upper inlet into the air velocity control duct **24** is designed to provide minimum disturbance to the fiber forming region. As shown in Figure 6.6b, the upper portion of the air control apparatus comprises a top horizontal isolating barrier **30** with an opening **31** through which the filaments **20** enter the air control duct **24**. The major air inlets **32** are at the front and back sides of the feeder **10** beneath the isolating barrier **30**. The side inlets **32** are several times larger than the opening between the feeder **10** and the top opening **31** through which the filaments pass. This difference in size assures that most of the air introduced into the duct **24** comes from the side inlets **32**. As the air enters the side inlets, it impinges on two curved air deflectors **34** which protect the filaments from horizontal air forces by directing the induced air in a downwardly direction parallel to the filaments.

The duct or primary air channel **24** has a constant cross-sectional area from the level of the inlet **32** to the general level of the high velocity exhaust fan **60**. This helps assure that the air velocity along the length of the ducts remains constant. In the embodiment shown in Figure 6.6 for use with a very fine fiber system producing fibers 0.00015 inch in diameter, the inlet end **33** of the duct **24** is about 8 inches by 2 inches. Near the bottom end of the duct **24** adjacent the fan exhaust outlet **62**, the duct is about 4 square inches. Thus, the total cross-sectional area of the duct is about 16 square inches all along its length.

To make the use of the duct **24** as convenient as possible, the front panel **50** is hinged at one side to provide access into the duct during the startup of the fiber-forming operation for cleaning purposes. The hinges and latches are preferably attached only to the outside of the duct **24**. If any of their parts extended into the duct, they could cause undesirable air turbulence in the duct.

A high velocity fan **60** is attached by means of a connecting air duct or channel **64** to the main duct **24**. The point of attachment **62** of the connecting channel **64** to the main duct **24** is near the level where the filaments are gathered into a strand **26**. Once the filaments have been combined into one strand, the air drag on the filaments is less critical because the exposed surface area is less and the combined strength of the strand is greater than that of the individual filaments. The cross-sectional area of the opening of the connecting channel **64** is at least as large as that of the main duct **24**. This helps assure an even flow of air through and out of the main duct.

Below the connecting channel's point of attachment **62** to the main duct, the main duct is tapered to a restricted outlet **28**. By restricting the outlet to as small an opening as possible, the majority of the air passing through the main duct **24** enters through the larger side openings **32** at the top of the duct.

The gathering shoe **22** shown in this embodiment also functions as the binder applicator. Both functions are combined to present a minimum of interference with the air flow. The gathering shoe-applicator **22** is hollow and is supported by a hollow tube **23** which also supplies the binder under pressure to the hollow applicator.

The binder flows from the tube **23** into the circumferential groove **25** of the applicator **32** through small opening formed in the axis of the applicator. As the strand passes across the groove, binder is applied to it.

The binder may be applied to the filaments or strand in ways other than the embodiment shown. For example, a conventional belt applicator or the spraying of a fine mist of binder into the duct **24** may be satisfactory. Whatever method of binder application is employed, it should be designed to complement the air flow through the duct.

The pulling or attenuating force on filaments may be applied by an mechanical means. A standard winder is illustrated in Figure 6.6.

The operation of a filament forming operation is very straightforward. With the access panel **50** open the fiber-forming operation is started in a conventional manner. The initial filaments **20** are formed and wrapped around the collet **44** to begin the operation. The access panel is closed and as the winder **40** comes up to speed, the exhaust fan **60** is also started. Conventional sensing and control means are employed for sensing the relative speed of the filaments and the air which surrounds the filaments in the duct **24**. In response to the control signals the damper **66** in the connecting channel **64** is positioned to assure that the air speed is matched to the speed of the filaments **20**.

Movable Blower for Improved Temperature Control

In the past filaments have been formed by using feeders for holding molten material. There is a plurality of orifices located in the lower region of the feeder and streams of molten material issue from the orifices. The streams of molten material pass between two blowers located beneath the feeder and jets of air from the blowers attenuate the streams of molten material into filaments. The filaments that are formed can be either collected in a loose bundle or formed into a strand and wound onto a package.

One of the most significant variables in the filament-forming process is the heat of the lower region of the feeder in the vicinity of the orifices. The temperature of this region of the feeder controls the rate at which the filaments can be formed and also is a significant factor in establishing the diameter of the attenuated filament. Normally, the temperature of the feeder area in the vicinity of the orifices has been controlled by controlling the supply of energy to the feeder. Also, the terminals that are used to supply the electrical energy to the feeder can be moved to somewhat change the heat pattern or temperature of the feeder in the region of the orifices.

However, the above methods of temperature control do not completely control the temperature or heat pattern in the lower region of the feeder. Frequently localized hot or cold spots develop on the lower region of the feeder. These temperature nonuniformities disrupt the filament-forming process.

G.A. Pelagio, C.J. Stalego and W.B. Holland; U.S. Patent 4,150,937; April 24, 1979; assigned to Owens-Corning Fiberglas Corporation provide an apparatus for attenuating filaments. In one of its more specific aspects, the process relates to attenuating filaments with blasts of air from a blower which is movable with respect to the fiber-forming orifices.

In Figure 6.7 there is shown a feeder **10** for supplying molten glass that is to be formed into filaments from either a bushing having tips or a tipless bushing. The process will be explained with reference to the former. The lower region **12** of the feeder contains a plurality of tips **14**. Each of the tips is adapted with an aperture **16** that opens through the tip and into the feeder.

Below the feeder is a pair of blower elements **20** that are positioned along each side of the longitudinal row of tips in the feeder. The blower elements are in spaced-apart relationship with respect to the tips on the feeder. An inlet **22** is positioned at one end of the blower for introducing a fluid into the blower. One surface of each blower element is adapted with a plurality of apertures **24**.

Positioned at each end of the blower elements is means **30** for moving the elements to vary the space between the blower elements. Connected to means **30** is means **32** for moving the elements in a direction perpendicular to the longitudinal axes of the filaments being attenuated. Connected to means **32** is means **34** for moving the blower elements in a direction parallel to the longitudinal axes of the filaments being attenuated.

Figure 6.7: Apparatus with Movable Blower

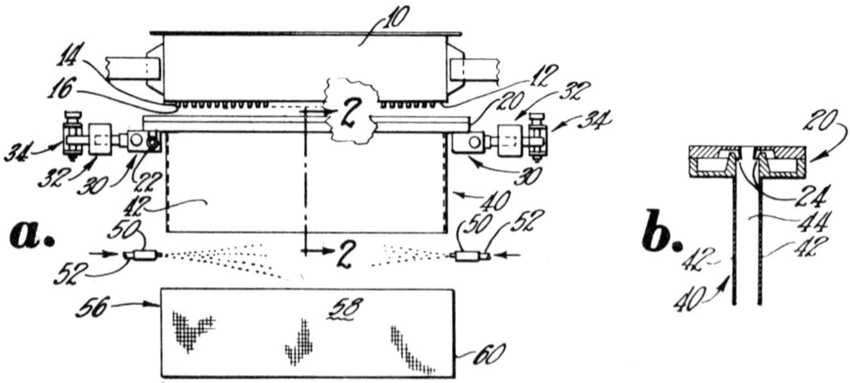

(a) Front elevation view
(b) Partial cross-sectional view along line **2–2** of Figure 6.7a

Source: U.S. Patent 4,150,937

Positioned beneath the blower elements **20** is skirt **40**, connected to the lower region of the blower elements and defining a generally rectangular-shaped passageway beneath the blower elements. The skirt has side walls **42** and **44**.

Located beneath the skirt is a plurality of spray nozzles **50** that are used to spray a coating onto the filaments. Conduit **52** connects the spray nozzles with a source of coating material, not shown.

Beneath the spray nozzles can be positioned container **56** for collecting the filaments that are formed. The side walls **58** and end walls **60** of the container are normally formed of a porous material.

In operation, molten material passes from the feeder through the orifices in the tips in the form of streams. The streams flow from the tips through the passageway between blower elements **20**. Fluid passing from apertures **24** in the blower elements engage the streams of molten material and attenuate the streams into filaments. The attenuated filaments pass into the rectangular passageway formed by the skirt **40** directly below the blower elements. The fluid exhausting from the apertures **24** creates a turbulent flow pattern in the skirt **40**. The turbulent flow causes the attenuated filaments to strike against the walls of the skirt and break.

As the short fibers exit the skirt **40**, spray nozzles **50** spray coating or sizing onto the filaments. The coating sprayed from the nozzles acts to protect the filaments from further abrasion. After the filaments have been coated they are collected in a container **56**. The porous walls of the container allow the air used to attenuate the filaments to escape from the container but retain the fibers within the container.

In operation, the position of the blower elements **20** with respect to the tips **14** will affect the temperaure of the tips which is an important factor in the formation and attenuation of the streams of molten material. To control the relationship of the tips and the blower elements, means **30** for varying the space between the elements, means **32** for moving the elements in a direction perpendicular to the longitudinal axes of the filaments being attenuated and means **34** for moving the elements in a direction parallel to the longitudinal axes of the filaments are provided.

Means **30**, means **32** and means **34** can be used to position the blower elements so that the proper temperature and heat pattern are established in the feeder. In addition, the blower elements can be moved relative to the feeder to maintain the desired temperature and heat pattern.

Cooling System

C.R. Nichols and K.A. Helberg; U.S. Patent 4,033,742; July 5, 1977; assigned to Kaiser Glass Fiber Corporation describe a cooling system for establishing and maintaining the running mode of a bushing used in production of glass fibers which embodies a heated orifice plate with closely spaced orifices and a bulk flow of upwardly directed gas (e.g., air), which cooling system comprises a series of opposing nozzles which provide a multiple air lance effect in starting up, clearing and maintaining the flow of individual glass fibers through each orifice.

Referring to schematic Figure 6.8, the glass is melted in furnace **1** and a reservoir of molten glass **2** is maintained in reservoir means or bushing **3**. The temperature of orifice plate **4** is controlled by electrical resistance heating means attached to the orifice plate.

Cooling air is introduced from at least two sides of the orifice plate through nozzles **6** and **6'**, which, in the apparatus shown, are connected to manifolds **7** and **7'**. The nozzles are arranged in generally opposing relationship on two sides of the orifice area. The glass fibers **8** are drawn from the orifices and are wound on collet **9**. Auxiliary equipment including sizing or dressing fluid applicators are not shown.

For establishing the running mode (i.e., startup or clearing), the air is introduced at an angle of from about 30° to about 60° to the orifice plate. This angle permits as much cooling air as possible to reach the orifice plate over the edge of the glass monolith as the monolith moves away from the glass plate while at the same time maintaining a flow of air at the cone area. The air is directed generally at the center of the plate but may vary somewhat because of the fanning effect as the air leaves the nozzle. If over-cooling occurs at the edge of the orifice plate while the center is undercooled, the impact centers of the cooling air are separated too far outwardly from the center. On the other hand, if the air streams overlap too much, the edges will tend to become too hot. The air leaving the nozzles will, of course, increase in cross section or fan out. The nozzles should be located relative to the orifice area so the cross-section of air flow at the orifice plate extends at least sufficiently to include the edge of the orifices nearest the nozzle. The air flow from all the nozzles should, of course, be of sufficient total cross section at the orifice plate that it encompasses the entire orifice area. A preferred angle for the nozzles is from about 40° to about 50° to the orifice plate, while a particularly preferred angle is about 45°.

The use of nozzles, as contrasted for example to the use of an elongated slit, is important in order to maintain impact pressure while avoiding the use of excessive air. Generally nozzle diameter will range from about 0.031 inch to about 0.188 inch and are spaced sufficiently close to each other to obtain a substantially even flow of air along the orifice plate. Nozzles having a diameter of 0.062 inch spaced on 0.5 inch centers positioned about 3 inches from the center of the orifice area have successfully been employed for a 2,000 hole orifice plate which was 1.65 inches wide and 8.3 inches long. The orifice area was 21 rows wide. Generally, the nozzles may be placed from about 2 to about 5 inches from the center of the orifice area. Representative air exit velocity at the nozzles ranges from about 100 to about 400 feet per second.

The apparatus may be employed to establish the running mode at startup accord- to the following procedure.

The orifice plate is heated sufficiently to cause flooding of the entire underside thereof and form a glass matrix or monolith. While it is not necessary, it is desirable to use cooling air assist the matrix to form more quickly. During this stage of startup the temperatures of the orifice plate will generally be from about 25° to about 150°C below normal operating temperatures and glass viscosities flowing through the orifices will be above about 100 poises. Normal operating temperature of an orifice plate for E glass, depending upon orifice diameter and desired throughput rate, will be within the range of from about 1150° to 1350°C. Operating temperatures for other types of glass will vary but such temperatures are well known to the art.

Inasmuch as the molten glass passing through the orifices must be fiberized, the viscosity of the glass in the fiber-forming cone beneath the orifices from this point forward in the process should be within the normal fiberizing viscosity range of from about 300 to about 1,000 poises, preferably from about 500 to about 700 poises.

After the monolith is formed, it is moved slowly away from the underside of the orifice plate. The movement of the monolith may be accomplished by mechanical means such as by tongs or a glass rod embedded in the monolith. Alternatively, but generally less desirably, the monolith can be permitted to fall of its own weight.

As the monolith moves away from the underside of the orifice plate, the clearing air is employed and begins to impact on the orifice plate itself to assist clearing. Generally clearing will occur beginning at the edges of the orifice which are nearest the clearing air sources. Because of normal manufacturing tolerances, minor variations in heat patterns and the like, however, some areas will fiberize, while other isolated areas will remain flooded.

Figure 6.8: Glass Fiber Production Equipment with Clearing Air Delivery Manifold

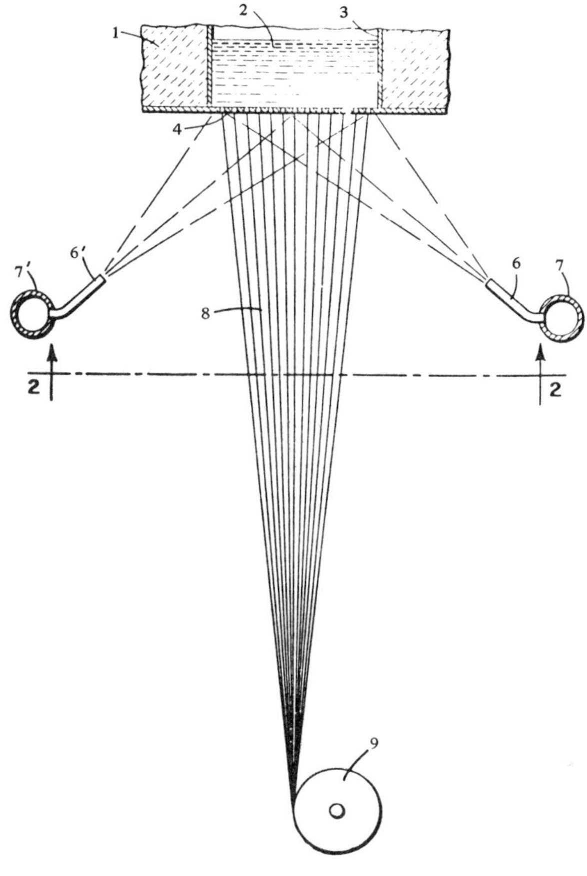

Source: U.S. Patent 4,033,742

At this point, the air flow may be reduced somewhat. The reduced air flow will permit the viscosity of the molten glass in the isolated flood areas to decrease and become more fluid. As a result, these floods will tend to flow into adjacent cones from which fibers are being formed. At about the same time the air flow is first reduced, the temperature of the orifice plate is increased slowly and the increase in plate temperature continues until normal operating temperatures and viscosities are achieved.

When the flooded areas spread to adjacent fiber-forming cones the cooling air is again increased. This increase in cooling air offsets the effect on the fiber-forming cones of the increasing plate temperature and also provides sufficient cooling to increase the viscosity of the fiber-forming cones so that enough tension can be applied to the fibers drawn from those cones to cause the adjacent flooded areas to fiberize.

In order to clear the remaining localized flooded areas, the air may be increased and decreased again with a typical cycle (i.e., one air increase to the next) taking from about 15 to about 20 seconds, although it should be understood that the time frequency may vary somewhat depending upon operating conditions. It can be expected that not all floods will be cleared with the first variation of air flow and that additional alternating increases and decreases in air flow can be employed. In the event a few small isolated flooded areas remain, a hand air lance may be employed to direct cooling air selectively to those areas as described below.

The above procedure has provided 80 to 90% clearing for a 2,000 hole bushing in a period of only about one and one-half minutes. Once this condition has been achieved, bulk air, including the bulk air configuration described herein, can be employed to maintain the running mode. A hand air lance which permits a stream of cooling air to be directed to specific areas can be used to separate the remaining few flooded areas after the bulk air is turned on.

While the above description has been directed to the startup of an orifice plate, essentially the same procedure can be employed if an operating orifice for some reason floods. In this latter instance, the temperature of the orifice plate is simply dropped 25° to about 150°C below operating temperatures and the above steps are repeated. By turning the operating air off after a flood begins to occur, the complete flooding desirable for restarting can be accelerated.

Example: A molten bath of E glass having a bulk temperature of about 2300°F was established over an orifice plate made of platinum and containing 2,068 orifices having a diameter of 0.052 inch and spaced centers that varied from 0.065 to 0.080 inch. The orifice area was 21 rows wide and occupied an area 1.65 inches wide and 8.3 inches long.

A clearing manifold was mounted on each of the long sides of the orifice area. Each manifold contained 16 nozzles having about a 0.063 inch inside diameter. Pairs of nozzles were mounted on 0.5 inch centers with 0.63 inch spacing between pairs to accommodate reinforcing ribs in the orifice plate. The nozzles were at an angle of about 45° to the orifice plate and the nozzle outlets were about 3 inches from the center of the plate. The centerline of the air stream of the nozzles was aimed at about the center of the orifice area in the orifice plate.

The orifice plate was allowed to flood and the temperature of the plate was reduced about 50°C (a reduction of 50° to 80°F measured about ½ inch up on a flange which defined a sidewall of the reservoir). The cooling air was employed at a manifold pressure of 4 to 12 psig.

The matrix of glass was pulled slowly from the plate and the air was cycled at about 15 second intervals as the orifice plate temperature was slowly increased to its normal operating temperature of about 2265°F. The matrix and then the fibers were initially hand pulled from the orifice plate until operating orifice plate temperature was reached and, thereafter, the fibers were pulled by a pull roll. About 80% separation was achieved after only about 3 cycles of the clearing air and after only about 1½ minutes. At that point bulk air was introduced onto the plate and the remaining flooded area were cleared with a hand air lance within about another 2½ minutes. The bulk air was introduced from 12¼ i.d. nozzles on 0.7 inch centers mounted 9 inches below the plate and at an angle of about 80° to the plate. The manifold pressure for the bulk air was about 1.5 pounds.

In clearing floods on numerous occasions, the air was often cycled from 2 to 4 times to provide 80% clearing of floods. On other occasions, the floods were entirely cleared with the auxiliary cooling system.

Laval Nozzle Generating Supersonic Flow

A.H. Walz and P.-J. Hahn; U.S. Patent 4,001,357; January 4, 1977 disclose a process for the manufacture of fibers from fusible materials and particularly materials of low viscosity. The melt and gaseous flow medium is passed through a Laval nozzle at a supersonic speed. The nozzle and speed are so selected as to cause an abrupt temperature drop in the flow medium within the nozzle which causes the melt to solidify into fibers. A Laval nozzle, known per se, is particularly suitable as a flow device.

If a pressure ratio p_3/p_1 of about 0.53 is set with respect to the pressure p_1 at the inlet side where the velocity, $U_1 = 0$ and where the medium to be separated in fibers is still being carried in liquid, single thread form, and to the pressure p_3 at the outlet side, then the speed of flow U_2 at the narrowest location **2** (see Figure 6.9) will be exactly equal to the speed of sound a.

The temperature ratio, with k = 1.4 for air:

$$T_2/T_1 = (p_2/p_1)^{(k-1)/k} = (p_2/p_1)^{0.286} = 0.83$$

The flow medium may be brought to supersonic speed behind this most narrow location if: (1) the pressure ration p_3/p_1 is less than 0.53; and (2) the cross section of the Laval nozzle behind this most narrow location will again be enlarged. As is known, U will always remain equal to a at the most narrow location, regardless of how small a value of p_3/p_1 is being selected.

The cross section F(x) behind the most narrow location of the Laval nozzle in accordance with the known laws of gas dynamics determines the speed U, the pressure p and, in accordance with the laws of thermodynamics, also the temperature T. One obtains, therefore, the temperature ratio $T(x)/T_2$ as the following relation:

$$F(x)/F_2 = f[U(x)/a] = f[M(x)]$$

wherein M = local Mach number, M = U/a.

In particular, when k = 1.4 for air:

$$T(x)/T_1 = [p(x)/p_1]^{0.286}$$

By means of a functional cross section F(x) fitted to the temperature, one may achieve a temperature drop of the blowing medium along a predetermined short segment Δx and within an associated short time:

$$\Delta t = \Delta x/U_{mean} \approx 2\Delta x/(U_2 + U_3)$$

The inlet orifice in the x-direction of a Laval nozzle **10** is denoted by **1** in Figure 6.9a, the location of the most narrow cross section is referenced by **2**, the range of the temperature drop by **3** and the outlet orifice by **3'**. The function of the cross section by means of which the temperature shock may be achieved is denoted by F(x).

In the embodiment shown in Figure 6.9b, melting occurs by means of directly heating a noble metal crucible consisting of coaxial cylinders **5'** and **5"** and of a crucible bottom **6**. The ohmic heat of the coaxial cylinder walls **5'** and **5"** traversed by electrical current and of the crucible bottom **6** provided with melt outlet orifices **7** is supplied to melt material **8**. The latter may be a ceramic substance in the embodiment shown.

A nozzle plate **9** is mounted at a spacing from and below the melt outlet orifices **7**, each of which is associated with a Laval nozzle in the nozzle plate **9**.

Nozzle plate **9** is the upper closure of an evacuation chamber **10'** operating in concert with an injector **11** which is provided with an annular chamber **12** into which lead drive medium conduits supplying a pressure medium such as compressed air from a compressed air source. Injector **11** is equipped with a Laval nozzle **16** shaped similar to an annular slit.

The melt jet is sucked through the Laval nozzles of the process, which are symmetric with respect to rotation, and in the process are markedly accelerated. After passing through the phase of a fiber bundle or melt film, the state of the fiber will be frozen by means of the temperature shock of the process.

Generation of an under-pressure p_3' in an under-pressure chamber **10'** common to all individual Laval nozzles **10** in Figure 6.9a is obtained by means of the supersonic injector **11**. With k = 1.4 for air:

$$T_3/T_1 = (p_3/p_1)^{0.286} \quad \text{and} \quad T_{3'}/T_1 = (p_{3'}/p_1)^{0.286}$$

Figure 6.9: Laval Nozzle Generating Supersonic Flow

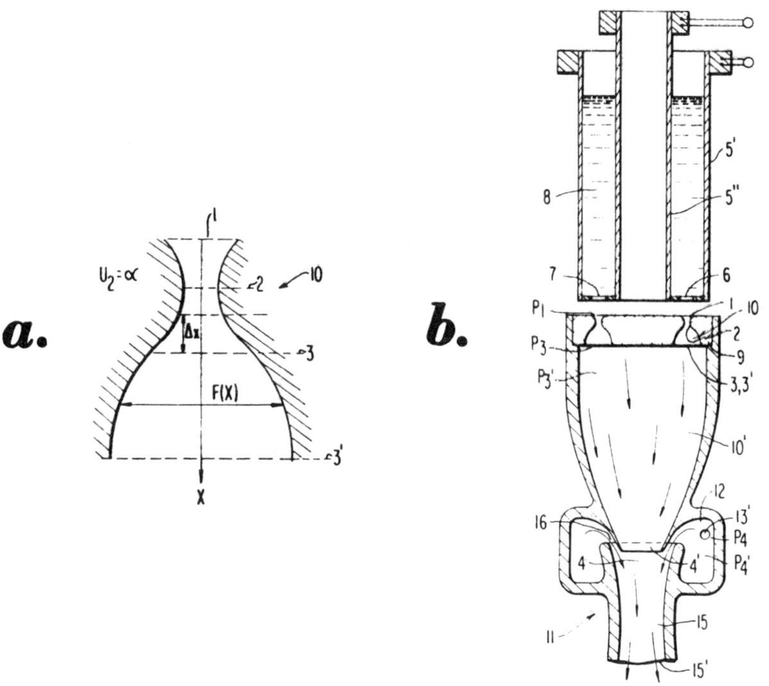

(a) Laval nozzle
(b) Device for manufacture of fibers by means of Laval nozzle

Source: U.S. Patent 4,001,357

Pressure p_4' will be equal to or less than p_3' in the range of supersonic decrease **4'** between the drive medium and the medium evacuated from the individual Laval nozzles **10.** To that end, the Laval nozzle **16** shaped nearly like an annular slot in accordance with the process will be such as regards cross section that it will correspond to the same gas dynamic relations known per se that p_4' will be less than p_3'. Furthermore, the pressure of the drive medium p_4 will be selected of such high value that the energy losses will be covered in mixing zone **4'.**

For a pressure p_4 = 6 atm = 6 kg/cm^2, a Mach number of about 1.7 may be achieved at location **3'.**

A subsequent subsonic diffuser **15** by means of conventional cross-sectional enlargement allows appreciable recovery of kinetic flow energy, so that the drive pressure p_4 to be applied remains relatively low and hence the overall efficiency of the device of the process becomes very advantageous. By means of sufficient length and enlargement of the cross section in accordance with the known laws of flow mechanics, one may raise p_{15}' at the outlet diffuser **15** to about atmospheric pressure p_1 (see U.S. Patent 3,283,039).

Example: *Rockwool* — The raw material was of the following composition:

	% by Wt
SiO_2	51.7
Al_2O_3	8.0
$Fe_2O_3 + FeO$	3.3
TiO_2	0.6
CaO	29.9
MgO	4.0
$Na_2O + K_2O$	3
SO_3	0.2

The thermoplastic range extends from about 800° to about 1450°C. The melting temperature in the Pt-Rh crucible (80 - 20 = 80% Pt, 20% Rh) ranges from about 1350° to 1450°C.

A fiber product is produced with a fiber diameter of about 0.5 to 3 x 10^{-3} mm and of a length of about 5 to 20 mm.

The table below lists the pressures, speeds and temperatures at the individual locations of the device.

	Location in Figures 6.9a and 6.9b						
	1	2	3	3'	4	4'	15
p, kg/cm^2 (atm)	1.0	0.528	0.300	0.200	6.0	p_4'/p_4 = 0.15	~1.0
U, m/s	0	340	412	472	0	–	few
M, U/a	0	1.0	1.4	1.7	0	1.9	~0
T, °K	300	249	215	189	300	–	300

Note: $\Delta x = x_3 - x_2 = 5$ mm = length of the path of appreciable separation into fibers (selected)

$\Delta t = 2x/(U_2 + U_3) = 5 \times 10^{-3}/376 = 1.33 \times 10^{-6}$ = seconds (time of separation into fibers)

$\Delta T = T_2 - T_3 = 34°C$ = temperature (slight for rockwool on account of large thermoplastic temperature range)

TEXTURED YARNS

Novelty yarns or strands are produced by a variety of processes in the textile field and find utility for many specific textile purposes, for example, in the manufacture of draperies, industrial cloths used as reinforcement for resin articles and the like. Many processes have been designed, especially in processing glass fiber strands, to produce decorative effects on textile strands to render them useful in providing bulk for cloth to be woven for decorative purposes as well as cloth woven for industrial purposes.

Slubby Yarn

A process is disclosed by *H.W. Barch and A.G. Bohy; U.S. Patents 4,096,685; June 27, 1978; 4,099,307; July 11, 1978; 4,100,723; July 18, 1978; and 4,100,726; July 18, 1978; all assigned to PPG Industries, Inc.* for preparing slubby strands in which a textile strand is passed either over a surface or through a slubbing tool.

The surface, if employed, may include a plurality of fluid passages positioned therein. As the strand passes over the surface or through the slubbing tool, high pressure fluid is introduced onto the surface of and through the strand. The high pressure fluid passes through the strand. This passage of fluid through the strand causes the strand to twist and bulk along its length to form a slub which is rapidly removed from the working surface of the slubbing surface or tool as a consolidated slub positioned on the strand. When a surface is employed, the surface may be a moving surface and when the surface includes a plurality of fluid passages, the passages may be vented at a point below the surface.

Figure 6.10a shows a schematic diagram of the operation of one embodiment of the process using conventional forming packages as the feed and wherein two glass strands are being consolidated into a single strand during their passage over the strand treating surface.

There is shown, positioned in a side-by-side relationship, two glass strand containing forming packages **1** and **2**. Each of these forming packages **1** and **2** have inserted therein a central pin **6** and **5**, respectively, which contains on its outer surface a circular ring member **7** and **8**, respectively. These ring members **7** and **8** are provided so that the strand can be ballooned out over the forming package for easy removal. The strands **3** and **4**, removed from the packages **1** and **2**, respectively, are consolidated into a single strand in an eyelet **9** and passed under an eyelet **10** for feeding to a rotating wheel **13**. Located on the surface **21** formed on the face of the wheel **13** are a plurality of gas passages **14** which are drilled into the body of the wheel member **13**. The surface **21** may be a curved, generally u-shaped, surface, as shown, or may be a relatively flat surface. Gas passages **14** communicate with the atmosphere surrounding the wheel member **13** through a plurality of holes contained on a circular plate member **15** affixed to the surface of the wheel member **13**.

The holes, **23, 24, 25, 26, 27, 28, 29** and **30** shown in Figure 6.10b on the plate member **15** are exit holes of varying diameter which are placed over the opening of the gas passages **14** and vent the passage **14** to the atmosphere. By adjusting the position of circular plate **15**, any of the holes **23** through **30** can be aligned with the passage **14**.

Figure 6.10: Production of Slubby Yarn

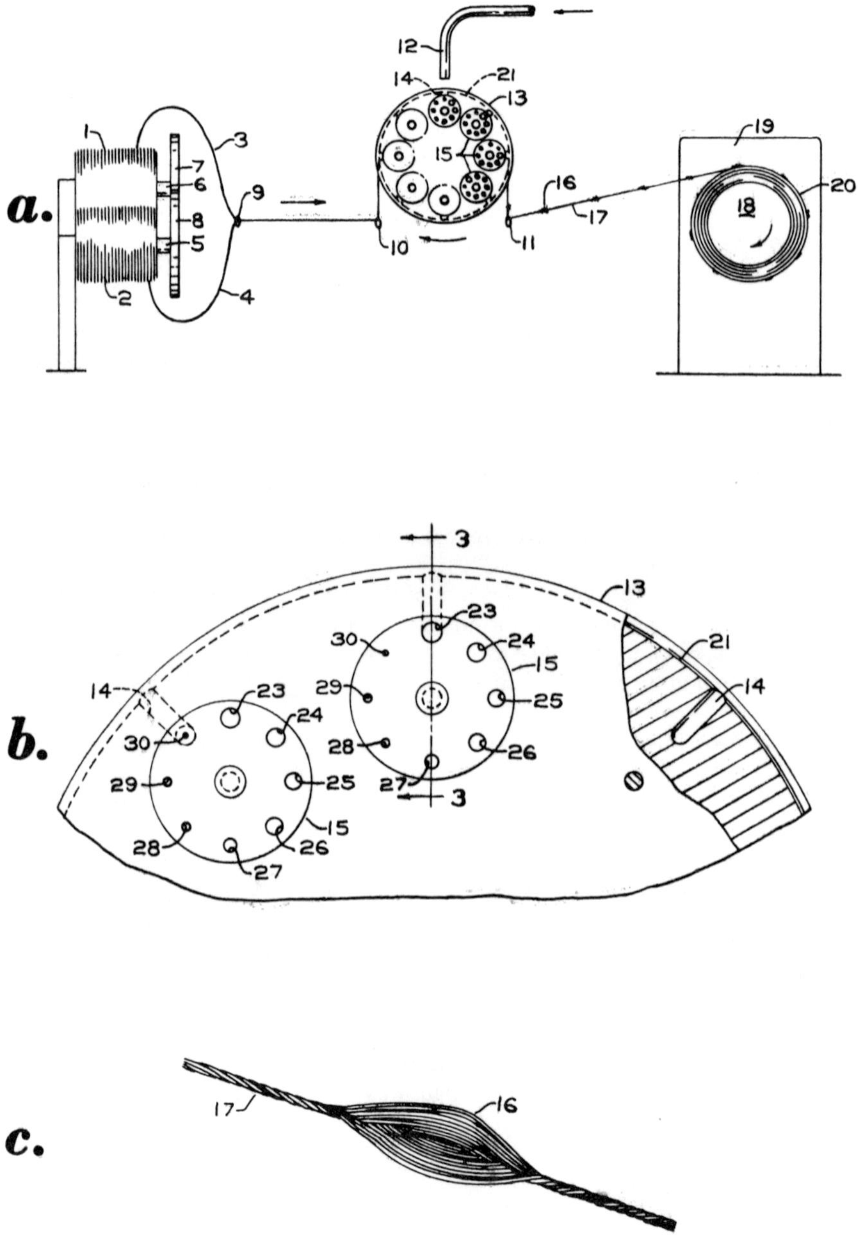

(continued)

Figure 6.10: (continued)

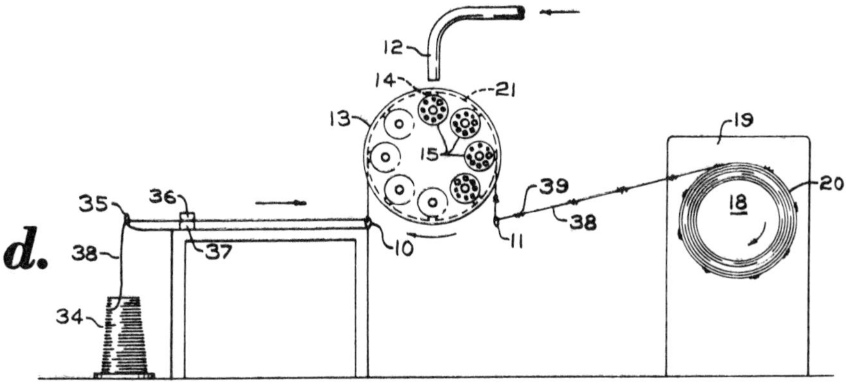

(a) Schematic diagram
(b) Enlarged view of wheel **13** of Figure 6.10a, partially broken away
(c) Glass strand produced by the process
(d) Schematic diagram of an alternate embodiment

Source: U.S. Patent 4,096,685

If desired, the solid plate **15** can be placed over any of the passages to thus close it to the atmosphere. Thus, as will be readily appreciated, the venting of the gas passages **14** to the atmosphere can be readily controlled. Adjustment of the venting provides a varying gas flow through the passage **14.**

As shown in Figure 6.10c, the process produces a slub **16** on strand **17.** The slub **16** is characterized by being interposed between counterdirectionally twisted strand on either side thereof thus rendering the slub physically stable during later processing. It is also characteristic of the slub **16** produced in accordance with this process that very few breaks occur on the strands. This is particularly useful in the production of fiber glass strands, since, due to their lack of elasticity, most texturizing systems utilized in the production of fiber glass strands result in numerous breaks of filaments and a consequent significant reduction in the tensile strength of the strand produced.

In an alternative method of producing strand containing slubs with particular reference to Figure 6.10d, a single end **38** is utilized to produce fiber glass strand having intermittent slubs **39** produced along the length thereof. In this embodiment, the single strand **38** is removed from a bobbin **34,** passed through eyelet **35** and through tension control members **36** and **37.** The tension control members are typically polished plates **36** and **37** through which the strand **38** is passed. The tension is normally controlled by varying the weight of the top plate

36 or by placing more than one plate **36** over the plate **37**. The strand **38** is then passed through eyelet **10** and over the wheel member **13** which is identical to the wheel of Figures 6.10a and 6.10b. Appropriate adjustments are made in the wheel member **13** to place over the apertures of the gas passages **14** located in the sidewall of the wheel **13** the appropriate hole sizes. This is accomplished by rotating device **15** to line up the appropriate hole therein with the exit of passage **14**. In this manner, slubs are produced from a single strand **38** in the same manner, as they are produced on the double strand shown in Figure 6.10a and again the strand produced characteristically is counterdirectionally twisted on either end of the slub **39** thus produced.

Example: In this example the equipment utilized was that generally depicted in Figure 6.10d and the runs will be described with reference to that figure. A series of runs were made with single end yarns **38** supplied from a bobbin **34**. The yarns **38** were drawn by the winder **18** from the bobbin **34** and across the wheel **13** prior to being wound into package **20**. The speed of travel of the yarn across the wheel **13** was varied from 244 feet per minute to 2,250 feet per minute (74.4 to 658.8 meters per minute). The air pressure during the run was 30 to 40 pounds per square inch gauge (206,842 to 275,790 pascal). The air jet had a nozzle having an opening of the "dog bone" pattern. The wheel **13** was provided with cylindrical air passages **14**, each of which had a diameter of 0.196 inch (4.98 millimeters). The exhaust control buttons **15** were set so that each air passage **14** had an exhaust hole 0.0086 inch (0.218 millimeter) in diameter. The yarns collected during these runs were tested for strength using the feed yarn as the standard and were also visually examined to determine slub formation.

Break strength of the slubbed yarns ranged from 3.88 to 22.01 pounds, compared to 3.34 to 26.24 pounds for comparable runs with no slubs.

Randomly Slubbed Yarn

A method and apparatus are disclosed by *A. T. McDonald; U.S. Patent 4,096,687; June 27, 1978; assigned to PPG Industries, Inc.* for producing randomly slubbed textile yarns. The method involves passing the yarn through a zone of fluid turbulence having a pair of counterdirectional fluid streams maintained under different fluid pressures.

The yarn passing between these counterdirectional fluid streams is alternately treated in opposing directions, with slubs being formed at the null points between the opposite directional treatments. Due to the uneven fluid pressures employed, the length of treatment in each direction along the yarn is not consistent, thus leading to slubs at various points along the length of the yarn. A fluid jet is also disclosed for accomplishing this result. The fluid jet includes a pair of interconnected and overlapping passageways which form a chamber through which the yarn passes, with each passageway having one or more fluid inlets arranged to direct fluid circumferentially around it and with the fluid streams being counterdirectional to one another. The fluid inlets for each passageway are fed fluid at uneven or different fluid pressures to accomplish the random slubbing effect.

Referring to Figure 6.11, a fluid jet **1** is shown having a main body **2** and a pair of interconnected and overlapping passageways **3** and **4**. The passageways **3** and **4** form a chamber **10** which has a generally figure-eight shape. While the chambers **3** and **4** are illustrated as being round, the passageways **3** and **4** could take any desired shape, such as elliptical, oval, triangular and the like, with the result that the chamber **10** has a shape which resembles generally a figure-eight.

As can best be seen in Figures 6.11b and 6.11c, each of the passageways **3** and **4** are connected in fluid transfer relation with one or more fluid inlets **9** along their lengths. The fluid inlets **9** connected to chamber **3** receive treatment fluid, and preferably gaseous fluid, such as air, oxygen, nitrogen and the like, through fluid intake **5**, which is preferably connected to fluid chamber **11**, which is in turn connected to the fluid inlets **9**. Optionally, each fluid inlet **9** could be connected directly to the fluid intakes. Likewise, the fluid inlets **9** connected to passageway **4** receive fluid from intake **7** which is preferably connected in fluid transfer relation to fluid chamber **12** and the inlets **9**.

The jet **1** may be formed of numerous materials, including plastic, ceramic, glass and metal. Preferably, the jet **1** is constructed of a metal, such as brass or stainless steel.

The jet **1** may be formed as a single piece, with the chamber **10** being formed by a pair of overlapped drillings into the walls of the jet **1**. The fluid inlets **9** and common fluid chambers or headers **11** and **12** are then drilled into the body **2**, with plugs employed to seal the inlets **9** and fluid chambers **11** and **12** at their ends.

Figure 6.11: Apparatus for Production of Randomly Slubbed Yarn

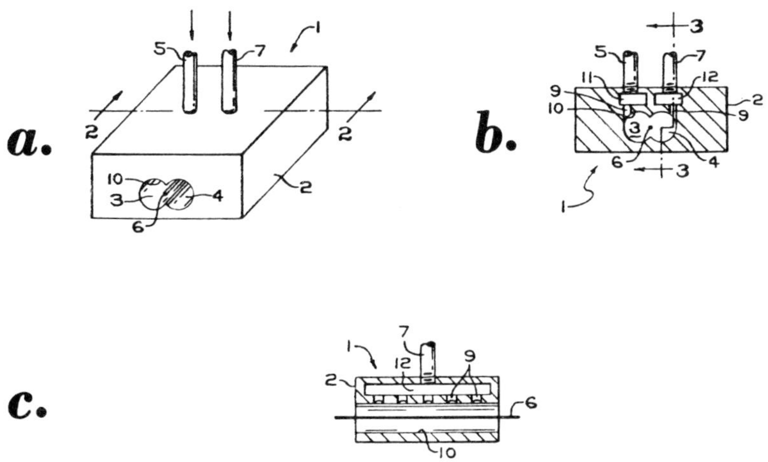

(continued)

Figure 6.11: (continued)

d.

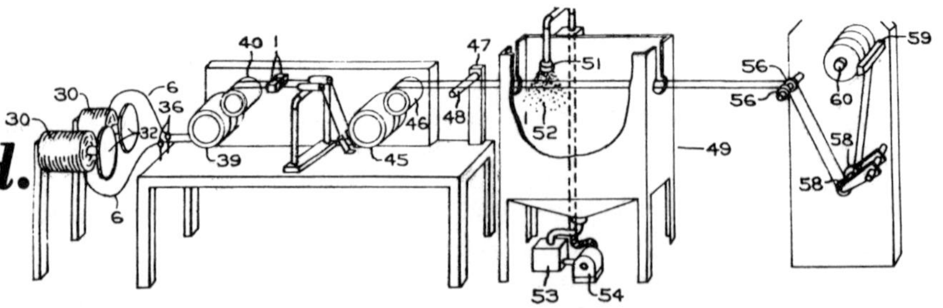

e.

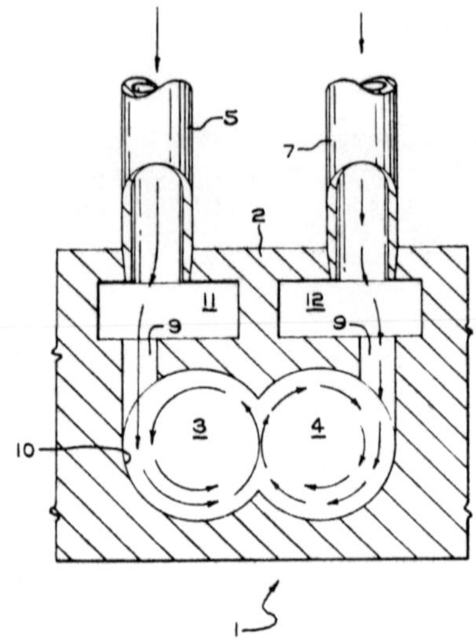

(a) Perspective view of the fluid jet
(b) Cross-sectional view of the jet taken through line **2–2** of Figure 6.11a
(c) Cross-sectional view of the jet taken through line **3–3** of Figure 6.11b
(d) Diagrammatic representation of a slub yarn-forming operation
(e) Expanded view of the chamber and fluid inlets, illustrating the fluid flow through the jet

Source: U.S. Patent 4,096,687

The jet **1** may also be formed of two or more sections. Thus, for example, two sections may be machined to be connected at the overlap and interconnection of the passageways **3** and **4** or at the connection of the fluid inlets **9** to the passageways **3** and **4**, or a combination of these may also be employed.

As previously mentioned, the chamber **10** is formed by the intersecting, overlapping passageways **3** and **4**. The chamber **10** is thus formed within the walls of the body **2**. At the intersection and overlap of the passageways **3** and **4**, there is a constriction in the chamber **10**. Thus, the height of the chamber **10** at the intersection of the passageways **3** and **4** is less than the height at the center of the passageways **3** and **4**. Also, the cross-sectional area of the intersection is smaller than the balance of either passageway **3** or **4**.

Again looking at the chamber **10**, the chamber may be formed of four arcs, two of which are approximately colinear and the other two arcs being mirro images of the first two. The central portions of the approximately colinear arcs intersect and overlap one another to form the constriction previously described.

As can best be seen in Figure 6.11e, the fluid inlets **9** are constructed and arranged to direct the treating fluid circumferentially around the passageways **3** and **4** respectively. Thus, the treating fluid within passageway **3** flows in a counterclockwise direction, while the treating fluid in passageway **4** flows in a clockwise direction. Yarn **6** passes through the chamber **10** and is alternately treated by the fluid within passageways **3** and **4**. In each passageway, the treating fluid entangles the filaments of the yarn with one another and imputes a false twist to the yarn in the direction of the fluid flow. At the null points, i.e., the points where the yarn **6** passes from passageway **3** to passageway **4** and the direction of twist reverses, puffs or slubs are formed in the yarn **6**.

To randomize the slubs in the yarn, the treatment fluid from intakes **5** and **7** are fed to the jet **1** under unequal pressures. Thus, the intake pressure for each of the intakes **5** and **7** may range from about 5 to 80 psig (0.35 to 5.6 kilograms per square centimeter gauge), but the pressures are not equal. Due to the pressure differential within the passageways **3** and **4**, the yarn **6** passing between the passageways **3** and **4** does not spend equal amounts of time in the passageways **3** and **4**, but rather remains for a longer portion of time in the passageway under the lower pressure. Because of this, the slubs in the yarn **6** are not formed at regular intervals, as they would be if the yarns **6** followed the generally figure-eight shape of the chamber **10** regularly, but rather are formed at irregular intervals.

Control of the randomness of the slubbing and degree of slubbing can be accomplished by varying the pressure differential and absolute pressures of the fluid intakes **5** and **7**, the speed of the yarn **6** passing through the jet **1**, and the tension on the yarn **6**.

Figure 6.11d illustrates an apparatus which may be employed to produce a slubbed yarn in accordance with this process. The yarns **6** are removed from forming packages **30**. The yarns **6** pass over the exterior of wheels **32** so that the yarns **6** can be removed from the outside of the packages without any snagging. The yarns **6** pass through yarn guides **36** and over the surface of a drive roll **39** coupled for rotation to a suitable drive source (not shown) and subsequently over a nip roller **40** journaled for rotation with its outer cylindrical surface in frictional contact with the outer cylindrical surface of roll **39**. Yarns **6** are then passed from the surface of the nip roll **40** and through the fluid jets **1** as shown in Figures 6.11a, 6.11b and 6.11c.

After emerging from the fluid jets **1**, yarns **6** pass through strand guides and over drive roll **45** which is coupled to a power source (not shown) for rotation. The yarns **6** pass from roll **45** over the surface of a nip roll **46** which is journaled for rotation with its outer cylindrical surface in frictional contact with the outer cylindrical surface of roll **45**. Yarns **6** are then passed over guide bar **48** mounted on a bracket **47** and the yarns **6** are passed under a binder spray head **51** which applies binder **52** to the yarns. Binder **52** is pumped to the spray head **51** by a pump **54** through pipe **50** from a binder reservoir **53**. Excess binder is collected continuously in reservoir **53** by a suitable drain arrangement in the bottom of the binder applicator zone.

The binder used can be any desired composition consistency and viscosity so long as it can be applied through the spray head **51**. Thus, binders containing starches, oils, resin, hot metals or solvent-type materials and the like including emulsions, suspensions, dilutions and the like can be utilized. It is not a requirement of the process that this binder be applied to the yarns **6**. However, it has been found that such aftertreatment of the yarns **6** helps set the slubs in the yarns **6** and can provide other beneficial effects to the yarns **6**.

Yarns **6** are passed to the winding operation after binder **52** is applied thereto by passing them over rolls **56**. The yarns **6** are then passed over tension rolls **58** which coact with a motor (not shown) driving mandrel **60** to maintain constant tension on the yarns **6** during winding and maintain constant takeup winding speed of the yarns **6**. The yarns **6** are wound in two packages on winder **60** which is equipped with a roller bail **59** to maintain the packages smooth on the surface and square ended. The texturizing of and binder application to the yarn **6** is more fully described in U.S. 3,730,137, and the winder **60** employed with the tension rolls **58** is the winder more fully described in U.S. Patent 3,814,339.

As previously mentioned, the degree of slubbiness in the yarn is controlled by the speed of passage of the yarn through the fluid jet **1**, the fluid pressures in the jet **1** and the tension on the yarn **6**. Fluid pressure differentials are controlled by controlling the fluid input through intake lines **5** and **7**.

As illustrated in Figure 6.11d, speed of passage of the yarns **6** through the jet **1** is controlled by controlling the speed of the mandrel **60**. The tension on the yarns **6** as they pass through the jets **1** is controlled by the relative speed of drive rolls **39** and **45**. In operation, drive roll **39** is operated at a speed somewhat in excess of the speed of drive roll **45**. Thus, the speed of the yarns **6** entering the fluid jets **1** is slightly in excess of the speed of exit from the jets **1**. This speed differential between drive rolls **39** and **45** is expressed as percent overfeed. This percentage of overfeed generally ranges from 1 to 10% or higher. At higher overfeed percentages, the tension in the yarns **6** as they pass through the jets **1** is essentially zero. This will produce larger and more frequent slubs in the yarn **6**. For different novelty effects, the percentage of overfeed can be adjusted to produce slubs which occur somewhat less often and which are somewhat smaller.

Textured Fiber Glass Utilizing Yarn from Forming Package

A process is described by *W.W. Drummond; U.S. Patent 4,003,111; January 18, 1977; assigned to PPG Industries, Inc.* for producing texturized textile yarns having improved characteristics and in which forming supply packages may be used as the feed yarn. The process has particular utility in the production of fiber glass yarns and produces yarns having a rounded appearance with good texture on the outer surfaces.

Yarn is removed from a dried forming package, passed through a fluid treating device which in its interior subjects the yarn to considerable agitation by circumferential whirling fluid. The whirling fluid acts to disturb the filaments within the strand and breaks any adherence between filaments that might exist. It also provides the yarn with a false twist between its exit from the fluid treating area and its entrance into the texturizing jets utilized to apply texture to the yarns.

Referring to Figures 6.12a and 6.12b, there is shown a blower **10** which is provided with a central bore or passageway **13** formed by the walls of the blower **10**. An elongated slot **19** is provided to permit passage of strand **14** into the chamber **13** from the exterior of the blower **10**. A plurality of fluid inlets **17** are positioned in a row on the lip of the blower **10** and the inlets **17** are machined to direct all fluid exiting from the inlets to the underside of the chamber **13** positioned above the inlet **17**. A similar row of inlets **18** is provided at the back of the chamber **13** and the inlets are machined to direct fluid emanating from the inlets **18** against the surface of the back wall of chamber **13** towards the inlets **17** and in a path around the chamber surface which is perpendicular to the strand **14** passing longitudinally through the chamber **13**. Inlets **17** are fed from a common header **16** located on the interior of the blower **10**. Inlets **18** are fed from a common header or chamber **15** located on the interior of blower **10**.

These chambers **16** and **15** are typically provided by boring a hole along the longitudinal axis of the blower **10**, boring holes **17** and **18** into the side of the chamber or header **16** and **15**, respectively, and sealing one end of the chambers **16** and **15** by brazing a plug therein. An inlet for fluid is provided on the open end of each chamber **16** and **15** so that fluid inlets **11** and **12** can be connected thereto to provide for the feeding of fluid under pressure to the chambers **16** and **15**, respectively. The strand **14** traveling through the device **10** is thus subjected to tangential contact with high velocity air as it travels through the blower. Typically, the central bore **13** of the blower **10** is of a diameter which is at least ten times the diameter of the strand passing through the bore.

Figure 6.12: Apparatus for Manufacture of Textured Fiber Glass

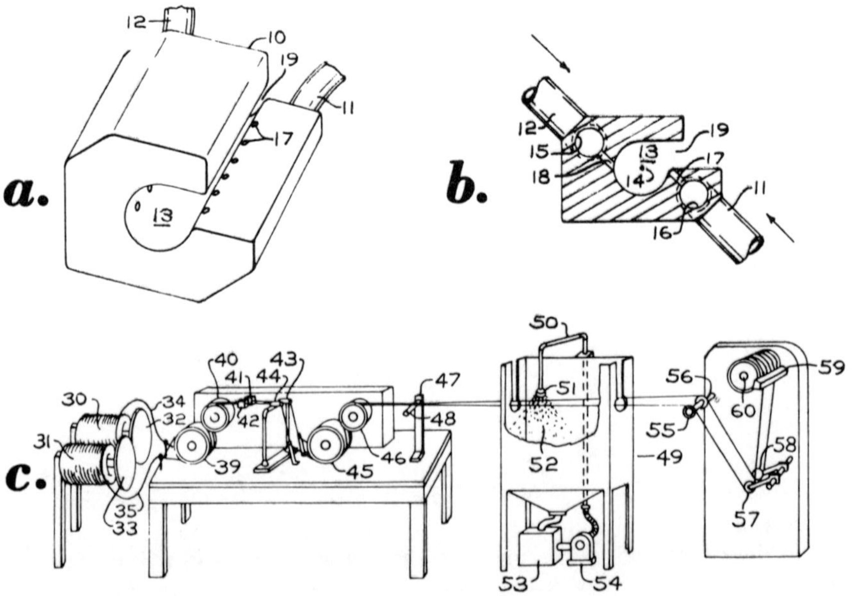

(a) Perspective view of one embodiment having a slotted strand entry and two rows of fluid inlets

(b) Cross section of Figure 6.12a showing the internal air chamber and the fluid inlet communications therewith

(c) Diagrammatic illustration of utilization of the blower of Figures 6.12a and 6.12b

Source: U.S. Patent 4,003,111

In the operation of the process as shown in Figure 6.12c, the feed supply strands or yarns **34** and **35** were supplied to the process from forming packages **30** and **31**. The forming packages **30** and **31** in the case of the treatment of fiber glass yarns are prepared by drawing molten glass from a molten glass source into a multiplicity of filaments in the conventional manner, applying a binder or size to the filaments, gathering the filaments into a strand and winding the strand at high speed on the surface of a winding machine which carries a forming package tube thereon. The forming packages resulting from this operation are normally dried to provide a forming package having a moisture content of less than 0.5% by weight. Forming packages having this weight percent of water or less thereon are thus utilized as the source of material for use in the process.

As shown in Figure 6.12c, the strands **34** and **35** are removed from the packages **30** and **31**, respectively. The strands **34** and **35** pass over the exterior of the wheel devices **32** and **33**, respectively so that the strand can be removed from the outside of the packages without any snagging. The strands **34** and **35** are then passed over the surface of a drive roll **39** coupled for rotation to a suitable drive source (not shown) and subsequently over a nip roller **40** journaled for rotation with its outer cylindrical surface in frictional contact with the outer cylindrical surface of a roll **39**. Strands **34** and **35** are then passed from the surface of the nip roll **40** through blowers **41** and **42**, respectively.

Air, the preferred fluid, is fed to these blowers through a suitable supply line (not shown) and the air is passed circumferentially around the interior cavity or channel running lengthwise of the blowers **41** and **42**. The air is fed from a pressurized air source (not shown) at about 20 to 80 pounds per square inch (1.406 to 5.624 kilograms per square centimeter) or more. The whirling air inside of the blowers **41** and **42** resulting from the air feed and its passage around the cavity of the blowers **41** and **42** through which strands **34** and **35** are passing tends to create in the strands a false twist leaving them in a more receptive state for the air texturizing that follows than that which is normally encountered by a twisted strand from a bobbin source being fed to a similar air texturizing system.

Strands **34** and **35** after emerging from the blowers **41** and **42** are passed through texturizing jets **43** and **44**, respectively. These jets are standard jets used to texturize yarn surfaces and are described in detail in U.S. Patents 2,783,609, 3,328,863 and 3,381,346.

After being texturized, strands **34** and **35** are passed over the roll **45** which is coupled to a power source for rotation. The yarns **34** and **35** pass from roll **45** over the surface of nip roll **46** which is journaled for rotation with its outer cylindrical surface in frictional contact with the outer cylindrical surface of roll **45**. Yarns **34** and **35** are then passed over guide bar **48** mounted on a bracket **47** and both yarns are passed under the binder spray head **51** which applied binder **52** to the strands. Binder **52** is pumped to the spray head **51** by a pump **54** through pipe **50** from a binder reservoir **53**. Excess binder is collected continuously in reservoir **53** by a suitable drain arrangement in the bottom of the binder applicator zone.

Strands **34** and **35** are passed to the winding operation after binder **52** is applied thereto by passing them over rolls **56** and **55**, respectively. The strands are then passed over tension rolls **58** and **57** which coact with the motor (not shown) driving mandrel **60** to maintain constant tension on the strands **34** and **35** during winding and maintain a constant takeup winding speed of the strand. The strands **34** and **35** are wound in two packages on winder **60** which is equipped with a roller bail **59** to maintain the packages smooth on the surface and square ended.

Continuous Yarn with Appearance of Spun Staple Yarn

W.W. Drummond; U.S. Patent 4,024,700; May 24, 1977; assigned to PPG Industries, Inc. discloses a bulky yarn having a low bulk density which comprises a spun ribbon of filaments or groups of filaments which may additionally be combined with a core strand. The spun continuous filaments form a plurality of loops which tangle with each other and with the core, if employed, to produce a continuous yarn having the appearance of a spun staple yarn.

Referring to Figure 6.13a, filaments or groups of filaments **10** are shown being passed over a ribbonizer **11** which aligns the filaments **10** into approximately parallel relationship with respect to each other prior to their being fed into the attenuator, generally designated by the numeral **12**. In the attenuator **12**, the approximately parallel filaments or groups of filaments, designated as **40**, are drawn around the belt **13** while still in parallel relationships, and are passed between the belts **13** and **14**. Belts **13** and **14** are driven by pulleys **15** and **16** connected through the shafts **17** and **18**, respectively. Shaft **18** is driven by a motor (not shown). The pulleys **15** and **16** drive the belts **13** and **14** at a radial velocity sufficient to propel the filaments **40** in parallel when the belts are reflexed over the pins **33** and **34** into a space at the velocity preferably from about 5,000 to 20,000 feet per minute (1,524 to 6,096 meters per minute). The filaments **40** are impinged upon the interior surface **19** of a rotatable, inverted open cone **20**.

In this embodiment of the process the rotatable inverted open cone **20** is shown as an open-end spindle, as known in the art. The open-end spindle **20**, operating at preferably about 10,000 to 50,000 revolutions per minute, is driven by means of a belt **21** in riding contact with a pulley **22** which is fixedly mounted to the shaft **23** of the open-end spindle **20**. The open-end spindle **20** is rotationally driven by the electric motor **24** at the aforementioned velocity with armature **9** and the pulley **8** fixedly mounted thereto. Thus, belt **21** rides on a pulley **8** and pulley **22**. As the attenuated filaments or groups of filaments **40** impinge on the surface of the inverted cone **19** or strand core yarn **25**, the filaments are displaced from their parallel form and are scattered into disarray as a plurality of connected loops within the open cone **20**.

As can be seen in Figure 6.13b, the loops of continuous filaments **40** are tangled with each other and continuously wrapped and tangled about the strand **25** which is traveling vertically through the cone **20**. Wrapping of the tangled filaments **40** about the strand **25** results in applying a twist to the yarn **28**. The strand core yarn **25** is traveling at a velocity of preferably about 200 to 2,000 feet per minute (609.6 to 6,096 meters per minute) as it passes through the cone **20**. The velocity of both the filaments **40** and the strand core yarn **25** may be adjusted to produce a variety of desired bulking characteristics and twists.

The continuous strand core yarn **25** is drawn from a source, not shown, such as a creel or a bobbin, over a guide rail **26** which guides the strand core yarn **25** through the aperture **27** of the open-end spindle **20**. After the bulky yarn **28** is formed, it is passed over a idler **29** which aids in the collection of the bulky yarn product. After passing over the idler **29**, the bulky yarn **28** in contacted with a traverse **30**, the traverse aiding the winding of the bulky yarn in random patterns on a rotating drum **31**.

Figure 6.13: Production of Bulky Continuous Yarn Including a Core

a.

b.

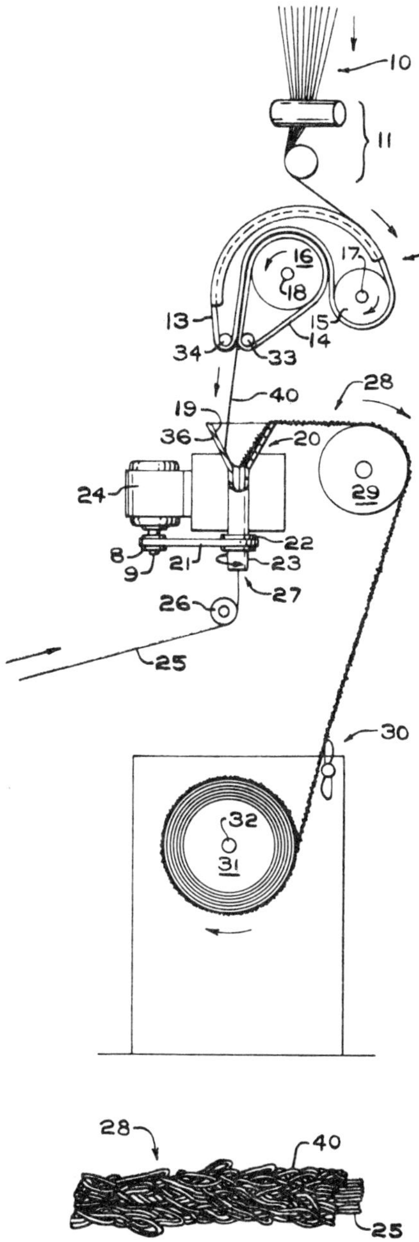

(a) Elevational view of apparatus, partly in section
(b) Bulky strand produced by apparatus (Figure 6.13a)

Source: U.S. Patent 4,024,700

The rotating drum **31** is driven by a shaft **32** connected to a suitable motor. The rotating drum **31** is driven at a velocity adequate to draw the strand core yarn **25** through the cone **20** at the desired speed.

Kinky Fibers from Glass

R.G. Russell; U.S. Patents 4,050,916; September 27, 1977; and 4,145,199; March 20, 1979; both assigned to the Owens-Corning Fiberglas Corporation discloses a method of and apparatus for processing heat-softenable materials, such as glass, to form fibers or filaments from streams of the material and particularly a method of and apparatus for effecting successive distortions, oscillations or vibrations of streams of the material at the region of formation or attenuation of the streams to fibers or filaments, and cooling or quenching the streams at the region of formation of the fibers or filaments for establishing successive kinks, bends or crimps in the attenuated fibers or filaments.

One embodiment of such means is illustrated in Figure 6.14. Disposed lengthwise of the stream feeder **12** is a manifold or member **40** having a hollow interior **41** which is supplied with a quenching fluid, such as water under pressure. Disposed lengthwise of the manifold **40** in spaced relation are tubes, pipes or tubular nozzles **44** joined with the manifold **40.** Figure 6.14c is a longitudinal sectional view of the tube, pipe or tubular nozzle **44**, the tubular nozzle **44** being of circular cylindrical cross section of substantially uniform diameter throughout its length.

As shown in Figure 6.14d, each of the pipes **44** is aligned with a transverse row of glass streams **20**. While Figures 6.14a and 6.14b illustrate six orificed projections **18**, each having an orifice **19** flowing a glass steam, the stream feeder may be provided with a larger number of transverse rows of orificed projections with twelve or more projections in each transverse row and upwards of fifteen or more rows.

The manifold **40** is connected with a supply of fluid, the manifold conveying the fluid to each of the pipes or tubular nozzles **44.** The fluid in the manifold **40** is under pressure and the fluid is projected as a coherent stream **50** from each of the tubes **44.** Water is preferred as a quenching fluid for the streams of glass but other suitable fluids may be used. Under certain operating conditions of the method as hereinafter described, the fibers or filaments attenuated from the glass streams may be vibrated or oscillated and the glass streams quenched by fluids such as air or other gases to form kinks or bends **35** in the fibers or filaments.

Arranged adjacent the opposite side region of the feeder **12** is a trough or receptacle **54** disposed lengthwise of the feeder **12** and positioned to receive and convey away the fluid after it has engaged the glass streams of a transverse row of streams. The receptacle or trough **54** is provided with an outlet pipe **56** which may be connected to a pump (not shown) for recirculating the fluid through the manifold **40** for reuse in the quenching operation.

Figure 6.14: Production of Kinky Fibers from Glass

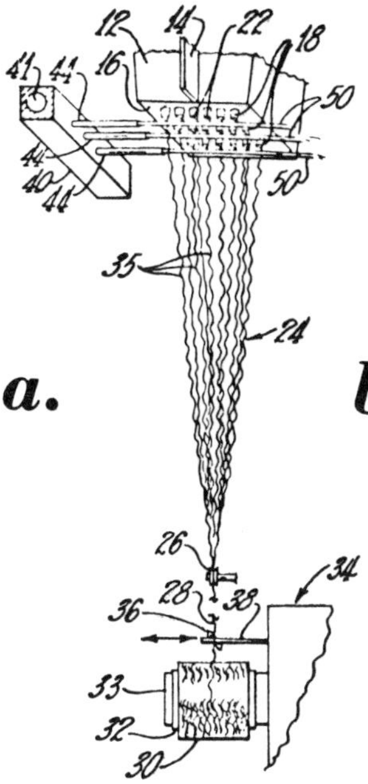

a.

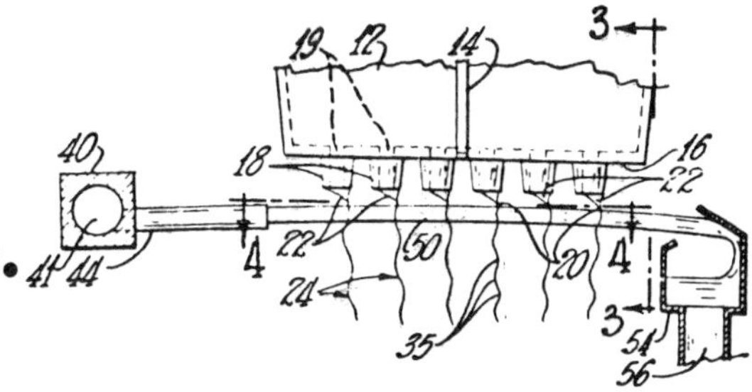

b.

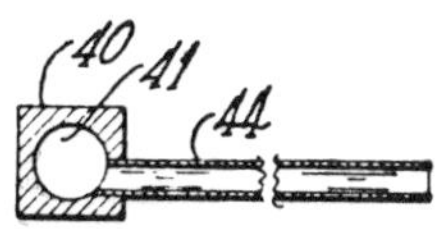

c.

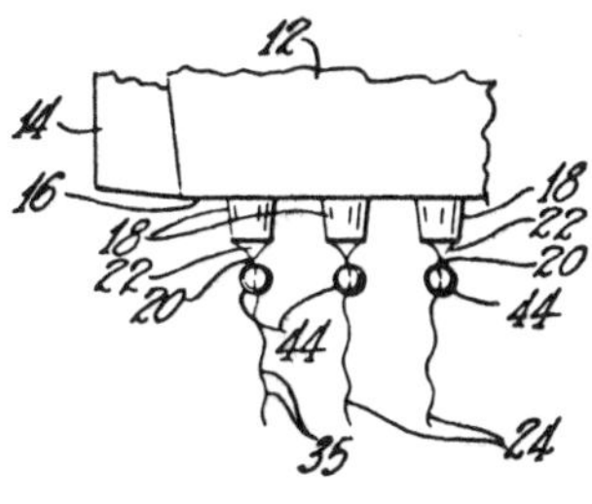

d.

e.

(continued)

Figure 6.14: (continued)

f.

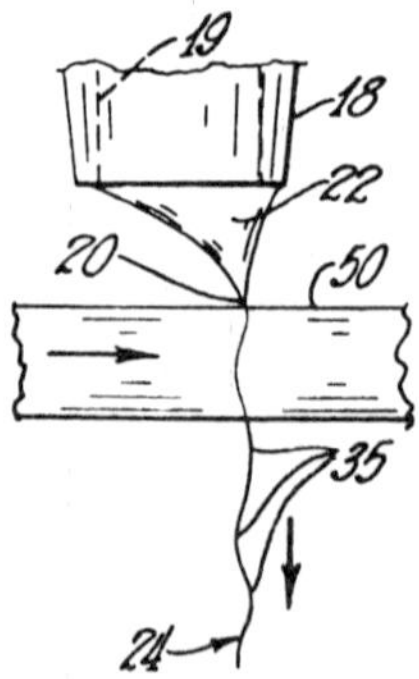

(a) Schematic isometric view of an arrangement for attenuating streams of glass from a feeder
(b) Elevational view illustrating the engagement of a stream of fluid with the streams of glass
(c) Longitudinal sectional view of a tubular nozzle for delivering a stream of fluid at the fiber-attenuating region
(d) Side view of the feeder
(e) Fragmentary sectional view taken substantially on line **4–4** of Figure 6.14b
(f) Enlarged view illustrating a stream of fluid impinging a stream of heat-softened glass

Source: U.S. Patent 4,050,916

It is desirable that the streams of water **50** or other quenching fluid be of sufficient velocity so that the streams **50** of the water impinge against the streams **20** of glass adjacent the tips of the cones **22** to distort, vibrate or oscillate the glass streams as shown in Figures 6.14b and 6.14f from their otherwise normal vertical paths of descent from the axes of the passageways or orifices **19.** It is found that the size or diameter of the stream of water, where water is used as the quenching fluid, must be adequate to impinge against all of the glass streams of a row.

A coherent stream of quenching fluid, such as water, projected from a tube, pipe or tubular nozzle **44** of about one-eighth of an inch internal diameter and of a length of above five or six inches has been found to be satisfactory in engaging all of the glass streams of a row with the water being under proper pressure. It is found that water under pressure of from thirty pounds per square inch to about sixty pounds per square inch will provide an eighth inch diameter stream of water with sufficient velocity to engage the glass streams of a row and that the stream **50** of water has substantially a rectilinear coherent trajectory with a minimum of curvature as illustrated in Figure 6.14b.

It is found that a coherent or concentrated stream of fluid, such as water, delivered from the tubular nozzle **44** under sufficient pressure may be impinged against at least fourteen streams of glass in a row and satisfactorily quench all of the glass streams. If the water pressure providing the quenching stream **50** is too high, then the velocity impact of the water of the stream against the glass streams at the attenuating region tends to tear or break the glass streams and glass beads tend to form. The streams of glass **20** of a row are aligned whereby the fibers as they are attenuated are drawn through the central region of the stream **50** of fluid as illustrated in Figures 6.14d and 6.14e.

In the method of operation of the arrangement shown, the attenuated fibers or filaments are converged by a member or gathering shoe **26** into a linear group or strand **28** and the strand wound into a package on a rotating forming tube **32** mounted upon a motor driven mandrel **33** of the winder **34**. The traverse member **36** mounted on the bar **38** is rotated and reciprocated in a manner conventional in glass textile winding machines. The traverse or guide **36** is preferably of a type for effecting a crossing of the individual convolutions of strand **28** as the strand is wound into the package **30** to facilitate withdrawing the strand from the package without breakage of the strand. If it is desired to wind the strand **28** into a package without effecting a crossing of the convolutions of strand, a nonrotatable conventional strand guide may be mounted on the reciprocating bar **38** in lieu of the rotatable guide **36**.

After the initiation of the flow of the streams of glass from the orificed projections **18** of the feeder by rotation of the forming tube **32**, the coherent streams of quenching fluid **50**, such as water, are projected from the tubes **44** and the water of the streams impinged against the rows of glass streams **20**. The streams of water impinging the glass streams at or adjacent the tips of the cones **22** of glass distort the cones **22** of glass and the glass streams.

Figure 6.14f is a greatly enlarged view and purports to illustrate the formation of kinks or bends **35** in a fiber or filament **24** being attenuated from the tip region **20** of the cone of glass **22**. The stream **50** of fluid, such as water, moving in the direction of the arrow substantially normal to the general direction of movement of the fiber or filament being attenuated, impinges against the glass stream, the fiber or filament as it is attenuated being drawn through the stream **50** of fluid.

The kinky fibers of filaments formed of glasses of different coefficients of expansion have different characteristics of hand or feel. Kinky fibers or filaments formed of a glass composition of a comparatively high coefficient of expansion have an improved feel in a mass, that is, are quite fluffy as compared with the feel or less fluffiness of a mass of kinky fibers or filaments formed of a glass having a lower coefficient of expansion.

During an attenuating and quenching operation in forming fibers or filaments having successive kinks or bends therein, where the filaments are under tension by winding a strand or linear group of the filaments into a wound package or by advancing the filaments by pull wheels or other means, the kinks or bends **35** permanently formed in the attenuated filaments are not visible to the eye by reason of the tension of attenuation in the fibers or filaments straightening out the kinks or bends **35**. When the filaments are relaxed or not under tension, the kinks or bends **35** are distinctively visible and such kinky fibers or filaments in a mass formation have the feel or characteristic similar to the feel or characteristic of a mass of cotton fibers.

Twisted or Distorted Filament

According to *R.G. Russell; U.S. Patent 4,144,044; March 13, 1979; assigned to Owens-Corning Fiberglas Corporation,* a method and apparatus for forming a twisted or distorted filament are provided by flowing streams of molten material and moving a member located within the stream of molten material at the zone of attenuation sufficient to produce a textured filament.

As shown in Figures 6.15a and 6.15b, the fiber-forming assembly **10** consists of a feeder **12** having depending tips or projections **13** through which heat-softenable material **15** is delivered as streams of molten material **16**. From the distal ends **14** of the tips **13** to the solidified portion of the filament extends a region or zone otherwise known as the zone of attenuation of the filaments. The filaments **17** are attenuated by the action of the winder **18** as the filaments are wound into package **19**. The winder or means for attenuating the filaments can be of the conventional constant tension type which are commercially available.

To form a textured filament **17** having a twisted geometric shape the fiber-forming assembly **10** further consists of motors **21** which are fastened to support means **23**, which can be insulated to protect the motors. Rotatable members or shafts **25** are engaged with motors **21** and are adapted to extend through the molten material **15** and feeder **12** to a point beyond or below the distal end **14** of the projections **13**.

In operation the molten material **15** flows completely over the portion of rod **25** which extends through projection **13**. As the rods are moved or rotated the streams of molten material **16** are also agitated or rotated as they are being attenuated such that the twists or deformations formed in the softened material are retained in the solidified filaments.

As can be more readily seen from Figure 6.15b, the shafts **25** have a tapered section **27** which generally conforms to the interior of the projection **13** but is spaced therefrom sufficient to allow the molten material **15** to pass between the angular void formed therebetween. The tapered sections, or chisel pointed sections, **27** extend beyond or below the distal end of projection **13** at the zone of attenuation of the streams of molten material. As such the tapered sections **27** are located or positioned within the interior of the streams of molten material.

Figure 6.15: Apparatus for Forming Twisted or Distorted Filament

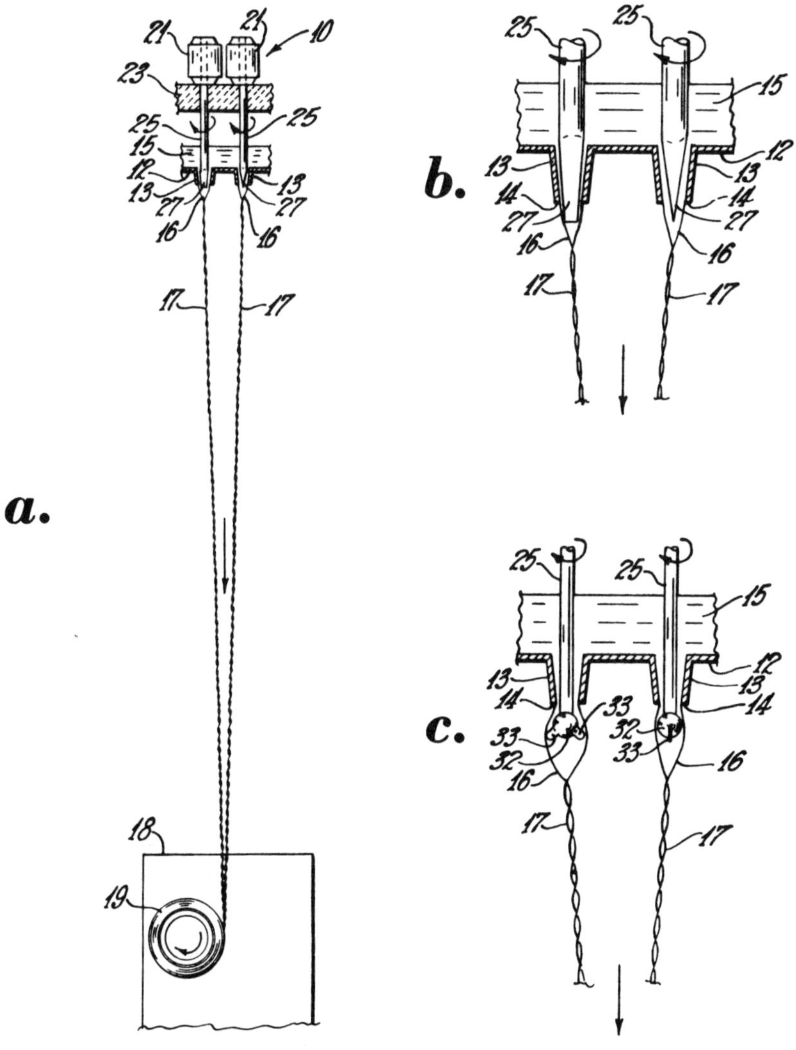

(a) Elevational view of a portion of a glass fiber-forming position including a winder for attenuating the filaments

(b) Enlarged cross-sectional elevational view of the feeder shown in Figure 6.15a

(c) Cross-sectional elevational view of an alternative embodiment of the agitating apparatus located at the zone of attenuation of the filaments

Source: U.S. Patent 4,144,044

It is believed that with such a system the surface tension of the molten material normally tending to form the material into its lowest energy state (i.e., a sphere) is offset by the centrifugal forces set up within the material due to the rotation of the shafts **25** within the streams at the zone of attenuation. As shown in Figure 6.15c, the rotatable or movable members are shafts **25** have an arcuate or spherical section **32** located below the distal end **14** of the projections **13** within the streams of molten material. Each of these spherical sections **32** has a plurality of extending elements or ears **33** to, among other things, magnify the effects of the rotation of the members **25** upon the streams of molten glass **16**.

It is to be understood that the relative speed at which the members **25** are rotated or moved with the respect to the speed at which the filaments are attenuated produces a filament having a varying degree of twists per unit of filament length.

Composite Texturized Yarn

G.E. Benson; U.S. Patent 4,058,968; November 22, 1977; assigned to Owens-Corning Fiberglas Corporation provides a method and apparatus for producing a composite bulked or texturized yarn having a generally uniformly bulked core yarn component of continuous filaments having groups of core yarn filaments extending outwardly from the core component in undulatory waves having substantially less filament density than the core component, and having a varyingly bulked effect yarn component of continuous filaments having groups of effect yarn filaments.

The effect yarn filaments extend outwardly from the effect yarn component in undulatory waves such that the groups of effect yarn filaments form slubs at regions along the length of the composite yarn. The filaments forming the slubs return to each respective nucleus portion of the segment on each side of the slub. Selected filaments within the waves have different lengths so that the waves are comprised of filaments based at various distances from the core and are spaced from one another within the wave. The product is further characterized by the absence of closed or crunodal loops and by the undulatory shape of the waves.

Referring to Figure 6.16a, a horizontal platform **14** supports a vertically extending housing **16** and a winder mechanism **18**. A yarn supply station **20** extends below platform **14**.

A pneumatic cylinder **22** and fluid jet **30** are fixed relative to housing **16**. The construction and operation of such fluid jets are well known in the art. A pair of double diameter, coaxial yarn feed and take-up rollers **36** and **37** are secured to rotatable shafts **38** and **39**, respectively. Shafts **38** and **39** are journaled for rotation on parallel axes in housing **16** by suitable bearings (not shown). The arrangement of feed and take-up rollers, fluid jet and winder mechanism is well known in the art.

Figure 6.16: Production of Composite Texturized Yarn

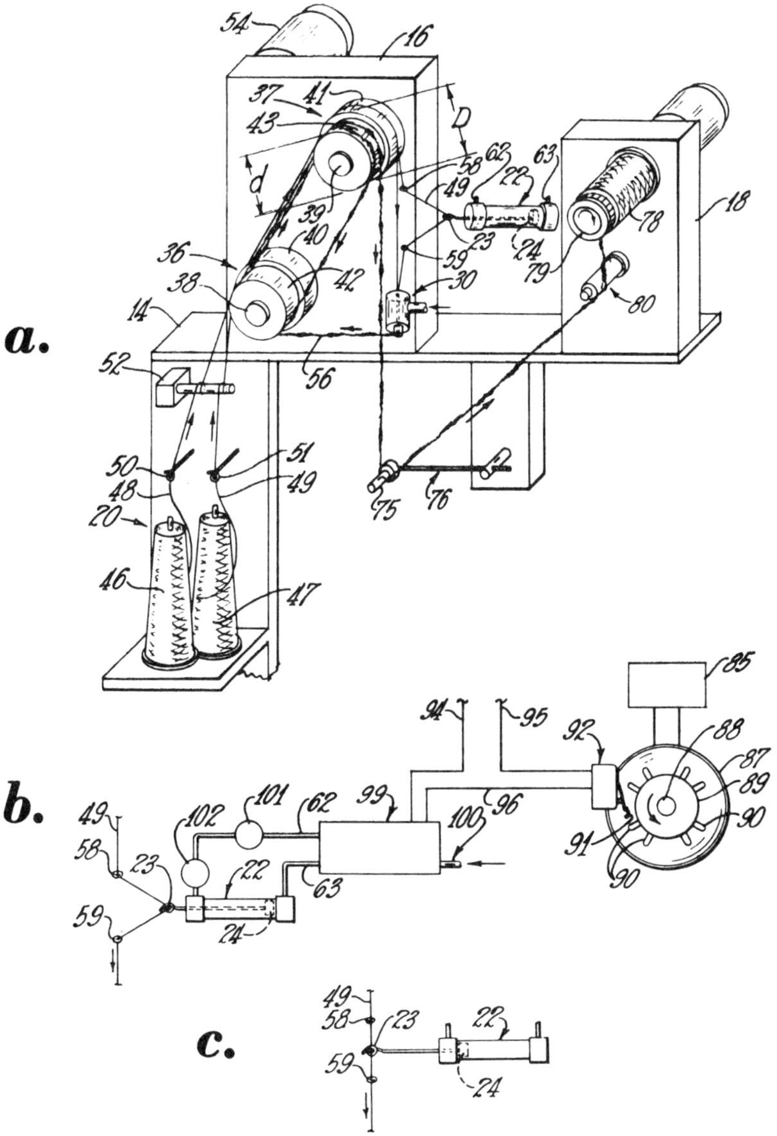

(a) Schematic view in perspective of the complete yarn bulking apparatus

(b) Schematic view of the effect yarn being deflected or accumulated before entering the texturizing jet and of a means for cycling or controlling the air cylinder and guide eye for deflecting or accumulating the effect yarn

(c) Schematic view showing the effect yarn in an undeflected state

Source: U.S. Patent 4,058,968

Rollers **36** and **37** can be rotated clockwise at the same speed, and assuming that each of the rollers **36** and **37** have identical dimensions, the surface or peripheral speeds of rollers **36** and **37** are the same for corresponding points thereon.

Each of the rollers **36** and **37** have first cylindrical sections **40** and **41**, respectively, having a diameter "D" and second coaxially secured cylindrical sections **42** and **43**, respectively, having a diameter "d" which is smaller than diameter "D". As is well known, when the first cylindrical sections **40** and **41** are used to feed a yarn or yarns to jet **30** and the second cylindrical sections **42** and **43** are used to withdraw the texturized or bulked yarn from jet **30** the yarn or yarns will be advanced toward the jet **30** at the same first velocity and the texturized or bulked yarn will be withdrawn from the jet **30** at a second velocity, which is less than the first velocity.

Such a difference in velocities produces a predetermined amount of overfeed; that is, because the first cylindrical sections **40** and **41** are coaxial with second cylindrical sections **42** and **43**, respectively, and are turned at the same angular speed, the peripheral speed of sections **40** and **41**, due to the larger diameter, will cause the yarn to be initially advanced to the jet **30** at a linear velocity in excess of the linear velocity at which the bulked yarn is withdrawn from the jet **30** by sections **42** and **43**, thus producing a predetermined amount of initial overfeed of the yarn or yarns advancing undisturbed toward jet **30.** The amount of initial overfeed is determined by the ratio of diameters, D/d, assuming the yarn or yarns are advancing undisturbed to the jet **30**.

An untexturized core yarn **48** and an untexturized effect yarn **49** are drawn from supply packages **46** and **47** respectively, through guide eyes **50** and **51** respectively, to a tensioning unit **52**. The tensioning units **52** and guide eyes **50** and **51** are fixed to platform **14.**

The core yarn **48** and effect yarn **49** are directed around and between the outer surfaces of cylindrical sections **40** and **41** a number of times sufficient to provide enough engagement with these surfaces to prevent slippage when the rollers are driven at high speed.

The core yarn **48** advances directly, relatively undisturbed, from the surface of cylindrical section **41** to fluid jet **30** at a constant predetermined first velocity. The core yarn **48** has a constant predetermined amount of overfeed since the combination texturized or bulked yarn **56** is withdrawn from jet **30** at constant predetermined second velocity different from the first velocity. The combination yarn **56** is directed around and between the outer surfaces of cylindrical sections **42** and **43** a number of times sufficient to provide enough engagement with these surfaces to prevent slippage when the rollers are driven at high speed.

As the core yarn **48** advances from the surface of cylindrical section **41** the effect yarn **49** advances therefrom at the same first velocity. However, the effect yarn **49** passes through an intermittent accumulation or deflection zone before entering jet **30.** In such zone the effect yarn **49** is acted upon to alter the velocity of the effect yarn entering the jet **30** such that the effect yarn **49** advances to jet **30** at randomly varying velocities for randomly varying periods thereby randomly varying the final overfeed of the effect yarn.

Effect yarn **49** advances through guide eyes **58** and **59**, which are fixed relative to housing **16**. Guide eye **23** is fixed to and reciprocates with piston **24** of air cylinder **22**, which is a dual acting type of air cylinder. With the piston **24** at an extreme end of its stroke, the guide eye **23** allows the effect yarn **49** to assume a straight or primary path between guide eyes **58** and **59**. Of course, air cylinder **22** could be located to provide a primary path for the effect yarn at some other point along the stroke of the piston **24**.

The operation of such air cylinders is well known in the art. Briefly, however, as compressed air is alternately supplied to and released from opposite ends of the bore of the air cylinder by lines or hoses **62** and **63** the piston **24** moves along the axial bore thereof, moving guide eye **23** with it.

The velocity at which guide eye **23** moves transversely to the undeflected or primary path of the effect yarn also controls the final velocity at which the effect yarn enters the jet **30**. As guide eye **23** moves transversely away from the undeflected path, it causes the effect yarn **49** to take a longer and longer path, which, so long as the guide keeps moving away, causes the effect yarn **49** to have a final velocity entering the jet **30** less than the constant predetermined velocity of the effect yarn entering the accumulation zone.

If the guide eye **23** stops, the final velocity of the effect yarn will be equal to the constant predetermined velocity of the effect yarn entering the zone. Conversely, if guide eye **23** is moving toward the undeflected or primary path the final velocity of the effect yarn entering the jet **30** will be greater than the constant predetermined velocity of the effect yarn entering the accumulation zone. The final velocity is a function of constant predetermined velocity and deflection velocity at which guide eye **23** transversely deflects the effect yarn **49**. Thus, the final overfeed of the effect yarn is increased and decreased.

The rate at which guide eye **23** transversely moves the effect yarn **49** toward its undeflected path determines the degree of bulkiness of the slub. Generally, the faster the rate of return the larger the cross-sectional dimensions of the composite yarn at the slub, assuming that the jet can assimilate all of the increase in overfeed.

After being withdrawn from jet **30** at right angles immediately upon exit therefrom and passing around cylindrical surfaces **42** and **43**, the combination yarn **56** is advanced downwardly to a yarn guide or roller **75** which is rotatably secured to the free end of a pivoted tension arm **76** which is a part of the constant tension winder mechanism **18**.

Constant tension winders such as those commercially available from a number of manufacturers include a tension sensing mechanism, such as tension arm **76**, which, through electrical or electrical-mechanical controls will vary the winding speed of the yarn package on its mandrel in accordance with variations in the tension of the yarn being wound. The composite texturized yarn, after being led around roller **75**, is taken directly to a composite texturized or bulked yarn package **78**, wound on mandrel **79** of the constant tension winder which includes a traverse mechanism **80**.

Example: A combination bulked yarn or glass filaments was produced using apparatus similar to that in Figure 6.16, using a core yarn of glass filaments (a DE100) and an effect yarn of glass filaments (a DE37). The linear velocity of the combination yarn leaving the jet was between 1,200 and 1,800 feet per minute. The pressure of the air delivered to the jet (approximately 75 psig) was within the normal range for such jets and yarns used in systems similar to those described in U.S. Patent 3,488,670, and the rate of initial overfeed was approximately 10%. However, systems having rates of initial overfeed within the range of 5 to 20% are believed to be suitable.

The constant tensile load on the combination yarn between roller **75** and traverse mechanism **80** was measured to be within the range of 90 to 115 grams. A constant tensile load within the range of 80 to 125 grams can be employed to produce a suitable combination yarn. The amount of the load is a factor in the characteristics of the final product. No sizing was applied to the combination yarn as it was being wound on the winder. Using the deflection or accumulation system similar to that in Figure 6.16b, a combination bulked yarn having slubbed sections of randomly varying length at randomly varying intervals along the length the otherwise generally, uniformly bulked yarn was produced. It appeared that such system ran for hours without producing a repeating pattern of slubs.

Glassy Fibers with Unmelted Protuberances

T. Otouma, S. Aoki, T. Minaki, K. Shibata and K. Mori; U.S. Patent 4,018,964; April 19, 1977; assigned to Nippon Asbestos Company, Ltd., Japan describe a method for preparing glassy fibers having protuberances studded on the surface of the fiber, characterized in that each protuberance comprises an unmelted component of the starting materials and is formed on the surface of the fiber by fiberizing the partially unmelted starting materials by a spinner or blowing technique.

Example: The following composition contains the components listed below.

	% by Wt
SiO_2	42.0
Al_2O_3	14.0
MgO	23.5
ZrO_2	13.5
B_2O_3	5.0
Na_2O	2.0

Silica stone, alumina, zircon-sand, magnesia clinker, soda ash and borax were mixed so as to give the above composition, and the resultant mixture was formed into bricks 50 x 50 x 80 mm. The bricks were then melted in a cupola furnace at 1320°C for about 7 minutes. The particle size of the zircon-sand used was 65 to 170 mesh. The melt was then taken out from the furnace as a stream having a diameter of about 30 mm and was fiberized by means of a spinner method. The fibers thus obtained had an average diameter of 5.2 μ, and had many protuberances. The protuberances had a size twice as large as the diameter of the fiber, and were located at intervals of 1 to 2 mm along each fiber. The unmelted particles included in the protuberances had a diameter of about 4.5 to 5.0 μ, and most of them were zircon-sand.

In comparison, starting materials having the same composition as above were melted in a three-phase arc-furnace until the materials were completely melted without leaving any unmelted part. The melt was then cooled to 1390°C, and was fiberized by means of a spinner method. The fibers thus obtained had no protuberances.

The above prepared two types of fibers were tested as reinforcing material in preparing calcium silicate thermal insulation. That is, eight parts of the above two types of fibers were respectively stirred in 1,000 parts of water to uniformly disperse them in the water. The dispersion was then mixed with 40 parts of diatomaceous earth, 40 parts of slaked lime and 12 parts of bentonite to produce a homogeneous slurry. The slurry was heated at 90°C for about 2 hours to form a gel and then the gel was dehydrated and molded into a formed product. The formed product was subjected to hydrothermal treatment in an autoclave for 7 hours under a saturated steam pressure of 9 kg/cm^2 thereby causing a hardening reaction. The product thus hydrothermally treated was dried to produce a calcium silicate thermal insulation.

The thermal insulations were tested for density, modulus of rupture in the ordinary state and modulus of rupture after they were heated at 650°C. The test results are shown in the table below.

	. . Thermal Insulation . .	
	A*	B**
Density (g/cm^3)	0.19	0.19
Modulus of rupture in the ordinary state (kg/cm^2)	8.8	6.7
Modulus of rupture after heating at 650°C (kg/cm^2)	5.3	4.6

*Reinforced by fibers of an average diameter of 5.2 μ having protuberances.

**Reinforced by fibers of an average diameter of 5.0 μ having no protuberances.

It is clear from the data shown that protuberances contribute to the increase in strength.

Glass fibers having protuberances studded on their surfaces can be used for reinforcing building materials and other industrial materials such as gypsum boards, calcium silicate boards and the like.

OPTICAL FIBERS

As is well known, the feasibility of transmitting communication signals over optical fibers has been successfully demonstrated in the laboratory. Attention is focused on the development of practical systems for use in the field. The first commercial uses of such optical communication systems will be in densely populated metropolitan areas, such as New York City, where existing cable ducts are straining at the seams and the cost of constructing new duct work is prohibitively expensive. Under these circumstances, optical cables are extremely attractive because of their inherently smaller cross section.

Helical Optical Fibers

E.A.J. Marcatili; U.S. Patent 4,028,081; June 7, 1977; assigned to Bell Telephone Laboratories, Incorporated has developed an optical cable which is relatively immune to the effects of bending and/or tension comprising a helical optical fiber loosely confined within a protective sheath. Apparatus is disclosed for fabricating the helical fiber. For example, as the fiber is drawn from a heated preform the fiber is subjected to periodic, pulsating jets of an inert gas, such as nitrogen, which physically deform the fiber into the desired helical configuration.

As shown in Figure 6.17a, one such configuration is that of a helix **31** having a period ℓ_0 which is long compared to the diameter **d** of the helix, that is to say where $\ell_0 \gg d$. It will be apparent that a fiber having such a configuration can readily be deformed without introducing stresses of sufficient magnitude to fracture the fiber and without introducing short local radii of curvature that induce microcracking losses. Of course, the size of the helix is greatly exaggerated in the figure to simplify the explanation of the process. In an optical cable, there will, of course, be a plurality of optical fibers contained within some suitable protective sheath or jacket.

Figure 6.17b shows one format of tapes or ribbons for such an optical cable. As shown, a plurality of optical fibers **32** are sandwiched between two layers of adhesive tape **33** and then a number of such sandwiches or tapes are stacked and stranded at the center of the cable. It will be readily apparent that when the cable is bent, for example, by inserting the cable into a duct, some of the fibers in the cable will be compressed and others will be under tension.

Figure 6.17c is a diagram of a known apparatus for pulling an optical fiber from a glass preform. Preform **40** is circumferentially heated by an oven **41** while simultaneously being pulled by a drum **42** driven by a motor **43**, thereby to yield an optical fiber **45**. In accordance with this process, this known apparatus is modified by the provision of a first nozzle **46** through which some suitable inert gas, such as nitrogen, from a source **47** is applied to the molten fiber. A second nozzle **48**, the longitudinal axis of which is orthogonal to the axis of nozzle **46**, also applies a jet of inert gas from source **47** to the fiber. The gas from source **47** passes through a pair of valves **51** and **52** which periodically interrupt the flow of gas to nozzles **46** and **48**, respectively. Thus, both nozzles supply a periodic jet of gas to the floor.

Figure 6.17: Manufacture of Helical Optical Fiber

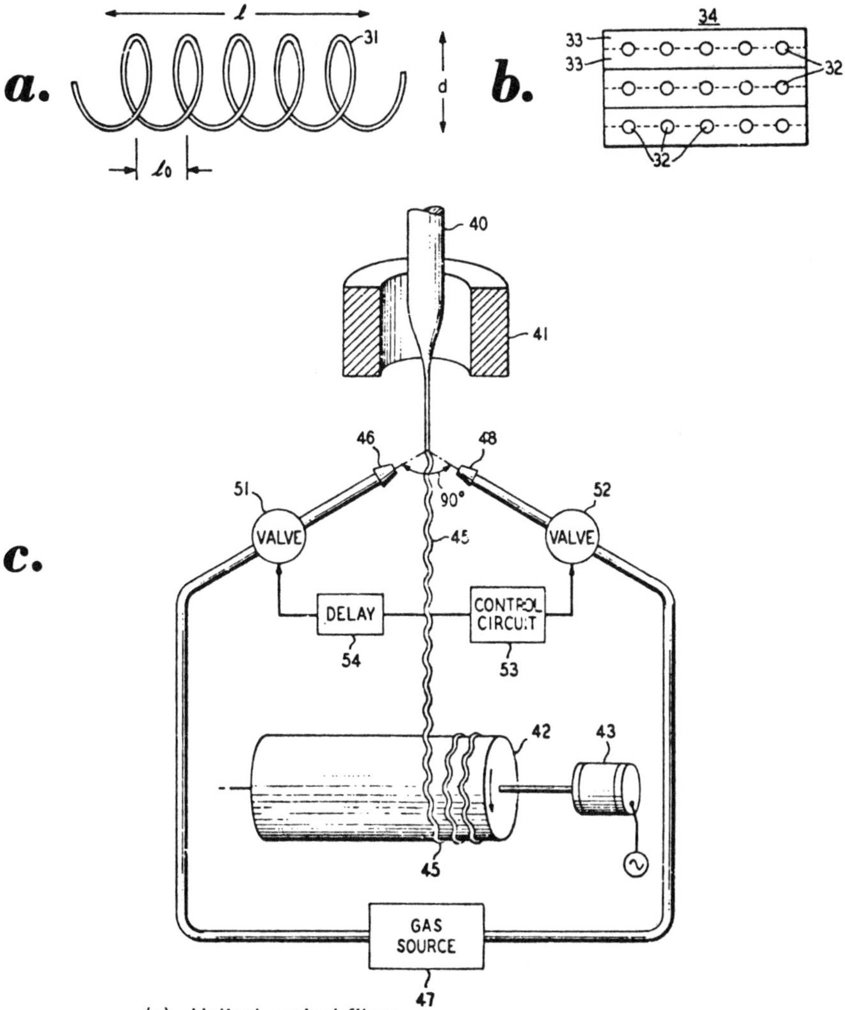

(a) Helical optical fiber
(b) Cross-sectional view of fibers arranged in a cable
(c) Partially schematic, partially diagrammatic view of associated apparatus

Source: U.S. Patent 4,028,081

Both valves are controlled by a control circuit **53**, but the control signal to valve **51** passes through a delay circuit **54** which delays the control signal. As a result, the exit of the gas alternates between nozzles **46** and **48**, thus, the gas jets have the same period but opposite phase. Both nozzles axes and the fiber axes are perpendicular to each other. The forces directed onto the molten fiber by nozzles **46** and **48** are of sufficient magnitude to physically deform the fiber and cause it to assume a helical configuration as it is drawn out of the preform. The periodicity of the gas jets must obviously be related to the rate at which the fiber is drawn from the preform and this period is advantageously selected such that the pitch of the helix which is formed is much larger than the diameter of the fiber.

Multicomponent Coaxial Rods

E. Roeder and E. Steinbeck; U.S. Patent 4,063,914; December 20, 1977; assigned to U.S. Philips Corporation provide a method of and a device for a simultaneous extrusion of multicomponent coaxial rods with very uniform, centered cores. A block of cladding material has core material coaxially contained in the geometric relationship desired for the ultimate product, and extrusion is by pressure opposite to the direction of material flow.

Using the method, composite bodies can be manufactured which have constant dimensions and a uniform geometry also over large lengths of approximately one meter. The method is advantageous notably for the manufacture of glass fibers for optical data transmission, because the high quality of the extruded product results in a low rejection percentage, so that the major part of the expensive kinds of glass can actually be used for the ultimate fiber manufacture.

Figure 6.18 is a cross-sectional view of a device for carrying out the process. A glass block **34** is arranged in a block holder **2** in a reservoir **1**. The block **34** is composed of a cylindrical body **3** consisting of cladding glass, wherein a rod-shaped body **4** of core glass, is concentrically arranged. The ratio between the diameters of the two glass bodies **3** and **4**, whose surfaces should be clean and polished, determines the type of glass fiber to be ultimately manufactured. After the glass block **34** has been arranged in the block holder **2** against a closed end **22** of the holder, it is heated by means of a high-frequency induction coil **7**. At the same time, the reservoir **1** is evacuated via a connection **9** so as to avoid gas bubbles at the interface of the two glass bodies, the gas bubbles being liable to cause dispersion of the light in the fiber at a later stage. If necessary or if desired, a protective gas, for example, an inert gas, can be introduced into the reservoir via a connection **11**.

Subsequently, the glass block **34** is extruded to form a strand **6**, a hollow extrusion ram **5**, comprising a discharge duct **51**, and a die **55**, comprising a discharge opening **56**, are pressed into the block holder **2** toward the closed end **22** such that the glass block **34** plasticized by heating is compressed in the direction opposite the discharge direction of the glass strand **6**.

Figure 6.18: Extrusion of Multicomponent Coaxial Rods

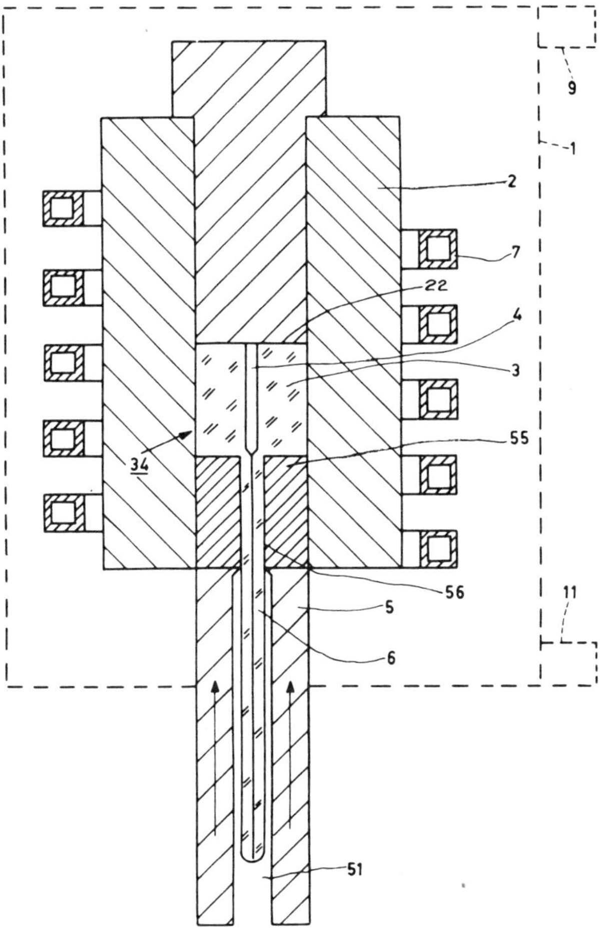

Source: U.S. Patent 4,063,914

Because the core glass in the form of a rod **4** is concentrically arranged in the cylindrical body **3** of cladding glass, so that it already has a geometric configuration corresponding to the desired configuration in the ultimate product (the glass strand **6**), deviations in the diameter ratio of core glass and jacket glass of the glass strand remain within narrow limits of 1% or less. Using such a device, a glass strand having a length of 90 cm and a diameter of 6.35 mm was extruded, the variations of the core diameter being limited to a maximum of ±1% over a length of 60 cm.

Core with Loose-Fitting Cladding Tube

H. Aulich, J. Grabmaier and K.-H. Eisenrith; U.S. Patent 4,133,664; January 9, 1979; assigned to Siemens AG, Germany describe a method and apparatus for producing a light-conducting fiber having a core disposed in a loose-fitting cladding tube characterized by a double crucible arrangement having a first crucible for a melt of the material for the cladding tube and a second crucible for the melt of the material of the core.

The principles of the process are particularly useful in an apparatus as shown in Figure 6.19a, having a double crucible generally indicated at **30** for producing a light-conducting fiber generally indicated at **31** in Figure 6.19b. The fiber **31** has a core **1000** in a loose fitting cladding tube or sleeve **2000.**

The double crucible **30** includes a first crucible **1** and a second crucible **2.** Each of the crucibles **1** and **2** may consist of a material selected from a group consisting of platinum, iridium and platinum-rhodium alloy. The double crucible **30** can also be formed of different materials which are both heat resistant and resistant to chemical reactions or attack.

As illustrated, the first crucible **1** is an exterior crucible with the second crucible **2** being disposed therein. The exterior or first crucible **1** has the base with a ring-shaped orifice **10** and the second crucible has a single orifice **20** which is of a given diameter that is less than the inner diameter of the ring-shaped orifice **10.** As illustrated, the orifice **20** is arranged within the ring-shaped orifice **10** with the center axis of the orifice **20** substantially coinciding with the center of the orifice **10.** The second crucible **2** is supported within the first crucible **1** by cross supports such as **3** so that the wall of the second crucible **2** forms an inner annular wall of the first crucible **1.** The double crucible arrangement can be heated with a heating system such as an induction heating system or coil **4.**

A closure element **5** is mounted by means for moving which includes a quartz rod **6.** The element **5** has an annular projection **7** for closing the orifice **10** and a second projection **8** for closing the orifice **20.** The movement of the rod **6** in a direction opposite to the direction of the arrow of Figure 6.19a causes the annular projection **7** and the projection **8** to simultaneously close both the orifices **10** and **20.**

The first crucible **1** serves for melting of a material such as a glass mass to form a mass of melted material or melt **11** which will be formed into a tube or sleeve **111** which is subsequently drawn into the cladding tube **2000** of the light conductive fiber **31.** The internal or second crucible **2** will melt a mass of glass to form a mass of melted material or melt **21** which will eventually form a glass strand **211** that is subsequently formed into a core **1000** of the fiber **31.** The amount of glass in each of the melts **11** and **21** can be continuously replenished by the addition of glass and solid bodies such as rods **110** for the melt **11** and the rod **210** for the melt **21.** It is obvious that the glass can also be replenished in other forms such as by the addition of solid masses such as in the shape of balls into the melts **11** and **21**. It is also possible to refill the crucibles **1** and **2** with a premelted glass melt.

Figure 6.19: Core with Loose-Fitting Cladding Tube

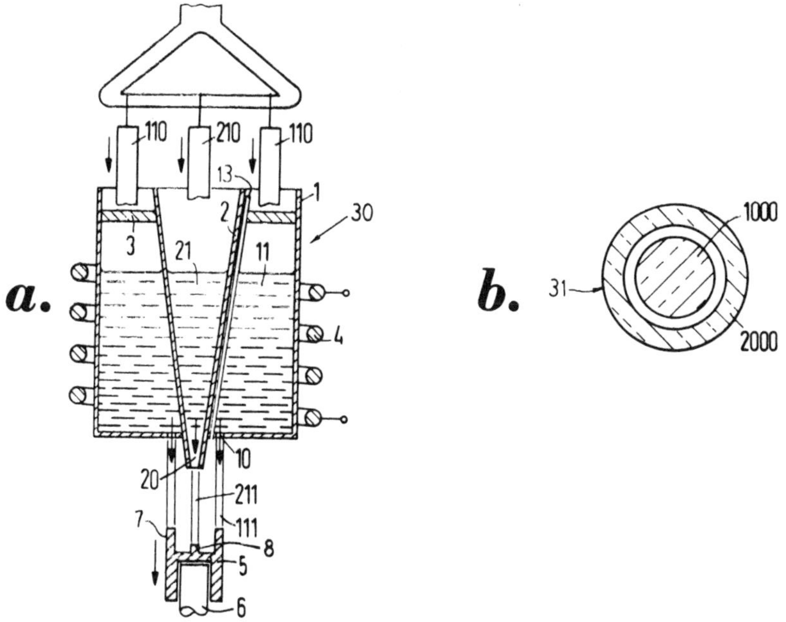

(a) Cross section of apparatus with portions in elevation
(b) Cross-sectional view of an optical fiber

Source: U.S. Patent 4,133,664

During the melting of the glass compounds in the crucibles **1** and **2**, the orifices **10** and **20** of the double crucible **30** are closed by means of the closure element **5**. After the melts are obtained, they are brought to the desired temperature for the drawing of the glass fiber which is usually accomplished by allowing them to cool to this temperature. After the melts **11** and **21** have reached the desired temperature for drawing, the closure element **5** is moved axially by the rod **6** in the direction of the arrow so that a glass tube **111** which adheres to the annular projection **7** and a glass strand **211** which adheres to the projection **8** will be drawn through the orifices **10** and **20**, respectively. The glass tube **111** will be subsequently formed into the cladding tube **2000** while the strand **211** will be subsequently formed in light-conducting fiber core **1000**.

After the rod **6**, which forms means for moving the closure element **5**, has moved the element a sufficient distance, the closure element may be removed and the drawn portion comprising the tube **111** with the strand **211** disposed therein is mounted on a conventional winding drum for glass fibers so that a light-conducting fiber with a loose fitting glass sleeve can be pulled from the double crucible **30.** The size of the cross section of the finished light conducting fiber can be determined by a selection of the pulling velocity.

In order to avoid collapsing of the tube **111** during the pulling of the light conductive fiber **31**, an inert gas, for example a noble gas, can be inserted between the glass tube **111** and the glass strand **211.** This insertion can be accomplished by applying the gas through a tube **13** which terminates with a port which is disposed between the ring-shaped orifice **10** and the orifice **20.**

As a result of drawing the fiber from the double crucible **30**, the fiber **31** will have a light-conducting fiber core **1000** which is received in a loose fitting cladding tube **2000** (Figure 6.19b). As illustrated in the figure, the core **1000** is loosely arranged within the cladding tube or sleeve **2000.** The cross section of the fiber core **1000** and the cross section of the tube or sleeve **2000** and the distance between the surface of the core and the inner surface of the sleeve or tube **2000** is determined by the viscosity of the glass melts **11** and **21**, the drawing velocity or speed and the geometry of the orifices such as **10** and **20.**

ADHESIVE COATING

P. Bourrain and P. Giroud; U.S. Patent 4,148,963; April 10, 1979; assigned to Rhone-Poulenc-Textile, France disclose an adhesive-coating composition for increasing the adhesion of heat-stable, high-modulus, modal, polyester-based yarns and glass fibers towards compounded rubbers based on natural rubber and on synthetic rubber, the composition comprising a solution of a precondensed resorcinol/formaldehyde resin, a latex based on vinylpyridine and polychloroprene latex, 28% strength ammonium hydroxide and a dicarboxylated butadiene/styrene resin. Yarns and fibers coated with the adhesive-coating composition are useful in the manufacture of tires, conveyors, conveyor belts, hoses and the like.

Example: An adhesive-coating bath is prepared, consisting of a resin solution A containing:

	Parts by Weight
Softened water	547
Sodium hydroxide	1
Precondensed resorcinol/formaldehyde resin (Penacolite B_1A)	57
Formaldehyde (40% strength)	15.5

A latex solution B, which is left to ripen for 3 to 4 minutes, and a latex solution C, are prepared separately, while stirring and at ambient temperature; these solutions respectively contain the ingredients as shown below.

	Parts by Weight
Latex Solution B	
Latex based on vinylpyridine	805
Polychloroprene	92
Ammonium hydroxide (28% strength)	80
Latex Solution C	
Softened water	200
Ammonium hydroxide (28% strength)	15
Dicarboxylated butadiene/styrene resin (Pliolite 4121)	350

Latex solution C is poured into solution B while stirring, and solution A is then poured into the mixture thus obtained, while stirring at ambient temperature.

	Percent
Solids content of the composition	30
Dry weight of resin/solids content	9.8
Dry weight of latex/solids content	86
Dry weight of latex/dry weight of resin	8.7/1
Dry weight of chlorinated latex/total of the latices	10

The bath is left to stand for twenty-four hours.

The adhesive-coating composition thus obtained, of pH 9.6 has a stable viscosity and a good mechanical stability.

Three cords of glass fibers each of 340 dtex, with each strand having a diameter of 9 μ, are impregnated by means of the composition thus prepared.

Each cord is drained on a roller equipped with a wiper blade to remove the excess bath, and is then dried continuously at a speed of 45 m/min in a 12 meter long oven maintained at 270°C; the treatment of crosslinking the resin is then carried out continuously in a 12 meter long oven at 270°C at a speed of 45 m/min.

Each cord adhesive-coated in this way is twisted at 140 turns/m in the Z-direction and the three cords are then twisted together at 140 turns/m in the S-direction.

Apart from very great flexibility, the twisted cord thus prepared has the following characteristics: adhesion towards compounded rubber based on polychloroprene, of the above composition, 8.5 kg/6 mm (mean of 20 measurements); proportion deposited, (adhesive-coated product – raw product)/raw product, 17%; and breaking load, 26.4 kg.

Carbon and Metal Carbide Fibers

CARBON FIBER PRECURSORS

Polybutadiene

According to *A. Nagasaka, H. Ashitaka, Y. Kusuki, D. Oda, and T. Yoshinaga; U.S. Patent 4,131,644; December 26, 1978; assigned to Ube Industries, Inc., Japan*, carbon fibers having excellent modulus of elasticity and tensile strength are produced in a high yield by a process where polybutadiene fibers are prepared by melt spinning polybutadiene having at least 85% of vinyl structure, a melting point of at least 100°C and a reduced viscosity of 0.3 to 1.2 dl/g. The fibers are then treated in a solution or suspension of a Lewis acid in an inert organic liquid where the polybutadiene is cyclized and crosslinked.

The polybutadiene fibers, treated as mentioned above, are further treated in a melted sulfur or a sulfur-containing solution at a temperature of 170° to 300°C, where the cyclized and crosslinked polybutadiene is dehydrogenated and aromatized. The fibers treated, as mentioned above, are carbonized by raising the temperature to 750° to 3000°C to form carbon fibers or graphite fibers. If so desired, the ordinary carbon fibers which have been produced at 750° to 1500°C can be converted to graphite fibers by heating them to a temperature of 1500° to 3000°C.

The sulfur treatment is carried out so that the ratio of the number of hydrogen atoms to carbon atoms in the treated polybutadiene (H/C) falls in a range of 0.2 to 1.0, preferably 0.5 to 0.9, and more preferably 0.65 to 0.85. If the value of H/C is less than 0.2, a relatively low yield of carbon fibers with relatively low tensile strength and modulus of elasticity is produced. Also, if the value of H/C is greater than 1.0, the resultant carbon fibers are poor in yield and quality.

Sometimes the sulfur treated fibers having an H/C greater than 1.0 tend to fuse or stick to each other in the subsequent carbonization step.

Example: Polybutadiene used herein contained 99% of vinyl structure and had a melting point of 190°C and a reduced viscosity of 0.62 dl/g. A polymer blend was prepared by blending 100 parts of polybutadiene, 1 part of acrylaminoethyl-(3,5-di-tert-butyl-4-hydroxybenzyl) ether, 0.2 part of tetrakis[methylene-3-(3,5-di-tert-butyl-4-hydroxyphenyl)propionate] methane, 0.2 part of distearyl-3,3'-thiodipropionate, and 1 part of a fatty acid diethanolamide (Dianol 300).

The polymer blend was melted at a temperature of 205°C and extruded through spinning orifices to form filaments each having a diameter of 22.4 μ. A bundle of 1,200 filaments so formed was drawn at a draw ratio of 2.8 at a temperature of 90°C to form a bundle of drawn filaments each having a diameter of 13 μ. The drawn polybutadiene filaments had an initial modulus of elasticity of 16×10^3 kg/cm^2 and a tensile strength of 1.9×10^3 kg/cm^2. The polybutadiene filament bundle was immersed in a Lewis acid solution which had been prepared by dissolving 2 g of aluminum bromide in 100 ml of benzene in a nitrogen atmosphere at a temperature of 20°C for 6 hours under a tensile stress of 90 kg/cm^2.

The immersed filament bundle was washed with benzene, methyl alcohol containing 20% of hydrochloric acid and then with methyl alcohol. Thereafter, the washed filament bundle was dried at room temperature under reduced pressure. The individual filaments thus treated had a tensile strength of 1.2×10^3 kg/cm^2. The filament bundle was then immersed in a sulfur melt at a temperature of 230°C for 5 hours under a tensile stress of 15 kg/cm^2. Thereafter, the filament bundle was washed with hot xylene and dried at room temperature under reduced pressure. The H/C of the filament bundle was 0.72.

Finally, the filament bundle was carbonized in an argon atmosphere by increasing the temperature of the filament bundle to 1000°C at a rate of 10°C/min without tensile stress. A carbon filament bundle was obtained with a carbonization yield of 89%. The resulting carbon filaments had a modulus of elasticity of 670×10^3 kg/cm^2 and a tensile strength of 8.8×10^3 kg/cm^2.

Comparative Example: The same polybutadiene filament bundle treated with the same Lewis acid solution by the same procedures as above was directly subjected to a heating process without sulfur treatment. The heating process was carried out in an argon atmosphere by heating the filament bundle which had been treated with the Lewis acid solution, to a temperature of 1000°C at a rate of temperature increase of 10°C/min. When the temperature reached 450°C, the filament bundle was decomposed and became liquid and, thereafter, when the temperature reached 1000°C, the liquid completely decomposed and evaporated without leaving any traces.

Polyimides from Wet Spun Polyamic Acid

F.S. Galasso, R.G. Bourdeau, and R.A. Pike; U.S. Patent 4,056,598; November 1, 1977; assigned to United Technologies Corporation describe a process for producing continuous filaments from polyamic acid suitable for conversion to polyimide filaments comprising a starting solution consisting essentially of equimolar amounts of p,p-diaminophenyl ether and bis(3,4-dicarboxyphenyl) sulfone dianhydride, the remainder a solvent selected from the group consisting of dimethylformamide, methylpyrrolidone and mixtures thereof, the starting solution having a viscosity ranging from 500 to 5,000 cp, preferably 1,500 to 3,000 cp, and forming the starting solution into filaments by extruding it in a coagulating bath consisting essentially of a liquid selected from the group consisting of propanol, methyl alcohol, ethyl alcohol, butyl alcohol, a mixture of ethyl alcohol and propanol, and a mixture of methyl alcohol and acetone.

The filaments formed according to the procedures above described are then dried and imidized. Imidization satisfactory to allow subsequent carbonization occurs upon heating to temperatures of approximately 250° to 300°C for at least approximately 15 minutes, preferably 270° to 290°C for at least approximately 30 minutes. Below 250°C, the fiber tends to deform (stretch) during any subsequent carbonization while above 300°C, the imidization leads to the bonding of individual fibers together and/or to the spool.

After the aforementioned thermal imidization, the polyimide filament may be carbonized in order to provide a carbon filament suitable for use as a substrate.

The carbon filaments produced are well suited for meeting the requirements for usage as substrates in various coating deposition processes. In one case, for example, a 1.2 mil carbon monofilament having a resistance of 5,000 ohms/inch was utilized as a substrate in an electrically resistive vapor deposition process in which a 5 mil silicon carbide fiber was produced.

Acrylonitrile Copolymer

According to *V.C.R. McLoughlin, R. Moreton, and W. Watt; U.S. Patent 4,107,408; August 15, 1978; assigned to National Research Development Corporation, England*, a suitable precursor for the production of carbon fiber is an acrylonitrile copolymer containing at least one nonacrylic moiety which is a carbonyl group, a hydroxymethylene group, or a two carbon moiety containing a carbonyl group or a hydroxymethylene group and in which the nonacrylonitrile moiety is present to the extent of 5 to 15 molar parts per 100 molar parts of acrylonitrile moiety.

Example: A copolymer of acrylonitrile-oxoethylene containing 12½ mols of oxoethylene per 100 mols of acrylonitrile was prepared by:

> Copolymerizing a mixture of acrylonitrile and tert-butyl vinyl ether dissolved in dimethyl sulfoxide using azobisisobutyronitrile as an initiator to form an acrylonitrile-butoxyethylene copolymer;
>
> Dissolving the copolymer in an aqueous solution of sodium thiocyanate and hydrolyzing the tert-butoxy grouping with hydrobromic acid to form an acrylonitrile-hydroxyethylene copolymer; and

Oxidizing this copolymer with dimethyl sulfoxide-acetic anhydride to form the acrylonitrile-oxoethylene copolymer.

The experimental details are as follows.

A solution was made containing 18.9 g of acrylonitrile, tert-butyl vinyl ether (2.48 g) and dimethyl sulfoxide (100 ml). To this solution was added 170 mg of azobisisobutyronitrile and the solution was heated at 50°C for 105 min under nitrogen. (The solution remained clear at all times.) The solution, which was by now viscous, was then added dropwise into methanol (500 ml) with stirring to precipitate the copolymer which was filtered, washed with methanol and dried. The yield was 6.35 g of a white solid having the approximate composition $O_{30}H_{36}N_8O$ and an inherent viscosity of 1.8 dl/g.

This copolymer (3.9 g) was dissolved in a 50% by weight aqueous solution of sodium thiocyanate with stirring at 20°C. To the solution was slowly added 1.25 ml of 48% by weight aqueous hydrobromic acid, and the solution was stirred for 2 hours at 20°C. The solution was then slowly added to a large volume of water and the resultant white precipitate filtered, washed and dried. The white solid produced had a composition of approximately $C_{26}H_{28}N_8O$ and an inherent viscosity of 2.15 dl/g.

A solution of the resultant solid (3.4 g) in dimethyl sulfoxide (360 ml) was treated with acetic acid anhydride (18 ml) with stirring. After 18 hours at 20°C, the product was precipitated in water (4 liters) to give a white copolymer having an approximate formula of $C_{26}H_{26}N_8O$ and an inherent viscosity of 1.9 dl/g.

This copolymer was converted to fibers by passing an 8% by weight solution of the polymer in a 50% by weight solution of aqueous sodium thiocyanate through a spinneret into a coagulation bath of 10% by weight aqueous sodium thiocyanate solution. The resultant fibers were then washed by passing them through a bath of distilled water and were stretched by a factor of 11.

A sample of the fiber was then converted to carbon fiber by heating in a nitrogen atmosphere, initially at 400°C for 6 hours and finally at 1000°C for ½ hour. During the heating the fiber was wound onto silica frames to restrain shrinkage of the fibers. The carbon fibers produced were then removed from the frames and a portion of these subjected to a further heat treatment in an argon atmosphere at 2500°C for ½ hour.

As a control, the batch of Courtelle fiber was also converted to carbon fiber in the manner described above.

Also as a control, a second batch of Courtelle fiber was converted to carbon fiber by heating the fibers initially at 220°C in air for 5 hours and finally at 1000°C for ½ hour in nitrogen. The heat treatment in air served as an oxygen treatment step. During both heating steps the fibers were wound onto silica frames in such a manner as to restrain shrinkage of the fibers. As before, a portion of the carbon fibers produced were removed from the frames and subjected to a further heat treatment in argon at 2500°C for ½ hour.

Courtelle fiber is a commercially available polyacrylonitrile based fiber containing about 6 mol % of methyl acrylate.

The properties of the carbon fibers produced from the acrylonitrile-oxoethylene copolymer are intermediate between fibers prepared from Courtelle fibers with and without the inclusion of an oxygen treatment step. Although the properties of the fibers are not as good as those produced from Courtelle fiber using the oxygen treatment stage described above prior to carbonization, the fibers still have useful properties and, as such, will find many applications. The main advantage of using the copolymer is the ability to produce fibers having useful properties without the need for the use of the time consuming and, therefore, expensive oxygen treatment stage.

Although in this example the fibers were heated to a temperature of 400°C for 6 hours prior to carbonization, this step may be omitted without detriment to the properties of the fiber. Restraining the shrinkage of the fibers during initial heating to 400°C is, however, believed to be necessary to obtain fibers having these properties.

Acrylonitrile Containing an Aminosiloxane

S. Kishimoto and S. Okazaki; U.S. Patent 4,080,417; March 21, 1978; assigned to Japan Exlan Company Ltd., Japan disclose a process for producing carbon fibers by heat treating an acrylonitrile fiber containing or impregnated with an aminosiloxane at a temperature of at least 150°C for 0.1 sec to 30 min so that the fiber contains 20 to 80 weight % undissolved matter upon immersing it in a 60% aqueous solution of sodium thiocyanate at 80°C for 20 min and then carbonizing the fiber. The productivity throughout the whole process, including the step of producing the precursor fiber and the step of producing the carbon fiber, is increased and a carbon fiber having high tensile strength and high modulus of elasticity is obtained in an industrially advantageous manner.

Example: A spinning solution was obtained by dissolving 17 parts of an acrylonitrile copolymer containing 97% acrylonitrile in 83 parts of a 50% aqueous solution of sodium thiocyanate. After this spinning solution was extruded into air through a spinneret, it was introduced into a 12% aqueous solution of sodium thiocyanate to form coagulated filaments. Thereafter, the fiber was washed with water and then stretched 3 times the length in boiling water and further stretched 2.5 times in superheated steam to obtain an acrylonitrile fiber in a swollen state having a water content of 160% and having a single filament fineness of 1.3 den.

The water-swollen fiber was then immersed in an aqueous emulsion consisting of 100 parts of an aminosiloxane (NH_2 content 0.5%) of the following formula:

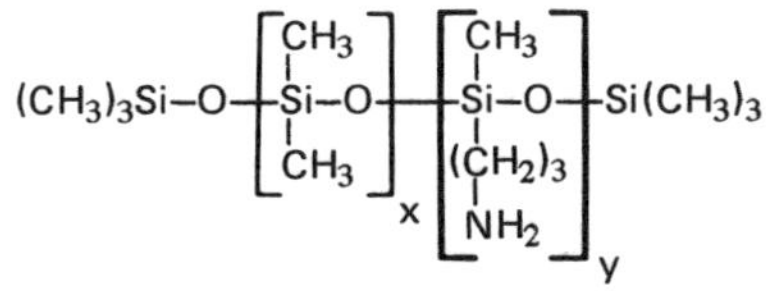

50 parts of POE (9) nonylphenyl phosphate and 4 parts of zinc acetate as the catalyst and, thereafter, the fiber was dried at 120°C for 3 min. In this way, an acrylonitrile fiber impregnated with 0.42% of the aminosiloxane was obtained. The fiber thus obtained was free from such defects as filament fluffiness, breakage and spreading. Subsequently, this fiber containing the aminosiloxane was heat treated under a definite length at 230°C for 1 minute.

Upon immersing the treated fiber thus obtained into a 60% aqueous solution of sodium thiocyanate maintained at 80°C for 20 minutes, the ratio of the undissolved matter in this solution was found to be 21%. On the other hand, when the same fiber dried without the aminosiloxane treatment was subjected to the abovementioned heat treatment, the concentration of the undissolved matter of the treated fiber thus obtained in a 60% aqueous solution of sodium thiocyanate solution was found to be 4%.

Each of the fibers after the heat treatment was subjected to a thermal stabilization treatment by passing it continuously under an air atmosphere through an electric furnace having a continuous temperature gradient from 200° to 280°C under a tension of 42 g/den for 25 minutes, and then it was carbonized by further passing it continuously under a nitrogen atmosphere through an electric furnace having a continuous temperature gradient from 250°C to 1300°C for 2 minutes.

It was found that, when the amount of the undissolved matter in a 60% aqueous solution of sodium thiocyanate is from 20 to 80%, carbon fibers having excellent physical properties can be obtained.

Splice Used in Thermal Stabilization of Fibrous Acrylics

It is well known that acrylic fibrous materials when subjected to heat undergo a thermal stabilization reaction where the fibrous material is transformed to a black form which is nonburning when subjected to an ordinary match flame.

Such modification generally has been accomplished by heating the acrylic fibrous material in an oxygen-containing atmosphere. It is believed that the resulting thermal stabilization reaction involves an oxidative crosslinking reaction of adjoining molecules as well as a cyclization reaction of pendant nitrile grounds to a condensed dihydropyridine structure. The cyclization reaction is exothermic in nature and must be controlled if the fibrous configuration of the acrylic polymer undergoing stabilization is to be preserved. The resulting thermally stabilized fibrous material is useful as a precursor in the formation of carbon fibers.

R.A. Breznak and J.A. Parker; U.S. Patent 4,048,277; September 13, 1977; assigned to Celanese Corporation have made improvements in a process for the thermal stabilization of a substantially flat multifilament band of an acrylic fibrous material where the ends of at least two discrete band segments are joined and continuously are passed in a successive manner in the direction of their length through a heated thermal stabilization zone while maintaining the original fibrous configuration substantially intact for a residence time sufficient to render the band black in appearance and nonburning when subjected to an ordinary match flame.

These improved results are accomplished when the ends of at least two of the band segments are joined by:

> Surrounding an end of each band segment with a fabric which is capable of withstanding the heated thermal stabilization zone with a portion of the fabric extending beyond each enclosed end of the band segment to form a substantially flat fabric extension;
>
> Sewing the fabric to each band segment in the area where the band is enclosed by the fabric with a thread which is capable of withstanding the heated thermal stabilization zone while maintaining the enclosed portion of the band in a substantially flat configuration;
>
> Overlapping a pair of the substantially flat fabric extensions with each of the band segments extending in an opposite direction; and
>
> Sewing together the overlapped substantially flat fabric extensions with a thread which is capable of withstanding the heated thermal stabilization zone to join the band segments of acrylic fibrous material in the absence of any substantial overlap of the acrylic fibrous material of each band segment.

The resulting splice is flat in configuration and enables the band segments including the splice to be continuously passed through a heated thermal stabilization zone without hang-up or impairment of movement resulting from the presence of the splice. The usual exothermic heat of reaction which is evolved during the cyclization portion of the thermal stabilization reaction effectively is dissipated from within the splice without harm.

Carbonaceous Mesophase Pitch

According to *L.S. Singer; U.S. Patent 4,005,183; January 25, 1977; assigned to Union Carbide Corporation,* high modulus, high strength carbon fibers having a highly oriented structure characterized by the presence of carbon crystallites, preferentially aligned parallel to the fiber axis, are produced from carbonaceous pitches which have been transformed, in part, to a liquid crystal or so-called "mesophase" state. When heated to graphitizing temperatures, these fibers develop the three dimensional order characteristic of polycrystalline graphite.

Example: An acenaphthylene pitch was prepared by heating acenaphthylene to form a polymeric mixture, and then pyrolyzing the mixture by heating it under reflux for 6 hours. At the end of this time, air was bubbled through the pitch for about 7 hours while the pitch was maintained at a temperature of about 250°C in order to remove acenaphthene and other volatiles. The resulting pitch had a density of 1.29 g/cc, a softening temperature of 234°C and contained 0.6% by weight quinoline insolubles. (Q.I. was determined by quinoline extraction at 75°C.) Chemical analysis showed a carbon content of 94.91% and a hydrogen content of 4.49%.

A portion of the pitch produced in this manner was added to an extrusion cylinder and heated in the extruder to 400°C over a 2 hour period under a nitrogen atmosphere. The temperature of the pitch was then raised from 400° to 436°C over a period of about 3.5 hours. When the pitch reached the latter temperature, a piston was used to apply pressure to the pitch while the molten pitch was extruded through a pinhole orifice (diameter 0.015") at the bottom of the extruder to produce a filament which was taken up by a reel at a rate of about 20 fpm. The filament passed through a nitrogen atmosphere as it left the extruder orifice and before it was taken up by the reel. A considerable quantity of fiber 20 to 30 μ in diameter was produced in this manner at a temperature between 436° and 440°C.

A sample of the pitch which had been heated to 438°C was found to contain 88% by weight pyridine insolubles, indicating a mesophase content of approximately 88%. (P.I. was determined by Soxhlet extraction in boiling pyridine.) Fiber spun from this pitch was examined by x-ray diffraction techniques and shown to be highly oriented. [The preferred orientation of the fiber, determined by microdensitometer scanning of the (002) band of the exposed x-ray film, was found to be 26° (FWHM).] The apparent stack height, L_c, of the aligned domains of pitch molecules in the fiber was determined by microdensitometer scanning of the width of the (002) diffraction arc to be 40 A.

Polarized light microscopy examination of the fiber indicated the presence of large, elongated anisotropic domains, having a fibrillar-shaped appearance, preferentially aligned parallel to the fiber axis.

A portion of the as-drawn fibers produced in this manner were heated to 343°C in oxygen over a period of about 1 hr and held at this temperature for about 6 minutes. The resulting oxidized fibers were totally infusible and could be heated at elevated temperatures without sagging.

The infusible fibers were heated to a temperature of 812°C over a period of about 100 minutes in an argon atmosphere and then to various temperatures up to 2000°C in about ½ hour. In each instance the fibers were held at the final heat treatment temperature for about 10 minutes.

Fibers having diameters of less than 30 μ produced in this manner exhibited tensile strengths in excess of 100 x 10^3 psi and Young's modulus of elasticity in excess of about 20 x 10^6 psi. Illustratively, fiber heated to 1200°C had a tensile strength of 129 x 10^3 psi and a Young's modulus of 23.1 x 10^6 psi. Fiber heated to 1400°C had a tensile strength of 134 x 10^3 psi and a Young's modulus of 26.3 x 10^6 psi, while fiber heated to 1600°C had a tensile strength of 128 x 10^3 psi and a Young's modulus of 34.8 x 10^6 psi.

X-ray diffraction studies of fibers prepared in a similar manner (by heating fibers produced from the same pitch to 350°C in oxygen at a rate of 10°C/min and subsequently carbonizing them by heating to a temperature of 1000°C in an argon atmosphere over a period of 1 hour) showed such fibers to have a preferred orientation (FWHM) of 33° and an apparent stack height (L_c) of 19 A. The 3000°C heat-treated fibers had a preferred orientation of about 8° and an apparent layer size (L_a) and stack height (L_c) in excess of 1000 A.

Polarized light microscopy examination of fibers prepared in this manner but heat treated to 1605°C indicates the presence of large oriented elongated graphitizable domains resembling those of the as-drawn fiber. Fibers prepared in the same manner but heat treated to 3000°C likewise were characterized by large oriented elongated domains preferentially aligned parallel to the fiber axis (although now graphitic rather than graphitizable).

Easily Handleable Pitch-Based Fibers

D.A. Schulz; U.S. Patent 4,138,525; February 6, 1979; assigned to Union Carbide Corporation discovered that the tensile strength and handleability of fibers spun from a carbonaceous pitch which has been transformed, in part, to a liquid crystal or so-called "mesophase" state can be significantly improved by oxidizing the fibers to an oxygen content of from 17 to 30% by weight, preferably from 18 to 22% by weight.

While those skilled in the art initially sought to limit oxidation of pitch fibers to the minimum amount required to thermoset them in the belief that excessive oxidation would reduce the strength of the carbonized and graphitized fibers produced therefrom, it has been discovered, quite surprisingly, that, not only does oxidation to the high level stated above greatly increase the strength of the spun filament, but also, that it has no deleterious effect on the strength of the carbonized or graphitized fibers produced therefrom.

Because of their greater strength and handleability, the highly oxidized fibers are less subject to breakage and damage during subsequent thermal processing. This allows such fibers to be processed at high speeds by means of conventional yarn-transport systems where the fibers are subject to higher tensions and rougher treatment than the lower oxidized fibers are capable of withstanding. Thus, such fibers can be rapidly transported through eyelets, over pulleys, through furnaces, and wound at high speeds while the lower oxidized fibers cannot. In addition, the high handleability of these fibers allows them to be utilized in textile type processes, such as weaving or knitting, where demanding high speed operations limit the use of the more fragile lower oxidized fibers. The cloth produced from these processes may, of course, then be further processed to produce carbon or graphite cloth by further heat treatment, thereby eliminating the difficulty of weaving or knitting cloth from fibers which have been stiffened to a high modulus by such thermal processing.

Example: A commercial petroleum pitch was employed to produce a pitch having a mesophase content of about 56% by weight. The precursor pitch had a density of 1.23 mg/m^3, a softening temperature of 120°C and contained 0.3% by weight quinoline insolubles. (Q.I. was determined by quinoline extraction at 75°C.)

The mesophase pitch was produced by heating the precursor petroleum pitch at a temperature of about 400°C for about 19 hours under flowing nitrogen. The pitch was continuously stirred during this time, and nitrogen gas was continuously bubbled through the pitch. After heating, the pitch exhibited a softening point of 341°C and contained 56.6% by weight pyridine insolubles indicating that the pitch had a mesophase content of close to 56%.

A portion of the pitch produced in this manner was then melt spun into fibers at a rate of 325 m/min through a 240 hole spinneret (0.07 mm diameter) at a temperature of 385°C. The filaments passed through a nitrogen atmosphere as they left the spinneret and were then taken up by a reel. A considerable quantity of fiber 9 to 12 μ in diameter was produced in this manner.

A portion of the spun filaments was placed in a stainless steel wire mesh tray and heated in a forced-air convection oven to a temperature of 315°C over a period of 45 minutes. This procedure was repeated a number of times with different portions of the spun filaments, except that varying hold times at 315°C were employed with each successive portion so as to vary the exposure time of each portion to the oxidizing atmosphere and the resulting oxygen content of the fibers of each lot. The oxygen content, tensile strength and modulus of the fibers produced in each run were then determined.

Samples containing greater than 20% oxygen were found to be highly handleable and could be woven into a cloth without difficulty. This cloth could be carbonized or graphitized by further heat treatment.

CONDUCTIVE FIBERS

Electrically Conductive Sheath/Core Filament

S. Kato; U.S. Patent 4,085,182; April 18, 1978; assigned to Teijin Limited, Japan describes a process for producing sheath/core type electrically conductive synthetic composite filaments, which comprises simultaneously melt-extruding in a sheath/core filament configuration an electrically conductive core composition of a thermoplastic synthetic polymer containing an electrically conductive carbon black dispersed therein and a nonconductive sheath composition of a thermoplastic fiber-forming synthetic polymer, and taking up the extruded sheath/core type synthetic filaments at a take up speed of at least 2,500 m/min, and an electrically conductive sheath/core type synthetic filament having a polyamide sheath.

Example: 25 pbw of a fine powder of electrically conductive carbon black (Columbia Carbon) was added to 75 pbw of polycapramide having an intrinsic viscosity, as measured on a m-cresol solution at 35°C, of 0.91. They were melt mixed in an atmosphere of nitrogen, and the mixture was extruded, cooled, and cut to form chips of the polymer containing the electrically conductive carbon.

The resulting chips as a core polymer and chips of polycapramide having an intrinsic viscosity, as measured on a m-cresol solution at 35°C, of 1.07 as a sheath polymer were cospun at varying take-up speeds to produce concentric core/sheath type composite filaments.

The spinning pack had a filtering area of 15.9 cm^2 both on the core side and the sheath side. A 20 mesh wire gauze was used in the filtering layer on the core side of the spinning pack. The filter layer on the sheath side consisted of an upper layer of 10 g of glass beads having an average particle diameter of 160 μ; an intermediate layer of 20 g of glass beads having an average particle diameter of 100 μ; a lower layer of 10 g of glass beads having an average particle diameter of 160 μ; a 325 mesh wire gauze; and a 50 mesh wire gauze placed beneath the lower filter layer.

The core-to-sheath ratio was 1:9, and the temperature of the polymers was 260°C. The number of the filaments was 3. The amount of the polymers melt extruded was adjusted so that the denier size of the filaments became 30 den/3 filaments under the take-up conditions.

The resulting filaments were mixed with a nylon 6 yarn (200 den/34 filaments) by a turbulent flow type air nozzle, and a knitted fabric was produced using the mixed yarn. The initial frictional voltage and the frictional voltage after pulling and rubbing of the knitted fabric were measured.

In two comparison composite filaments, the frictional voltage after pulling and rubbing was very much increased, ranging from 1,080 to 2,724 V compared to 187 to 249 V for the filaments prepared by the process. A microscopic examination of the surface of each of the knitted fabrics used in these tests showed that most of the composite filaments were broken and partly dropped off.

All composite filaments, of this process and comparison, were dissolved in formic acid, and the form of the electrically conductive carbon black was examined. It was found that in all of the filaments, the carbon black retained more than 90% of its chainlike structure.

Comparative Example: Composite filaments were prepared using the same apparatus and under the same conditions as in the Example except that the take-up speed was adjusted to 1,000 m/min, and the denier size of the filaments after drawing was 30 den/3 filaments in accordance with a conventional process.

The resulting undrawn filaments were drawn at a draw ratio of 2.5 at a drawing speed of 800 m/min. The electric resistance of the filaments was measured.

The undrawn filaments had an electric resistance of 6.5×10^9 ohms/cm, but the drawn filaments had an electric resistance of 1.1×10^{12} ohms/cm and, thus, exhibited poor electric conductivity.

The properties of the undrawn and drawn filaments were as follows.

	Undrawn Filaments	Drawn Filaments
Tenacity, g/den	1.40	2.56
Elongation, %	295.3	71.5
Young's modulus, kg/mm^2	61	192

The undrawn and drawn filaments were dissolved in formic acid, and the form of the electrically conductive carbon black contained in the filaments was examined. It was found that the carbon black in the undrawn filaments retained more than 90% of its chainlike structure, whereas about 90% of the chainlike structure of the carbon black contained in drawn filaments were destroyed and changed into a fine grainlike structure. The filaments can be crimped, scoured or bleached, and mix-spun or mix-woven with other fibers.

Electrically Conductive Side-by-Side Biconstituent Fiber

An electrically conductive side-by-side biconstituent melt spun fiber is provided by *N.W. Boe; U.S. Patent 4,129,677; December 12, 1978; assigned to Monsanto Company,* having a conductive constituent and a nonconductive constituent. Both constituents comprise a relatively nonconductive polymer. The conductive constituent additionally contains an electrically conductive material such as carbon black, and wraps around the nonconductive constituent.

Example: This example illustrates the preparation of fiber using nylon 66 polymer to form the nonconductive constituent and nylon 6 polymer containing uniformly dispersed therein 31% by weight of conductive carbon black to form the conductive constituent. The carbon black had fixed carbon of 98.5%; volatiles of 1.5%; particle size of 23 mμ; surface area of 125 m^2/g; and very low electrical resistivity.

Using the side-by-side biconstituent fiber melt spinning apparatus described and illustrated in U.S. Patent 3,969,559, the nylon 66 polymer (containing about 0.05% by weight moisture), in the form of chips of cubelike shape, was fed to a 1½ inch screw melter and the carbon-black-containing nylon 6 polymer (containing 0.110% by weight moisture), in the form of cylinders measuring about 3 mm in diameter and about 3 to 6 mm in length, was fed to the other 1½ inch screw melter. Each polymer was delivered via a separate metering pump to the spinneret where the polymers converge in single laminal streams in the lower portion of each capillary of the spinneret. The following conditions were employed.

	Nylon 66	Carbon Black-Containing Nylon 6
Percent by volume	94	6
Melt temperature, °C	300	260
Moisture, % by wt	0.05	0.110
Spinneret temperature, °C	275	–

The extruded biconstituent fiber was spun in a conventional melt spinning chimney having a crossflow of cooling air with a delivery temperature of 18°C. A known finish was applied to the fiber, and the fiber was collected at a speed of 411.48 m/min. The fiber was then withdrawn overend from a package and forwarded to a pair of feed rolls that delivered the fiber at a speed of 216 m/min to a hot roll (225°C), driven at a periphery speed of 217 m/min. Between the feed roll and hot roll the fiber made one wrap around a draw pin (6.4 mm diameter) at ambient temperature. From the hot roll the fiber was forwarded to a cold draw roll driven at a periphery speed of 751 m/min. The fiber made five wraps around the draw roll and its associated separator roll. The fiber was then collected using a conventional ring traveler take-up device.

The resulting drawn fiber had denier of 24.5; tenacity of 2.7 g/den; elongation of 71.7%; and electrical resistance of 3.87×10^6 ohms/cm.

Photomicrographs were made of the cross-section of the fiber. From the photomicrographs it was determined that the conductive constituent extended along 29.17% of the periphery of the fiber and inwardly a depth equivalent to 7.69% of the diameter of the fiber. The L:D (length to distance) ratio of the conductive constituent was calculated to be 11.9:1 using the formula:

$$\frac{L}{D} = \frac{0.2917 \times \text{diameter} \times \pi}{0.0769 \times \text{diameter}} = 11.9168$$

In the second run a fiber was prepared using the same procedure and conditions as described above except that the spinning speed was 822.96 m/min instead of 411.48 m/min. In this instance the drawn fiber had denier of 24.7; tenacity of 3.27 g/den; elongation of 66.5%; electrical resistance of 1.89×10^6 ohms/cm; and a L:D ratio of 7.855.

For purposes of comparison another run (run 3) was made by which a fiber was prepared using the same procedure and conditions as in the first run except that the nylon 6 contained 39% instead of 31% by weight carbon black, and the carbon black-containing nylon 6 polymer was dried to a moisture content of 0.046% instead of 0.110%. In this instance the fiber has denier of 24.7; tenacity of 2.2 g/den; elongation of 50.82%; electrical resistance of 2.01×10^6 ohms/cm; and an L:D ratio of 0.393:1.

The fiber made in run 3 was made in accordance with U.S. Patent 3,969,559. As can be seen from the L:D ratio of the fiber of run 3, the nonconductive constituent had a lower melt viscosity than the conductive constituent and, therefore, encapsulated or wrapped around the conductive constituent. In contrast, in runs 1 and 2, the conductive polymer, being of a lower melt viscosity than the nonconductive polymer, wrapped around the more viscous nonconductive polymer.

Heat Conductive Polytetrafluoroethylene

A. Wimmer; U.S. Patent 4,163,825; August 7, 1979; assigned to Chemiefaser Lenzing AG, Austria describes a method of making threads or fibers of polytetrafluoroethylene with improved heat conductibility which uses thread of fiber cores enclosed by envelopes which contain fine particle polytetrafluoroethylene, a heat-conducting pigment, and a binder. Such threads or fibers can be used to form a sealing packing.

Example 1: 4.20 kg of powdered graphite with a powder density of 375 g/l and 0.21 kg of emulsifying agent Triton in 2.59 kg of water were added to 3.00 kg of an aqueous dispersion of polytetrafluoroethylene with a content of 60% by weight of solids. The result was 10 kg of a conductible polytetrafluoroethylene dispersion with a total solid content of 60% by weight and a viscosity of 100 cp.

A polytetrafluoroethylene basic yarn with a titer of 10,000 dtex was immersed in the dispersion, and the surface of the yarn was stripped with a fork. The impregnated yarn was dried. The dried product had a titer of 20,000 dtex, a very good heat conductibility, and the advantage of being interlaceable to form sealing packages.

Example 2: As described in Example 1, 10 kg of a polytetrafluoroethylene dispersion were produced by mixing 6.43 kg of an aqueous polytetrafluoroethylene dispersion with a solid content of 35% by weight, 2.25 kg of electrode graphite, 0.09 kg of emulsifying agent and 1.23 kg of water. The dispersion mixture had a total solid content of 45% by weight and a viscosity of 15 cp. It can be used as described in the abovementioned examples.

In Figure 7.1a there is a perspective illustration of a single polytetrafluoroethylene thread with a titer of about 20 dtex. The cross-section of the thread **1** is rectangular since such threads, according to the split-peeling method, are milled off a rotating roller consisting of solid polytetrafluoroethylene. The thread **1** is embedded in an envelope **2** consisting of fine particle polytetrafluoroethylene, a heat-conducting pigment, such as soot or graphite and a binder.

Figure 7.1b shows a bundle consisting of enveloped single threads, the embedding mass again being denoted by **2** and the single threads by **1**.

Figure 7.1: Heat Conductive Polytetrafluoroethylene

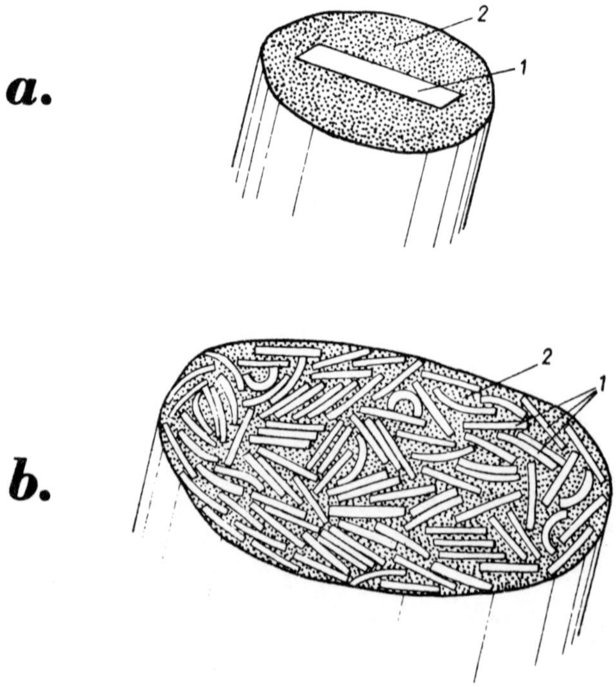

(a) Cross-section of a single thread
(b) Bundle of threads

Source: U.S. Patent 4,163,825

REINFORCING YARNS

Carbon Yarn Coated with Resorcinol-Formaldehyde-Latex

M. Toki, N. Kusatsu, Y. Kohno, and K. Matsugasako; U.S. Patent 4,044,540; August 30, 1977; assigned to Toray Industries, Inc., Japan, disclose multifilament reinforcing carbon yarn coated with a resorcinol-formaldehyde-latex (RFL) composition for use in reinforced rubber articles, such as tires. The reinforcing yarn is prepared from a multifilament carbon yarn having a limited denier of about 500 to 4,000, having a twist factor of the yarn of about 0 to 250, and having individual filaments which have a denier of 1.4 to 2.0. The reinforcing yarn is impregnated with RFL, so that the amount of coated RFL is at least about 10 wt % based on the yarn. Treatment of the carbon filament yarn with an epoxy resin, prior to being subjected to impregnation with RFL, is also disclosed.

Examples 1 through 4 and Comparative Examples 5 through 8: These examples illustrate the effectiveness of construction of original multifilament carbon yarns in increasing the tenacity of RFL-coated multifilament carbon yarns.

In these examples, each original multifilament carbon yarn is prepared from Torayca Type 300 carbon yarn having a tenacity of 18 g/den per individual filament. Construction of each of the original multifilament yarns are stated in Table 2 together with total denier, denier per filament and twist factor.

Each original multifilament yarn except in Comparative Example 8 is impregnated with a liquid containing RFL having the formula shown in Table 1, and the pick-up weight of RFL is adjusted as shown in Table 2.

Table 1

Ingredient	Parts by Weight	Dry
Resorcinol	8.5	8.5
Formalin (37%)	4.1	1.5
Vinylpyridine-styrene-butadiene terpolymer latex (40%)	46.8	18.7
Styrene-butadiene copolymer (40%)	15.6	6.3
Water	25.0	–
Total	100.0	35.0

The product was matured for 7 days at room temperature. Its solid content was 35% by weight.

In Comparative Example 8, the original multifilament yarn was impregnated with a liquid prepared by diluting the compound given in Table 1 with water to a solid content of 17.5%.

Each impregnant was dried by heating to 150°C in air for 2 min and baked by heating to 200°C in air for 1 min. The results are shown in Table 2.

Table 2

Example Number	Total Denier of Yarn	Denier per Filament	Twist Factor	RFL Pick-up (%)	Tenacity of RFL Coated Yarn g/den
1	900	0.6	45	25	9.4
2	1,800	0.6	64	20	7.4
3	3,600	0.6	90	17	7.1
4	1,800	1.8	64	21	7.0
5	4,500	0.6	100	13	5.8
6	1,800	0.3	64	9.8	6.5
7	1,800	0.6	297	15	5.5
8	1,800	0.6	64	9.6	5.2

Assembly of Metal-Coated Carbon Fibers

K. Nakamura, Y. Fukube and T. Osaki; U.S. Patent 4,132,828; January 2, 1979;

assigned to Toho Beslon Co., Ltd., and Nihon Shinku Gijutsu KK, both of Japan describe an assembly, useful in producing a carbon-fiber-reinforced metal, of a plurality of carbon fibers each coated with a matrix metal layer, the fibers having bonded points at the metal layers to form a two-dimensional network structure or both a two-dimensional network structure and a three-dimensional network structure, produced advantageously by an ion-plating process or a vacuum deposition process.

Example: In a vacuum system, nontwisted tows each consisting of 6,000 carbon filaments having a diameter of 9.2 μ, a tenacity of 200 kg/mm^2 and a Young's modulus of 19.5 tons/mm^2 were aligned in a tape form in one row with the width being adjusted to 18 mm and the average number of fibers in the height direction being adjusted to 3. High purity aluminum was placed in the system, and the system was evacuated to a vacuum of less than 1 x 10^{-4} torr. Then argon was introduced into the system and, while maintaining the argon pressure at 2 x 10^{-2} torr, a voltage of -1.0 kV was applied to the carbon fibers to perform plasma-etching of the carbon fibers for 5 minutes.

After etching, aluminum was evaporated by resistance heating, and deposited by ion-plating on the carbon fibers held at 100°C at a deposition rate of 1.0 μ/min. Thus, a tapelike assembly of carbon fibers having an aluminum coating with a thickness of 4 μ was obtained. The proportion of the carbon fibers in the assembly was 30% by volume.

Twenty such assemblies were stacked in one direction and heated under pressure at 560°C and 900 kg/cm^2 for 1 hour at a vacuum of 2 x 10^{-5} torr to produce a carbon-fiber-reinforced metal having a carbon fiber content of 31% by volume. The metal partly escaped from the system at the time of heating under pressure. The resulting carbon-fiber-reinforced metal had a tenacity, in the direction of the fiber axis, of 46 kg/mm^2 and good electric conductivity.

Gut for Tennis Racket

A gut is provided by *T. Kanemaru, Y. Hisatomi and T. Nishimoto; U.S. Patent 4,084,399; April 18, 1978; assigned to Nippon Carbon KK, Japan* for a sports racket, such as a tennis racket. The gut is made with carbon fibers. The gut may be made solely of carbon fibers, or there may be a combination of carbon fibers and organic and/or inorganic fibers. A coating may be provided for the resultant product, which may be a synthetic resin or a metal. The carbon fiber is above 150 kg/mm^2 in tensile strength.

Figures 7.2a and 7.2b show a gut in accordance with one embodiment. Element **1** is a carbon fiber multifilament (18 tons/mm^2 in Young's modulus, 300 kg/mm^2 in tensile strength and 1.7% in elongation) which is prepared such that 2,000 carbon fiber monofilaments, each being 5 μ in diameter, are twisted together in a combination of 200 turns/m. Element **2** is a filament of nylon 6 No. 1 (0.16 mm in diameter) which is 34 kg/mm^2 in tensile strength and 200 kg/mm^2 in Young's modulus. One of the aforegoing carbon fiber multifilament **1** and three of the aforegoing nylon filaments **2** are twisted together in a combination of 50 turns/m, and the resultant intermediate product is coated with a nylon layer **3** by passing the same through a molten nylon bath.

Figure 7.2: Gut for Tennis Racket

a.

b.

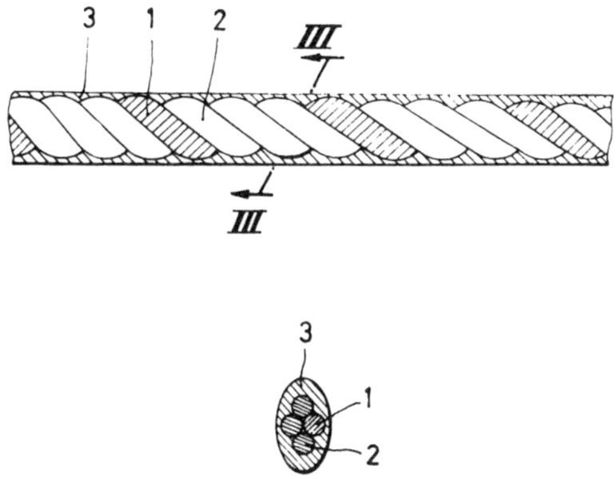

(a) Sectional side view
(b) Sectional view along line **III–III** in (a)

Source: U.S. Patent 4,084,399

The gut thus obtained is 1.4 mm in diameter and various properties thereof have been measured. This product which is identified as A, is compared with conventional ones in the table below.

	Wire Density (g/m)	Gut Diameter (mm)	Breaking Strength (kg/gut)	Modulus (kg/unit)
Product A	1.35	1.4	70	270
Whale gut product	1.65	1.25	35	290
Sheep gut product	1.54	1.4	48	260
Nylon gut product	1.69	1.45	56	230

METAL CARBIDES

Flexible Metal Carbide Fabrics

Flexible metal carbide fabrics are produced by *B.H. Hamling; U.S. Patent 4,162,301; July 24, 1979; assigned to Union Carbide Corporation* by a process that involves:

Impregnating a preformed organic polymeric fabric with a solution of a metal compound;

Heating the impregnated fabric to evolve volatile decomposition products and to leave a carbonaceous relic containing the metal in finely dispersed form; and

Further heating the relic to 1000° to 2400°C in a nonoxidizing atmosphere to form the metal carbide. Boron carbide and silicon carbide fabrics produced by this process are attractive for high temperature structural applications.

Example 1: 7.1 g of 1.5 den regular viscose fabric is preswelled by immersion in water for 1 hr, and then immersed in 100 ml of a uranyl chloride (UO_2Cl_2) aqueous solution having a molarity of 4.20 and a specific gravity of 2.11. Since the water in the preswollen region dilutes the solution, the latter is replaced with a new 100 ml of 4.2 molar UO_2Cl_2 solution. After 10 min of immersion in the solution, the fabric is removed and centrifuged to remove excess solution. The fabric contains 3.3 g of solution per gram of rayon. It is dried in a warm air stream and after drying weighs 22.2 g, including 1.57 g imbibed uranium per gram rayon.

The fabric is next placed in a tube-type furnace and heated in a vacuum of about 1 μ Hg at a rate of 10°C/hr to 900°C. The furnace is then allowed to cool to room temperature under the vacuum. The resulting carbonaceous fiber relic weighs 14.9 g and contains finely dispersed uranium with a carbon to uranium atom ratio of 3.7. The fabric is black, lustrous and strong; there is no fiber breakage.

These carbonaceous fiber relics have approximately the correct carbon to uranium ratio for producing monouranium carbide fibers by heating to temperatures of about 1500° to 2000°C in a nonoxidizing atmosphere. For example, UC fibers by the reaction $UO_2 + 3C \rightarrow UC + 2CO$ can be prepared by heating uranium impregnated rayon fabric to about 1700°C and maintaining this temperature level for a period of 1 hr in an argon atmosphere.

Example 2: The starting rayon fibers were 1.5 den viscose tow (133,000 continuous filaments in the tow). The tow weighing 6.8 g was preswollen in water for 15 min prior to immersion in 2.84 molar $ZrOCl_2$ solution for 4 hr. It was next centrifuged on unimbibed solution from between the fibers and dried in air. The salt-loaded rayon tow weighed 12.5 g and contained 0.84 g zirconium salt per gram rayon.

The tow was placed in a tube furnace and heated under a vacuum for 1 to 10 μ Hg pressure at a rate of 50°C/hr to 400°C, then 100°C/hr to 1000°C. The pyrolyzed fibers were black, had a high luster and weighed 6.0 g. The fiber relics contained 33.7% carbon and 43.0% zirconium by weight.

In order to carry out the carburization reaction between the carbon and the zirconium, the fiber relics were placed in a graphite crucible and heated at 1900°C for 2 hr in a hydrogen atmosphere employing a high frequency induction furnace. The resulting fibers weighed 4.2 g, had a black-gray color, showed no evidence of sintering together and had a diameter of 4 to 5 μ. They were very flexible and had a tensile strength between 100,000 and 150,000 polycrystalline face-centered-cubic zirconium carbide (ZrC). No graphite lines were evident in the x-ray patterns. The fibers were over 1 foot long with crystal grain size of less than 0.2 μ.

Example 3: A one-foot length of 20 den viscose rayon tow (10,000 continuous filaments in the tow) was immersed in an aqueous solution containing 25% by weight boric acid for 1 hr. The solution was kept at 90°C during the imbibment to keep the boric acid in solution. The tow was then centrifuged while still hot to remove unimbibed boric acid from between the fibers. After drying the tow in air, it was subjected to the same pyrolysis and carburization treatment as described in Example 2.

The reacted fibers were black, free from each other, and had appreciable flexibility and strength. X-ray diffraction analysis showed the fibers to be composed of crystalline boron carbide (B_4C). The fiber diameter was about 20 μ, and the crystal grain size was less than 0.2 of such diameter.

Calcining Acrylonitrile plus Powdery Metal

K. Oohara, T. Shizuki, H. Mitamura, and M. Sugino; U.S. Patent 4,126,652; November 21, 1978; assigned to Toyo Boseki KK, Japan disclose a process for preparation of a metal-carbide-containing molded product which comprises heating a molded composition comprising at least one powdery metal selected from the group consisting of B, Ti, Si, Zr, Hf, V, Nb, Ta, Mo, W, Cr, Fe and U, and having an average particle size of not more than about 50 μ and an acrylonitrile polymer at a temperature of about 200° to 400°C and then calcining the resulting product at a temperature of about 900° to 2500°C in an inert atmosphere.

Example 1: Dimethylformamide (70 parts), metallic silicon (average particle size, 3 μ or less, purity 96% by weight) (15 parts) and polyacrylonitrile (molecular weight about 100,000) (15 parts) are mixed together under stirring at 50°C for several hours to obtain a dispersion having a viscosity of about 300 poises. This viscous liquid is spun up into a hot air-spinning tube at 200° to 250°C, and the solidified yarns are reeled off at a speed of 50 m/min, stretched about 3 times at 125°C under pressure of steam and dried to obtain a metallic silicon-containing polyacrylonitrile fiber. The strength of the fiber is 4.1 tons/cm^2, and the elongation is 10%.

The thus obtained fiber is treated with the hot air of 200° to 350°C for 1 hr and, after gradual elevation of the temperature in the argon atmosphere, a heat treatment at 1300°C is effected for about 1 hr to afford a fiber substantially consisting of silicon carbide. The strength of the fiber is 20 tons/cm^2, and the modulus of elasticity is 2,500 tons/cm^2.

Example 2 (Comparative): The same procedure as in Example 1 is repeated but using metallic silicon having an average particle size of 80 μ. The obtained liquid shows a viscosity of about 200 poises. When the liquid is subjected to spinning, snapping of yarns takes place frequently, and reeling cannot be effected in practice.

Composite Material of Silicon Carbide and Beryllium

According to *S. Yajima and M. Hamano; U.S. Patent 4,141,726; February 27, 1979; assigned to The Research Institute for Iron, Steel and Other Metals of the Tohoku University, Japan,* beryllium composite material reinforced with continuous silicon carbide fibers is obtained by bonding tightly continuous silicon carbide fibers obtained by baking spun fibers of organosilicon high molecular weight compound, with beryllium and its alloys as a matrix. The silicon carbide fiber-beryllium composite material is excellent in the mechanical strength, heat resistance and oxidation resistance, and is useful as a material for aerospace instruments and as a material for the nuclear industry.

Example 1: Silicon carbide fiber-beryllium composite materials were produced from beryllium metal as a matrix and silicon carbide fibers. The silicon carbide fibers were produced from a polycarbosilane synthesized from polysilane which was obtained from dimethyldichlorosilane by thermal polycondensation. Bundles of silicon carbide fibers obtained by baking at 1250°C in N_2 gas and having a thickness of 10 to 20 μm and a length of 50 mm were set in an alumina crucible, and the crucible was hung at the upper portion of a heating chamber which was connected to a vacuum line of 1×10^{-3} mm Hg.

Beryllium metal was charged into an alumina vessel, and the vessel was placed at the lower portion of the heating chamber. Beryllium metal in the vessel was heated from the exterior of the vessel and melted at about 1300°C. The crucible was brought down, dipped in the melted beryllium for 1 min, subjected to 5 atmospheres of argon gas pressure for 5 min, and then brought up. The resulting silicon carbide fiber-beryllium composite material was worked into a rod which was used as a test piece.

Example 2: A 30 x 30 mm square foil 0.05 mm thick consisting of a 99:1 wt % Be:Ag alloy and a layer of unidirectional SiC crude fibers (10 to 20 μ thick, 30 mm long) were alternately superposed to produce a laminate 2 mm thick containing 15% by weight of SiC fibers. The laminate was hot pressed at 1250°C for 30 min at a pressure of 200 kg/cm^2 in argon to obtain a composite of the SiC fibers and the Be alloy.

The composite has a thickness of 1.5 mm and a density of 2.0 g/cm^3. Microscopic observation showed that the alloy had been softened and permeated into spaces between the layered SiC fibers. This composite was somewhat superior to a composite containing 10% of SiC fibers in mechanical and other properties.

Core Yarns and Other Composite Fibers

TIRE CORDS AND BEARINGS

Aramid with Polyester or Nylon

P.D. Shepherd and R.S. Bhakuni; U.S. Patent 4,155,394; May 22, 1979; assigned to The Goodyear Tire & Rubber Company disclose a cord composite suitable for rubber tire reinforcement comprised of a plurality of plies of yarns selectively cabled in a manner so that, upon application of longitudinal stress, at the initial elongation of the cable, the primary load bearing ply is a polyester or a nylon yarn and so that after appreciable elongation of the cable, the primary load bearing yarn is an aramid yarn. The disclosure further relates to a pneumatic rubber tire, industrial belt or hose having a carcass containing a fabric of such a cord as a reinforcing member. For further understanding of the process, reference should be made to the following drawings.

Figure 8.1: Tire Cord Composite

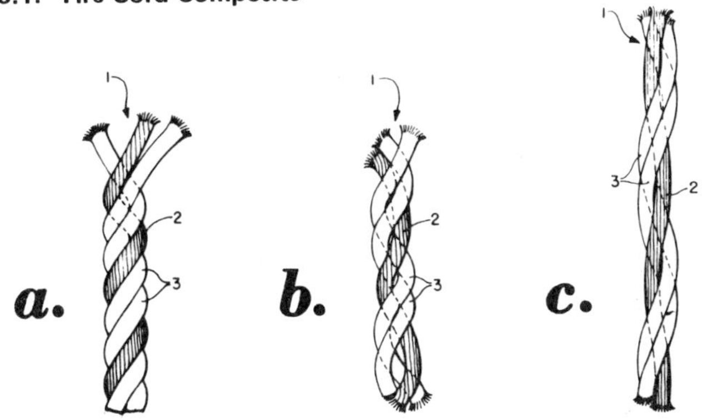

(continued)

Figure 8.1: (continued)

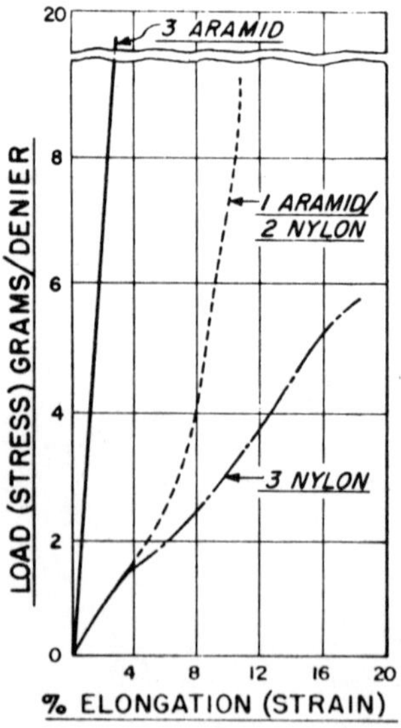

d.

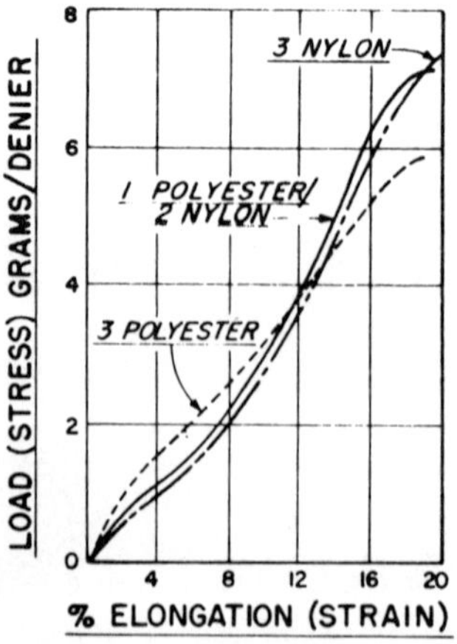

e.

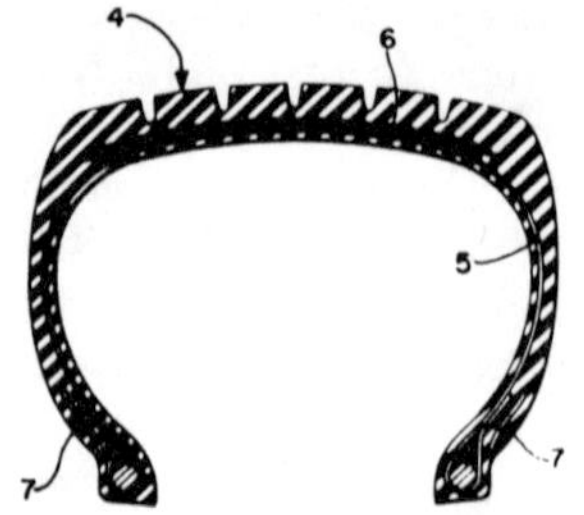

f.

(a)(b)(c)	Three-ply cable of twisted dissimilar yarns, as it is longitudinally stretched
(d)	Stress/strain graph showing stress/strain relationships of various three-ply cords of cabled aramid, aramid/nylon and nylon yarns
(e)	Stress/strain graph showing stress/strain relationships between three-ply cords of cabled polyester, polyester/nylon and nylon yarns
(f)	Cross-sectional view of a pneumatic tire showing cord fabric reinforcement

Source: U.S. Patent 4,155,394

Referring to the drawings, in Figure 8.1a a cord **1** of helically cabled dissimilar yarns is shown having a twist of about eight turns per inch comprised of one ply or yarn of aramid **2** having a residual twist of about six turns per inch and two plies or yarns of polyester or nylon **3** having a residual twist of one-half turn per inch.

In Figure 8.1b, as the cord **1** is initially stretched, the load is primarily taken by the polyester or nylon cord **3** as they elongate by stretching in an axial fashion whereas the essentially inextensible aramid yarn **2**, instead of stretching, simply elongates by beginning to straighten out.

In Figure 8.1c, as the cord **1** is stretched further in the longitudinal direction, the polyester or nylon yarn **3** continues to elongate by stretching, until the aramid yarn **2** essentially reaches the limit of its elongation by straightening and then becomes the principal load-bearing member of the cord.

To more clearly illustrate the problems involved in cabling together the yarns to achieve the cooperative effort toward providing a cord suitable for tire reinforcement purposes, reference is made to Figures 8.1d and 8.1e. In Figure 8.1d, it can conveniently be seen that a three-ply aramid yarn is substantially inextensible and comparatively undergoes very little elongation, whereas a three-ply nylon cord under similar longitudinal load has a substantially much greater elongation. Uniquely, forming a cord of helically cabled yarns containing one aramid ply and two nylon plies by the method of this process is shown in Figure 8.1d to yield a cord having intermediary elongation for a comparable load. In this instance, although the two plies of nylon stretch somewhat axially, the aramid ply, instead of stretching, elongates by tending to straighten out along the centerline of the cord. Through this longitudinal load-bearing phenomena the process is achieved.

In Figure 8.1e it is shown that the load versus elongation of polyester and nylon three-ply cabled yarns are substantially the same. This drawing amplifies the convenience of the heretofore teaching to simply adjust the twist of the individual polyester and/or nylon yarn in the cable to achieve a cooperative effort for yarns having substantially equal moduli instead of the teaching of the process. Here, both the polyester and nylon yarns cooperate to bear the load practically throughout the application of longitudinal stress. This is practically possible and desirable because of the similarity of their moduli.

Figure 8.1f, a cross-sectional view of a pneumatic rubber tire, shows a tread **4** and a positioning of a tire reinforcing fabric in the carcass ply **5**, belt **6** and chipper **7** of a cord which is comprised of the plies of helically cabled dissimilar yarns.

Sheath/Core Weft Yarns in Reinforcing Fabric

E.R. Barron; U.S. Patent 4,024,895; May 24, 1977; assigned to E.I. DuPont de Nemours and Company describes a reinforcing fabric comprising a web of tire cords maintained in spaced relationship with sheath/core weft yarns having a cotton sheath about a core of spin-oriented polyester filaments. The weft yarns have a high initial modulus and a high break elongation which is substantially retained after heat-aging the yarn at 450° or 475°F. The fabric is particularly useful as reinforcement in radial tire constructions.

Example: Conventional cotton system ring spinning technology is used to prepare the sheath-core yarn. Seven ends of a commercial 50-grain carded cotton sliver are processed on a Saco-Lowell DE-7C Versamatic Drawing Frame into a 58 grain sliver (1 pass, 6.0 X draft). The drawn sliver is then processed into a 1.6 hank roving on a Saco-Lowell FS-2 Roving Frame (11.1 X draft). The sheath/core yarn is prepared on a Saco-Lowell spinning frame, a single roving being fed to each spinning position. The spin-oriented polyethylene terephthalate core yarn, a commercially available 100-denier, 34-filament yarn with about 120% elongation at break, is fed to the last drafting roll by appropriate guides and a tensioning device to maintain very low tension. Spinning conditions are: 29.3 X draft, 19.2 Z tpi, No. 3 traveler, and 6,000 rpm. Spin tubes are twist set using the following conditions: 90 minutes at 180°F (82°C) dry bulb and 170°F (77°C) wet bulb. Twist-set yarns are coned on a Schlafhorst Auto-Coner; slub catchers are left open and a wax applied. This yarn has the following properties: yarn size, 23.7/1 cc (224 denier); sheath, carded cotton (124 denier); core, 100-34 spin-oriented polyester; composition, 55/45 cotton/polyester; twist multiplier, 4.0.

Tenacity, initial modulus and break elongation are measured according to the ASTM designation D-2256-69. Results on the as-spun yarn were: tenacity, 0.92%; break, 1.17 gpd; initial modulus, 19.1 gpd; elongation, 106%.

The core-spun yarn described is used as a pick yarn to prepare a 2,000 denier (1,000/½ cord construction), polyester carcass fabric with 2 picks/inch (0.8 pick/cm). A 63-inch (160-cm) wide fabric is woven on a Draper X-3 loom. The resulting fabric is processed with a standard two-step hot stretching process at 80 ypm (73.2 mpm). After application of the first dip, fabric is cured 1 min in an oven at 475°F (246°C). Then a second dip is applied. The fabric is dried 1 min in a second oven at 360°F (182°C) and finally cured for 1 min in a third oven at 460°F (238°C).

The treated polyester carcass fabric is 54 inches (137.2 cm) wide and has 29 cords/inch (11.4 cords/cm). Elongation of pick yarns removed from the processed fabric is 120%, when tested at 0.29 inch (0.74 cm) gauge length. This fabric is made into a two-ply polyester carcass, steel-bolted radial passenger tire. Carcass cord-to-cord spacing problems (e.g., undulations in the side walls), are considerably reduced in these tires as compared with tires made with carcass fabrics which contain conventional cotton pick yarns.

Steel Wire Cord from Particulate Iron Oxides

A method for making steel wire cord from particulate iron oxides with the aid of a fiber-forming acrylic polymer is disclosed by *E.J. Dobo; U.S. Patent*

4,023,989; May 17, 1977; assigned to Monsanto Company. A plurality of continuous filaments are first formed by wet-spinning an acrylic polymer spin dope in which particles of iron oxide are dispersed. The resulting filaments are drawn to increase their tenacity and then twisted to form a precursor cord. This is followed by exposing the precursor to a reducing atmosphere (e.g., a gaseous mixture of hydrogen and carbon monoxide) at a temperature in the range of from about 900° to 1150°C for a period of about 3 to 8 minutes. Under these conditions, the iron oxide particles are reduced to the metallic state and the polymer in the precursor is pyrolized to carbon and by-product gases. The carbon diffuses into the resulting iron, and the individual metal particles sinter to form continuous steel wire cord having an essentially ferritic/pearlitic structure. When the product is converted to a tempered martensite structure, tensile strengths of at least 280,000 psi are attainable.

Example: 1,000 g of hematite (Fe_2O_3) and 212.1 g of a copolymer consisting of 93% by wt of acrylonitrile and 7% by wt of vinyl acetate were intimately mixed in a rod mill for 10 hours. Thereafter, a solvent mix consisting of 850 cc of dimethyl acetamide, 0.5 cc ethylene glycol and 1.2 cc of sorbitan monopalmitate was chilled to 10°C and placed in a large Waring blender. The mixture of iron oxide and polymer was then also transferred to the blender and stirred in by hand to give a fairly uniform mixture. The solvent was prechilled to reduce its solvency so that the polymer could be dispersed mechanically with limited amounts going into solution. The blender was then brought to high speed and further blending of the oxide and complete solution of the polymer took place. The blender was turned off when a final temperature of 42.5°C was attained as sensed by a thermocouple in the mixture. The heat for the temperature rise resulted from the degradation of mechanical energy supplied by the blending device. During the mixing period, a vacuum of 22 inches of mercury was pulled on the contents of the blender to reduce the amount of air entrapment in the mixture.

The contents of the blender were transferred to the dope pot of a wet spinning line where the precursor dope was subjected to a vacuum of 22 inches Hg for one-half hour and then pressurized to 35 psi for one-quarter hour. This step was undertaken to again reduce entrained air that could cause voids in the precursor fiber.

A positive displacement pump delivered 15.6 cc/min of the dope through a cup spinnerette which had 5 holes each of 20 mils in diameter. Upon emerging from the spinnerette, the dope thread lines entered a coagulation bath which was at a temperature of 24°C. The coagulation system employed consisted of a mixture of 50.2% by volume of ethylene glycol and 49.8% by volume of dimethyl acetamide. An acrylic plasticizer (N,N-dimethyl lauramide) was also present in the bath in an amount of 0.1% by weight of the coagulating mixture. The thread line was taken up at the first godet (thread advancing rolls) at 20 ft/min and washed with the bath solution to continue the gentle coagulation process. The second godet received the thread line at the rate of 20 ft/min. Here the thread line was washed with water to complete the coagulation. Then the thread line was stretched in boiling water to orient the filaments. This step occurred between the third and fourth godet which rotated at the rate of 40 ft/min. On leaving the fourth godet, the filamentary thread line was taken up on a standard cap twister with the spindle rotating at 720 rpm to provide 1.5 tpi in the thread line.

The take-up bobbin from the filament spinning line was placed in the feed position of a furnace conversion system, and a thread line was fed into the furnace at a rate of approximately 11 in/min on a belt moving at the rate of 4.0 in/min. The dwell time in the furnace was about 6 min at a temperature of 1100°C. The difference in the rates of movement between the belt and the precursor feed is accounted for by the shrinkage of the thread line which occurs during the conversion operation. To coordinate the feed rate of the thread line with the belt movement, the thread line position before entering the furnace was sensed by a photoelectric relay. A reducing gas mixture consisting of 92.3% by vol of hydrogen, 4.4% by vol of methane and 3.3% by vol of carbon monoxide was introduced into the furnace near the end opposite to that of the feed entry.

The steel cord product obtained was of an essentially pearlitic-ferritic structure having a carbon content of 0.7% by wt. In spite of the closeness of the individual filaments due to the twist imparted, the cord product showed no married strands.

To convert into a tempered martensite, the steel wire cord was heated continuously in a furnace to 831°C and then quenched in oil at 102°C to give a martensitic structure. Post-tempering at 300°C in oil for 5 min gave a tempered-martensitic structure having a tensile strength of 280,000 psi and an elongation of 2.1%.

Polyurethane-Impregnated Polyamide Cable

According to *E.P. Babayan; U.S. Patent 4,095,404; June 20, 1978; assigned to Hitco*, a high strength, low weight, electromechanical cable is manufactured from aromatic polyamide multifilament yarns impregnated with a hydrolytically stable polyurethane resin to form a protective coating. The coating provides load adjustment from fiber to fiber, eliminates abrasive self-destruction of the fibers during flexing of the yarn under load, protects the fiber to some extent from ultraviolet radiation, aggressive chemicals or abrasive particles and makes it possible to preform the yarn.

The coating comprises a reaction product of a liquid tetramethylene glycol, an aliphatic/cycloaliphatic diisocyanate and a diamine coupling-curing agent. The impregnated yarns are dried, twisted together, heated above the softening point of the resin to fuse the coatings of adjacent yarns, and then cooled to form a set twisted helix of the yarns.

The high modulus fibers are a synthetic, organic polymer having very high tensile strength and resistance to stretch and having light weight, good toughness and environmental stability.

The preferred material is a continuous yarn known as PRD-49 or Kevlar 49 (DuPont) which is an aromatic polyamide. A fabrication technique is schematically shown in Figure 8.2a and Figure 8.2b.

Figure 8.2: Manufacture of a Polyurethane-Impregnated Polyamide Cable

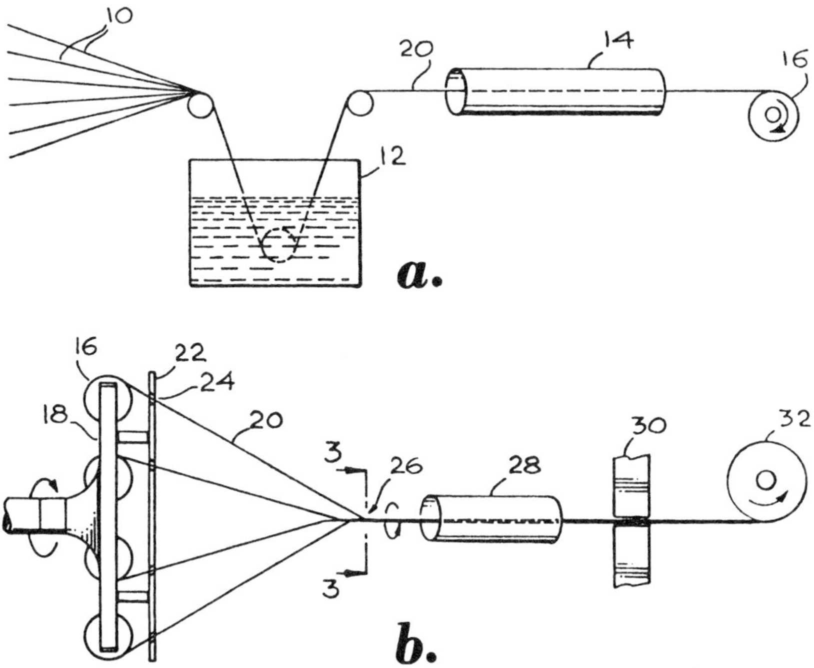

(a) Impregnation stage
(b) Composite formation stage

Source: U.S. Patent 4,095,404

Referring to the figure, the individual yarns **10** are precoated with resin in bath **12** and dried in oven **14** at a temperature from 150° to 200°F. The precoated yarns **20** with coating thereon are then wound on spools **16** mounted in a rotatable frame **18**. The coated yarns are passed through a template **22** which rotates with the frame. A series of concentric holes **24** are drilled in the template [the same 1, 6, 12, 18. . .6(n−1) pattern used in winding stranded ropes], and each yarn **20** is passed through its own individual hole. The yarns are pretensioned and then twisted together in a conveying helix **26** as they pass through a preheater **28** at a temperature of 200° to 300°F (to soften the resin matrix to a nearly fluid state), pulled through a heated sizing die **30** at a temperature of 170° to 220°F, and cooled to room temperature before being wound on a storage reel **32**.

The result is a tightly twisted helix **26** of filaments **20** which retain an infinitesimal coating of resin for lubrication and structural bonding. The helix angle (lay length) is controlled by fixing the ratio of turns per unit of advance of the precoated yarns. Diameter of the strength member becomes primarily a function of the number of filaments, and is only weakly sensitive to lay length, yarn tension-initial resin fraction or die temperature. The sizing die acts primarily to control circularity and to wipe away excess resin.

The simultaneous twisting/heating process also reduces void content to a negligible fraction ($\leqslant$1%) by wringing trapped air and solvent vapor out of the filament helix. Those minute voids which remain are confined to a thin annulus of resin between the filaments and the outer surface, and do not degrade the properties of the member. Packing fractions for the filaments in the composite member have been running between 66 and 69%.

The polyurethane lacquer is impregnated onto the fibers in an amount of from 5 to 95% by wt, suitably from 5 to 40%. Optimum physical properties are provided in the range of 20 to 35% by wt. The polyurethane is the reaction product of a stoichiometric mixture of an aliphatic/cycloaliphatic diisocyanate with a liquid polytetramethylene glycol which is further cured with an aliphatic diamine coupling-curing agent. The final polyurethane is a soluble thermoplastic capable of solution-coating of the fibers and capable of heating to fusion after application. The polytetramethylene glycol has a molecular weight from 500 to 3,000 and is suitably a Polymeg 650, 1,000 or 2,000.

The preferred diisocyanate is an alkylated, isocyanatoalkyl cyclohexyl isocyanate of the formula:

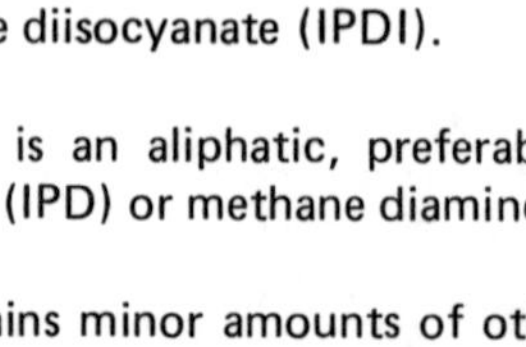

where R_3 is lower alkyl. When all the R_3s are methyl and R^1 is methylene, the compound is isophorone diisocyanate (IPDI).

The coupling-curing agent is an aliphatic, preferably cycloaliphatic, diamine such as isophorone diamine (IPD) or methane diamine.

The composition also contains minor amounts of other additives such as 0.1 to 0.5 phr of a curing catalyst such as dibutyltin dilaurate, 1 to 5 phr of a drying agent such as a molecular sieve. Colloidal or amphoteric silicate fillers can be added in an amount from 1 to 10 phr to increase the strength of the coating. Minor amounts of other additives such as ultraviolet absorber, antioxidants or dyes and pigment can be added if desired.

The reactive ingredients are combined in a solvent system which is a solvent for the ingredients and for the polymer. Preferably, the Polymeg, molecular sieve, catalyst and IPDI are first reacted in xylene to form a prepolymer. The diamine dissolved in part of a mixture of isopropanol and methyl ethyl ketone (MEK) is slowly added to the prepolymer until the pH is from 7 to 8. Isopropanol provides a retardant effect avoiding gelling and xylene and MEK contribute to chain build of the polyurethane.

A preferred formulation for the polyurethane lacquer is provided in the following table:

Ingredient	Range	Preferred
	.... (pbw)	....
Part A		
Polymeg 650	100	100
Molecular sieve	1-5	2
Dibutyltin dilaurate	0.1-0.5	0.2
Xylene	50-150	92
IPDI	stoichiometric	69.3
Part B		
Isopropanol	150-400	244
IPD	stoichiometric	26.9
MEK	150-400	237

Part A is mixed and prereacted to form a prepolymer. Part of the isopropanol and MEK are added to Part A and the IPD is dissolved in the remaining solvent and slowly added until the pH is 7 to 8. If the final pH is above this range, the composition turns yellow on aging and the properties degrade. The lacquer is stable and does not contain any reactive isocyanate groups. Test specimens were cast and the solvent evaporated. The films exhibited a tensile strength of 5,000 to 6,000 and an elongation from 400 to 500.

Strength members for cables were prepared from 380 denier PRD-49 impregnated with the polyurethane lacquer according to the procedure of Figures 8.2a and 8.2b. The composite members exhibit excellent tensile strength and very low specific gravity, the significance being most apparent when the properties of the strength members are compared to commercial cabling steels.

Low-Friction Fabric Bearing

According to *R.J. Matt; U.S. Patent 4,074,512; February 21, 1978; assigned to Textron, Inc.*, improved wear and load-carrying properties result for a resin-impregnated fabric bearing wherein the bearing surface includes, as the most significant fabric component, a yarn which is a compounded bundle of sized TFE (tetrafluoroethylene) filaments and sized filaments of high-temperature nylon, in the proportion of at least no more than 50% TFE by volume. The yarn is loosely twisted, to the extent that after construction of a fabric with the yarn, the yarn is fully washable, to remove the sizing, down to the innermost filaments. This enables full impregnation of the fabric with synthetic resin, and curing in substantially intimate and continuous and void-free relation of the resin to all filaments within the yarn of the fabric.

Example: A satin-weave fabric, with 4:1 face, is produced using a multicomponent warp yarn and a filling yarn of continuous-filament, type 430 Nomex (a high-temperature nylon). The filling yarn is strictly described, for each end, as 200/100/3S, of the type 430 Nomex; i.e., 200-denier Nomex yarn, consisting of 100 filaments, the yarn being twisted 3 times per inch. The warp yarn is strictly described as one end of 400/60/0 Teflon (TFE filament) and two ends of 200/100/0 type 430 Nomex; in other words the TFE end comprises 400-denier TFE yarn with 60 filaments, and with 0 twist, while the two Nomex ends are each as described for the filler, but with 0 twist. The components of the warp yarn are then twisted 3 turns per inch. The sectional areas of the TFE filaments and Nomex filaments in the warp yarn are chosen to present about 25% TFE at the bearing surface, the rest of the bearing-surface filaments being Nomex.

After weaving, the fabric is thoroughly washed in boiling water to remove the sizing, should the sizing happen to be water-soluble. Then after appropriate rinsing, drying and pressing, the fabric is cut and applied as the liner **20** of a plane spherical bearing, as shown in Figure 8.3.

Figure 8.3: Simplified Longitudinal Sectional View of a Complete Bearing

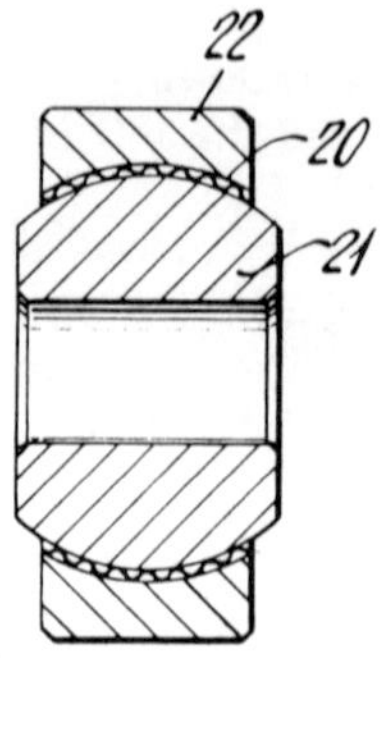

Source: U.S. Patent 4,074,512

In making the bearing of Figure 8.3, the inner ring **21** is a spherical ball which is first coated with a suitable mold-release agent. The outer ring **22** is deformable and ultimately has a spherical bore, as shown, but this is initially a straight cylindrical bore into which is inserted a cutting from the washed fabric, suitably coated with an uncured synthetic-resin adhesive; the adhesive may be a phenolic, epoxy, fluoroethylene-propylene (FEP), or polyimide. Ring deformation, bonding and curing may proceed generally as described in U.S. Patent 3,303,557.

The completed bearing exhibits superior performance, exceeding that of "all-Teflon" surfaced bearings in the important respects of maximum load, life (i.e., low wear) under load, and in general a higher upper temperature limit for satisfactory operation, the latter feature being a function of the synthetic resin and of the non-Teflon yarns or yarn components.

Composite Article Providing Seamless Fabric-Lined Bearings

In a process developed by *N.W. Butzow and B. Harris; U.S. Patent 4,107,381; August 15, 1978; assigned to Rexnord Inc.* a fabric sleeve having lengthwise yarns of a low-friction material and circumferential threads of a bondable material is secured on a mandrel utilizing in particular the heat shrinkage of the circumferential threads but also the axial shrinkage of the low-friction material. After cooling, the sleeve is impregnated with a liquid resin and may be stored for future use utilizing the ends of the mandrel for support and handling. When required, the resin is cured to form a rigid tube. The tube has an internal bearing surface a predetermined size larger than the mandrel and may be readily withdrawn from the tube and then cut to selected lengths to comprise the desired bearings.

Figure 8.4: Production of Bearings

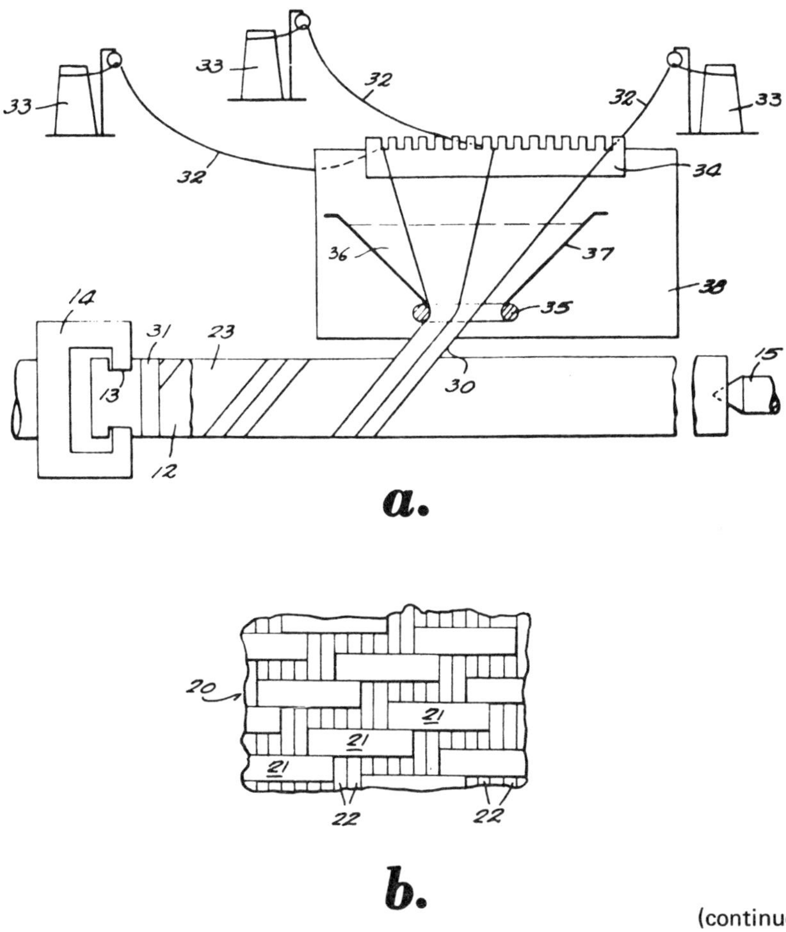

(continued)

Figure 8.4: (continued)

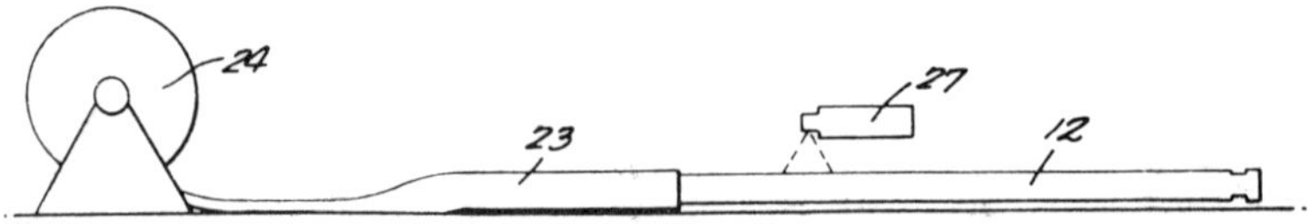

c.

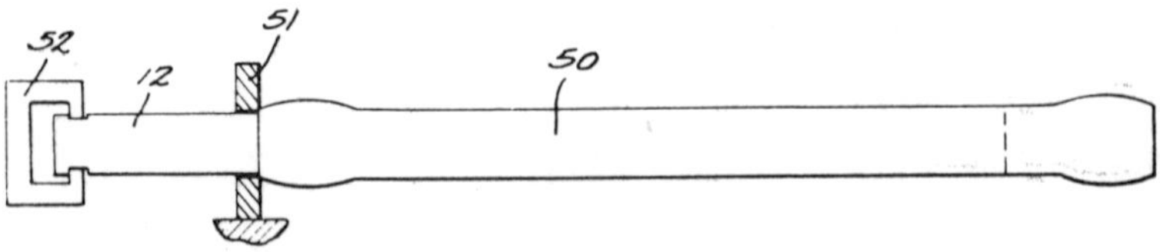

d.

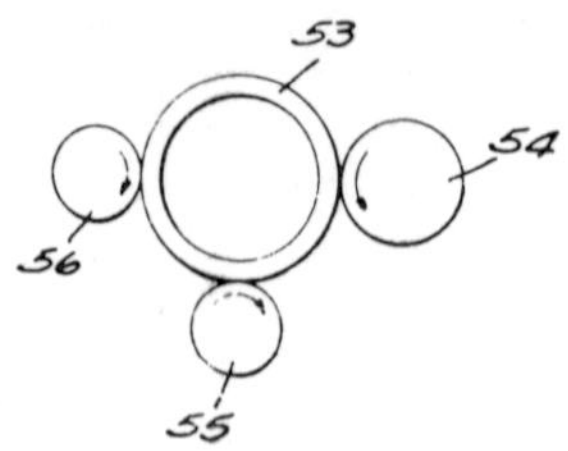

e.

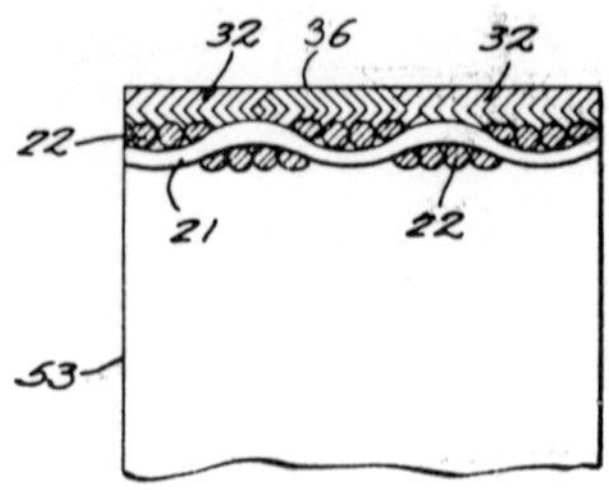

f.

(a) Side elevation of the mandrel as supported at its ends in a winding machine; the tape as it is wound over the sleeve; and, diagrammatically, the roves as they are drawn from creels through the liquid resin and through the ring of the winding machine.

(b) Enlarged view of a small section of fabric as it may be woven into a sleeve.

(c) Fabric sleeve being drawn from a reel onto the mandrel.

(d) Removal of the mandrel from the tube.

(e) Centerless grinding the outside of each bearing to the selected cylindrical diameter.

(f) Enlarged section through one-half of a completed bearing.

Source: U.S. Patent 4,107,381

In Figure 8.4a, the mandrel **12** is of stainless steel provided with a smooth polished surface and a diameter which, with correction factors, will determine the size of the finished bearing. For convenience, the mandrel is machined at both ends **13** to fit the driving head **14** of the winding machine, not otherwise shown, and the supporting arbor **15**.

The fabric **20** shown in Figures 8.4b and 8.4c includes a warp of Teflon yarns **21** and a fill of Dacron yarns **22**. The fabric is woven in the form of the sleeve **23** having a size slightly larger than the circumference of mandrel **12** and may be stored on the reel **24**. The satin weave shown is one of several which may be found suitable. Two filler yarns **22** of Dacron filaments are used, and they must, of course, be unshrunk. The Teflon yarns **21** have a denier approximately twice that of the filler yarns and may have a twist suitable to their warping. The denier of yarns **21** and **22** may be adjusted to provide optimum bearing properties with adequate securement of the fabric.

The mandrel is first coated with a parting-agent such as a silicone material available in an aerosol-pressurized spray can **27**. Sleeve **23** is drawn from the reel and cut to a length less than that of the mandrel and drawn over the mandrel. The mandrel is then placed in an oven to heat the sleeve momentarily to shrink yarns **22**.

The mandrel is then supported between the driving head and arbor of the winding machine. At this point it may or may not be desirable to apply to the sleeve a preliminary coating of resin depending upon such factors as room temperature and the resin curing cycle.

The end of the prepared tape **30** is then secured as with band **31** to the end of the mandrel, as shown in Figure 8.4a. The tape consists of a number of parallel roves **32** of glass filaments and the resin carried thereby. The roves are drawn as from the creels **33** and over the comb **34** in the usual manner.

Roves may be dipped in a pan of resin before passing through the ring **35** disposed above the mandrel. Alternatively, as shown, the roves pass downwardly through the resin **36** carried in the cone **37** and through the ring **35** which comprises the lower end of the cone. Means such as a loose plug, not shown, are provided to close the lower end of the cone sufficiently to retain the resin supply.

Comb **34** and cone **37**, including the ring, are supported by the reciprocating carriage **38**, which may also be provided with means, not shown, to keep the cone supplied with resin.

With rotation of the mandrel and reciprocation of carriage, the sleeve is wrapped with the tape in the prior art manner. Some attention must also be given the tension applied to the tape in winding. A tension in the order of 2 to 4 pounds per rove has been found adequate to keep the tape tight and also to be sure of driving or forcing the resin through the sleeve to the mandrel. With such a tension, a pressure of 2 to 4 pounds per square inch would be applied to the sleeve. It is the adjustment of this pressure, by varying the tension referred to, which can be utilized within limits to adjust the ratio of exposed Teflon and resin at the bearing surface.

When the wrap is completed, the tape is cut and the end is similarly secured with the band. The mandrel is then removed from the winding machine and is immediately ready for curing of the resin which has been applied.

The mandrel is handled at its ends only and is placed in the oven. The mandrel is supported horizontally by and between a rotating head and arbor so that the mandrel rotates in the order of 30 rpm during the curing cycle.

Referring to Figures 8.4d and 8.4e, upon hardening of the resin rigid tube **50** is formed and when cooled may be removed from the mandrel. The cooling facilitates handling, and the contraction of the mandrel relative to the tube facilitates removing the tube from the mandrel. For that purpose the end of the tube may be set against block **51**, and the end of the mandrel projecting therethrough may be pulled as with gripper **52** to remove the same.

The tube may be stored and, any time as required, the individual bearings **53** of the desired lengths may be cut therefrom. The exterior of each bearing is finished first by grinding. Centerless grinding as by placing the bearing between wheel **54** and rollers **55** and **56** is entirely satisfactory for smaller sizes. Larger sizes would be chucked, and expecially large sizes in the order of 12 inches in diameter and length might be ground in a lathe. The edges of the bearing may be smoothed with sanding.

The finished bearing **53**, as shown in Figure 8.4f, comprises the Teflon yarns **21** extending axially at the surface, the Dacron yarns **22** extending circumferentially at the bearing surface, the surrounding layers of glass filaments of roves **32** extending in opposite spirals from end to end of the bearing, and the hardened resin **36** which forms the body of the bearing.

POLYESTERS, ACRYLICS, AND VINYLS

Side-by-Side Bicomponent Polyester Filaments

Side-by-side bicomponent polyester filaments, and process for making same, are disclosed by *G. Barbe and R. Habault; U.S. Patent 4,117,194; September 26, 1978; assigned to Rhone-Poulenc-Textile, France*, wherein the filaments are of polyethylene terephthalate and polybutylene terephthalate, with the polybutylene terephthalate being sparingly crosslinked by at least about 0.20 mol percent of crosslinking agent. The sparingly crosslinked polybutylene terephthalate has a melted state viscosity VF of at least about 2,000 poises at 260°C. The polyethylene terephthalate partly envelopes the sparingly crosslinked polybutylene terephthalate, in cross section, so as to cover at least about 45% of the periphery of the sparingly crosslinked polybutylene terephthalate.

Figure 8.5 is a cross-sectional view of a spinneret head assembly which can be used in the process. The sparingly crosslinked polybutylene terephthalate, in the molten state, is passed through channel **4** of distribution plate **1**, and then is directed by chamfer **2** of bore **3** through the bore and through orifice **5** to the atmosphere. The polyethylene terephthalate component is passed through the distribution plate and spinneret plate **9** to the bore, wherein the polyethylene terephthalate envelopes the sparingly crosslinked polybutylene terephthalate totally and eccentrically. The polymers then pass through the bore and orifice to the atmosphere. It will be noted that seals **11** are used to separate the polymers from each other until they have passed through the distribution plate.

Figure 8.5: Cross Section of Spinneret Assembly

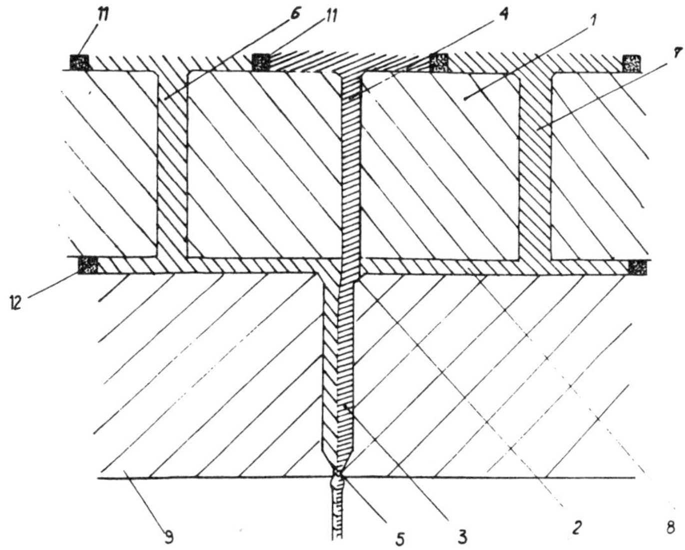

Source: U.S. Patent 4,117,194

Example: Four lots of different polymers were prepared as follows:

Polymer A
: Polyethylene terephthalate having an intrinsic viscosity of 0.66 dl/g (measured on a 25°C solution in orthochlorophenol containing 1% by wt/vol of polymer) and having a molten state viscosity VF of 2,300 poises at 290°C.

Polymer B
: Polybutylene terephthalate crosslinked by trimethylol propane in an amount of 0.3 mol %, based on the amount of terephthalic acid units in the polymer, having a molten state viscosity VF of 4,000 poises at 260°C.

Polymer C

Polybutylene terephthalate crosslinked by trimethylol propane in an amount corresponding to 0.3 mol %, based on the number of terephthalic acid units in the polymer, having a molten state viscosity VF of 8,000 poises at 260°C.

Polymer D

Polybutylene terephthalate crosslinked by trimethylol propane in an amount of 0.3 mol %, based on the terephthalic acid units in the polymer, and having a molten state viscosity VF of 10,000 poises at 260°C.

The polymers were separately melted, with polymer A melted at 280°C and polymers B, C and D melted at 245°C. Then bicomponent filaments were spun from polymer couples A-B, A-C and A-D, in proportions of 50% by vol of polymer A and 50% by vol of polymers B, C or D, respectively. The polymers were spun at 275°C through a spinneret of the type illustrated in Figure 8.5, where channel **4** had a diameter of 1.5 mm, bore **3** had a diameter of 2.5 mm, channels **6** and **7** had a diameter of 2.5 mm, the axis of channel **4** was displaced from the axis of bore **3** at a distance of 1 mm, and chamfer **2** had a length of 0.75 mm, the space **8** between distribution plate **7** and spinneret plate **9** was 1 mm. The spinneret had 32 spinning orifices therein, with each orifice having a diameter of 0.50 mm.

The extruded bicomponent filaments were passed over delivery rollers at a speed of 550 m/min and maintained at a temperature of 90°C, and then passed over drawing rollers which were maintained at a speed of 2,190 m/min, producing a draw ratio of about 4. The filaments were then wound up at a speed of 2,000 m/min, and then steam relaxed (at a pressure of 10 bars) at a temperature of 160°C.

The resulting filaments had a bicomponent structure where the polyethylene terephthalate component covered from 50 to 70% of the periphery of the sparingly crosslinked polybutylene terephthalate component.

For comparison, the polymer couples A-B, A-C and A-D were spun in accordance with the above procedure, but in place of the assembly described in Figure 8.5, these comparative filaments involved an assembly according to U.S. Patent 2,386,173 where the spinneret had 32 round orifices each 0.34 mm in diameter.

Yarns produced according to the process exhibit better crimp frequencies, and especially much higher half-uncrimping force values and half-recrimping force values, as a result of the filament crimp being more stable than the crimp of the comparative filaments.

Dyed Multicomponent Polyester Filaments

H. Möller and J. Wiegand; U.S. Patent 4,002,427; January 11, 1977; assigned to Hoechst AG, Germany disclose a method of making crimped, dyed multicomponent filaments of at least 2 polyester components including spinning the filaments together in a side-by-side relationship or in an eccentric sheath-core relationship, stretching the spun filaments and developing the crimp. These crimped filaments are wound on a bobbin with a Shore hardness of from 15° to 70°, then dyed on the bobbin. The crimped, dyed multicomponent filaments can be processed further to yield textile fabrics directly without an intermediate rewinding process.

The crimp is determined by the following method: The crimped filament is loaded with 1 mg/dtex; after a loading time of 5 minutes, its length l_0 is measured. Subsequently, the same preload is applied to the filament which is heated to 180°C for 2 minutes in a shelf dryer. The filament is withdrawn from the shelf dryer and loaded with 100 mg/dtex for 1 min. Under this preload the length l_1 is attained. The initial crimp value is then found by calculation:

$$K_1 = \frac{l_1 - l_0}{l_0}$$

Example: Polytetramethylene terephthalate with a specific viscosity of 1.38 was spun in side-by-side position to polyethylene terephthalate which had been modified with 0.6% of triethoxy silane ethyl-phosphonic acid diethyl ester and which had a specific viscosity of 0.45. The output was 37 g/min per component, the nozzle had 32 orifices. The spinning filament was withdrawn at the rate of 1,500 m/min and subsequently stretched on a draw-twister, Zinser 16 S, 1:3.205 at a temperature of 90°C of the godet and at 130°C of the heating plate (flat iron). The latent crimp of the stretched filament was initiated by means of a device such as described by British Patent 1,198,035, and the crimped filament was wound on a bobbin, the tube of which had holes drilled through. The bobbin weighed 1.5 kg; Alucolor DSB 1000 was used as the winding device. The cross angle was 17°, the Shore hardness of the winding was 40°. The K_1 value of the off-white filament was 30.2±1.1%; the error as specified corresponds to the 95% reliability range.

The off-white bobbin was then dyed first green at 125°C, then once more yellow at 125°C; this double-dyeing process inflicts especially hard conditions on the crimp of the filament. The K_1 value of the blue filament was 30.3±2.3%, which means the crimp underwent the dyeing process without any crimp loss. The Shore hardness was measured by means of a Zwick-densimeter.

Hydrophilic Bicomponent Acrylic Threads

U. Reinehr, C. Pieper and P. Kleinschmidt; U.S. Patent 4,163,078; July 31, 1979; U. Reinehr, H.-J. Jungverdorben and T. Herbertz; U.S. Patent 4,180,617;

December 25, 1979; both assigned to Bayer AG, Germany describe a process for the production of a hydrophilic bicomponent filament or fiber from two different fiber-forming, synthetic polymers, which are placed eccentrically to each other in defined areas, according to a solution spinning process, which comprises adding to at least one spinning solution 5 to 50% by weight, based on the solvent and polymeric solid, of a substance which is: (a) a nonsolvent for the polymers to be spun and (b) readily miscible with the spinning solvent and with water or another liquid used as washing liquid for filaments or fibers so spun.

The production of bicomponent fibers is carried out in known manner by means of conjugate spinning of at least two different polymer solutions in a side-by-side or in a core/sheath structure. Acrylonitrile polymerizates are preferably spun in this way. These polymerizates preferably contain at least 50% by weight, and most preferably at least 85% by weight, of acrylonitrile units.

In addition, they may contain one or several of the copolymerized comonomers which are known in the acrylic fiber industry. Examples of these include acrylic acid and methacrylic acid as well as their derivatives, such as esters, preferably alkyl esters, such as methacrylic acid methyl or ethyl ester, substituted or unsubstituted amides such as methacrylamide, N-methyl methacrylamide, N,N-dimethyl methacrylamide, also vinyl esters such as vinyl acetate. Comonomers with groups which have an affinity for dyes, preferably acid groups, may also be copolymerized. These groups preferably include methallyl sulfonic acid, vinyl sulfonic acid, styrene sulfonic acid and their salts, preferably alkali metal salts.

In order to obtain fibers with the best possible hydrophilic nature, the spinning treatment is selected so that as little as possible of the liquid added during the dry spinning process in the spinning duct is evaporated or extracted together with the evaporating spinning solvent. Spinning duct temperatures which are as low as possible and which are scarcely above the boiling point of the spinning solvent to be evaporated, short spinning ducts and high spinning outlets and thus short residence times in the spinning duct have proved to be exceptionally advantageous. For these reasons, the maximum temperature in the spinning duct should be 80°C, preferably 5° to 30°C, above the boiling temperature of the spinning solvent used.

The hydrophilic nature of the bicomponent fibers produced in this way, and having a core/sheath structure in at least one component, is also influenced by the type and manner of post-treatment.

If, for example, acrylic fibers from a dimethylformamide-glycerine mixture after spinning according to the processs are stretched in steam or water and only then washed, dried and finished, the original compact coating surface of the fibers or threads becomes strongly microporous by means of diffused glycerine, whereby acrylic fibers with particularly good hydrophilic properties are obtained.

Drying conditions which are as mild as possible with a maximum of 160°C, preferably 80° to 140°C, and short residence times of a maximum of 1 to 3 minutes in the dryer, give core/sheath fibers with very good hydrophilic properties.

According to these methods, bicomponent threads and fibers with core/sheath structure in at least one component may be obtained, although these are spun side-by-side. However, they show the typical eccentric side-by-side structure internally, which is responsible for the permanent crimping. They have a moisture absorption of at least 1.5% and a water-retaining power of at least 10%.

Example: 5.7 kg of an acrylonitrile copolymerizate of 93.6% of acrylonitrile, 5.7% acrylic acid methyl ester and 0.7% of sodium methallyl sulfonate were dissolved in a mixture of 19.8 kg of dimethylformamide and 4.5 kg of glycerine at 90°C. 7.5 kg of another acrylonitrile copolymerizate mixture, comprising 5.5 kg of acrylonitrile homopolymerizate and 2 kg of an acrylonitrile copolymerizate of 91% of acrylonitrile, 5.6% of acrylic acid methyl ester and 3.4% of sodium methallyl sulfonate were dissolved in dimethylformamide at 100°C. Both solutions were lead to a bifilar nozzle and dry spun side-by-side in the ratio 1:1. The fibers were combined in a cable, stretched 1:3.6 fold in boiling water, rinsed, prepared, dried under tension at 110°C, crimped, cut and fixed in steam in 1.5 minutes.

The fibers had an individual titer of 3.3 dtex, a strength of 1.9 p/dtex with an elongation of 48%. The fibers possessed a pronounced core/sheath structure with mushroom-shaped cross section, as shown in light-microscopic images of the cross sections. The width of the hem of the sheath amounts to approximately 2 μm and is compact in relation to the fine-pored core. The bicomponent fibers were spun into Nm 16/4 yarns which were rope dyed for 1 hour in a boiling dyeing bath and dried without tension.

The moisture absorption of the dyed yarns amounted to 2.0% at 65% relative atmospheric humidity and 21°C, and 9.5% at 90% relative atmospheric humidity and 21°C. The water-retaining power amounted to 26%. Fibers from the yarn had approximately 10 crimping bows per centimeter and a crimping of 11.2%. The fibers had a crimping reversibility of 0.2%, a porosity of 21.4% and a specific surface of 8.8 m^2/g. The dyed yarns were plump with a wool-like feel.

U. Reinehr, T. Herbertz and H.-J. Jungverdorben; U.S. Patent 4,185,058; January 22, 1980; assigned to Bayer AG, Germany provide hydrophilic fibers and filaments with good coloring response to dyes from filament-forming hydrophobic synthetic polymers having a sheath/core structure with a highly microporous core and a substantially compact sheath, and having a water-retention capacity of at least 10% wherein the pores in the core have an average pore diameter, measured in the direction of the cross section of the fiber, of at most 4000 A. Also described is a process for the production of those filaments and fibers according to a dry-spinning process wherein a spinning solution is spun below the boiling point of the spinning solvent used.

Example: 52 kg of dimethylformamide (DMF) were mixed with 12 kg of tetraethylene glycol in a vessel with stirring. 36 kg of an acrylonitrile copolymer of 93.6% of acrylonitrile, 5.7% of methyl acrylate and 0.7% of sodium methallyl sulfonate were then added with stirring at room temperature. The suspension was heated to 135°C in a heating apparatus. After leaving the heating apparatus, the spinning solution was filtered and transferred to the spinning duct. The total residence time of the suspension, from the heating apparatus to the spinneret, was approximately 5 minutes.

The spinning solution was dry-spun from a 72-bore spinneret. The temperature of the duct was 30°C, air temperature 40°C. The quantity of air supplied was 40 m^3/hr. The fibrous material, which had a denier of 2,440 dtex, was collected on bobbins and doubled to form a tow with an overall denier of 1,708,000 dtex. The tow was then drawn in a ratio of 1:4.0 in boiling water, washed, treated with an antistatic dressing and dried under conditions permitting 20% shrinkage. It was then crimped and cut up into staple fibers 100 mm in length.

The individual fibers, which had a final denier of 11 dtex, had a water-retention capacity according to DIN 53 814 of 49%. They had a pronounced core/sheath structure. The cross-sectional surface area of the sheath amounted to approximately 5% of the total cross-sectional area. The average pore diameter was approximately 1000 A and the internal surface area, measured by the BET-method, was 57.1 m^2/g.

The fibers were dyed in a concentration series ranging from 0.1 to 4% of a blue dye represented by the following formula:

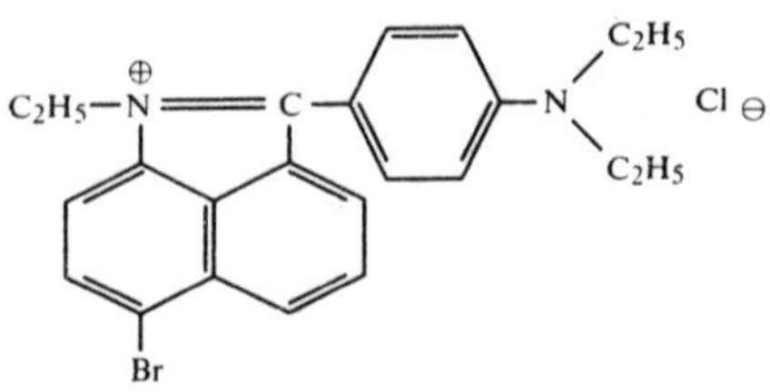

A commercial dry-spun acrylic fiber of the same denier and the same composition was used for comparison. The dyeings obtained were assessed visually and compared with each other by remission measurements. The additional amount of dye used, compared with that used by the ordinary commercial acrylic fibers, was 40%.

U. Reinehr and H.-J. Jungverdorben; U.S. Patent 4,185,059; January 22, 1980; assigned to Bayer AG, Germany describe a process for the production of hydrophilic filaments and fibers which comprises wet- or dry-spinning a fiber-forming synthetic polymer from a spinning solvent containing from 5 to 50% of a substance which has a higher melting or boiling point than the spinning solvent which is miscible with the spinning solvent and with a liquid suitable for use as a washing liquid and which is a nonsolvent for the polymer to be spun and from 0.1 to 20% by weight of at least one substance which decomposes into a gas under heating.

Example: 19.8 kg of dimethylformamide (DMF), 4.1 kg of glycerol and 0.2 kg of ammonium acetate were mixed in a reaction vessel with stirring. 5.7 kg of an acrylonitrile copolymer of 93.6% acrylonitrile, 5.7% methyl acrylate and 0.7% sodium methallyl sulfonate were then added with stirring. The mixture was stirred at 80°C for 1 hour and filtered, and the spinning solution thus obtained was dry-spun through a 180-aperture die into a spinning shaft by the methods known in the art.

The temperature of the shaft was 175°C. The viscosity of the spinning solution, which had a solids concentration of 19% and a glycerol content of 14% by weight, based on DMF plus polyacrylonitrile powder, was 65 falling-ball seconds. The proportion of the substance which decomposes into gaseous constituents was 3.5% by weight, based on the dry weight of acrylonitrile polymer.

The spun product, having a titer of 3,470 dtex, was collected on spools and doubled to form a spinning band having a total titer of 104,100 dtex. On leaving the spinning shaft, the spinning band still contained 12.3% by weight of glycerol.

The glycerol content in the spinning band was determined by gas chromatographic analysis. The fiber cable was then stretched in a ratio of 1:3:6 in boiling water, washed in boiling water for 3 minutes under a light tension and then treated with an antistatic dressing. It was dried in a sieve drum dryer at 140°C under conditions permitting 20% shrinkage and cut up into staple fibers 60 mm in length.

The individual filaments had a final titer of 6.7 dtex, a moisture-absorption capacity of 3.2% and a water-retention capacity of 84%. The ultimate tensile strength was 2.3 p/dtex; elongation on tearing was 40%.

On leaving the spinning shaft, the fibers had a pronounced core/sheath structure with irregular, mostly trilobate cross section. The width of the sheath in cross section was about 4 μm. More than 100 fiber cross sections were quantitatively analyzed to determine the ratio of core-to-sheath in the fibers. According to these measurements, on average 32% of the cross-sectional area of the fiber consists of sheath.

The proportion of residual solvent in the fiber was less than 0.2% by weight, and the proportion of glycerol left in the fiber was 0.6% by weight. The fibers could be deep dyed to an intense shade with a blue dye having the formula shown on the following page. The extinction was 1.31 for 100 mg of fiber per 100 ml of DMF (570 mμ, 1 cm cuvette).

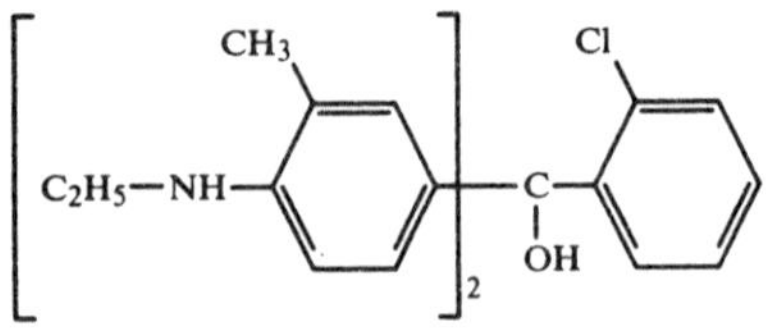

Foamed Vinyl Plastisol on Filament Core

In a method developed by *J.P. Okie and J.D. Worrall; U.S. Patent 4,144,371; March 13, 1979; assigned to Engineered Yarns, Inc.*, a fabric is prepared from a latent foam-encapsulated filament yarn, the fabric being either woven or nonwoven. The fabric is taken to a temperature high enough to activate the foaming agent within the thread under conditions such that thread junctions are bonded together and the thread itself is flattened. The degree of flattening is controlled in order to achieve a selected percentage of open area in the resultant fabric.

Figure 8.6: Foamed Vinyl Plastisol on Filament Core

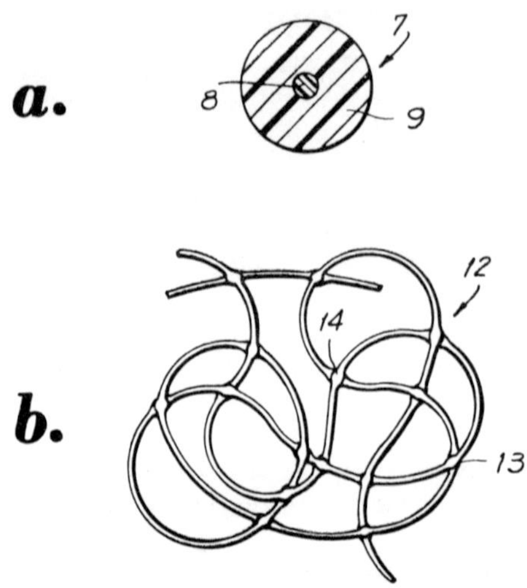

(a) Sectional view of an encapsulated monofilament yarn
(b) Nonwoven flattened fabric

Source: U.S. Patent 4,144,371

Figure 8.6a shows a thread indicated generally by the reference numeral **7**, the thread comprising a central filament **8** of a material such as rayon, nylon, glass or polyester encapsulated in a plastisol **9**. The plastisol is applied to the filament by any convenient means such as dipping, etc. The plastisol generally is a plasticized vinyl compound containing 50 to 60% of polyvinyl chloride resin and 50 to 40% of plasticizers, fillers, pigments, stabilizers, etc. In addition, the plastisol incorporates a blowing agent. A blowing agent is selected which has an activation temperature substantially higher than the curing temperature of the plastisol. The quantity of blowing agent present depends upon the degree of expansion desired. For instance, conveniently, the quantity of blowing agent may be such that a cured thread which is 0.028 inch in diameter expands to approximately 0.054 inch in diameter. The yarn is preferably of polyester.

The plastisol-coated yarn is drawn through a circular die to give it the shape shown in Figure 8.6a, after which it is cured or partially cured at a temperature below the activation temperature of the blowing agent. Preferably the sequence of steps, namely coating, sizing by means of a die and curing is repeated once or, if desired, more than once. The purpose, of course, is to increase the final diameter of the cured thread. The cure may be either partial (elimination of tackiness and providing adequate mechanical strength to withstand subsequent processing), or full. Exactly the same process may be used with a multifilament yarn.

After curing, thread either of the monofilament or multifilament type can be formed into a nonwoven fabric such as is shown in Figure 8.6b, the fabric being represented generally by the reference numeral **12**. After forming into the nonwoven fabric, the fabric is heated to a temperature sufficient to activate the blowing agent and weld the thread at the junctions. Flattening occurs at such junctions. While soft, the fabric may be further calendered so that the junctions formed by crossing threads, such as at **13** and **14** are further flattened. Also, in the calendering process, the threads are flattened between junctions.

Example: A continuous filament polyester yarn of 1,000 denier is double-coated to a diameter of 28 mils with a latent foam vinyl coating. This yarn has a weight of approximately 1 lb/1,000 yd and is generally circular in cross section with the core yarn well protected as an inner core. The cured thread may be woven into a construction of 6 picks per inch in both fill and warp. The fabric is then passed through a tenter frame at which time the fabric is held taut in both the fill and warp directions. Hot air at approximately 375°F is passed through the fabric, causing the vinyl to soften and the blowing agent to release.

Under these conditions, the yarn expands from 28 mils to about 54 mils. At the same time, the tensions being applied to the fabric cause the thread to flatten at the points where warp and fill cross, bonding taking place at each of these points. The result is a very stable fabric of foamed, flat vinyl-coated yarns lying in the fill and warp directions. Flattening is greatest at the cross-over points, due in part to the pressure on the thread at such points. The width of the yarn is increased to 65 mils, the yarn becoming smaller in thickness simultaneously.

Further flattening can be accomplished by calendering the fabric while it is still soft. A round 28-mil yarn in a woven construction can be flattened to 90 mils in width while still protecting the core yarn and retaining 30% of the foam structure.

The product has great resistance to outdoor weathering and can be produced in a multitude of colors. The product also has good abrasion resistance and the core yarn is protected from deterioration by sunlight. It can be used for outdoor fencing around gardens, patios, miscellaneous enclosures, and especially for fencing material to cut down the glare of lights between parallel opposing lanes of traffic on a highway. It can also be used for barrier fences in ski areas, for home shade cloth and protection against wind. Depending on the open area, both controlled visibility and air flow can be achieved.

Multicomponent Fiber and Subsequent Removal of a Component

M. Okamoto, K. Watanabe, Z. Izumi, T. Aya and H. Kitagawa; U.S. Patent 4,008,344; February 15, 1977; assigned to Toray Industries, Inc., Japan describe a multicomponent fiber which is drawable without fusing at a low temperature (of not more than about 85°C), which has excellent spinning stability, and one component which is easily removable by dissolution with a solvent.

The polymer consists mainly of a copolymer of styrene and an acrylic vinyl compound having a heat deformation temperature (HDT) measured by British standard method 2782 of 40° to 75°C, an elongation in hot water at 70°C of at least 100%, and a shrinkage in hot water at 85°C of at least 15%.

Example 1: Polyethylene terephthalate copolymerized with 5 mol percent of isophthalic acid was used as the island component of an islands-in-a-sea type composite fiber (see British Patent 1,171,843). The product obtained by copolymerizing 80 parts of styrene and 20 parts of 2-ethylhexyl acrylate was used as the sea component. The fiber was spun at a ratio of islands/sea of 52/48 and 16 islands/filament to obtain a 10 denier x 84 filament undrawn yarn. The spinning process was very stable and no traces of gel could be recognized after dismantling the pack. Using the undrawn yarn, drawing was carried out while varying the drawing conditions.

The shrinkage was greatly affected by the drawing temperature. In order to make a highly contractible fiber, the fiber had to be drawn at a low temperature, and the object could be achieved easily and effectively only by combination of the polymers of this process.

Example 2: As a control, Example 1 was repeated except using a styrene (80 parts)/ethyl acrylate (20 parts) copolymer as the sea component in spinning.

After spinning for 5 hours, the pack was dismantled, the spinneret was taken out and cooled. Thereafter it was immersed in trichloroethylene. Even after 20 hours the sea component polymer had not dissolved and the spinneret was still blocked with swollen gel.

Example 3: A total 150 denier x 48 filament woolly false-twisted yarn of polyethylene terephthalate was used as warp and a spun yarn from an islands-in-a-sea type staple fiber whose island component was polyethylene terephthalate and whose sea component was a copolymer of 72 parts styrene and 28 parts of 2-ethylhexyl acrylate, was used as weft in weaving a fabric. The islands-in-a-sea fiber had the following characteristics:

Number of islands	16
Island component ratio	55 % by wt
Sea component ratio	45 % by wt
Denier	3.0
Fiber length	51 mm
Number of crimps	8-12 per inch

Draw ratio at the time of drawing was 3.2. This staple fiber was processed favorably through carding, drafting, roving and spinning steps, and coiling of the staple around the card, drafting rollers and spinning rollers took place very seldom, and good spinning could easily be carried out.

Further, the yarns were used as warp and weft, respectively, and they were woven into a 5-ply satin fabric to produce a weaving density of 100 warps per inch and 60 wefts per inch. At this time, an edge of about 1 centimeter was made and a basket weave was selected as the woven pattern.

The fabric was washed with hot water at 80°C. After it was dried, the fabric was thoroughly washed 5 times with trichloroethylene, and the sea component of the weft was removed. After it was dried, the fabric was passed through hot water containing an ordinary oiling agent for napping, and dried.

The resulting fabric was passed 7 times through a rotary card cloth napping machine to nap the surface, to obtain a napped fabric in which superfine fibers were thoroughly napped. This fabric had smooth touch. However, it had poor pilling resistance and hand, and the firmness that resulted when it was bent longitudinally was greatly different from that which resulted when it was bent transversely. When it was bent transversely creases were left behind. The fabric was out of balance.

This fabric was impregnated with a 10% dilute polyurethane emulsion consisting mainly of polypropylene glycol, toluene diisocyanate and hexamethylene diamine prepared according to known methods of polymerization, and dried. The amount that adhered, calculated as pure polyurethane, was 21 g/cm^2. The resulting fabric was napped (buffed) by No. 150 mesh sandpaper. The front surface was buffed thrice and the back surface was buffed once by a belt sander buffing machine. As a result of this buffing, the fabric became a suedelike fabric whose surface was covered with compact superfine nap. It was heat treated at 170°C for 3 minutes in hot air.

Next, the fabric was dyed to a brown color by a circular type pressure dyeing machine, washed with water, a finishing oiling agent applied, and the fabric dried in hot air at 90°C according to known methods. The back surface was buffed to remove ugly naps raised here and there, and the surface brushed and finished. When the naps were brought down to one side and the fabric was so dried that the surface could contact the surface of a dryer roll at 90°C, the fabric developed excellent surface gloss and brilliance.

The resulting fabric was a suedelike fabric, the texture of which was hardly visible, having a soft and smooth surface touch and soft hand, and so well balanced that any directional difference in hand when it was bent longitudinally and transversely was almost unnoticeable.

Because the fiber was well drawn in this fabric, superfine naps were in evidence, having almost no tendency to be entangled, and they had excellent gloss.

Core and Sheath with Subsequent Removal of One Component

According to *T. Ishikawa, K. Yamashita, N. Okubo, M. Sakashita, A. Okamoto and T. Kusunose; U.S. Patent 4,085,174; April 18, 1978; assigned to Asahi Kasei Kogyo KK, Japan,* synthetic fibers capable of absorbing aqueous media are manufactured by simultaneously extruding the fiber material and a second material in a core-and-sheath arrangement. Turbulence is produced in the zone where these two materials combine, which results in undulations in their cross sections. The second material is then dissolved away with a solvent to leave a fiber with an undulating surface. When the second material is extruded as the core, the fiber is hollow with undulations on the inside; when the second material is extruded as the sheath, the solid fiber has an undulating outer surface.

Example 1: Polycapramide dissolved in 99% sulfuric acid and having a relative viscosity: 2.6 (intrinsic viscosity: 1.6) and polyethylene phthalate, $[\eta]$: 0.90, were fused and fed from respective first and second spinning pumps and finally through a spinneret of the structure as shown on the following page in Figures 8.7a and 8.7b. Polycapramide was used as sheath and polyethylene terephthalate was utilized as core. The volumetric ratio between the sheath material and the core material was set to 1:3.

Referring to the figures, the spinneret had 13 extrusion orifices **10**, each having 0.5 mm ϕ. Zone **5** was constituted by 20 sand particles per extrusion orifice, having passed through a sieve of 20 mesh. The spinneret temperature was set to 300°C and the winding speed of the produced conjugate fiber was adjusted to 1,000 m/min.

The thus formed and unstretched yarn was stretched as conventionally to its 30-times length into a 50 d/13 fil polyamide-polyester sheath-core conjugate yarn which was then treated with an aqueous caustic soda solution for dissolving out the core component. In this way a hollow yarn, as shown in Figure 8.6c, was obtained, representing random and heavy undulating inside surface **17'**.

The dissolving-treating conditions were: core-dissolving bath, aqueous 20% caustic soda solution; bath ratio, 1:20; bath temperature, 90°C; treating period, 30 minutes.

Figure 8.7: Process for Spinning Modified Synthetic Fibers

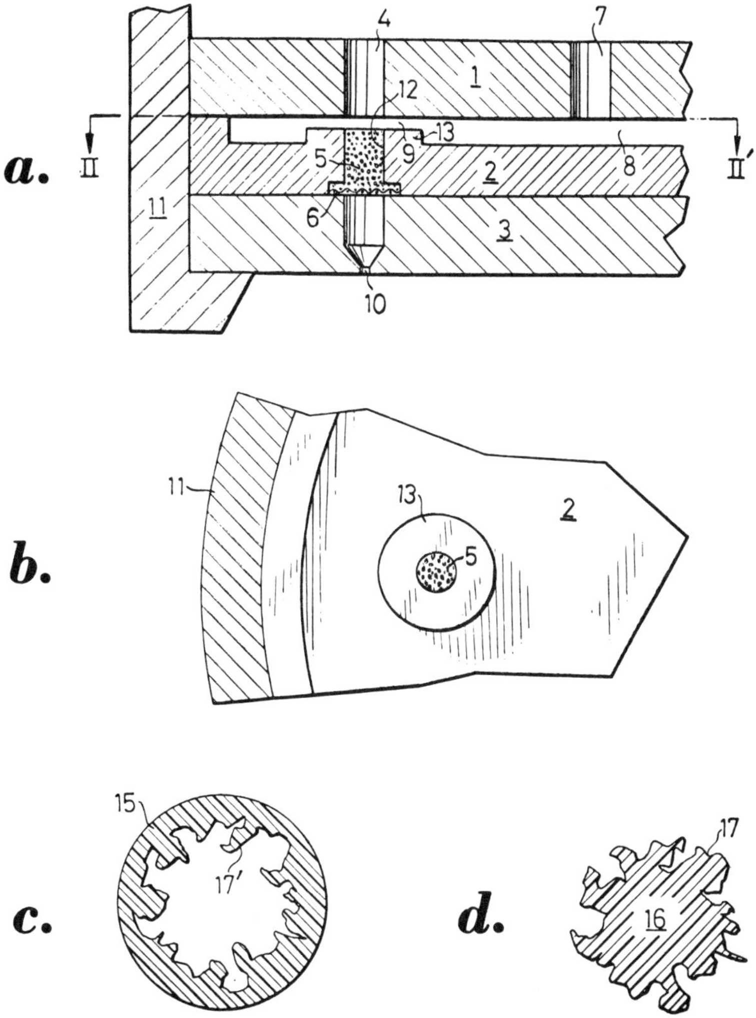

(a) Partial sectional view of spinneret
(b) Sectional plan view along section line II–II′ in (a)
(c) Cross section of filament
(d) Cross section of filament

Source: U.S. Patent 4,085,174

The main physical properties of this yarn were: strength, 4.1 g/d; degree of elongation, 37%; Young's modulus, 514 kg/mm^2; with good dyeing ability for acid dyestuff. The sweat-absorbing power of the fiber was found to be about 10 times that of corresponding plain nylon multifilament. Appearance and feeling were silky and linenlike, respectively.

Example 2: ϵ-Caprolactam/hexamethylene diadipate copolymerized in weight ratio of 3:7 and dissolved in 95% sulfuric acid and having a relative viscosity of 2.6 (intrinsic viscosity: 1.2), and polyethylene terephthalate, $[\eta]$: 0.70, were fused separately and fed at a volume ratio of 1:2 from respective spinning pumps to the spinneret, having a similar structure as shown in Figures 8.7a and 8.7b, the spinneret having 10 extrusion orifices, each being of 0.25 mm ϕ, so as to provide sheath-core conjugate yarns. Polyamide copolymer was used for sheath, while polyethylene terephthalate was utilized for the core. Zone **5** was filled with 8 ball-like members, 1.5 mm ϕ, made from a length of stainless steel rod, 10 cm long, 0.5 mm ϕ. The temperature of the spinneret was set to 275°C and the winding-up speed was set to 600 m/min. The unstretched conjugate yarns were stretched to 3.2 times their lengths, to provide yarns of 100 d/10 fil.

These yarns were treated with a sulfuric acid bath, under the following conditions: bath, 70% aqueous H_2SO_4 solution; bath ratio, 1:30; bath temperature, 50°C; for a 5-min treating period. In this way the sheath was dissolved away and the thus-obtained yarns took a cross-sectional configuration similar to that in Figure 8.7d. Sweat absorption was 9 times that of conventional comparative plain nylon multifilaments. Appearance was silky and hand feeling was linenlike.

TWISTED AND WRAPPED YARNS

Twisted Core Yarn

J. Umiastowski; U.S. Patent 4,099,370; July 11, 1978 describes a twisted yarn which has an elongated flexible core at least part of which is twisted in a given direction. A plurality of surface fibers are carried by the core at the twisted part thereof and are helically wound around the twisted part of the core in a direction opposite to the direction of twist of the part of the core so as to maintain the latter part of the core twisted in the given direction.

Referring to Figure 8.8a, on the following page, there is shown at the upper part a fiber drawing means formed by the pairs of rollers **2** and **3**. As is well known, rollers **3** rotate at a speed somewhat greater than rollers **2** for the purpose of drawing the sliver **1**, which is schematically illustrated, travelling from drawing rollers **2** toward drawing rollers **3**.

The yarn is fed along a predetermined path coinciding with the axis of the yarn by way of a feed means which includes the pair of feed rolls **12** which act on the yarn to pull it with a predetermined controlled tension from drawing rollers **3** toward the feed rolls. In this way the yarn is fed along a predetermined path extending through a rotary means which surrounds this path as well as the yarn fed therealong by the feed means. This rotary means takes the form of the illustrated twister **10** which is acted upon by a rotating means formed in part by belt **9** so that the twister can be rotated at high speed.

Figure 8.8: Twisted Core Yarn

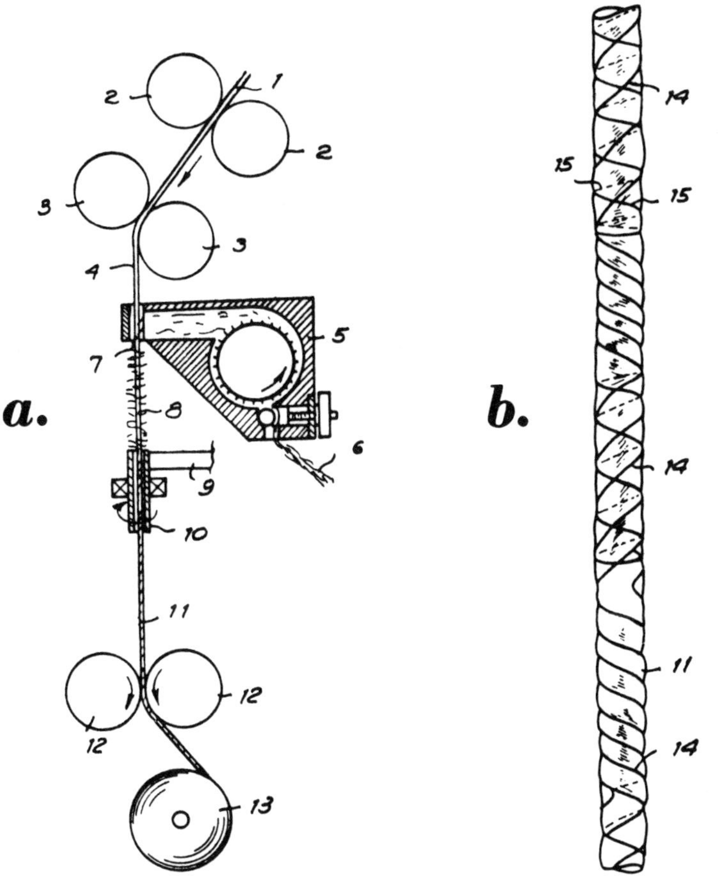

(a) Schematic, partly sectional elevation, of manufacturing process
(b) Fragmentary elevation of the yarn

Source: U.S. Patent 4,099,370

Upstream of the rotary means, considered in the direction of travel of the yarn from rollers **3** toward the feed rolls, is a depositing means for depositing fibers on the exterior surface of the drawn sliver. This depositing means takes the form of a lickerin device **5** which operates to supply additional staple fibers to drawn sliver **4**, approximately at location **7** between drawing rollers **3** and the twister. Sliver **6** is supplied to the lickerin device which acts in a known way to separate the fibers of the sliver and to direct them in open separated condition to the surface of sliver **4** to which these additional staple fibers cling.

In this way, downstream of location **7** the yarn has the composite structure **8** of a core, in the form of fibrous drawn sliver, for example, from which additional fibers laterally extend, these additional fibers being provided at the exterior surface of the core by way of the depositing means **5**. This composite fiber structure travels through the twister and is continuously advanced by suitable yarn-pulling rollers **12** which serve to impart positive and controlled tension to the yarn.

Subsequent to the feed means **12**, the yarn **11** is wound by a suitable winding means which includes the illustrated collecting roll **13** which is rotated about its axis in a known way while the completed yarn is wound thereon.

The abovedescribed apparatus operates in the following manner: The rotary means formed by the twister imparts false twist to the yarn passing through the twister and, as a result, the drawn sliver **4** rotates in the direction illustrated by the curved arrow in Figure 8.8a. This rotating sliver is capable of collecting the additional staple fibers provided by the lickerin device **5** and these additional staple fibers remain loosely attached to the drawn sliver as the drawn sliver advances to the twister. Thus, the composite fiber structure **8** has many free fiber ends projecting laterally from the surface of the core where they are held more or less at right angles to the yarn axis, with these additional staple fibers projecting laterally from the core in part as a result of centrifugal force of rotation. It is important that the twister rotate faster than the composite fiber structure which enters into the twister so that there will be a substantial slippage between the twister and the composite fiber structure.

The upstream annular end surface at the entrance end of the rotary means formed by the twister is roughened or otherwise treated so as to have a relatively high coefficient friction with respect to the fibers which project laterally from the yarn core. As a result, when the fiber ends of the structure come into contact with this end surface of the twister, the fiber ends become twisted or helically wound around the core of the fibrous structure, this core in the figure consisting primarily of drawn sliver **4**, and the direction of helical winding of the fibers is opposite to the direction of twist imparted to the drafted sliver.

It is to be noted that the false twist is imparted to the core by way of the action of the rotary means on the fibers projecting from the core. In other words, the rotary means acts on these fibers to place them under a certain tension while they are helically wound along the core and are displaced at least partly into engagement with the exterior surface thereof, and it is this pulling of the fibers during the helical winding which results in false twisting of the core itself with the initial twist in the core which is introduced during the false twisting being in an opposite direction to the helical winding of the fibers. Moreover, since the twist is imparted to the core by pulling on the fibers projecting from the core, these fibers are twisted to a greater extent than the core itself.

In accordance with the false-twist principle, at the exit end of the twister the twist initially introduced into the drawn sliver is removed and this action causes the helically wound surface fibers at the exterior of the core to be tightened along the surface of the core as the latter is released to tend to untwist itself, so that in this way complete twist removal is prevented in the core at those regions where there is an abundance of the helically wound surface fibers. However, in those regions of the core where there is a lack of surface fibers or where the surface fibers are only situated in relatively small numbers or are not sufficiently tightly twisted for any reason, the core will undergo twist reversal so that at parts of the core where there are no surface fibers, for example, the twist of the yarn will be opposite to the twist which is retained therein by the surface fibers.

Referring to Figure 8.8b, which shows part of a yarn **11**, it will be seen that at the upper portion of the figure there are surface fibers **15** helically wound against the exterior surface of the core **14**. As is apparent from the upper portion of the figure, the direction of twist of the core is opposite from the direction of helical winding of the surface yarns and the extent to which the surface yarns are twisted is greater than the extent to which the core is twisted.

Just below this upper portion of the figure where the surface yarns **15** are illustrated there is a portion of the yarn **11** where the core **14** has no surface fibers and it will be seen that this part of the core is twisted in the same direction as the helically wound fibers **15**, which is to say in an opposite direction from the direction of twist at that part of the core where the fibers are located. A further region of the yarn **11** where there are no surface fibers is shown just above the lowermost part of the figure, while between the latter region and the upper region where there are no surface fibers is illustrated an additional intermediate region where there are surface fibers retaining in the yarn the direction of twist which is opposite to the direction of twist where there are no surface fibers.

Thus, it is apparent that the process is based upon the premise that not all fibers in a yarn need be twisted in the same direction in order to lend strength to the yarn.

Elastic Yarn

According to *F.B. Northup; U.S. Patent 4,150,529; April 24, 1979; assigned to Hanes Corporation,* a balanced elastic yarn includes an elastomeric core twisted in one direction and a single cover strand helically wrapped about the elastic core in a direction opposite to the twist direction of the elastic core. After a predetermined twist has been inserted in the elastic core, the core is tensioned with the elongation being positively controlled between spaced points intermediate to the yarn take-up and the elastic core supply and a high number of wraps per inch of the cover strand are helically wound about the elongated core while the core draft or stretch is positively controlled. The elastic core take-up package is subjected to heat to relieve the stresses in the cover strand, resulting in a balanced elastic yarn in the relaxed state.

As shown in Figure 8.9, on the following page, an elastic core **E** is supplied from a spool **10** and has a cover strand **C** wrapped thereabout in a stretch or draft zone **X** before being wound upon take-up roll or package **12**. The elastic core may be any suitable elastic thread or filamentary material formed of rubber, spandex, or the like, having desired stretch, strength, durability and other required characteristics.

The inelastic cover thread, or strand **C**, is supplied from a spool **14** and is helically wrapped about the core in the draft zone. While the cover may be of various materials or blends of materials, preferably the cover strand is of thermoplastic, continuous, multifilamentary form.

The elastic core upon spool **10** is untwisted, as it is received from a producer. To apply a twist in a first direction to the elastic core, this spool is positively driven by belt **16** and the core twists about its own axis as it is drawn upwardly over the end of the spool. The twisted core is directed through a guide **18** and around a star wheel **20** before advancing axially through the tubular spindle **22** of spool **14**, guide **24** and between nip rolls **26, 28** before reaching the take-up roll.

The star wheel and the nip rolls are positively driven to positively and accurately control the degree of elongation or draft applied to the twisted elastic core independent of the cover. The star wheel and the rolls are driven at variable, selected speeds by independently controlled power means, not shown.

The desired tension or elongation upon the core is maintained by directing the elastic core between the nip rolls at a selected higher rate than the rate at which the core is advanced by the star wheel. Therefore, the core stretch or draft is uniformly constant and exact since it is accurately controlled within the draft zone, which is spaced intermediate to and unaffected by the rotating core supply roll and/or take-up roll. The uniformity of the core elongation within this intermediate zone substantially eliminates variables in the elastic core elongation, thus resulting in a uniform control over the core stretch characteristics.

A certain amount of elongation is imparted by the air resistance as the core unwinds from the supply spool. Without the star wheel, the elongation would be solely a function of spindle speed, hence severely limiting the range of stretch of the covered yarn. Therefore, the rpm of the core supply spindle can be selected to achieve a balanced cover yarn while at the same time the desired or selected elongation of the covered yarn can be achieved by means of the star wheel. It is to be understood that suitable means, other than the star wheel and nip rolls may be provided for accurately controlling the core elongation in the zone **X**.

Figure 8.9: Method of Making Elastic Yarn

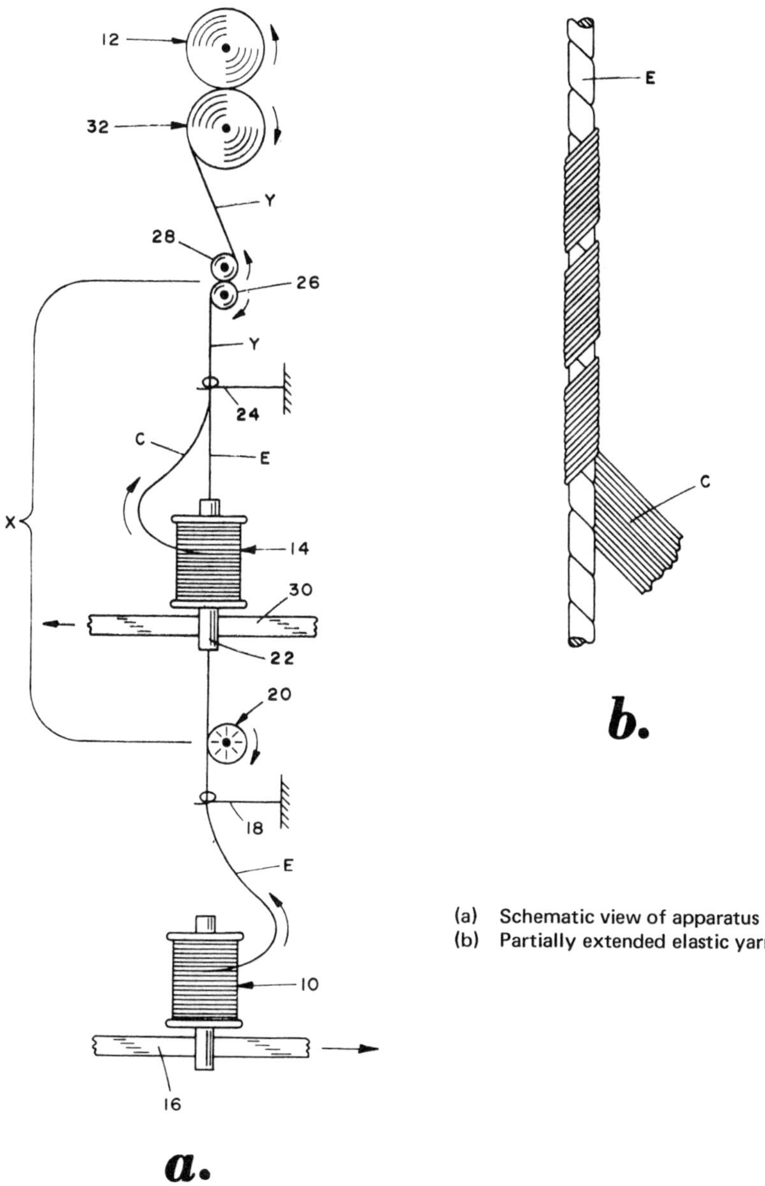

(a) Schematic view of apparatus
(b) Partially extended elastic yarn

Source: U.S. Patent 4,150,529

The cover strand is withdrawn upwardly over the end of spool **14**, which is driven at a desired, selective speed by belt **30** and helically wrapped about the elastic core which passes axially through the spool. The cover is wrapped helically about the core when the core is within the draft zone.

The cover strand on spool **14** has negligible twist, that is a small twist of 0.5 to 2.5 turns per inch, inserted while producing the yarn. As the cover is withdrawn from the spool it is not twisted about its own axis but it is wrapped about the core, due to the rotation of the driven spool, in a direction opposite to the axial twist of the elastic core, as shown by Figure 8.9b. While the helical winding of the cover in one direction tends to balance the twist of the core in the opposite direction, the elastic yarn at this point is not completely balanced.

The take-up roll is rotated by a friction drive drum **32** which is driven by a controlled, variable speed power means, not shown. After the composite core and cover **Y** emerge from the nip rolls, the tension in the core is at least partially relaxed, permitting partial retraction of the elastic core due to the surface speed differential between the nip rolls and the drive drum. The speed of rotation of the independently controlled nip rolls and the drive drum must be correlated in such a manner to wind the elastic yarn **Y** upon the take-up roll with a desired controlled tension or substantially slack with no tension, as desired.

Example: A twist-free 40-denier Lycra core is axially twisted in one direction as it is drawn from the supply spool **10** which is driven by bolt **16** at approximately 14,000 rpm. The twist-free cover consists of a 20-denier 14-filament nylon strand supplied from spool **14** which is driven at about 21,000 rpm to wrap helically the 14 filaments about the core in a direction such that the filaments tend to balance the twist in the core. The speeds of the core through the draft zone and the elongation of the core in the draft zone are such that when the elastic yarn **Y** is in the generally relaxed state, the core has approximately 46 wraps per inch of the cover strand when extended.

The finished yarn has an extensibility of 175 to 225%, and the core is more completely covered due to the high number of wraps. Heating the yarn with atmospheric steam does not shrink or set the cover strand but merely relieves the stresses in the nylon cover. The resultant single cover elastic yarn in the relaxed state is torque free, has an excellent covered, well-protected core, and is easy to knit.

Wrapped Yarn

F. Maag and F. Unger; U.S. Patent 4,018,042; April 19, 1977; assigned to Hoechst AG, Germany describe a yarn composed of an untwisted staple sliver and at least one filament wound around this sliver. The denier of the winding filamentary yarn is under 50 dtex, the elongation at break of the winding filamentary yarn is at least as great as the elongation at break of the staple fibers. The strength at a specific load of the winding filamentary yarn at 4% of elongation is at least 10 g. Moreover, the shrinkage factor of the winding filamentary yarn is within the same range as the shrinkage factor of the staple fibers. Also, at least 85% of the staple fibers are longitudinally orientated in the sense of the yarn direction.

Various yarns were manufactured by means of the device shown diagrammatically in Figure 8.10.

Figure 8.10: Manufacture of Wrapped Yarn

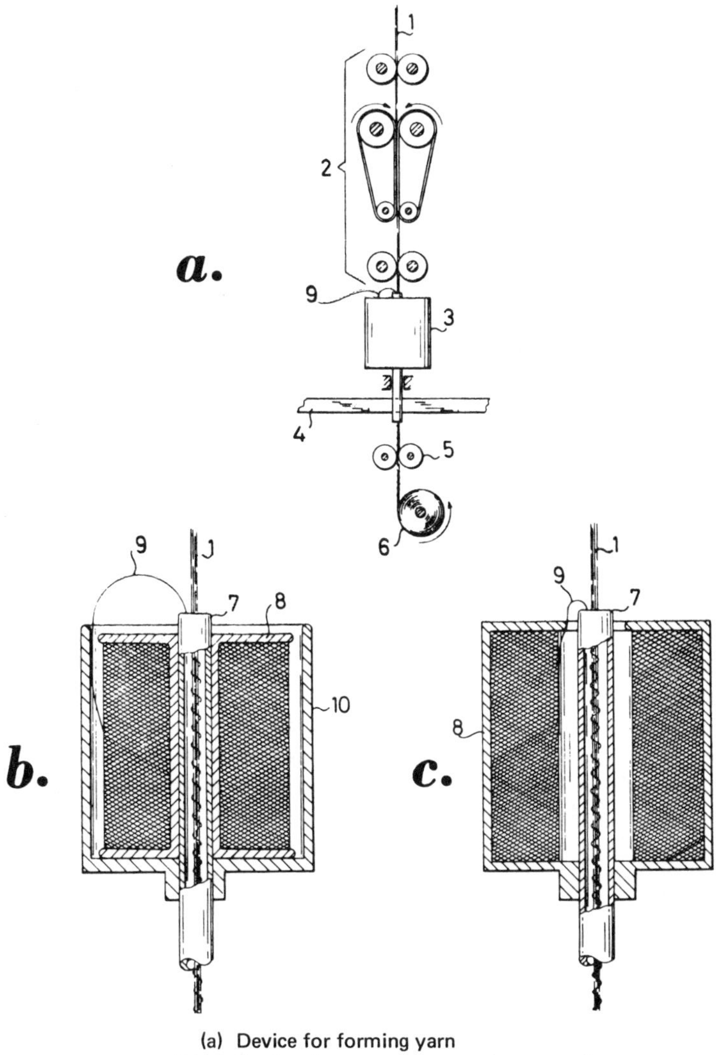

(a) Device for forming yarn
(b)(c) Devices for wrapping yarn about a roving

Source: U.S. Patent 4,018,042

Referring to Figure 8.10a, **1** is the fed-in roving, **2** is a double-belt drawing system, **3** is an enwrapping device, **4** represents a tangential belt, **5** are feed rolls and **6** is the winding bobbin. Figures 8.10b and 8.10c show specifically the enwrapping device **3**. The fed-in roving is led through a tube **7** inside the hollow shaft of the bobbin **8**, on which the yarn **9** is wound. At the same winding speed of rotation as the bobbin there is rotated simultaneously a balloon limiting device **10**.

Example: A roving having a denier of 10,000 dtex made of staple fibers of modified polyethylene terephthalate, according to German Patent 1,720,647; filament denier, 3.3 dtex; staple length, 60 mm; strength of the individual staple fibers, 29 g/tex; elongation at break, 35%; and shrinkage at 200°C being 6%, was introduced as roving into the device shown in Figure 8.10a.

The filamentary winding yarn had the following properties: 66 nylon (15 dtex); strength, 57 g/tex; elongation at break, 8%; shrinkage at 200°C of 8%; strength at a specific load at 4% elongation, 20 g.

48,000 rpm of the bobbin result in a wrapped yarn with the following properties: 275 dtex; 930 coils/m; twist coefficient (α), 154; degree of orientation of staple fibers, 98; maximum force, 523 g; strength, 19.0 g/tex; elongation, 10.3%; effectivity of the substance, 66%; output, 51.6 m/min. The maximum force is determined according to German Industrial Standard 53-815. The output indictates the speed at which the finished wrapped yarn is wound on the bobbin. The effectivity of the substance means the ratio of

$$\frac{\text{wrapped yarn strength (g/tex)}}{\text{total staple fiber strength (g/tex)}} \times 100\%$$

Synthetic Composite Resistant to High Temperatures

J.G. Sokaris; U.S. Patent 4,159,618; July 3, 1979; assigned to Albany International Corp. discloses a high-temperature-resistant, composite yarn useful in the manufacture of woven and knitted fabrics. The yarn comprises a core of a twisted multifilament or spun yarn and an outer coating or sheath of a high-temperature-resistant, synthetic, polymeric resin. The multifilaments or staple fibers of the core are selected from high-temperature-resistant synthetic materials. The outer coating comprises a plurality of coating layers intimately bonded to each other and to the twisted core so as to give the composite yarn an outward appearance and other characteristics of a monofilament. The composite yarn combines the properties of a monofilament and a multifilament or staple yarn.

As seen in Figure 8.11, on the following page, the twisted multifilaments **12** are encapsulated and covered by a coating **22** of a high-temperature-resistant synthetic polymeric resin.

Figure 8.11: Synthetic Composite Resistant to High Temperatures

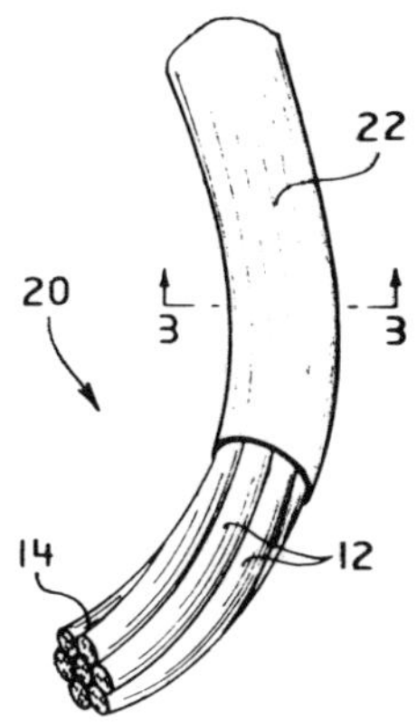

Source: U.S. Patent 4,159,618

Example: A 25/1 staple aramid yarn (Kevlar, supra) spun on the cotton system is provided having an outer diameter of 0.005 inch. The provided yarn is coated with a polyimide resin (RC-5057, DuPont) using five passes on a horizontal enameling machine and employing felt wipe-type applicators. After each coating pass the coated product is dried and the coating cured. The finished product has a diameter of 0.009 inch. The final product has a monofilament-type appearance and is flexible enough to be woven.

Similarly, repeating the above procedure but replacing the polyimide as used therein with a polysulfone or an organopolysilicone or a polyphenylene sulfide or a polyepoxide or a polyester or a polyester-imide or a polyquinoxaline, a composite yarn is obtained.

The pseudo-monofilament exhibits better flex fatigue resistance than comparable size metal wires and improved heat resistance at high temperatures over, for example, polyester monofilaments.

Increased Coherence Due to False Twisting

According to *B.C. White; U.S. Patent 4,180,968; January 1, 1980; assigned to Courtaulds Limited, England,* a composite filament yarn made by plying untexurized cellulose acetate yarn with polyamide or polyester yarn is made more coherent by false-twist texturizing. The increased coherence prevents bunching of the acetate yarn during knitting, avoiding the associated shade variation. The acetate yarn is spun-dyed and the polyamide or polyester yarn is package-dyed after pretexturizing. The acetate and synthetic yarns may be dyed to matching shades.

Example: Two ends of untexturized spun-dyed cellulose diacetate yarns, each of count 220 dtex and comprising 32 filaments, and one end of false-twist texturized package-dyed polyester yarn (polyethylene terephthalate) of count 167 dtex and comprising 32 filaments, were twisted together on a downtwister which inserted 80 turns per meter (tpm) folding twist. The cellulose diacetate yarns had been spun-dyed a carmine color and the polyester yarn had been package-dyed to a matching shade of carmine.

The 607/96 composite yarn so formed was then false-twist texturized on an A.R.T.C. double-spindle, double-heater, false-twist texturizing machine using the following machine conditions:

First heater temperature	175°C
Inserted twist, first spindle	1,285 tpm
Twist direction, first spindle	S
Overfeed between first and second feeds	-7 %
Second heater temperature	200°C
Inserted twist, second spindle	999 tpm
Twist direction, second spindle	Z
Overfeed between second and third feeds	0 %
Overfeed onto take-up package	+6 %

The texturized composite yarn was knitted into fabric which had a uniform carmine color. On examination the component yarns were still coherent and there was no apparent bunching of the cellulose diacetate yarns.

APPARATUS AND METHODS FOR PREPARING CORE YARNS

Fleece-Type Fibers Surrounding Strong Core Filament

In a process developed by *E. Hamel; U.S. Patent 4,070,818; January 31, 1978; assigned to Ofa AG, Switzerland*, a slubbing is formed into a thin band by a plurality of pairs of rollers. A core filament is thereafter applied to each face of this band and the two core filaments and band are then spun together to form a core yarn comprising a coherent mass of fibers in which is imbedded a pair of parallel and spaced-apart core filaments with some of the fibers of the mass between the spaced-apart filaments.

Referring to Figure 8.12 on the following pages, a core-type yarn **10** is shown in Figure 8.12a. It comprises a coherent mass **16** formed of fibers such as wool, cotton, rayon, or staple fiber, in which is imbedded a pair of like and spaced-apart filaments **12, 14** of high-strength synthetic-resin material such as polyester. The mass has a cross-sectional area that is a multiple of that of the filaments. The filaments can be monofilaments or a tightly wound thread, but in any case have a longitudinal strength which is considerably greater than the longitudinal strength of the coherent mass. These filaments are spaced from each other with a portion **16'** of the mass **16** between them.

Figure 8.12: Fleece-Type Fibers Surrounding Strong Core Filament

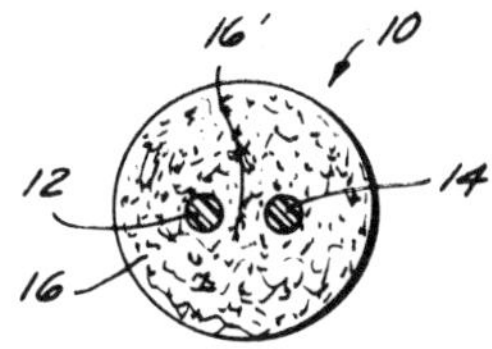

a.

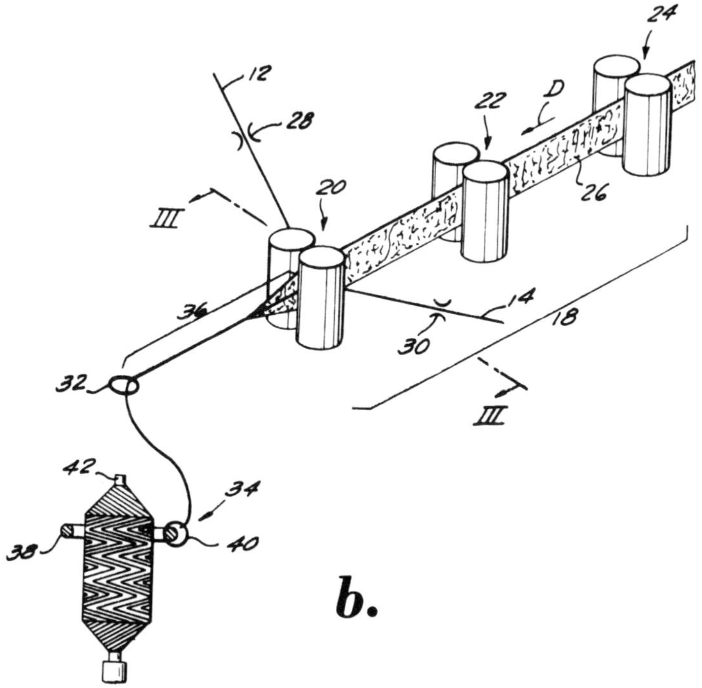

b.

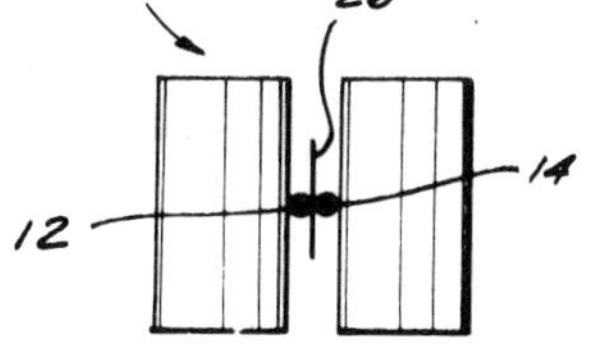

c.

(continued)

Figure 8.12: (continued)

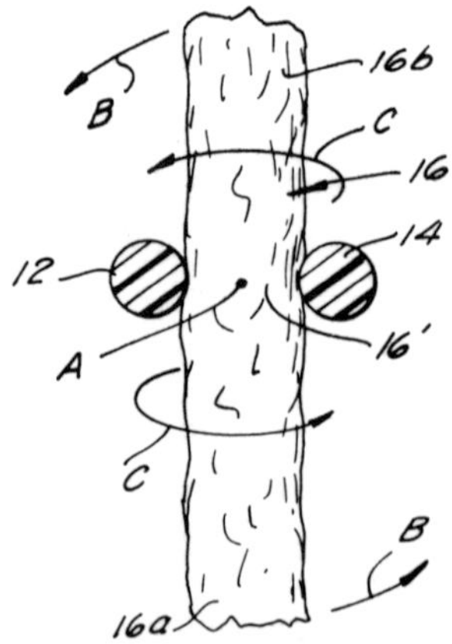

d.

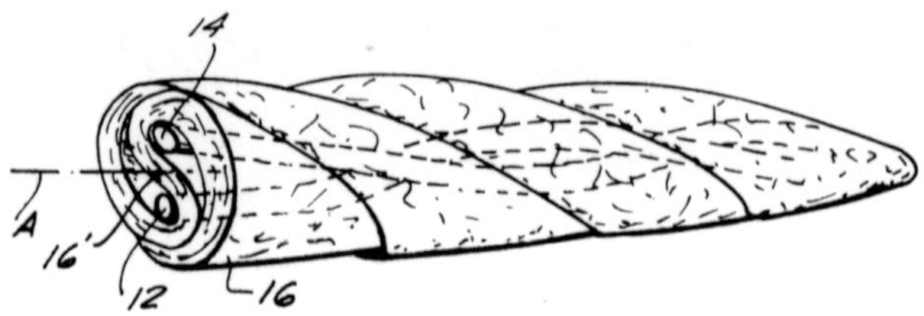

e.

(a) Cross section in enlarged scale through a yarn
(b) Partly diagrammatic perspective view showing a system for making the yarn
(c) Section taken along line **III–III** of Figure 8.12b
(d) Detail view diagramming the spinning action
(e) Perspective view showing results of spinning

Source: U.S. Patent 4,070,818

As shown in Figures 8.12b and 8.12c, this yarn is made in an apparatus having a former **18** constituted by three aligned pairs of cylindrical rollers **20, 22, 24** which advance a band **26** of slubbing in direction D. Rollers **20** rotate faster than rollers **22** which in turn rotate faster than rollers **24** so that the slubbing is formed into a thin upright band (as illustrated in Figure 8.12c). A pair of guides **28, 30** feed in the filaments **12** and **14**, respectively, to the combining station formed by rollers **20**. Figure 8.12c illustrates how these filaments are fed to opposite sides of the band directly across from each other and both lying in a plane perpendicular to the plane of the band of the combining station.

In region **36** between the combining rollers **20** and an eye **32**, the slubbing band and filaments are spun together and then wound up on a ring-type spinner **34** having a spool **42** that is rotated at high speed with the yarn to be spun passing through a traveler **40** on a ring **38** surrounding the spool as described on pages 376 and 377 of *The Way Things Work* [Simon and Schuster (1967)].

As can be seen from Figure 8.12d, during the spinning operation the upper and lower portions **16b** and **16a** of the fleece rotate about the axis **A** (arrows **B**) while the filaments are simultaneously twisted (arrows **C**) therearound. The overlapped portions of the fleece (Figure 8.12e) have their fibers intertwine so that the ultimate cross section is that shown in Figure 8.12a.

Fiber-Fleece Slubbing Material Covering Core Threads

E. Hamel; U.S. Patent 4,100,727; July 18, 1978; assigned to Ofa AG, Switzerland discloses a process for producing a core yarn of the type in which core threads are received within and surrounded by a fiber-fleece material which originally is in the form of at least one slubbing strip.

Referring to Figure 8.13 on the following pages:

- (a) Diagrammatic perspective view of a stretching arrangement illustrating a portion of the ring spindle downstream thereof for twisting the core yarn
- (b) View in the direction of arrow **II** in Figure 8.13a
- (c) Section taken along line **III–III** of Figure 8.13b
- (d) Side elevational view, also in diagrammatic form, illustrating the apparatus of Figure 8.13a
- (e) Detail elevational view of the feeding of the core threads to a device of the type illustrated in Figure 8.13a

In the drawings, the core yarn has been shown in section in Figures 8.13b and 8.13e, only the twist of the core threads being apparent therein. However, as will be evident from Figure 8.13a, the entire assembly of core threads and slubbing strips is twisted in practice so that, in a cross section as shown in Figure 8.13c, the core threads **18** and **20** are fully embedded or encased in the slubbing material.

In Figure 8.13a, there can be seen a stretching device which consists of three roller pairs **10, 12, 13**, through the nips of which the slubbing strips **14** and **16** are passed, stretched and flattened. (See Swiss Patent CH-PS 395,819.)

Figure 8.13: Method of Making a Core Yarn

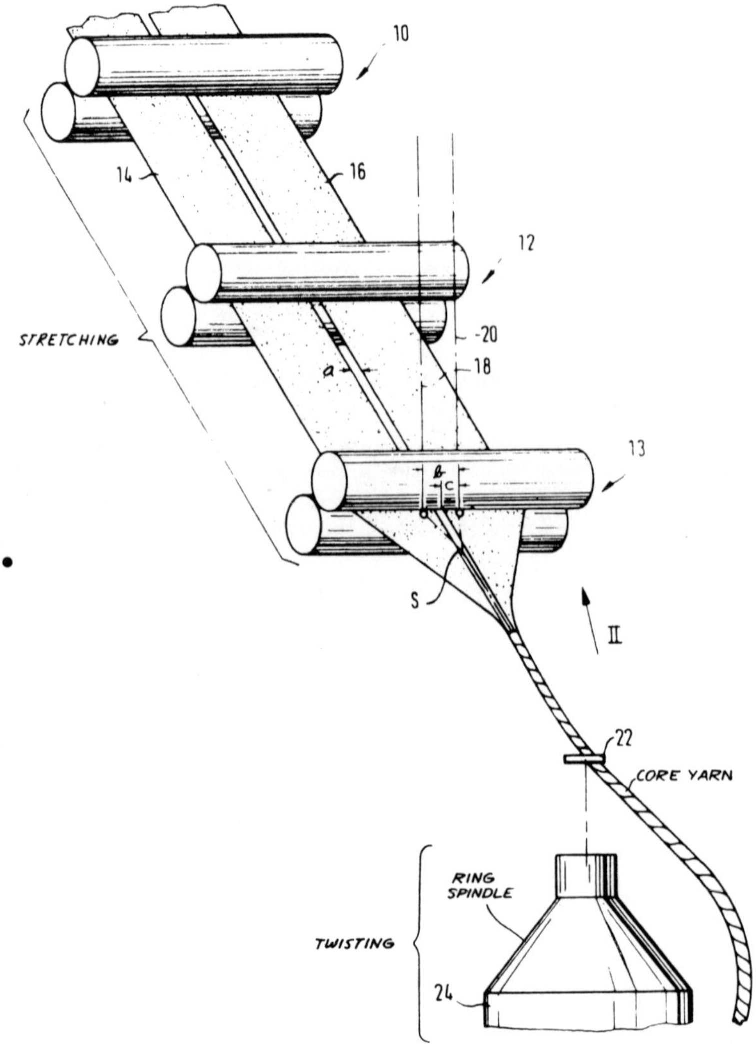

a.

(continued)

Figure 8.13: (continued)

b.

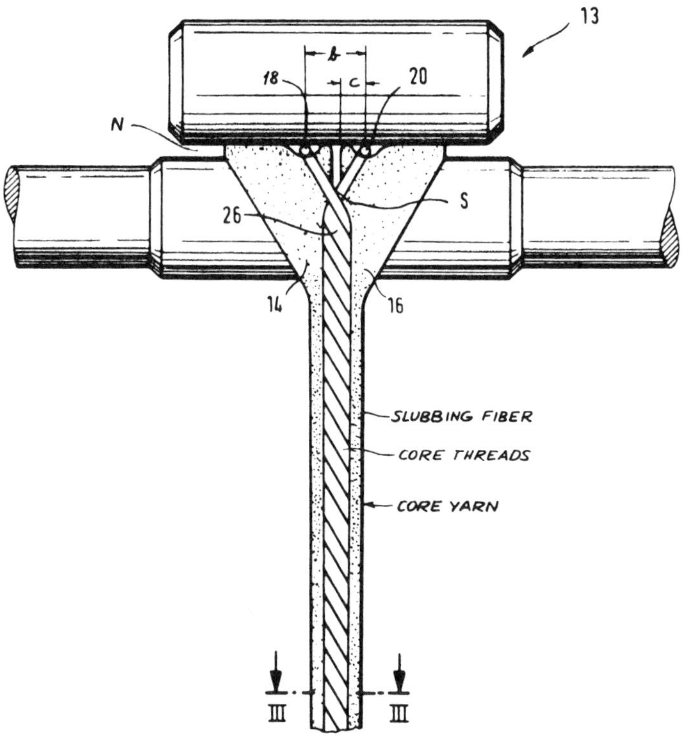

c.

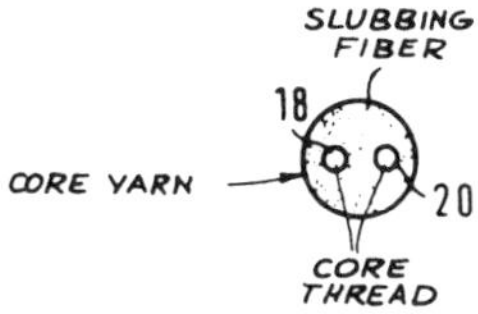

(continued)

Figure 8.13: (continued)

d.

e.

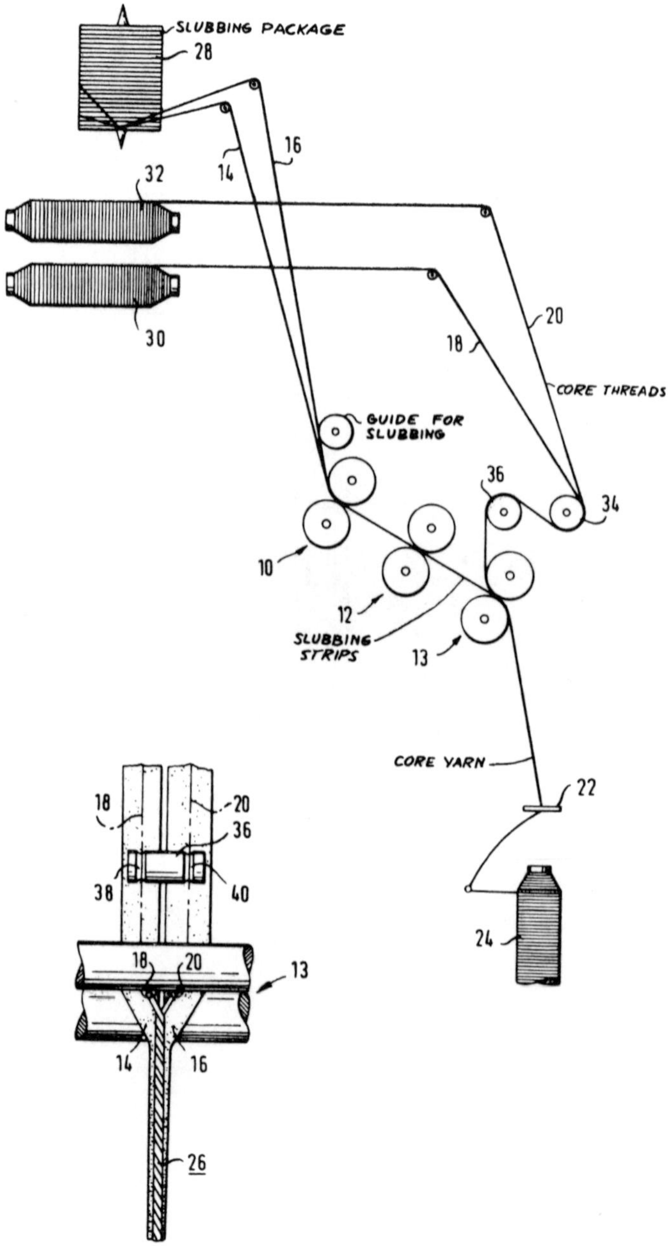

Source: U.S. Patent 4,100,727

As shown in Figure 8.13a, the mutually confronting (justaposed or proximal) edges of the fiber-fleece slubbing strips **14** and **16** are, shortly before the nip of the last roller pair **13**, spaced apart by a relatively small distance **a** which should not exceed 2 mm and preferably is less than 1 mm, even less than 0.5 mm. Advantageously, this distance is no greater than the thickness to which the slubbing strips are rolled. In fact, although such an arrangement is not shown, the proximal edges of the slubbing strips entering the last pair of rollers can be in contact with one another or even overlap slightly.

In the last stretching roller pair, a pair of core threads **18, 20** is introduced, the distance between these core threads being represented at **b** which is only slightly greater than the distance **a**. Thus, each of the core threads overlies a respective slubbing strip within the roller pair and has a spacing **c** from the proximal edge of the respective strip which is of the magnitude previously described. The core threads immediately downstream of the roller pair **13** run together in the gap **S** between the slubbing strips.

Downstream of the last roller pair of the stretching device is provided a thread-guide eye **22** which can be axially aligned with a ring spindle **24** on which the core yarn is wound, the ring spindle spinning the assembly of slubbings and core yarns together.

From Figure 8.13b it will be immediately apparent that the location **S** represents the vertex of the spinning triangle formed by the core of the core yarn directly adjacent the nip **N** of the roller pair **13**. The slubbing strips merge together to form the slubbing fiber layer disposed between and around the core threads **18, 20**, as shown in Figure 8.13c. More particularly, the portions of the slubbing disposed laterally of the respective core threads **18, 20** are partially sandwiched between these core threads and partially wound around them so that the two slubbing strips blend together and completely surround the core.

The core yarn has been found to have not only a high sliding resistance between the fiber-fleece slubbing material and the core threads, but also an optimum covering of the core threads by the fiber-fleece material.

In Figure 8.13d are seen the two slubbing packages **28** in the form of rovings of, for example, wool. These rovings are flattened in the first roller pair **10** of the stretching device. When cotton slubbing is to be used, it is advantageous to provide a separate spool for each of the rovings or slubbing strips of cotton to be fed to the stretching device. The core threads **18, 20** are drawn from a pair of bobbins **30, 32** and can pass around a guide roller **34** and over a further roller **36** which can be formed with spaced-apart circumferential grooves **38, 40** to properly maintain the gap between the core threads (Figure 8.13e).

If desired, a guide roller pair, also provided with appropriate grooves, can be disposed immediately upstream of the last roller pair of the stretching device to maintain the desired spacing of the slubbing strips.

The core yarns produced by this method have been found to be particularly advantageous for the production of knitted wear and this is especially the case when the core yarn is of a textured type. Knitwear produced with this core yarn has a high stretchability, high breaking strength and excellent self-flattening characteristics. Upon washing, the knitted wear has only limited shrinkage and remains substantially wrinkle free.

Core-and-Effect Fancy Yarn

C. Beraud, R. Guigal, R. Lehmann and A. Lyonnet; U.S. Patent 4,095,403; June 20, 1978; assigned to Institut Textile de France and ANVAR, France disclose fancy yarns and a method of making them in which at least two yarns, of which one forms a core and one an effect yarn, are fed to a junction point and then fed together through a heat treatment zone and assembled into the fancy yarn in a false twisting step, the core yarn having been provided with a thermoplastic binder upstream of the junction point by which the effect yarn is adhered to the core yarn.

As is shown schematically in Figure 8.14a (on the following page), a device for carrying out the process essentially comprises: feed means **1** for a core yarn **2**, for example, a feeder of the strap type or of the pressure-roller type; feed means **3** for an effect yarn **4**, which may or may not be of the same type as feed means **1**, these feed means **3** being optional; a heat treatment element **5**, such as a contact oven, heated by electricity or in any other way; a false-twist spindle **6** of, for instance, the mechanical, magnetic or friction type; draw-off means **7** for the fancy yarn formed and of a type similar to feed means **1**; and a conventional wind-up device **11**. The thermoplastics binder is supplied in the form of a yarn **8** which is brought in at feed means **1** to run parallel with the core yarn.

Example: Referring to Figure 8.14a, a fancy yarn is formed by the false-twist assembly of a core yarn **2** and an effect yarn **4**, both consisting of polyamide 66 of 78 decitex gauge, comprising 23 strands, twisted at 20 turns in the Z direction, nonshrunk, semimatt and of round cross section.

The thermoplastic binder yarn **8** consists of polyamide 11, of 32 decitex gauge, comprising 10 strands twisted at 20 turns in the Z direction, nonshrunk, semimatt and of round cross section.

The speed of the false-twist spindle **6**, which is mechanical, is 75,000 rpm. This spindle is of the crossed-bars type (type SKF, diameter 8). The draw-off speed of the core yarn and of the fancy yarn formed is 32 m/min (0 overfeed). The speed of the effect yarn is about 38 m/min; in this case, the effect yarn is not fed positively, that is to say, it does not pass through feeder **3**. This effect yarn winds itself around the core yarn and the binder yarn by the false-twist effect. The effect yarn is subjected to a very low tensile force of the order of 0.1 cN/tex and directed at right angles to the core yarn and to the binder yarn, the junction point **12** being stabilized by means of a fixed guide eyelet, which is not shown.

The temperature of the oven **5** is 200°C. The length of the oven is **1** m. The resulting fancy yarn is about 225 decitex gauge. The fancy yarn obtained is illustrated schematically in Figure 8.14b and exhibits a fibrous appearance, with the effect yarn **4** completely enveloping the core yarn **2** and being firmly bonded to the latter at point **14**.

Figure 8.14: Core-and-Effect Fancy Yarn

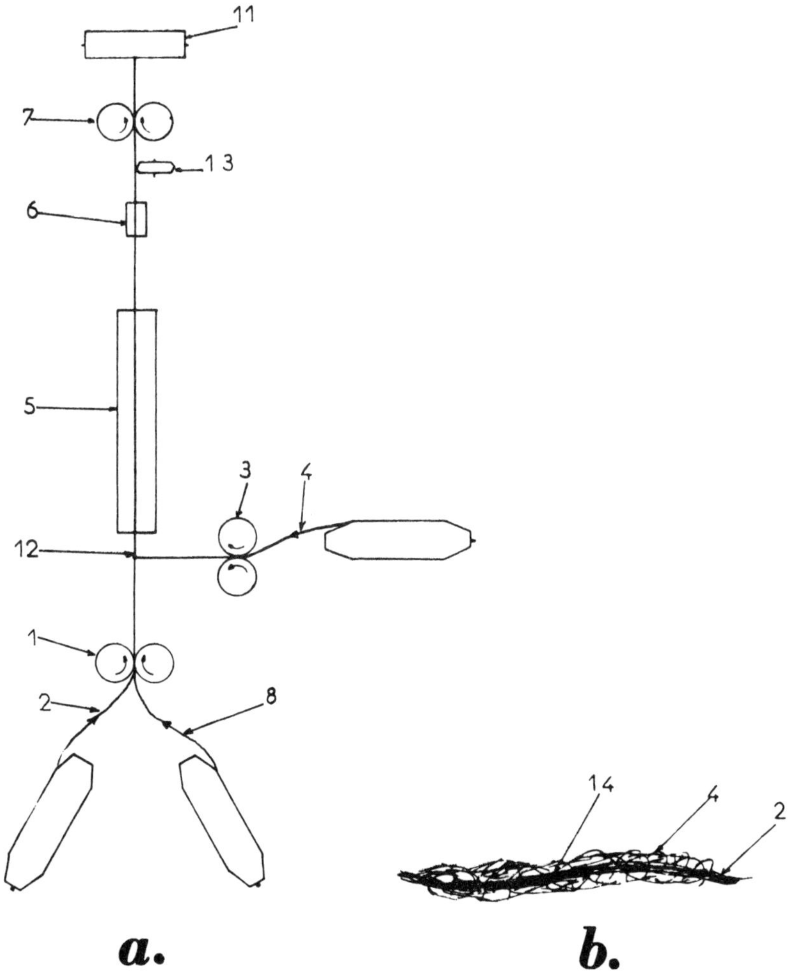

(a) Production of fancy yarn
(b) Schematic illustration of yarn obtained by Figure 8.14a

Source: U.S. Patent 4,095,403

Sintered Bimetallic Filamentary Composites

According to *E.J. Dobo; U.S. Patent 4,089,921; May 16, 1978; assigned to Monsanto Company,* bimetallic filamentary composites are produced by first forming a conjugate precursor filament comprised of an organic polymer together with particles of a first reducible metal oxide and particles of a second reducible metal oxide with the metal component of each of the two metal oxides being sinterable at a temperature which is below the melting point of the other. The structure of the precursor is characterized by a first longitudinally extending layer along its length which contains the particles of the first reducible metal oxide, and an adhering second essentially distinct longitudinally extending layer extending along its length which contains the particles of the second reducible metal oxide. The essentially discrete layers may be in a sheath-core arrangement or in side-by-side relationship.

Conversion to the bimetallic filamentary product is accomplished by exposing the precursor filament to a reducing environment at a temperature and dwell time sufficient for effecting a reduction and sintering of the metal particles. The temperatures employed are in a range which is below the melting point of the metal particles and above the vaporization or decomposition temperature of the nonmetal components of the precursor filament.

Example: A first film-forming dope was prepared containing 39.4 g NiO uniformly dispersed in a solution which contained 5.3 g of a copolymer consisting of 93% by wt acrylonitrile and 7% by wt vinyl acetate dissolved in 30 cc of dimethyl acetamide. A second film-forming dope was prepared with 30.0 g of Fe_2O_3 dispersed in a similar solution. The first and second film-forming dopes were then cast one on top of the other. The joined films were coagulated in water to produce a composite film from which precursor filaments were slit.

The precursor filaments were then converted to a bimetallic filamentary structure by heating at a temperature of 1000°C for a period of 3 min in a reducing atmosphere consisting of 93.5% hydrogen and 6.5% carbon monoxide. The bimetallic filament obtained contained two distinct longitudinally extending components. One component consisted of nickel and the other steel. Steel was formed by the diffusion of carbon present in the system into the iron oxide.

Avoidance of Sticking of Melt-Spun Sheath-Core Filaments

According to *R.A. Ross and D.R. Strachan; U.S. Patent 4,020,139; April 26, 1977; assigned to E.I. DuPont de Nemours and Company*, the sticking between an array of melt-spun eccentric bicomponent filaments converged together into yarn is avoided by selecting the filaments being converged so that the thin sheath regions of the filaments do not contact one another and conducting the convergence in two steps if the number of filaments required for the yarn is too great for convergence in one step while avoiding the aforesaid contact.

Referring to the following figure, Figure 8.15a shows an eccentric bicomponent filament **2** in which a core **4** of one component of the filament is eccentrically disposed, e.g., in the D-shaped cross section shown, within a sheath **6** of the other component of the filament. Because of the eccentric dispostion of the core within the sheath, the cross section of the sheath varies from a thin region **8** to a thick region **10** relative to one another. The thin region preferably has a thickness of only about 1% of the filament diameter.

Figure 8.15: Arrangement for Avoiding Sticking of Melt-Spun Filaments

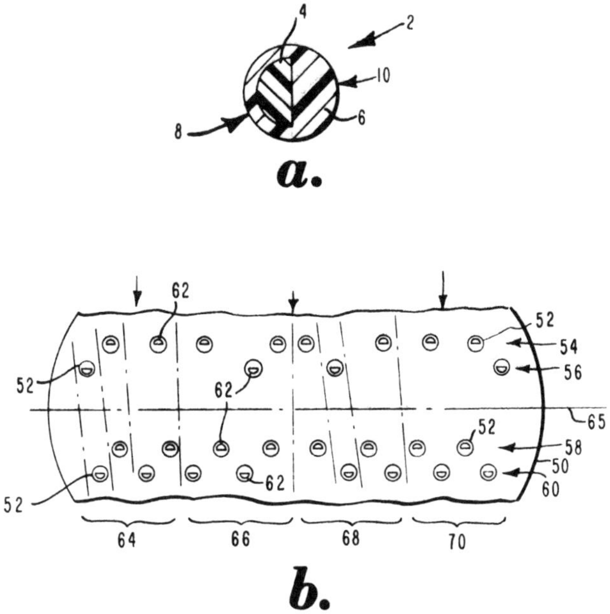

(a) Cross section of eccentric bicomponent filament
(b) Plan view of portion of spinneret plate and cross section of an array of filaments being melt spun

Source: U.S. Patent 4,020,139

An embodiment incorporating filament selection and stepwise convergence of the selected melt-spun filaments into yarn is shown in Figure 8.15b depicting a portion of the face of spinneret plate **50** from which is melt-spun a plurality of eccentric bicomponent filaments **52** in an array of four rows **54, 56, 58, 60** corresponding to the four rows of the same total number of orifices in the spinneret assembly of U.S. Patent 3,704,971, to make four 7-filament yarns. Since the same number of orifices is involved, this embodiment represents no loss of production capacity from that of the patent. In this embodiment, the difference from the patent spinneret assembly is the location of the orifices in the rows, which aids filament selection to avoid thin-sheath to thin-sheath contact.

The eccentric core **62** is shown for each filament (cross-section shading of filaments **52** is omitted for clarity) to show the opposite orientation of the thin-sheath region of each filament from row to row, which is the same orientation as produced by the spinneret assembly of the abovementioned patent. The thin-sheath regions in row **54** face the quench gas flow coming from the direction indicated by the arrows in the drawing, which is from the back of the apparatus, and the thin-sheath portions in row **56** face in the opposite direction. This orientation is repeated for rows **58** and **60** of filaments.

Example: The following general procedure was used for spinning filaments from different spinneret orifice arrangements. The eccentric bicomponent filament had a cross section like that of Figure 8.15a and had the following composition: The core was a copolymer of 30% by wt epsilon caproamide units and 70% by wt hexamethylene dodecanediamide units and the sheath was 100% polyhexamethylene dodecanediamide units. The core copolymer had a relative viscosity of 44.2±2.2 and the sheath polymer had a relative viscosity of 39.2±2.1, the relative viscosities being measured on samples taken from the melt of each polymeric material just before entering the spinneret assembly. The filament was composed of 60% by wt core copolymer and 40% by wt sheath polymer.

The melt temperature measured at the point where the relative viscosity samples were taken was 268°C for both polymeric materials. The windup speed was 421 yd/min (385 m/min) and the as-spun denier per filament was about 27. The 60-inch (1.52-m) quench chimney used cross-air flow at 49°±1°F. Before windup into packages, the yarn was passed through a 75.6-inch (1.92-m) long chamber containing saturated steam at atmospheric pressure for dimensional stabilization. The yarns were withdrawn from their packages and drawn at a draw ratio of 4.285X over an unheated draw pin located between the feed and draw rolls of a conventional draw twister to a denier per filament of 7.1 (7.9 dtex). The draw roll peripheral velocity was 790 yd/min (722.4 m/min).

After drawing, the yarn was immediately passed through a 5.8-inch (14.7-cm) long tube through which hot air was jetted cocurrently at 0.7±0.1 ft^3/min (19.8±2.8 l/min) to provide an exit air temperature of 150°±3°C. At the point of air entry, the tube was 0.08 inch (0.20 cm) in diameter, increasing gradually to 0.20 inch (0.51 cm) at a distance of 2.8 inch (7.1 cm). There the tube diameter increased abruptly to 0.25 inch (0.63 cm) and remained constant for the last 3 inches (7.6 cm) of tube length.

After exit from the tube the yarn passed in zig-zag fashion over three snub pins at a total contact angle of 390°±15°. The resultant annealed yarn was immediately packaged under tension. The ring and traveler windup following annealing was at a spindle speed of 7,710 rpm to provide 0.271 turns per inch (0.107 turns per centimeter) of inserted twist. The resultant drawn yarns were then knitted into the tubes and crimped, and these tubes were visually inspected for defects.

In the control experiment of this example, using the foregoing general procedure, the orifices of the spinneret had the four-row arrangement of the spinneret assembly of U.S. Patent 3,704,971. In a second experiment using the general procedure of this example, the orifices of the same spinneret assembly were arranged in accordance with the embodiment of Figure 8.15b with the centerline distance between rows of each pair being about 0.6 cm and between each pair of rows about 2.0 cm.

In the control experiment, the yarn was made by converging filaments selected from a pattern of two from one row, two from the adjacent row, two from a row on the opposite side of the spinneret centerline and one from the adjacent row thereto. These seven filaments were converged in one step at a convergence guide positioned between the quench chimney and steam chamber into seven-filament yarn.

In the second experiment, the filaments were converged into bundles of two filaments, two filaments and three filaments selected in accordance with the figure prior to entering the steam chamber, and at the exit of the steam chamber these bundles of filaments were converged into four seven-filament yarns. At least 50 packages of yarn were wound up and knitted into test tubing for each experiment.

The results of the control experiment showed that 12½% of packages were defective. Results of the second experiment showed that none of the packages were defective.

Production of Fibrils and Microfibers

POLYOLEFINS

The manufacture of synthetic papermaking fibers, commonly referred to as fibrils, by precipitating an olefin polymer from a hydrocarbon solution is known. A number of different processes are reported for manufacturing such fibrils and fibrils are offered for commercial sale in the United States by at least two suppliers.

Although fibrils made by different processes and/or obtained from different sources are essentially indistinguishable from each other by physical examination, it has been noted that water-laid sheets prepared from different types of fibrils differ significantly from each other in important properties. These observed differences are believed to result from small differences in fibril structure which are difficult to detect by physical observation.

Treating with Polyvinyl, Ultrasonic Vibrations

A process is disclosed by *D.R. Hayes, Jr. and R.R. Whitney; U.S. Patent 4,134,931; January 16, 1979; assigned to Gulf Oil Corporation* for treating high molecular weight olefin polymer fibrils in the presence of an aqueous solution of polyvinyl alcohol while subjecting the solution to ultrasonic vibrations.

Fibrils prepared by this process have excellent properties. The process does not require refining under conditions of high shear stress.

Example: A lot of fibrils were prepared following the procedure of Example 1 of U.S. Patent 4,013,751. The ethylene polymer from which the fibrils were prepared had a weight average molecular weight of about 1,500,000. The hydrocarbon-wet fibrils were refined two times for 120 seconds in a Waring Blendor employing 100 pbw of 91% isopropanol per part of fibrils. The fibrils were filtered and pressed to expel the maximum quantity of isopropanol.

Ten grams of the fibrils and 650 grams of water were stirred for 15 seconds in a Waring blendor to prepare a fine dispersion of fibrils in the water. This slurry and an additional 350 ml of water having 1 gram of polyvinyl alcohol (Vinol 540) dissolved therein were transferred to a 2-liter beaker equipped with a magnetic stirrer. An ultrasonic probe (Sonifier Model S-125 supplied by Branson Sonic Power) was placed in the slurry and operated at 10 A (dc output) of power. Stirring was continued for 15 minutes. The fibrils were filtered, washed with water, refiltered and dried for 16 hours at 120°C.

A first control A was prepared in the same manner, except that the ultrasonic probe was not used. A second control B was prepared by stirring 10 grams of fibrils with 1,000 ml of water containing 1 gram of polyvinyl alcohol for 60 seconds in a Waring blendor. Previous experience has shown that stirring these fibrils for more than 1 minute in the Waring blendor degrades the fibrils.

Films were pressed from the product of this example and both the controls. The polyvinyl alcohol content was measured by IR analysis with the results being shown in the table below.

Sample	Percent Polyvinyl Alcohol
Control A	5
Control B	4
Example	9

From the above data, it is apparent that the process gives close to a quantitative sorption of the polyvinyl alcohol.

The high load melt index of the ethylene polymer employed in the experiments, and the fibrils prepared in this example and controls A and B were measured. The results were shown in the table below.

Sample	High Load Melt Index
Starting polymer	0.0079
Control A	0.0094
Control B	0.0109
Example	0.0044

Neither of control A-B lowered the melt index of the starting polymer, whereas the product of the process reduced the melt index by nearly 50%, that is, (0.0079 - 0.0044)/0.0079 x 100.

Water-laid sheets were prepared from the product of the example and controls A and B. By hand tear tests, the product of the example gave significantly stronger sheets than either of the controls.

Heat-Softened Polymer Dispersion Forced Through Orifices

In a process by *D.G.H. Ballard, R.T. Murray and G.M.F. Jeffs; U.S. Patent 4,007,247; February 8, 1977; assigned to Imperial Chemical Industries Limited, England* fibrils of thermoplastic polymer, particularly crystalline thermoplastic polymers such as polyethylene and polypropylene, are prepared by forcing a dispersion of heat-softened polymer particles in a liquid medium through a pair of opposed coaxial orifices. The liquid medium is preferably aqueous and is pressurized to prevent boiling. The resulting fibrils have some degree of molecular orientation, and may be useful in the production of reinforced paper.

The apparatus used comprised a stirred pressure vessel leading to a fibrillator, as illustrated in Figure 9.1, which is a cross-sectional elevation of the apparatus. The fibrillator **6** is shown on a larger scale than the remainder of the apparatus for the sake of clarity.

Figure 9.1: Cross Sectional Elevation of Pressure Vessel and Fibrillator

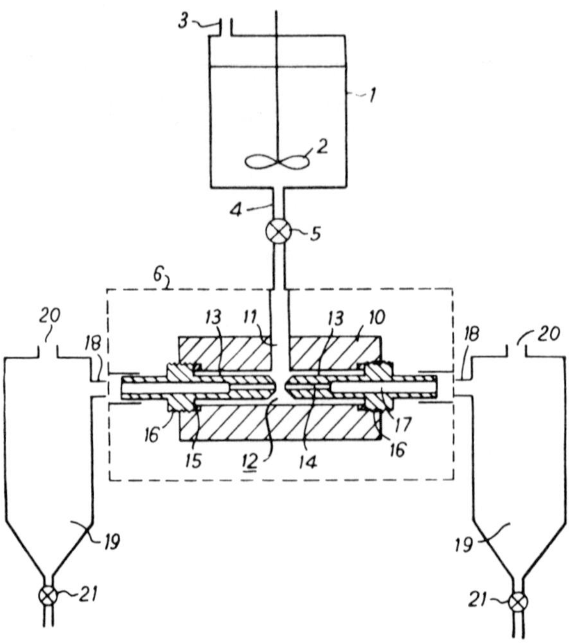

Source: U.S. Patent 4,007,247

Steel pressure vessel **1** was provided with stirrer **2**, electrical heating means (not shown), and inlet and outlet ports **3** and **4**, respectively. Inlet port **3** led to a pressurized nitrogen supply (not shown) and outlet port **4** led to fibrillator **6** via valve **5**. Fibrillator **6** comprised a rectangular metal pressure vessel **10**, heated by electrical means (not shown). A supply passage **11** led to a cylindrical inner chamber **12** which extended at right angles to passage **11** between a pair of opposite faces of the vessel and opened to the faces of the vessel by a pair of internally screw-threaded ports **16** of diameter larger than that of the chamber **12**. Two screw-threaded jets **13** were screwed into the ports **16** so as to enter the inner chamber **12**, the narrow central bores **14** of the jets constituting a pair of mutually opposed coaxial orifices. The spacing between the jets was adjustable by the insertion of spacers **15** of varying thickness in the ports **16**.

Within each jet, the narrow central bore **14** opened out into a passage **17** of larger diameter, the length of the narrow bore **14** constituting the effective length of the jet. The passages **17** communicated with the inlet ports **18** of expansion vessels **19**, provided with outlet ports **20** for the escape of the diluent' vapor and valves **21** for withdrawal of the suspension of fibrils in cooled liquid medium.

In use, a dispersion of particles of polymer in liquid medium was heated to a temperature a few degrees below the softening point of the polymer, and then forced through the supply passage **11** into the chamber **12** of the fibrillator which was at a much higher temperature, so that the temperature of the polymer was raised above its softening point. The dispersion of heat-softened particles was then forced through the bores **14** of the jets **13**, the flow of the liquid medium giving rise to a fluid dynamic field having a high longitudinal velocity gradient. This caused the particles of heat softened polymer to elongate into the form of fibrils which passed through bores **14** and issued from the passages **17** into the expansion vessels **19** as a slurry in super-heated liquid medium. Flash evaporation occurred, giving rise to vapor which escaped through outlet ports **20** and cooling the liquid medium to a temperature below the softening point of the polymer.

Example: An aqueous slurry of high density polyethylene (HDPE) in distilled water was prepared by stirring 25 grams of HDPE (melt flow index 0.02, MP 135°C) with 2,500 ml of distilled water and 0.25 gram of surface active agent sodium dioctyl sulfosuccinate (DOSS) in a conical flask. The HDPE was in the form of particles having sizes in the range 10 to 1,000 μm.

The slurry was then transferred to the stirred pressure vessel of the apparatus illustrated in Figure 9.1, heated to 126°-129°C and pressurized by nitrogen and steam to 225 psi.

The valve at the base of the pressure vessel was then opened and the slurry forced through the fibrillator which had been heated to 250°C, over a period of 60 seconds.

The jets of the fibrillator were of 0.80 mm internal diameter, 55 mm long and their adjacent ends were separated by 0.9 mm.

The product collected from the expansion vessels at the exits from the fibrillator jets was subjected to optical and scanning electron microscopy (SEM) and found to contain fibrils having diameters in the range 5 to 10 μm and aspect ratios in the range 100-200.

SEM examination showed that the fibrils had a rough surface and examination using crossed polar and x-ray diffraction techniques indicated considerable molecular orientation.

A proportion of the polymer feed was recovered as unmodified particles. Melt flow index was measured using 2.16 Kg load at 190°C, using procedure defined in ASTM D1238/65T.

Polymer Dispersions Extruded in Presence of Monomer

G. Zanella and L. Mancini; U.S. Patent 4,105,727; August 8, 1978; assigned to Montedison SpA, Italy provide a method for obtaining polypropylene fibrils from solutions or dispersions of the polymer which have a particularly uniform morphology even when at relatively low extrusion temperatures.

In the preparation of polypropylene fibrils from solutions or dispersions of the polymer in n-hexane by the flash-spinning method carried out in combination with the action of a gaseous fluid, it is possible to obtain more homogeneous fibrils having optimum dimensions, especially in regard to length, and even at relatively low extrusion temperatures, when the n-hexane solution or dispersion is extruded in the presence of propylene monomer dissolved in n-hexane in amounts of from 2 to 25 mols per 100 mols of the hexane-propylene solution.

Example: 500 grams of polypropylene (density = 0.9053, isotacticity index = 93, crystallinity = 58%), 10 liters of n-hexane and 10 grams of a surfactant consisting of ethoxylated stearylamine were introduced into a 24 liter autoclave, equipped with a heating jacket and with a comb stirrer. By means of heating, and by introducing contemporaneously gaseous propylene, under stirring, the following conditions were obtained in the autoclave: 185°C; 28 kg/cm^2 gauge total pressure; propylene concentration in n-hexane, 17 mols for 100 mols of hexane-propylene solution (corresponding to a propylene overpressure of 15 kg/cm^2 gauge).

The polymeric composition, consisting in this instance of two liquid homogeneously dispersed phases, was extruded, obtaining, as a product, individual fibrils having a surface area $>$1 m^2/g, 85% of which had a length comprised between 1 and 3 mm, and a diameter between 15 and 25 μ, while the remaining 15% had a length in part $<$1 mm, and in part between 3 and 5 mm, and a diameter between 25 and 40 μ.

Precipitation at Critical Temperature and Velocity

J. Aboulafia, R. Le Fustec and C. Schranz; U.S. Patent 4,069,287; January 17, 1978; assigned to Compagnie Francaise de Raffinage; Groupement Europeen de la Cellulose, France describe a process and apparatus for manufacturing polyolefin fibers from a composition of the polyolefins dissolved in a diluent, by subjecting the composition to flow at a velocity above a critical value in a tube or bundle of tubes, the temperature of the composition being within the range of critical temperatures of the composition over at least a portion of the length of the tube or bundle of tubes and separating the fibers from the liquid diluent.

The critical temperature range is below the temperature of solution of the polyolefin in the molten state in a diluent and is more specifically the range of temperatures in which the polyolefin remains in dissolved condition when the solution mixture is maintained at rest but precipitates when the mixture is subjected to certain disturbances such as, for instance, shearing. This typically is between 70° and 130°C. The lower limits of the critical velocity and the critical temperature are determined by the appearance of undesired powder coprecipitating with the fibers.

Referring to Figure 9.2, an enclosure **1** containing a polyolefin in a diluent under a nitrogen pressure of 3 bars is connected, via the lines **2** and **3**, to a straight tube **4** having ends **A** and **B**. The tube **4**, which is the place of formation of the fibers, has a cross section which is less than that of the lines **2** and **3**. Downstream of the tube **4** there is an enclosure **5** for the recovery of the fibers, which enclosure has a grid **6** on which the fibers deposit.

Figure 9.2: Schematic Representation of Apparatus for Production of Polyolefin Fibers

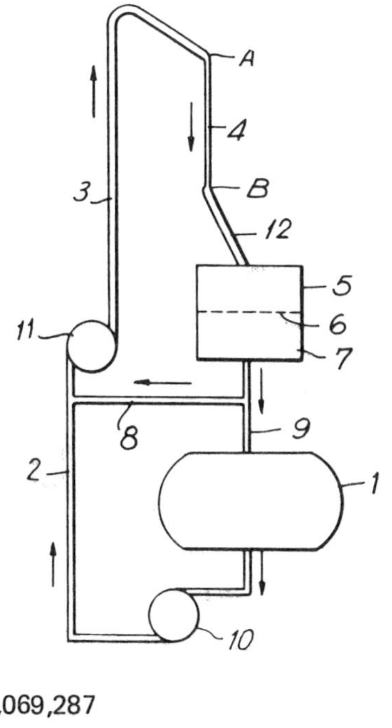

Source: U.S. Patent 4,069,287

The diluent is recovered in the lower portion **7** of the enclosure **5**. A part of the diluent recovered is recycled via the line **8** while another part is forwarded, via the line **9**, into the enclosure **1** when the latter is used as a solution enclosure for the polyolefin. The recycling through the line **8** can be eliminated when the diluent recovered at **7** does not contain any polyolefin.

The lines **2** and **3** contain pumps, **10** and **11** respectively. The pressure in the enclosure **5** is 3 bars. The loss of heat in the tube depends upon the geometry of this tube and the material of which it is made. The formation of the fibers in the tube takes place in a quasi-isothermal manner. The temperature of the composition at the point **A** is therefore substantially the same as that at point **B**, which means that the temperature of the composition at the point **A** is within the range of critical temperatures. The enclosure **5** is advantageously at the same temperature as the tube. The line **12** extending between the point **B** and the enclosure **5** may be eliminated, the tube then discharging directly into the enclosure **5**. Likewise, the end **A** may be connected directly to the outlet of the circulating pump **11**.

The carrying out of the process may require the use of a cooler (not shown in the figure) in the path of the line **2** or of the line **3** in the event that the temperature of the composition at the point **A**, in the absence of such cooler, would be above the upper limit of the range of critical temperatures and there is no, or practically no, heat loss in the tube.

Example: 28 grams of polyethylene (Vestolen 6016), having the properties: density = 0.960 g/cc, $\overline{M}_n$ = 11,500, $\overline{M}_w$ = 80,000, are dissolved in an enclosure of a volume of 1 liter at a temperature of 130°C in 600 ml of cyclohexane, corresponding to a concentration of 5% expressed by weight. The enclosure is connected, via a reciprocating pump, to a copper coil of a length of 10 m and a diameter of 4 mm placed in a thermostatic oil bath in such a manner that the flow in the coil is isothermal. The temperature of the coil is therefore fixed. For a pumping rate of 70 l/hr (which corresponds to a maximum linear velocity of the composition in the coil of 3 m/sec), fibers appear in the composition while it passes through the coil when the temperature of the composition at the entrance to the coil is between 82° and 78°C. These two values of the temperature define the range of critical temperatures.

When the temperature is above 82°C, all other things being equal, one does not note any precipitation of polyethylene in any form whatsoever. When the temperature is below 78°C, all other things being equal, there is noted the precipitation of a mixture of fibers and powder, the proportion of powder being higher the lower the temperature.

Within the same temperature range when the rate of the pump is fixed at 35 l/hr (which corresponds to a maximum linear velocity of 1.5 m/sec of the composition in the coil), the fibers obtained are shorter, and they are obtained mixed with powder. The velocity is therefore below the critical value.

Release of Pressure on Mixture of Molten Polymer and Solvent

A process is provided by *J.-P. Pleska and M. Marechal; U.S. Patent 4,010,229; March 1, 1977; assigned to Solvay & Cie, Belgium* for the manufacture of fibrils of short length by suddenly releasing the pressure acting on a two-phase liquid mixture of molten polymer and solvent and which is at a high pressure and a high temperature. The two-phase liquid mixture is ejected through a pressure release orifice so as to vaporize the solvent instantaneously and solidify the polymer, and the flow path of the two-phase liquid mixture is perturbed at the instant when it enters the pressure release orifice.

Spinnerets are provided which have a perturbation chamber containing a supply orifice and a pressure release orifice.

As is shown in Figure 9.3a, a spinneret **1** possesses a pressure release orifice **2** and a pre(pressure release) device. This pre(pressure release) device, which consists, according to the embodiment represented, of a partition **3** possessing a central orifice **4**, is by no means indispensable. Its role is to cause a pressure drop in the mixture of molten polymer and solvent which is in the form of a single-phase liquid passing through the spinneret, in order to bring about the formation of a two-phase liquid mixture.

Figure 9.3: Device for Manufacture of Short Fibrils

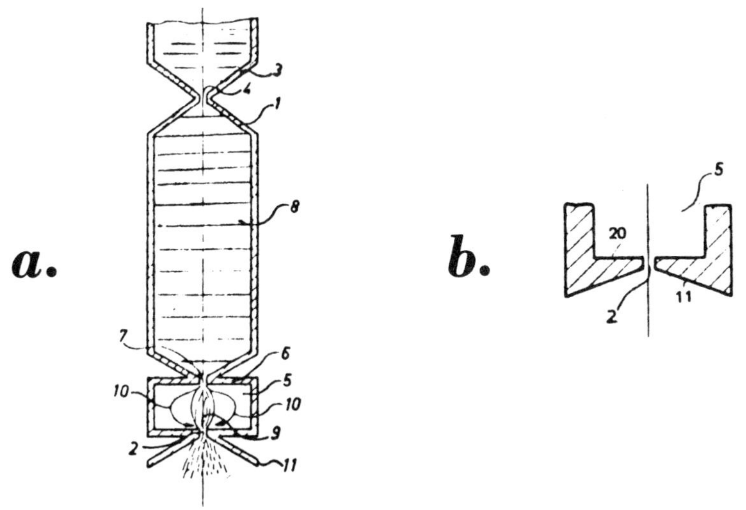

(a) Complete device
(b) Terminal part of the device

Source: U.S. Patent 4,010,229

As shown in the figures, the spinneret also comprises a perturbation chamber **5** defined especially by the wall **20** possessing the pressure release orifice **2** and by a flat upper wall **6** possessing the supply orifice **7** at its center. The supply and pressure release orifices are opposite and coaxial. The perturbation chamber is cylindrical with a circular cross section. The orifices are positioned at the centers of the bases. The distance between the supply and pressure release orifices (height) is generally less than 10 cm and preferably less than 7.5 cm.

Example: A two-phase liquid mixture is produced by bringing a mixture comprising 15% by weight of Eltex A 1050 (Solvay & Cie) and 85% by weight of technical hexane to a temperature of 195°C and to a pressure of 63 kg/cm^2. These conditions correspond to the start of the appearance of two liquid phases. Eltex A 1050 is a high density polyethylene with a melt index of 5.

The pressure acting on this mixture is released by passing the latter through a spinneret possessing a perturbation chamber as represented in Figure 9.3b and having the following geometric characteristics: ratio of the diameter of the perturbation chamber to the diameter of the supply orifice **7** (Figure 9.3a) equal to 16; ratio of the height of the perturbation chamber to the diameter of the supply orifice **7** equal to 5; ratio of the height of the perturbation chamber to the diameter of this chamber equal to 0.313; and diameters and heights of the supply orifice and of the pressure release orifice of 1 mm.

The diverging component **11** extending from the orifice for effecting sudden pressure release opens out at an angle of 150°. The two-phase liquid mixture is discharged at the rate of 21.4 kg of polymer per hour. Short fibrils of average length 1.7 mm are obtained directly. The specific surface area of these fibrils is as much as 35 m^2/g.

Polymer, Water, and Solvent Mixture Flashed Through a Nozzle

J.H. Kozlowski, P.C. Litzinger and F.J. Steffes; U.S. Patent 4,054,625; Oct. 18, 1977; assigned to Crown Zellerbach Corporation have developed a process of manufacturing fibers by forming a mixture of a solvent, a polymer and water and flashing the mixture through a nozzle. Water is present as a discontinuous phase in the mixture. The polymer may be any polymer capable of forming fiber, preferably a crystalline or a partially crystalline polymer. The preferred polymers are crystalline or partially crystalline polyolefins, preferably polypropylene. The discrete fibers thus formed can be easily refined and used for making paper webs.

Example: The dissolution vessel employed in this example was an 800 gallon, glass-lined, baffled tank having a centrally disposed 7.5 horsepower agitator having two 4-bladed turbine-type stirrers located thereon. The lower turbine was 27 inches from tip-to-tip and the upper turbine was 19 inches from tip-to-tip. The stirrer was operated throughout the run. Sixty gallons of water containing 1% by weight polyvinyl alcohol (Monsanto Gelvatol, viscosity of 4-6 cp and 87.7-89% hydrolyzed) based on the weight of the polyethylene to be employed was added to the tank, followed by the addition of 45.4 kg of high density polyethylene having an intrinsic viscosity of 1.4 dl/g and melt index of 5.5 (Mitsui 2200P). The vessel was sealed, and 300 gallons of n-hexane added. The water-to-hexane volume ratio in the mixture was 1:5, and the polyethylene concentration was 40 g/l of hexane. The contents of the vessel were then heated to 130°-135°C and a pressure of 120 psig.

A valve in the bottom of the vessel was then opened and the mixture fed through a 3 inch line 5 inches in length to a Valtek angle-type control valve having a port with a diameter of 9.5 mm and a length of 7 mm. A nipple-shaped plug in the valve port formed an annulus with the wall of the port. The valve was operated about one-quarter open.

The mixture flowed through the annulus which formed the flash nozzle at the rate of 180 grams of polyethylene per minute. The mixture flashed into a line having a diameter of 1 inch. At a distance of about 6 inches past the nozzle, low pressure (10 psig) steam was introduced into the 1 inch line to strip the solvent which had not previously flashed from the fibrous noodle. At a distance of 18 ft from the nozzle the 1 inch line opened into a vapor-separation vessel, the vapors being removed from the top of the vessel. Hot dilution water at 85°C was introduced at the top of the vapor-separation vessel at a rate of about 5 gal/min. The water and fibrous noodle dropped to the bottom of the vapor-separation vessel, and was fed at temperature of 80° to 90°C and a consistency of about 1% to a Jones double disc refiner (12 inch diameter discs) equipped with brushing-type plates (Jones type 1, 1, 1.5, +10°). The discs were operated at 2117 rpm at a clearance of 0.004±0.004 inch.

The fibers were subjected to a total of four passes through the refiner, the second through fourth passes being at a stock temperature of 40°-50°C. An additional 1% by weight Gelvatol 20-30 polyvinyl alcohol (based on the polyethylene) was added to the stock tank and the stock given a fifth refiner pass with the plates set at 0.010±0.004 inch. The polyethylene fibers had the following fiber fractionation when tested according to TAPPI Standard T233 SU-64:

Mesh	Weight Percent
+20	10.3
+35	33.8
+65	23.8
+150	17.4
+270	5.0
-270	9.7

The fibers had a drainage factor of 0.24 sec/g. One portion of the resulting fibers was made into handsheets in accordance with TAPPI Standard T-205 M-58 with modified wet pressing (400 psi) and a heat-bonding step (121°C at minimum pressure). Another portion of the fibers was blended 50/50 with bleached alder kraft pulp having a Canadian Standard Freeness of 410 cc and made into handsheets by the same procedure. The resulting handsheets were tested and the results are as follows.

Property	100%	50% Blend
Basis weight, g/m^2	61.2	61.2
Density, g/cc	0.36	0.48
Opacity, %	83.4	78.9
Brightness, % (Elrepho no. 8)	88.9	84.4
Breaking length, m	139	2416
TEA, kg-cm/cm^2 *	0.06	34.3
Internal bond, Scott units	14	90
Tear, g/sheet	6	28

*Tensile energy absorption.

Hydrophilic Fibers Containing Inorganic Pigment

In a process developed by *W. Gordon, H.J. Leugering and H. Schaefer; U.S. Patent 4,013,617; March 22, 1977; and W. Gordon; U.S. Patent 4,098,757; July 4, 1978; both assigned to Hoechst AG, Germany* hydrophilic polyolefin fibers containing inorganic pigments are obtained by means of flash evaporation of a superheated suspension, which is under at least autogenous pressure. The suspension comprises a dispersion of a hydrophobized inorganic pigment in an emulsion prepared from a solution of a polyolefin in an easily boiling solvent and an aqueous solution of a hydrophilizing agent. The suspension is passed through a nozzle into a low pressure zone to form pigmented fibers.

Example 1: Referring to Figure 9.4, into a pressure vessel **A** which has a volume of 70 liters and which is equipped with a multistate, countercurrent agitator with 5 flat paddles **B**, is placed 0.4 kg of polyethylene with a density of 0.960 g/cm^3 having a reduced specific viscosity of 1.4 dl/g and a molecular weight distribution $\bar{M}_w/\bar{M}_n$ of 6, as well as 20 liters of hexane, 15 liters of water, 60 grams of polyvinyl alcohol with a solution viscosity of 4 cp (4% in water at 20°C and a saponification degree of 98%), as well as 1.6 kg of hydrated aluminum silicate corresponding to the formula: $Al_2O_3 \cdot 2SiO_2 \cdot 2H_2O$, 90% of the particles of which are smaller than 10 microns and with a 1% hydrophobization which had been carried out in accordance with Example 1 of German Patent 847,486; the mixture was emulsified and suspended at 140°C with an agitator speed of 600 rpm.

The total pressure in the pressure vessel is adjusted by means of nitrogen to 16 kg/cm^2. When the valve **C** at the bottom of the pressure vessel is opened, the emulsion flows through the pipe-shaped nozzle **D** which has an inside diameter of 4 mm and a length of 1.20 m, into vessel **E** in which vacuum pump **F** provides a vacuum of approximately 100 mm Hg and where the resulting fibers are collected. The hexane residues which have remained in the fibers are removed under vacuum by passing steam over the fibers from the steam supply **H**. The fibers containing water are removed from the container **E** through the discharge connection **G** which can be closed.

After partial dehydration by mechanical squeezing to approximately 30% of their volume, the fibers produced contain 76.3% of the hydrophobized aluminum silicate, i.e., the retention during the flash spraying is 95.5%. The produced

Figure 9.4: Process for Manufacturing Fibers Containing Pigment

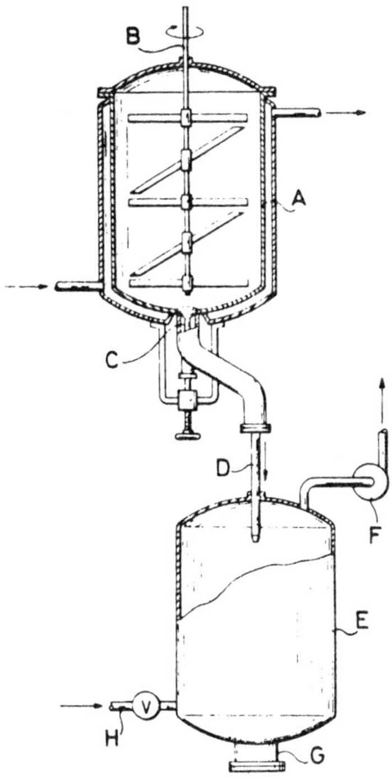

Source: U.S. Patent 4,013,617

fibers are hydrophilic and can be dispersed in water without difficulty. When 2 grams of these fibers are dispersed in 800 ml of water in a 1 liter measuring cylinder and the fiber suspension is allowed to settle for exactly two minutes, the fibers sink only slightly, so that after two minutes the supernatant water volume which is free of fibers amounts to 40 ml.

When the fibers obtained are classified in a membrane classifier of the Brecht-Holl type for 10 min with the sieve at 0.5 atm above atmospheric pressure and maximum stroke, there remains of a 2 gram weighed test sample 10% on a sieve with a 0.40 mm mesh size, and 57% on a sieve with a 0.12 mm mesh size, while 33% of the weighed sample passes through the latter sieve. This indicates that the fibers are so uniform and so short that they may be employed, for example, for the manufacture of paper without further shredding.

If a sheet of 160 g/m^2 is manufactured from the fibers obtained above on a Rapid-Kothen sheet-forming device, the pigment contained in the sheet amounts to 74.2%, i.e., the pigment retention during the fiber processing stage is 97.3%. On the other hand if one attempts to produce a sheet containing pigment from 75% hydrophilic pigment and from 25% of comparable polyethylene fibers which do not contain any pigment, it is found that the pigment retention amounts to only 19%. However, the pigment retention achieved through the fiber formation described is 92.8% up to the fiber processing stage.

Example 2: The same procedure outlined in Example 1 is used, with the pigment consisting of 1.6 kg of hydrated aluminum silicate which has not been made water-repellent, corresponding to the formula $Al_2O_3 \cdot 2SiO_2 \cdot 2H_2O$, with 90% of the particles being smaller than 10 microns.

Subsequent to partial dehydration by mechanical squeezing to approximately 30%, the produced fibers contain 35.5% pigment, i.e., the pigment retention amount to only 44.3%.

When these fibers are classified, 13% of the fibers remain on the 0.40 mm mesh sieve, 84% of the fibers remain on the 0.12 mm mesh sieve and 3% of the fibers pass through this latter sieve. Although these fibers are considerably hydrophilic, they are not freely dispersible in a diluted suspension, but continue to cling together.

Only after having been shredded in known manner in a 12 inch disk refiner, the fibers are reduced to a length that is comparable to that mentioned in Example 1. Classification indicates 9% retention on the 0.40 mm sieve, 64% retention on the 0.12 mm sieve and 27% pass through the 0.12 mm sieve. Following partial mechanical dehydration as indicated above, the pigment content amounts to only 26%.

When the thus refined fibers are used for forming in a Rapid-Kothern sheet-forming device a sheet of 160 g/m^2, the pigment content in the sheet is only 19%, i.e., the pigment retention between the manufacture of the fibers and the processing of the fibers is only 24%. It appears that there is no possibility of producing by this method hydrophilic fibers containing more than 50% of pigment. The amounts of pigment which have not been retained must be recovered and recycled at considerable cost.

Crystalline Fibrils from Polymer Solution Flowing Through Gauze

F.C. Frank, A. Keller and M.R. Mackley; U.S. Patent 4,020,266; April 26, 1977 describe a process in which crystalline polymer fibrils having a highly ordered crystalline core surrounded by less ordered crystalline polymer are produced by causing a solution of the polymer to flow through a gauze. The solution is at a temperature below the dissolution temperature of the polymer but above the temperature at which the polymer will crystallize spontaneously in an unstirred solution. The flow rate is selected in relation to the gauze dimensions to impart a sufficiently high longitudinal velocity gradient to the solution flowing through the gauze, preferably about 10^2 sec^{-1} or greater.

Referring to Figure 9.5, the apparatus comprises a closed vessel **10** with means (not shown) for maintaining it at a required elevated temperature. Extending downwardly into the vessel is a spindle **12** driven by a motor **14**, and having at its lower end portion within the vessel a radial arm **16**. At its end the arm carries an open-ended sleeve **18** across the interior of which extends a fine stainless steel gauze **20**. The vessel is charged with polymer solution **22**, and the sleeve **18** is arranged so that on rotation of the spindle **12** the solution flows substantially axially through the sleeve.

In an experiment using this apparatus, a solution was made containing 1.5% by weight polyethylene (Marlex 6002) in xylene, this being the maximum concentration which was readily obtainable. 500 ml of this solution was introduced into the vessel and maintained at a temperature of about 106°C. The gauze was made up from stainless steel wires 0.02 mm thickness and with 0.1 mm spacing, giving 40% of the gauze area obscured by wire.

The vessel was 12 cm in diameter, the arm **16** was 1 cm in length, and the sleeve was 2 cm in diameter. The spindle was rotated at 24 rpm, giving an estimated fluid velocity outside the gauze of 2.5 cm per second, and an estimated velocity through the gauze of 6.25 cm per second, and an estimated longitudinal velocity gradient of 375 sec^{-1} in the fluid along the axis of each opening in the gauze. These conditions were chosen in an attempt to obtain a suitably high velocity gradient with substantially laminar flow of fluid in the region of the gauze.

When the gauze was removed from the apparatus and examined, a network of fibrils were seen to have formed on the gauze. Each fibril consists of a central core from which extend transverse platelets.

Figure 9.5: Diagrammatic View of Crystallization Apparatus

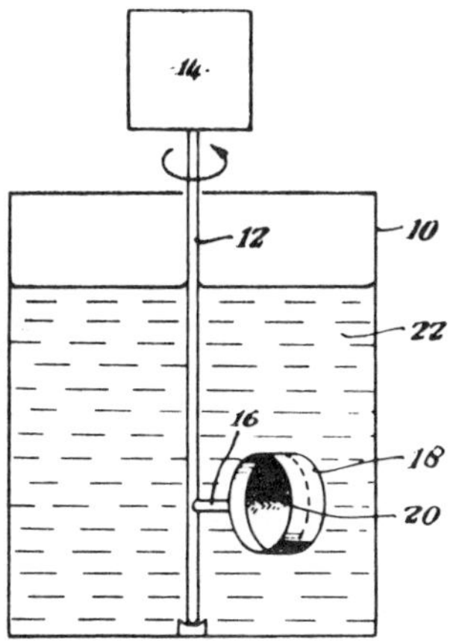

Source: U.S. Patent 4,020,266

Tests on the material indicated the presence of two modes of crystallization. The core is believed to be more lightly ordered and made up from extended polymer chains, whereas the platelets are thought to be relatively less ordered and made up from folded chains. The folded chain structure can be considered as the normal crystallization mode of the polymer, while the extended chain core is the result of the production of longitudinal velocity gradients in the solution at the point of crystallization.

Cooling While Agitating with Vibrations of Sonic Frequency

L.B. Keller and R.K. Jenkins; U.S. Patent 4,127,624; November 28, 1978; assigned to Hughes Aircraft Company have prepared a class of polymeric fibers and masses which are produced by precipitation from solutions of the polymer. The fiber mass is a coherent, three-dimensional network of interconnected crystalline polymeric fibers and fiber bundles randomly arrayed in all directions. The fibers are smaller in diameter than can be consistently produced by conventional fiber drawing and spinning techniques.

The fiber masses are produced from solutions of fiber-forming linear organic polymers having regularly repeating chain structures. The polymers are precipitated from these solutions by simultaneously cooling and agitating the polymer solution at vibrations in the sonic frequency range. The fiber masses may then be removed from the solvent or, in the case of a polymerizable solvent, a composite may be formed.

As shown in Figure 9.6, a frequency generator **1** is connected to an amplifier **2** used to vary the frequency of a shaker-head **3** which drives a shaft **4** connected to holder **5** which is capable of retaining container **6** in which a solution **7** is held by means of stopper **8**.

Figure 9.6: Device Used to Impart Vibrations to a Fiber-Forming Solution

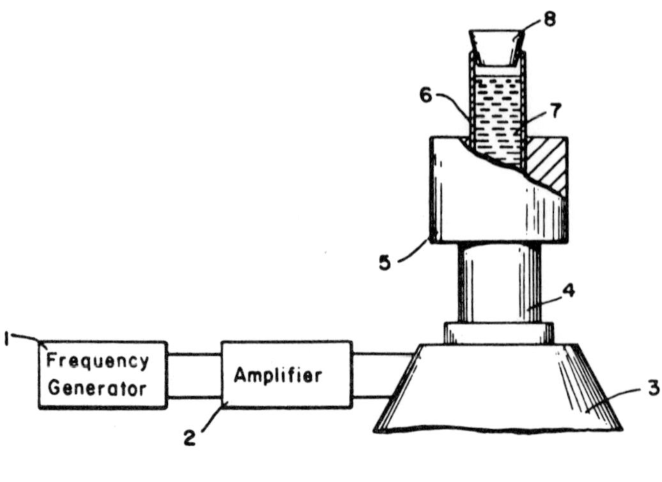

Source: U.S. Patent 4,127,624

Example: 20 ml of a 5% solution of isotactic polypropylene in styrene inhibited with hydroquinone was heated to 120°C to dissolve the polypropylene. Then the test tube with solution was placed into the holder of the vibrator system, and the solution was subjected to vibrations from about 800 to 90 Hz during which time resonance frequencies were visually located.

The sample was held at several resonance frequencies of notable agitation until such time as the entire sample had become a fibrous mass. During this period, the system was allowed to cool toward room temperature. The mass was removed from the test tube by inverting and tapping the tube. The fibrous mat, due to its enormous surface area, held the solvent styrene within its interstices, thus requiring solvent extraction (acetone) for complete removal.

Stretchable Spun-Bonded Polyolefin Web

R.N. Sabee and C.C. Sabee; U.S. Patent 4,013,816; March 22, 1977; assigned to Draper Products, Inc. describe a stretchable spun-bonded web suitable as top-liner for a disposable diaper, pad, bandage and the like, and a process for making the same. In particular, the disclosure relates to a polyolefin web where the crossover points of the fibers do not rupture when the web is stretched to 50% of its original length, but maintains approximately its original structure, which makes it extremely suitable for such applications. The web is made by melt blowing polypropylene of less than 1.2 intrinsic viscosity at a filament velocity of at least 15 meters per second on a smooth collecting surface having a temperature less than 65°F.

Figure 9.7: Graph Showing Web Elongation

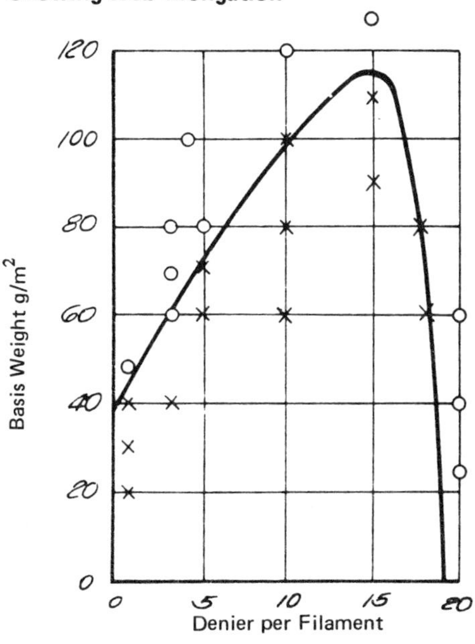

Source: U.S. Patent 4,013,816

Experiments show that the filament velocity has to exceed a certain minimum, and the birefringence be below a certain maximum, in order to obtain a web of good bond strength and a minimum web tensile elongation of 50%. In addition to these variables, a combination of filament size (dpf) and basis weight is required to obtain stretchability. The data are plotted in Figure 9.7.

Figure 9.7 is a curve which shows the percent web strip elongation at break for webs of differing basis weight and denier per filament. Only a web which falls on the curve or within the curve has the desired web elongation at break of larger than 50%.

In Figure 9.7 the **X** indicates a web having an elongation at break greater than 50% and the **O** indicates a web having an elongation at break of less than 50%.

According to Figure 9.7, the maximum basis weight is 116 g/m^2 at 16 dpf, and the maximum allowable dpf is about 20, above which cooling is apparently too slow even in a single filament to eliminate brittleness of the web.

Pillowed Web of Blown Microfibers

A blown microfiber web developed by *L.D. Humlicek; U.S. Patent 4,103,058; July 25, 1978; assigned to Minnesota Mining and Manufacturing Company* having a network of compacted high-density regions and pillowed low-density regions exhibits a unique combination of low density and good tensile strength and integrity. Such a web may be collected on a perforated screen so that microfibers deposited on the land area of the screen form the compacted high-density regions, and microfibers deposited over the openings of the screen form the pillowed low-density regions.

Figure 9.8 schematically shows illustrative apparatus for forming polymeric microfiber webs. The microfiber-blowing portion of the apparatus, for forming microfibers and directing them in a gaseous stream toward a collector, can be a conventional apparatus such as described in the article "Superfine Thermoplastic Fibers," in *Industrial Engineering Chemistry,* vol 48, p 1342 (1956) and Prentice, U.S. Patent 3, 704,198.

Such apparatus generally includes a die **15**, which has an extrusion chamber **16** through which liquified polymeric material is advanced; a set of polymer orifices **17** arranged in line across the forward end of the die; and gas orifices **18** adjacent to the polymer orifices. A gas, usually air, is supplied at high velocity through the orifices **18** and directed toward the path of polymeric material extruded from the polymer orifices **17**. The high-velocity gaseous stream draws out and attenuates the polymeric material extruded through the polymer orifices, whereupon the attenuated polymeric material solidifies as microfibers during travel to a collector **20**.

The collector **20** comprises an appropriate collection screen **21**, which in this illustrative apparatus is wrapped around parallel discs **22**. Supporting structure

Figure 9.8: Microfiber-Blowing Apparatus

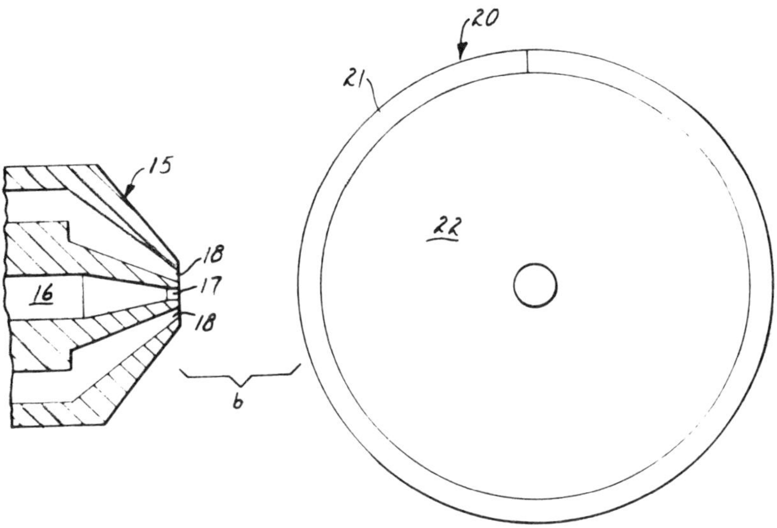

Source: U.S. Patent 4,103,058

for the collection screen may extend between the discs; for example, a small-mesh screen may be used to support a flexible collection screen that has sufficient depth to permit formation of the desired pillows in a collected web. A low-pressure region may be developed within the interior of the drum formed by the screen **21** and discs **22** to improve the withdrawal of the gas stream and to aid the penetration of microfibers into the openings of the collection screen. Once a microfiber web has been collected on the collection screen **21**, it is usually removed from the collection screen at a point remote from the deposition area and wound into a storage roll, whereupon it can be later used as is; cut into desired configurations; added to other structure, as by lamination to another sheet, or otherwise processed.

For certain uses of the microfiber web in which low-density high-volume pillows are needed, for example, when the web is to be used as a collector for fluids, the pillows have a density less than about 0.02 g/cm^3. For other uses, where the webs are to be used, for example, as filter media, thermal insulation, and acoustic barriers, the density of the pillows may be lower, such as about 0.004 g/cm^3.

The webs prepared are especially useful for presenting a three-dimensional arrangement of particles in which the particles can interact with (e.g., chemically or physically react with, or physically contact or be modified by) a medium to which the particles are exposed. The particles are physically entrapped within the interstices of the web and no binder material is required to hold them in place for typical useful functions of the web.

Reinforced Pillowed Microfiber Webs

According to *D.L. Krueger; U.S. Patent 4,042,740; August 16, 1977; assigned to Minnesota Mining and Manufacturing Company* webs of blown microfibers having a network of compacted high-density regions and pillowed low-density regions are reinforced by a mesh of filaments used to collect the web.

Example: A reinforced microfiber web was made by supporting a mesh of reinforcing polypropylene filaments about 5 cm in front of an extrusion apparatus. The mesh included four filaments per 5 cm of length and width of the mesh. Polypropylene microfibers were extruded through the apparatus and collected on the mesh using process conditions as follows:

Temperature of molten polypropylene*	250°C
Temperature of die	370°C
Temperature of air	510°C
Rate of polymer flow, g/min**	15.7
Air pressure, kg/cm^2***	0.7
Rate of movement of mesh, cm/sec	7.9

*Having a melt flow of 12 g/10 min.
**Using 200 polymer extrusion orifices having a width of 0.3 mm.
***The air orifice had a slot thickness of 0.3 mm.

A microfiber web was collected having a nominal thickness including the height of the pillows of 1.7 cm, with the microfibers having an average diameter of about 5 micrometers. Samples 5 by 10 cm were cut and evaluated for oil sorbency using heavy mineral oil (having a viscosity of 75 to 80 cs). The oil sorbency ratio was determined to be between 65 and 100 g oil per gram of web. Test samples were used to sorb oil, then squeezed to remove oil, and then used again to sorb oil several times. Data are given in the following table.

Saturated Web (g)	Squeezed Weight of Web (g)	Percent Oil Removed	Oil Sorbency Ratio	No. of Uses
20.9	1.4	-	79.4	1
13.5	1.25	92.5	50.9	2
10.8	1.2	91.1	40.5	3
10.9	1.1	92.2	40.9	4
10.8	1.1	92.1	40.5	5
10.3	1.1	92.6	38.6	6

Filter of Electrically Charged Fiber Material

J. van Turnhout and J.C. Rieke; U.S. Patent 4,178,157; December 11, 1979; assigned to N.V. Verto, Netherlands disclose a method for forming a filter of electrically charged electret fiber material consisting of a high molecular weight nonpolar polymeric substance, comprising drawing, charging, and fibrillating a foil of the high molecular weight substance. The fibrillated foil is wound and then crimping is effected.

As shown in Figure 9.9, each of the extruders **31** and **32** supply a component of the high molecular weight, polymeric substance, each of them differing in molecular weight, to a common nozzle **33** with the strip-like extrusion mouth **34**. The thus obtained foil **1**, in which both components are present in a laminate-shaped nonhomogeneous distribution, is, after passing the rollers **3** and **4**, subjected to a first drawing between the roller pair **5** and **6** and the roller pair **9** and **10** rotating at a higher speed, in which it is drawn over an edge **7** of a block **8** provided with a schematically indicated heater **15**.

Between the roller pair **9** and **10** and the roller pair **13** and **14** rotating at a still higher speed the foil **1** is subjected to a second drawing while it is advanced over a bent plate **12**, provided with a schematically indicated heater **16**. At the same time, the foil **1** is electrically charged by the spray device **18**, comprising corona wires which are connected to a not-shown high-tension source and of which the plate **12** forms the antipole. The splittable electret foil **1** obtained in this manner subsequently is advanced via the support roll **20** over the needle roll **29**, which fibrillates the foil **1** to a fibrillated fiber band **35**. Subsequently, the latter is widened by the spreading device **36** to a net-like fiber fleece **37** and wound on to the roll **38**.

When in this way a fiber mat of the desired thickness is wound, it is taken from the roll **38** by cutting it in the direction of the axis of the roll or by pushing it from the roll **38** without cutting. The fibers of which the fiber mat exists now are not held anymore in straightened shape and crimp spontaneously as a result of the different measure of relaxation-crimp in both components of the high-molecular substance of which they are made. The fibers mesh with each other and form a filter mat with a homogeneous and firm structure, which is further still improved by a heat treatment of about 15 minutes in an oven not shown in the drawing, the temperature of which is kept at about 70°C.

Figure 9.9: Method for Manufacture of Electret Filters

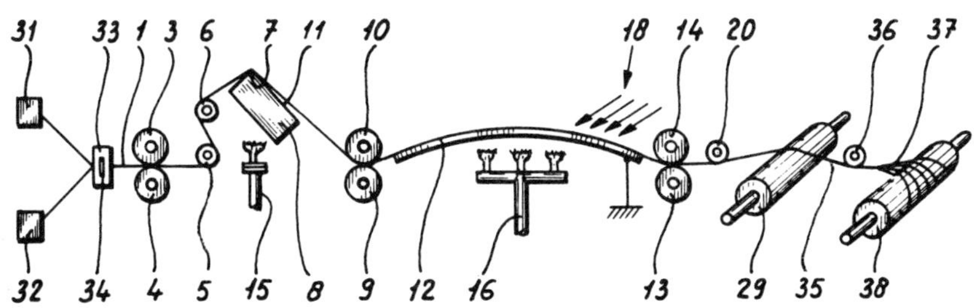

Source: U.S. Patent 4,178,157

Example: According to the method described with the aid of Figure 9.9 a laminate having a thickness of 50 μm and a width of 6 cm consisting of polypropylene with a melt flow index of 1.5 and polypropylene with a melt flow index of 8 is drawn with an initial speed of 13 m/min over block **8**, having a temperature of 100°C, in a draw ratio of 1:6 and subsequently over the plate **12** having a temperature of 150°C in a draw ratio of 1:1.5. The corona wires are 5 mm above the foil and have a charge of -10 kV with regard to the plate **12**. The foil is fibrillated with a needle roller with 60 rows of which the needles are positioned at a mutual distance of 0.5 mm. The obtained split fiber band is spread to a width of 90 cm and is wound in 170 layers to fiber mat with thickness of 3 mm. The fiber mat is taken from the roll and is placed for 15 minutes in an oven at 70°C.

Electrostatic Spinning

G.E. Martin, I.D. Cockshott and K.T. McAloon; U.S. Patent 4,127,706; Nov. 28, 1978; assigned to Imperial Chemical Industries Limited, England provide a method of preparing a porous sheet product which comprises the step of introducing a spinning liquid comprising an organic fiber-forming polymeric material into an electric field whereby fibers are drawn from the liquid to an electrode and collecting the fibers so produced upon the electrode. Polytetrafluoroethylene (PTFE) and other fluorinated polymer mats produced by the electrostatic process are useful as electrolytic cell diaphragms, battery separators etc.

The process of electrostatic spinning involves the introduction of a suitable spinning liquid into an electric field whereby fibers are drawn from the liquid to an electrode. While being drawn from the liquid the fibers harden, which may involve mere cooling (where the liquid is normally solid at room temperature, for example, and is melted to enable spinning to take place), chemical hardening (for example by treatment with a hardening vapor or by crosslinking) or by evaporation of solvent (for example by dehydration). The resulting fibers may be collected on a suitably located receiver and subsequently stripped from it conveniently in the form of a sheet or mat.

In one embodiment, the surface to which the fibers are drawn is a continuous surface, as of a drum, over which passes a belt which may be withdrawn from the region of charge, carrying with it the fibers which have been formed and which have become attached thereto.

Such an arrangement is shown in Figure 9.10, in which **1** is an earthed metal syringe needle supplied from a reservoir with spinning liquid at a rate related to the rate of fiber production. Belt **2** is of gauze driven by a driving roller **3** and an idler roller **4** to which is fed an electrostatic charge from a generator **5** (in the apparatus illustrated a Van de Graaff machine). Removal of the fiber mat **6** from belt **1** is by any convenient means, for example by suction or by air jet, or it may be removed by juxtaposition of a second belt carrying sufficient electrostatic charge to effect detachment of the mat from belt **2**. In the figure the mat is shown being picked up by a roller **7** rotating against the belt.

Figure 9.10: Diagrammatic View of Electrostatic Spinning Apparatus

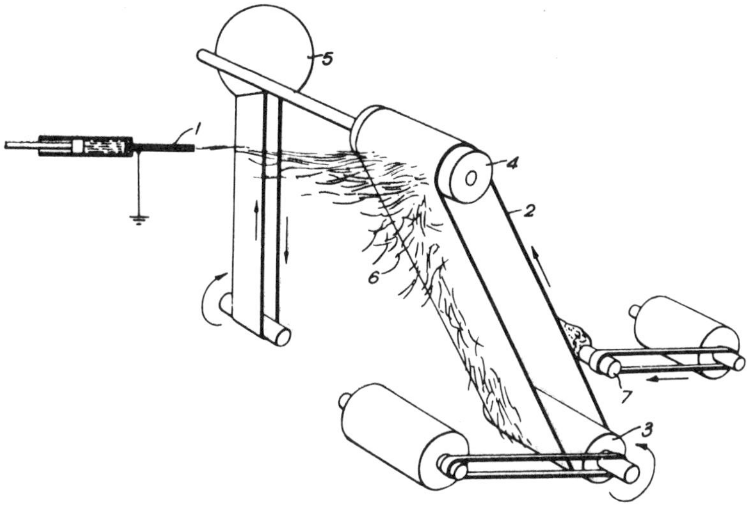

Source: U.S. Patent 4,127,706

Example 1: The apparatus employed was as shown in Figure 9.10; the belt was of "Terylene" net 20 cm wide. The spinning liquid was prepared by mixing 80 parts w/w of an aqueous polytetrafluoroethylene dispersion having a PTFE solids loading of 60% and containing 2% (w/w on PTFE) of Triton X-100 surfactant (Rohm and Haas) with 20 parts w/w of a 10% solution of polyethylene oxide "Polyox" WSRN 3000 in water. The PTFE was of number average mean particle size 0.22 micron and standard SG 2.190. The surfactant may be any of the range capable of stabilizing PTFE of which Triton X-100 and "Triton DN65" are examples. The spinning liquid was spun from 20 x 1 ml syringes on to the net (the charge on the roller being 20 kV) situated 20 cm from the earthed needle tips.

The fibers were deposited over a width of about 16 cm and a sheet 0.4 mm thick was obtained. This sheet was then removed, placed on a stainless steel gauze support and sintered at 360°C for 5 minutes. A tough, porous, white, slightly rough sheet of uniform thickness was produced, consisting of fibers of average diameter 2-3 microns apparently bonded together into a reticulum having 78% free volume.

Example 2: A sheet obtained as described in Example 1 was treated as follows with (a) a 10% w/w aqueous solution of sodium hydroxide at 18°C for 24 hours, (b) 10% hydrochloric acid at 18°C for 24 hours, (c) a 10% w/w aqueous solution of sodium hydrogen phosphate at a boil for 1 hour, and finally with (d) a constantly agitated 10% w/w suspension of titanium dioxide (average particle size 0.2 micron) in isopropyl alcohol for 5 hours.

The PTFE sheet impregnated with the titanium dioxide was washed with isopropyl alcohol to remove excess solid and then mounted in a vertical diaphragm cell for the electrolysis of sodium chloride.

OTHER MICROFIBER MATERIALS

Cellulose Acetate Fibers by Emulsion Flashing

Discrete cellulose acetate fibers are produced by *P.C. Litzinger; U.S. Patent 4,040,856; August 9, 1977; assigned to Crown Zellerbach Corporation* by flashing an emulsion of cellulose acetate, water, and a solvent which is miscible with the water at the temperature at which the emulsion is flashed. The fibers can be used directly to make paper according to conventional papermaking methods.

Example: A 22 liter steam-jacketed vessel was charged with 2.8 liters of acetone and 5.2 liters of water. As the contents were stirred, 0.8 kilogram of cellulose acetate having an acetic acid content of about 55% was added. The vessel was then sealed, purged with nitrogen, and heated from ambient temperature to 143°C with stirring to form an emulsion of acetone-plasticized cellulose acetate in an acetone-water medium. The emulsion was flashed through a nozzle having a diameter of 1.02 mm and a length of 2.86 cm into a receiving vessel. During flashing, the pressure in the stirred vessel was maintained between 180 and 188 psig. The receiving vessel was maintained at subatmospheric pressure by a steam ejector which facilitated vaporization and removal of the acetone.

The product was a slurry of discrete fibers in water. The fibers were screened from the water and reslurried in fresh water at a consistency of 1%. The slurry was passed through a disc refiner. Handsheets were made from the refined slurry and from a blend of 50% cellulose acetate fibers and 50% conventional papermaking wood fibers. The properties of the handsheets and the cellulose acetate fibers are reported below.

Properties of 100% Cellulose Acetate Handsheets

Basis weight	62.3 g/m^2
Caliper	0.126 mm
Density	0.49 g/cc
Tear resistance	6.8 g/sheet
Tensile strength	1.83 kg/15 mm
Breaking length	1.96 km
Stretch	1.92 %
Tensile energy absorption	0.012 $kg\text{-}cm/cm^2$
Internal bond	65.0 Scott units
Brightness	81.3 %
Scattering coefficient	980.0
Opacity	95.3 %

Properties of 50% Cellulose Acetate Fiber/50% Wood Pulp Handsheets

Basis weight	61.0 g/m^2
Caliper	0.102 mm
Density	0.594 g/cc
Tear resistance	26.0 g/sheet
Tensile strength	3.10 kg/15 mm
Breaking length	3.39 km
Stretch	2.42 %
Tensile energy absorption	0.036 $kg\text{-}cm/cm^2$
Internal bond	168.0 Scott units
Brightness	77.4 %
Scattering coefficient	719.0
Opacity	91.9 %

Properties of Cellulose Acetate Fibers

Surface area	22 m^2/g
Classified fiber length	0.63 mm
Fiber size distribution	
on 20 mesh screen	0.48 %
on 35 mesh screen	7.59 %
on 65 mesh screen	34.58 %
on 150 mesh screen	21.13 %
on 270 mesh screen	11.14 %
through 270 mesh screen	25.07 %
Coarseness	18.2 decigrex
Drainage factor	75 sec^{-1}

Photographs, at 20 times magnification, of the fibers collected on the 35, 65 and 150 mesh screen respectively, show that although the fibers might appear to be interconnected, they are actually only physically intertwined, with substantially no interstrand bonds being present. Thus, the fibers are neither continuous nor plexifilamentary.

Preparing Fibrous Materials by Wet Method

J. Maehara; U.S. Patent 4,141,942; February 27, 1979; assigned to Idemitsu Kosan Co., Ltd., Japan describes a process for preparing fibrous materials by wet method. Fibrous materials with excellent properties can be obtained by feeding the starting solution to the interface area of two phases of the coagulation bath.

Examples of a starting solution for fibrous materials are an aqueous solution of cellulose sodium xanthogenate for rayon fibrous materials; acetone or methylene chloride solution of cellulose acetate for cuprammonium rayon fibrous materials; rhodanate, nitrate, N,N-dimethylformamide, N,N-dimethylacetamide, dimethylsulfoxide or acetone solution of polyacrylonitrile for acrylic fibrous materials; and aqueous, alcohol or urea solution of polyvinyl alcohol for vinyon fibrous materials.

Other examples of starting solutions are acetone-carbon disulfide or acetone-benzene solution of polyvinyl chloride for polyvinyl chloride fibrous materials;

N,N-dimethylformamide, N,N-dimethylacetamide or dimethylsulfoxide solution of polyurethane for polyurethane fibrous materials; aqueous alkaline or dimethylsulfoxide solution of soy casein, milk casein or yeast protein for proteinous fibrous materials; and methylene dichloride solution of poly(L-glutamate-γ-methyl) or aqueous solution of poly(L-sodium glutamate) for polyamino acid fibrous materials.

Example: The starting solution of specific gravity of 1.10 was made by dissolving 10 parts by weight of polyacrylonitrile fiber (Patria) to 90 parts by weight of dimethylsulfoxide with stirring at room temperature and allowing the solution to stand for one day to remove gas.

In a treatment tank of 100 liter content and having a length from nozzle to outlet of fibrils of 2 meters, 30 liters of ethyl acetate (specific gravity: 0.90, coagulation value: 67.8) as the upper phase and 30 liters of water (specific gravity: 1.00, coagulation value: 8.3) as the lower phase, were introduced. Then the starting solution of fibrils described above was fed to the interface of the above two phases at a rate of 120 ml/min at 15°C from the slit nozzle with 50 mm in width and 0.2 mm in interval which was equipped in the one end of the tank. At the same time, a shear force was applied at 10 mm from the tip of the nozzle by reciprocal mixing with a linear velocity of 1.6 m/sec in the direction of a right angle to the extruding starting solution and parallel to the interface of the biphasic liquid in the coagulation bath.

By the above procedure, fibrils like absorbent cotton were formed exclusively in the interface area of the two phases of the bath. These fibrils were recovered on screen meshes and were washed with water at 15°C. After being treated with steam at 110°C, they were treated with hot air at 95°C. It was noted that the fibrils thus obtained were 3 to 5 μ in width and 10 to 30 mm in length.

Fiber Fleece of Electrostatically Spun Fibers

In a process developed by *W. Simm, C. Gosling, R. Bonart and B. von Falkai; U.S. Patent 4,143,196; March 6, 1979; assigned to Bayer AG, Germany* for the production of fiber filters, the fiber material is sprayed electrostatically from the liquid state and deposited as a fiber fleece on a conductive support. The starting materials used are solutions of high polymers in readily volatile solvents.

Figure 9.11a shows an installation for the production of fiber fleeces under the action of powerful electric fields and Figure 9.11b shows the annular spray electrode. The whole installation consists in principle of a spray electrode **1** from which the spray liquid is sprayed and precipitation electrodes **2** on which the electrically charged fibers are deposited. The voltage between the electrodes is between 50 and 200 kV. The electric supply to the installation is from the high voltage source **3**. The deposited fiber material **4** is carried out of the spray zone by movable electrodes, e.g., conveyor belts **5**. The stripper **6** serves for the continuous removal of the deposited fiber material.

Figure 9.11: Electrostatic Spinning Apparatus

a.

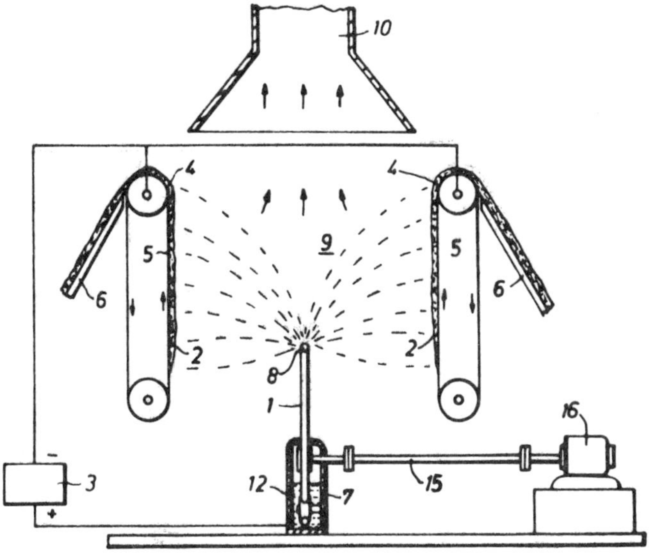

b.

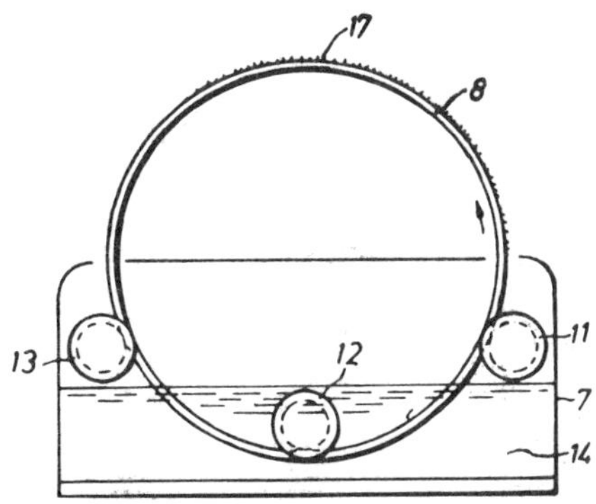

Source: U.S. Patent 4,143,196

The solutions which are to be spun are contained in the storage tank **7** from which the spinning liquid is continuously removed by the annular electrode **8** and carried into the spray chamber **9** where the spinning process takes place. The solvent vapors produced are discharged through an exhaust vent **10**.

Figure 9.11b shows the annular electrode in side view. The annular electrode is rotatably mounted on rollers **11**, **12** and **13** and dips with its lower part into the spinning liquid **14**. The drive is provided by the insulating shaft **15** via the transmission **16** (Figure 9.11a).

On rotation in the direction of the arrow, a thin layer of liquid on the surface of the ring is carried to the outside where it can be sprayed under the influence of the electric field. Numerous liquid points **17** which merge into thin spinning threads at the end are formed on the exposed part of the ring by the action of the field on the liquid film. The spinning threads become detached and solidify by evaporation of the solvent. When they detach themselves, the threads are highly charged and owing to their charge they move in the electric field towards the precipitation electrode where they are deposited to form a fleece.

Example: A spinning liquid is prepared from a 10% solution of polystyrene (MW approximately 180,000) in methylene chloride. The electric conductivity of the solution is $\chi = 2.10 \times 10^{-7}$ ohm^{-1} cm^{-1}. The spray electrode is an annular electrode as shown in Figure 9.11b having a diameter of 1 meter.

The distance from the precipitation electrodes arranged on the two sides of the ring is in each case 0.5 meter. There is a voltage of 120 kV between the ring and the precipitation electrodes, the ring being positive. The ring rotates at a rate of 30 revolutions per minute during the spray process. A fleece of 2 to 3 mm in thickness of dry, porous polystyrene fibers which have different thicknesses, the average thickness being about 20 μm, is formed after a few minutes on the precipitation electrodes which are covered with thin, highly permeable cellulose fleeces which have been produced by conventional methods.

An air filter which is immediately ready for use and which is practically impermeable to dust particles in the air which have a particle diameter of more than 0.5 μm is obtained by removing this coated support fleece from the electrodes and placing two layers together with the surfaces which are not covered by the support fleece in contact.

Formaldehyde Resin Fibers

Formaldehyde resin fibers, in particular urea-formaldehyde resin fibers, are produced by *P. Snowden; U.S. Patent 4,178,336; December 11, 1979; assigned to Imperial Chemical Industries Limited, England* by introducing a liquid formaldehyde resin and a catalyst into a spinning cup in the presence of cold humid air which inhibits drying and reaction, centrifugally spinning fibers from the cup into the path of hot air currents, at between 50° and 100°C, which dry the fibers without curing them, and curing and chemically stabilizing the dry fibers by heating at above 100°C until they are insoluble in cold water.

Referring to Figure 9.12a, aqueous urea-formaldehyde (UF) resin of viscosity 5 to 300 poises, preferably 10 to 100 poises, more preferably 15 to 75 poises, is introduced at **1** into a mixer **2**, where it is mixed with an aqueous solution of a resin-curing catalyst introduced at **3**. (The addition of a spinning aid, such as polyethylene oxide solution, and/or of a surfactant, such as "Lissapol", to the mixer **2** is advantageous).

Figure 9.12: Spinning Urea-Formaldehyde Fibers

a.

b.

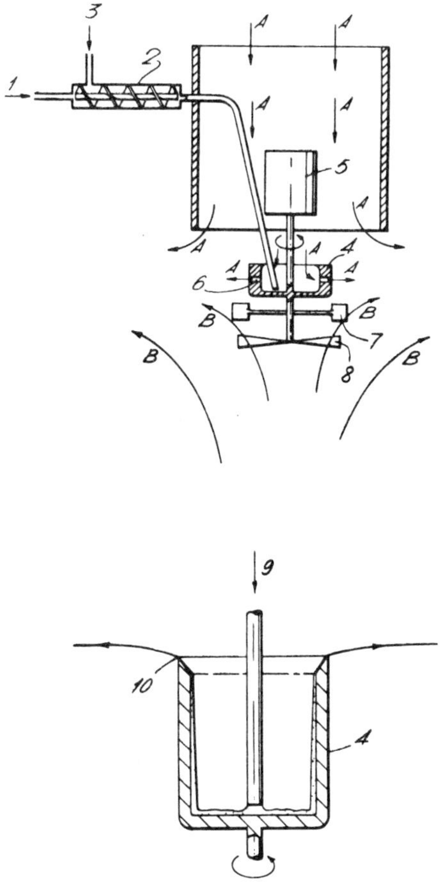

(a) Schematic illustration of spinning process
(b) Rotating cup

Source: U.S. Patent 4,178,336

From the mixer **2**, the UF resin mixture is introduced onto the base of a rotating cup **4** driven by a motor **5**. The resin spreads over the base and the wall of the cup **4** as a thin film, and is spun from apertures **6** in the wall of the rotating cup under such conditions as to produce a plurality of individual, separate fibers. These are allowed to attenuate or stretch to the diameter desired therefor, without drying or curing or disturbance from turbulent air, by first spinning them into a region of low temperature and high humidity. Such a region is provided by a downwardly-directed flow of cold, humid air which partly flows through the apertures **6** with the fibers and which is partly deflected outwardly by the cup, to form outwardly directed currents of cold humid air as shown by the arrows **A**. The pumping action caused by the rotation of the cup **4** causes the cold, humid air to be deflected outwardly with, and in the same manner and direction as, the fibers, thereby reducing the relative velocity between the cold, humid air and the resin fibers during the attenuation and stretching of the resin fibers. In most cases, ambient air will be suitable.

When the desired fiber diameter has been attained, the resin fibers are dried and transported to a fiber-collecting point by suitable outwardly-directed currents of dry, hot air indicated by arrows **B** in Figure 9.12a. These currents **B** may be created by a radial propellor **7** and an axial fan **8** fitted to the bottom of the cup **4**. The hot dry air is at a temperature such as to heat the fibers to between 50° and 100°C, for example about 65° or 70°C.

After the spun fibers leave the cup, they continue to be drawn out and attenuated or stretched into fibers of smaller diameter, but they are physically stabilized by the heat of the dry, hot air currents **B**, during their free flight from the cup, after they have attained the desired diameter but before they have a chance to develop into droplets or shot.

After collection, the fibers are cured and chemically stabilized by heating, during which the catalyst not only cures them but renders them insoluble in cold water. Suitable catalysts comprise acids or acid salts, for example sulfuric acid, formic acid, or their ammonium salts (e.g., ammonium sulfate), or mixtures thereof. The curing is carried out at above 100°C, preferably above 120°C.

One suitable cup design for use in the process is illustrated diagrammatically in Figure 9.12b. Liquid resin (e.g., a UF resin solution) is fed through **9** with a liquid catalyst to the bottom of the rotating cup **4**; it flows radially across the cup and then up the walls of the cup, where flow irregularities are smoothed under the centrifugal forces operating in the rotating cup. At the correct flow rate, fibers are spun outwardly from the lip **10** of the cup. The height of the cup is such as to allow the flow rate to be smoothed and depends upon the diameter of the cup, its rotational speed, and the viscosity of the liquid resin being spun. The diameter of the cup and its rotational speed can be varied over quite large ranges, and are adjustable to give the flow rates required by the process.

Example: "Aerolite 300" UF resin supplied by Ciba-Geigy was used. "Aerolite 300" is an aqueous UF resin prepared by condensing a mixture of urea and formaldehyde in a F:U molar ratio of about 1.95:1, followed by concentration to a solids content of about 65% by weight. It has a viscosity, depending upon its age, of about 40 to 200 poises at room temperature, and a water tolerance of 180%. The resin was adjusted to a viscosity of 75 poises, by the addition of water, and then fed to the bottom of a 3" diameter cup shaped according to Figure 9.12b, and rotating at a speed of 3,000 rpm. At a feed rate of 75 ml/min, good quality fibers were produced, having an average diameter of 15 μ.

At a feed rate of about 200 ml/min, the resin was spun from the rim of the cup as a continuous two-dimensional sheet and gave poor quality fibers.

Ettringite Fibers

J.-P. Caspar and J. Gaillard; U.S. Patent 4,140,540; February 20, 1979; assigned to Lafarge SA, France describe an improvement in the process of U.S. Patent 4,002,484, making it possible to obtain ettringite, or calcium trisulfoaluminate, in fiber form.

The process is carried out in the same way as in U.S. Patent 4,002,484 but with enough water to assure having a suspension containing a maximum of 25% dry matter at the end of the reaction, subjecting the suspension to moderate stirring of 1 to 5 times the force necessary to maintain a homogeneous suspension. The fibers serve as a replacement for asbestos fibers, notably as a fireproof product.

Example 1: In the previously-mentioned U.S. Patent 4,002,484, Example 7, a suspension of an aluminous cement of the following composition in parts by weight: 72 Al_2O_3, 27 CaO, 0.07 Fe_2O_3 and 0.24 volatile matter was reacted with gypsum of the composition in parts by weight: 35 to 36 CaO, 39 to 40 SO_3, 18 H_2O and 6 CO_2 with lime of the composition: 75% CaO, 1 to 2% CO_2, 23 to 24% H_2O and water, in the following proportions in parts by weight: 12.6 aluminous cement, 40.6 gypsum, 13.3 lime and 33.6 water to obtain stoichiometry and an additional amount of water to obtain a paste containing 15% by weight of ettringite or 567 parts of water. The whole mass was treated in a jar mill rotating at a velocity of 70 rotations/minute.

In this example the same aluminous cement as that described hereinabove was used and was reacted wth a calcium sulfate having a composition in parts by weight of: 33.2 CaO, 43.4 SO_3, 2.7 CO_2, 12.5 insolubles and 89% commercial quicklime and water in the following proportions: 22 parts aluminous cement, 59 parts calcium sulfate, 19 parts lime and 1,400 parts water, corresponding to obtaining a suspension comprising 10% by weight of ettringite.

The mixture was treated in a jar mill identical to that used in the abovementioned example, but maintained at 70°C and regulated to rotate at 6 rotations per minute, or 50% faster than the minimum velocity of rotation to maintain the solid matter in suspension. After 24 hour's stirring, a suspension of fibers having an average length of 60 μ and an average diameter of 1-2 μ (determined by a microscope with a systematic exploration system) was obtained.

Example 2: A fire resistance trial at 1100°C, using the ISO Standard method, was conducted with 4 cm panels, certain of which were used as controls, these being made of plaster prepared from a semihydrate, the others being made of a mixture prepared from 70 parts of the same semihydrate and 30 parts of ettringite fibers obtained as previously described. Fire resistance was 2 hours for the plaster panels and 4 hours for the mixed panels.

Protein Texturization by Centrifugal Spinning

In a process developed by *W.M. Hildebolt; U.S. Patent 4,062,987; December 13, 1977; assigned to Campbell Soup Company* a bland texturized monofilament protein product having a texture and mouthfeel simulating animal meat is prepared by forming an aqueous slurry of a protein material, injecting high pressure steam into the slurry to propel it through a confined treatment zone, centrifugally spinning the protein material into monofilaments, and recovering the textured monofilaments in a collection zone. Apparatus for this process comprises means for injecting steam into a protein slurry, means defining a confined treatment zone wherein the injected steam and protein are contacted, and means for centrifugally spinning the steam-treated protein material into monofilaments.

One embodiment of the apparatus will be described by reference to Figure 9.13. The slurry of protein to be texturized and water is formed in any suitable mixing means **1** such as a Hobart mixing bowl. Discharge from the mixing means is forwarded by a feeding means **2** communicating through a feed line **3** with a steam injection zone **4**. The feeding means can be any device capable of forwarding a relatively high consistency slurry. Positive displacement pumps such as the "Moyno" pump are well suited for this application. The feed line can contain a pressure gauge **5** and a check valve **6** which is effective to prevent backflow in the feed line.

Steam injection zone **4** comprises a mixing valve assembly in which the protein slurry from the feed line and high pressure steam are mixed. Any valve assembly which can effect a rapid and intimate mixture of steam and protein slurry without clogging can be employed. One suitable valve is the Schutte & Koerting Model No. 320 1¼" NPT. Steam enters the injection zone via inlet line **7**. Communicating with the discharge from the injection zone is a confined treating zone **8**. In the preferred embodiment this treating zone comprises an elongated cylindrical chamber or cooking tube. The dimensions of this chamber are not critical.

In practice, lengths of from 7 to 13 feet have been satisfactorily employed. The diameter should be large enough to prevent clogging and small enough to facilitate the maintenance of proper pressure conditions. Diameters in the range of about 1½ to 6 inches have been found to satisfy these conditions although a maximum of 3 inches is preferred. The chamber may be provided with temperature sensing element **9**, and a discharge pressure gauge **10**.

The discharge end of the cooking tube is provided with means for centrifugally spinning the protein material, generally shown as **12**. The centrifugal spinning means comprises a rotating multiorifice spinnerette head **13** which is driven by a motor **14** or other suitable means. Cold water spray nozzles **15** may also be provided to quench the emerging monofilaments. These nozzles are preferably arranged concentrically outside and above the spinnerette head. The centrifuge assembly is also provided with a product recovering zone **16** in which the monofilaments and water mixture may be collected. The cross section of the spun fibers may be varied by employing spinnerette orifices of different shape.

Figure 9.13: Protein Texturization by Centrifugal Spinning

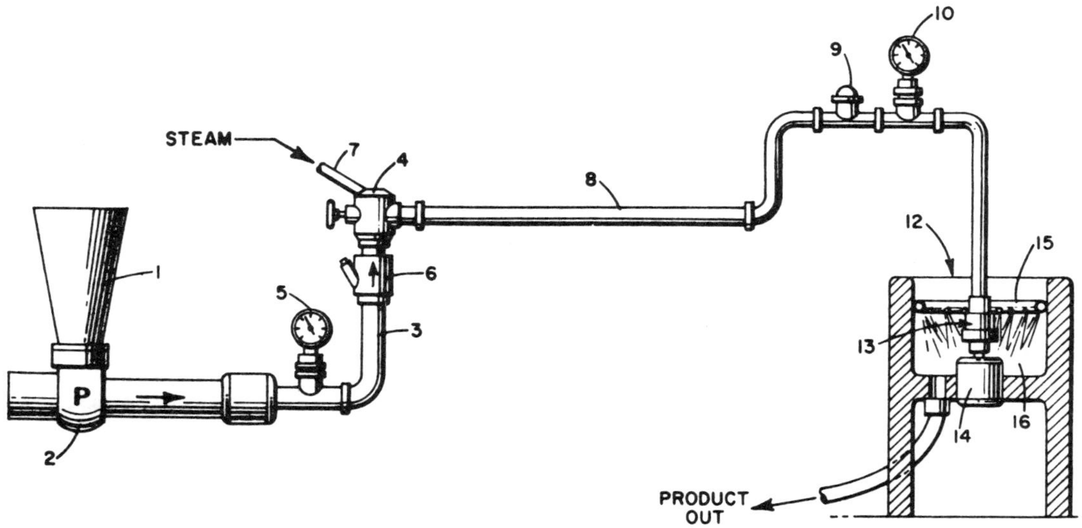

Source: U.S. Patent 4,062,987

Example: Textured soy protein monofilaments are produced according to the process as follows: 4,282 grams of protein mixture comprising 50% Promine R, (a soy isolate containing about 95% protein, Central Soya, Inc.) and 50% Soy Flour 200 W, (a soy flour containing about 50% protein, Central Soya, Inc.) is slurried with water to provide a solids content of 40%. Twenty grams of sodium bisulfite is added to the slurry to decrease its vicosity. This slurry is texturized in the apparatus as shown in Figure 9.13. The slurry is formed in Hobart bowl mixer equipped with pastry hooks and forwarded by a Model 3L4 Moyno pump. The steam mixing valve is Schutte & Koerting Model No. 320 1¼ npt. The steam supply is 110 psi and in this apparatus the resulting steam velocity is in the range of about 50 ft/sec.

The temperature in a 1½ inch diameter, 10 ft long cooking tube is maintained at 320°F. The back pressure in the cooking tube is about 70 psi. The orifices in the spinnerette head are set at 0.070" and the head is rotated at 1,800 rpm. The product issuing from the spinnerette head consists of 2 to 3 inch fibers which are quenched by a water supply. The fibers are very white and exhibit good texture qualities.

Fibers Containing Hair Particles

The method of *N. Orentreich; U.S. Patent 4,112,040; September 5, 1978* for making fibers suitable for making a hair brush includes a matrix or pad made of a flexible material and a plurality of relatively rigid, single natural-type bristles having their lower ends anchored by flexible, sleeve-like members projecting upward from the matrix. Also disclosed is a method of forming the bristles which includes comminuting natural bristles at a low temperature and then forming a thick slurry of the comminuted bristles and a binder, which is then molded and cured to form bristles having improved characteristics.

In accordance with this method, a quantity of human or animal hair is processed to form bristles in a manner to be described. The type of hair utilized depends on the desired coarseness of the bristle to be fabricated. That is, if very fine bristles are to be fabricated, the hair of an animal such as a hamster is suitable; for bristles of average coarseness, human hair would be used to advantage; and horsetail hairs could be used in the event a coarse bristle is desired.

These pieces of natural animal or human hair are first cleaned, preferably by being rinsed in a hexane bath for about 30 minutes, while being shaken to remove external contamination from the hair.

After being cleaned, the hair is then cut into small pieces, preferably below one centimeter, by any suitable cutting means. The cut pieces of hair are then ground or comminuted into small particles, less than 5 microns in diameter, of essentially keratin, a fibrous protein of which animal and human hair is constituted.

A presently preferred manner of comminuting or grinding the hair into those fine keratin particles is to place the pieces of hair into a container of liquid nitrogen at a temperature of between -195° and -190°C. The container may include a centrally located, movable hammer member and a pair of fixed metal members or anvils at each end. A pair of electromagnets is placed external to the container and ac current is supplied thereto to cause the hammer member to reciprocate into and out of engagement with the anvils for a period of about two minutes. As the hammer strikes against the anvils, the frozen pieces of hair are ground down to the desired particle size.

Thereafter, the ground keratin particles are placed in air at room temperature for a period of about 30 minutes until they are defrosted.

The defrosted fine hair particles are then mixed with a binder material in a suitable mixing apparatus. The binder is added in the amount necessary to form a thick slurry or paste of the keratin particles and binder that has the consistency of a pudding. A suitable binder material that may be used in this process to fabricate soft hair bristles is a silicone rubber such as Dow Corning Adhesive "891". For medium bristles, a suitable binder material is a nylon or polypropylene elastomer, and to fabricate a hard bristle, a suitable binder material for use in forming the slurry is 2 part epoxy.

The binder-keratin slurry is then introduced into a mold in which are formed openings of the diameter and length of the desired bristles. Typical dimensions of these openings to form bristles is about 0.1 to 0.3 mm in diameter and 0.5 to 3 cm in axial length. The bristles are pressure extruded in the mold and the extruded bristles are thereafter allowed to harden or cure in air at a temperature of about 60°C.

Thus, in a hair brush, natural-type bristles are employed which will not break off when the brush is used. The use of these bristles in the brush provides the advantages of natural bristle brushes such as the elimination of static electricity and the gentle action of a natural rounded tip natural bristle, without the formation of the natural bristles into clumps or bunches, which have the tendency to remove the user's hair.

Drawing, Stretching and Heatsetting

POLYESTERS AND POLYAMIDES

Processes for the production of polyamide and polyester filaments useful for textile and industrial purposes are known. Such products can be prepared by extruding molten polyamide or polyester polymer through a spinnerette, quenching the filaments and winding up the quenched filaments. To obtain useful properties, it is known that the filaments must be drawn several times their original length, thereby orienting the filament molecules.

Simplified Processing of Polyamide and Polyester Fibers

According to *W.N. Russell, J.I. Bruton and I.C. Twilley; U.S. Patent 4,113,821; Sept. 12, 1978; assigned to Allied Chemical Corporation,* the processing equipment has been simplified which results in a simple yet flexible process for producing high quality polyamide or polyester fibers. It has been found that a specific unexpected relationship exists with reference to the various functions of spinning and drawing which can be combined in order to yield an operable process at high speeds while producing high quality yarn.

The continuous high speed process comprises extruding molten synthetic linear polymer through a spinnerette into undrawn filaments at a temperature between about 235° and about 380°C, pretensioning the filaments continously at a constant tension level greater than about 0.005 gram per denier (undrawn) and less than a tension level required to draw the filaments, heating the filaments substantially immediately above their second order transition temperature, drawing the filaments substantially instantly subsequent to the heating at a temperature in the range of from about above their second order transition temperature to about 5°C below their melting point when winding up the drawn filaments.

Subsequent treatments such as bulking, interlacing, sizing, etc. are also usually required to produce a product acceptable for specific product applications.

Example: This example illustrates the combined spinning and drawing utilizing polyethylene terephthalate polymer. Polyethylene terephthalate polymer having an intrinsic viscosity of 0.95 was melted in an apparatus at 295°C and pumped through a filter to a 192 hole spinnerette. The spinning temperature was maintained within plus or minus 2°C by use of a Dowtherm liquid vapor temperature control system. The throughput rate of 45 lb/hr was maintained by an inlet pump pressure of 800 psi and an outlet pressure of 7,500 maximum.

The extruded filaments were quenched in ambient room temperature air. The filaments prior to convergence were lubricated by use of a rotating finish roll. From the convergence point, the yarn bundle passed over in three wraps a yarn pretensioning roll system maintained at ambient room temperature and operating at 1,383.4 fpm with a denier of 7,500. The yarn then passed over in ten wraps a feed roll system maintained at ambient room temperature and operating at 1,391.1 fpm. The predraft tension was maintained at 0.03 g per denier based upon the speed difference in the yarn pretensioning roll system and the feed roll system. The yarn bundle was then fed through the draw point localizing zone wherein the hot fluid (steam) temperature was 450°C and the pressure was 122 psig. The yarn then passed over the draw roll system maintained at a temperature of 130°C and operating at a surface speed of 8,198 fpm. The yarn bundle was then allowed to relax prior to being wound up by passing the yarn over a heated roll system controlled at a surface temperature of 160°C and surface speed of 7,813 fpm. The yarn was then wound up.

The properties of yarn produced under the abovedescribed conditions (A), yarn spun and drawn under prior art conditions, (B), and yarn sold in the trade for the same use, (C) are shown below. All yarns had deniers of 1,315-1,320; and fiber intrinsic viscosities of 0.88-0.90. Tensile modulus gives the stress at 100% elongation on the extrapolation of the initial straight line portion of the stress-strain curve. B.Q.I. (Beaming Quality Index) indicates the defects (broken filaments, strip backs, nubs, etc.) per million yards in beaming. The B.Q.I. is tested on an Ultra Yarn Inspector, Model 1007, Serial 2594, Lindly Corp.; major sensitivity set at 4.0%; minor sensitivity set at 3.0%.

Sample	Tenacity (gpd)	Elongation at Break (%)	Toughness	Thermal Shrinkage (%)	Pre-tension (gpd)	Tensile Modulus (gpd)	B.Q.I.
A	9.2	18.7	0.90	5.7	0.03	100	20
B	9.0	10.5	0.48	12.0	–	121	350
C	8.7	13.7	0.74	6.3	–	113	210

It is apparent that the properties and mechanical quality of the yarn produced from this process are superior to the yarns produced by the prior art.

Continuous Spindrawing for Polyethylene Terephthalate Yarns

A process is provided by *W.A. Chantry and R.M. Conrad; U.S. Patent 4,003,974; Jan. 18, 1977; assigned to E.I. Du Pont de Nemours and Company* in which a continuous filament polyethylene terephthalate yarn having an HRV (defined below) in the range of 24 to 28 is heated to a temperature in the range of 75° to 250°C and drawn to a total denier in the range of 50 to 500 at a draw ratio in the range of 5.2:1 to 6:1, passed over a draw roll having a temperature in the range of 225° to 250°C, relaxed at a tension in the range of 0.09 to 0.15 gram per denier and wound up at a minimum speed of 2,000 yards per minute (1,829 meters/minute). The yarns prepared by this process have a tenacity in the range of 7.5 to 9 grams per denier, a load-bearing capacity in the range of 3 to 5 grams per denier at 7% elongation, a maximum dry heat shrinkage of 4% at 177°C and an elongation at break in the range of 12 to 20%.

Yarn tenacity and break elongation are determined by means of an Instron Tensile Tester which extends a 10-inch (25.4 cm) length yarn sample to its breaking point at an extension rate of 6 in/min (15.2 cm/min) at a temperature of about 25°C. Extension and breaking load are automatically recorded for each sample.

T_7, a measure of yarn modulus, is the load-bearing capacity in g/den, of the yarn at 7% elongation. T_7 may be obtained from the stress-strain curve produced by the Instron Tensile Tester in measuring tenacity and break elongation, as above.

Dry-heat shrinkage at 177°C (DHS_{177}) is determined by exposing a measured length of yarn under zero tension to dry heat for 30 min in an oven maintained at 177°C and measuring the amount of retraction. The amount of shrinkage is expressed as a percentage of the original length.

HRV is a sensitive and precise measurement indicative of polymer molecular weight. HRV is the ratio of the viscosity of a solution of 0.8 gram of polymer dissolved at 49°C in 10 ml of hexafluoroisopropanol containing 80 ppm H_2SO_4 to the viscosity of the H_2SO_4-containing hexafluoroisopropanol itself, both measured at 25°C in a capillary viscometer and expressed in the same units. The use of hexafluoroisopropanol as solvent is important in that it allows dissolution at the specified temperature and thereby avoids the polymer degradation normally encountered when polyesters are dissolved at elevated temperatures. HRV values of 24 and 28 correspond roughly to intrinsic viscosity values of 0.68 and 0.74, respectively, when the intrinsic viscosity is measured at 25°C in a solvent composed of a mixture of trifluoroacetic acid and methylene chloride (25/75 by volume).

The tension on the yarn leaving the hot draw rolls may be measured by means of a Check Line Master Series Tensiometer obtainable from the Electromatic Equipment Company of Cedarhurst, NY.

Example: This example illustrates the preparation of a 210 denier polyester industrial yarn at a draw roll speed of 3,500 ypm.

Polyethylene terephthalate is melt spun through a 50-hole spinneret using a spinning block temperature of 285°C to give a yarn product having an HRV of 27. Immediately below the spinneret, the extruded filaments pass through a heated-wall delay baffle having a length of 7½ inches and having a wall temperature maintained at 300°-325°C. Below the delay baffle the filaments pass through a quench zone where the filaments are quenched in cross-flow air at room temperature. The quenched filaments pass around unheated feed rolls operating at 625 ypm (571 meters/min), then through a draw jet supplied with steam at about 60 psig (5.08 atm) at a temperature of 275°-300°C.

The yarn is then wound (7½ wraps) around a pair of draw rolls having a surface temperature of 245°-248°C and operating at 3,500 ypm (3,200 meters/min) for a residence time of 0.09 second on the draw rolls. The draw ratio is 5.6:1; the total drawn denier is 210. The drawn yarn then passes to and around a pair of unheated let-down rolls operated at 3,318 ypm (3,034 meters/min) which allows 5.2% retraction in length at a tension of 0.12 g/den. From the let-down rolls the yarn passes to a conventional surface driven package windup at a drive roll speed of 3,300 ypm (3,011 meters/min).

Conventional oil-in-water emulsion finishes are applied to the yarn at three points in the above process to provide lubrication and antistatic protection. A coconut-oil-based finish is applied just below the quench zone and between the draw jet and the draw rolls; a butyl-stearate-based finish is applied between the let-down rolls and the windup for a total finish on yarn of 0.57%. The yarn is also interlaced after the second of two let-down rolls to provide bundle coherency, using an interlacing jet in the manner described by Bunting and Nelson in U.S. Patent 3,110,151.

The 27 HRV yarn produced has a tenacity of 7.6 g/den, a break elongation of 16.0%, a T_7 value of 3.2 gpd and a dry-heat shrinkage at 177°C of 3.5%. The process operates efficiently without deleteriously affecting the advantageous properties of the yarn.

High Modulus Polyester Yarn

W.N. Knopka; U.S. Patent 4,082,731; April 4, 1978; assigned to Avtex Fibers Inc. describes a method for producing a high modulus yarn of filaments of a polyester resin consisting of at least 15 up to 50 mol percent of bibenzoate units and from 85 to 50 mol percent of terephthalate units, the yarn having been uniformly oriented and then heat-set while free-to-relax.

Example: Resin preparation was accomplished by transesterifying predetermined quantities of dimethyl terephthalate, diethyl bibenzoate and ethylene glycol in the presence of a catalyst, followed by polycondensation at elevated temperatures under vacuum. When the polycondensation was finished, the resin was spun into yarn containing 10 filaments.

Yarn processing was accomplished by drawing the yarn at a temperature of about 20°C higher than its glass transition temperature, at 75% of its break-draw ratio and then heatsetting the yarn while completely relaxed.

In Figure 10.1a, stress-strain curves for various fibers are shown. Curve **1** is for the fiber of a homopolymer of ethylene terephthalate having an intrinsic viscosity (IV) of 0.6, a tenacity of 3.57 g/d and an initial modulus of 49.1 g/d after heatsetting.

Curve **2** is of the fiber of a copolymer of 95 mol % ethylene terephthalate units and 5 mol % of ethylene bibenzoate units having an IV of 0.52, a tenacity of 3.15 g/d and an initial modulus of 54.4 g/d after heatsetting.

Curve **3** is of the fiber of a copolymer of 10 mol % of ethylene terephthalate units and 10 mol % of ethylene bibenzoate units having an IV of 0.53, a tenacity of 3.05 g/d and an initial modulus of 54.9 g/d after heatsetting.

Figure 10.1: Physical Properties of Yarns

a.

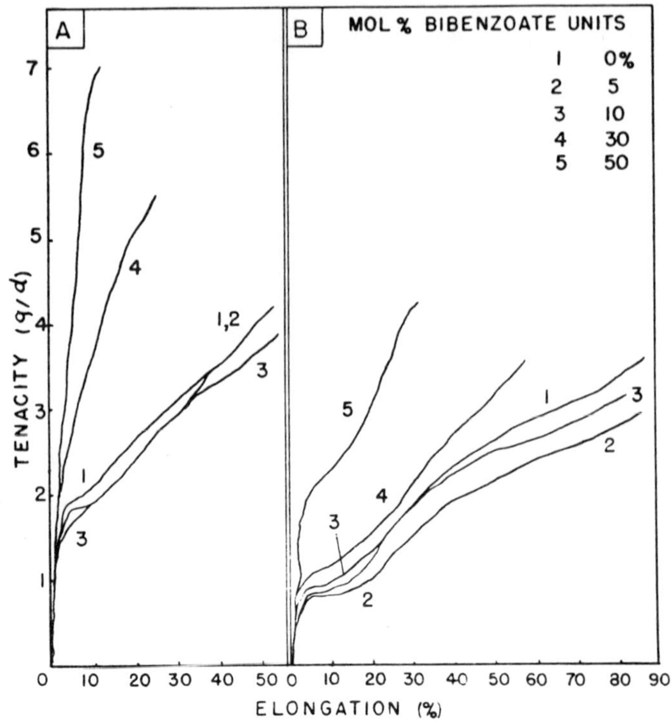

Stress-strain curves for fibers before (A) and after (B) heat-setting

(continued)

Figure 10.1: (continued)

b.

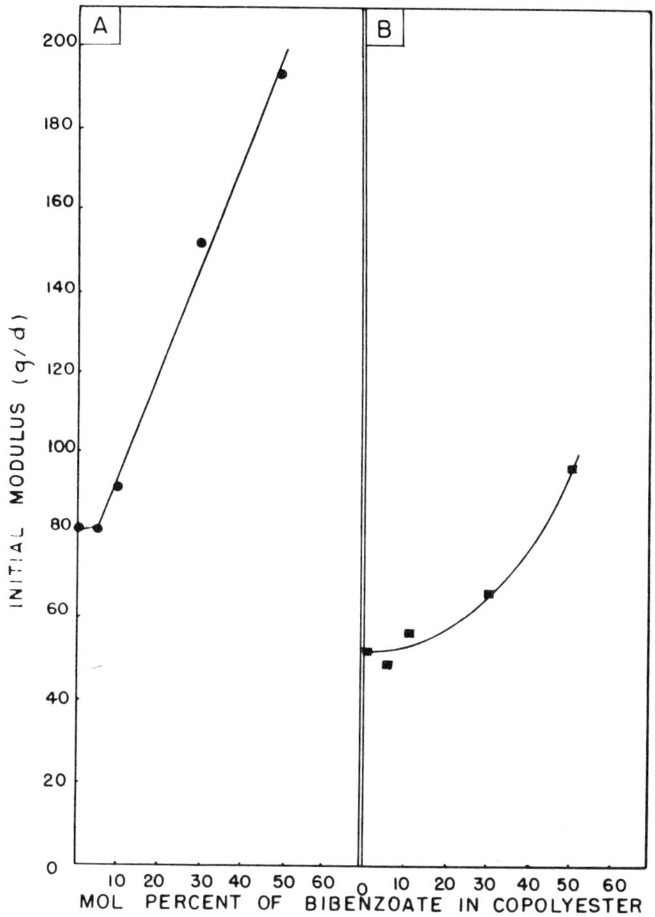

Initial modulus curves for fibers before (A) and after (B) heat-setting

Source: U.S. Patent 4,082,731

Curve **4** is of the fiber of a copolymer of 70 mol percent of ethylene terephthalate units and 30 mol percent of ethylene bibenzoate units having an IV of 0.65, a tenacity of 3.61 g/d and an initial modulus of 63.0 g/d after heatsetting.

Curve **5** is of the fiber of a copolymer of 50 mol percent of ethylene terephthalate units and 50 mol percent of ethylene bibenzoate units having an IV of 0.65, a tenacity of 4.17 g/d and an initial modulus of 93.9 g/d after heatsetting.

It can be seen from Figure 10.1a, graphs A and B, that the initial moduli (as represented by the stress-strain curves) of the fibers of this process, (curves **4** and **5**) are much higher both before and after heatsetting, while fibers which do not come within the parameters of this process, (curves **1**, **2** and **3**) have relatively lower initial moduli.

It can be seen from Figure 10.1b that a significant desirable change in initial moduli, after heatsetting while free-to-relax, does not occur until at least 15 mol percent of bibenzoate units are present in the polymer chain.

Molecular weight (as measured by intrinsic viscosities) has a distinct influence on the tenacity of the fibers, as would be expected since shorter polymer chains have less strength. Specifically, an IV of at least 0.55 is required so as to produce yarn which has at least as high tenacity as yarn of the homopolymer.

High Speed Stretching of High Birefrigence Filaments

A process developed by *H. Yasuda, K. Tani and J. Nagatamo; U.S. Patent 4,105,740; August 8, 1978; assigned to Toyo Boseki KK, Japan* involves stretching polyester tows at a speed greater than 150 m/min, wherein more than 85 mol percent of the recurring units are units of ethylene terephthalate, carried out at a stretch ratio satisfying the formula

$$-1.58 \log (\Delta n \times 10^3) + 4.18 \leqslant DR \leqslant -2.32 \log (\Delta n \times 10^3) + 5.93$$

wherein Δn represents the birefringence of the unstretched filaments and is a value of from 0.020 to 0.100, and DR is the stretch ratio.

Example: A polyethylene terephthalate having an intrinsic viscosity of 0.6 (measured in a mixed solvent of phenol and tetrachloroethane in the ratio of 6:4, at 30°C) was extruded at a spinning temperature of 290°C, through a spinnerette having 300 extrusion orifices, each 0.25 mm in diameter, provided in circular arrangement. The extruded filaments were cooled rapidly just below the spinnerette by a cool air current blown against it at the flow rate of 0.4 m/sec, from the entire surface surrounding the filaments. Thereafter, the cooled and solidified filaments were wound on a winder. By varying the extrusion amount of the polymer and the winder takeup speed, six kinds of unstretched yarns (Samples A to F) shown below were obtained.

Sample	Extrusion Amount of Polymer (g/min)	Winder Takeup Speed (m/min)	Birefringence Δn of the Unstretched Yarn
A	178	1,000	0.006
B	208	1,550	0.012
C	250	2,000	0.021
D	250	2,500	0.028
E	284	2,900	0.037
F	291	3,200	0.042

These unstretched yarns were then each bundled to form a tow of one million deniers, respectively, and the tows were stretched at the stretching speed of

160 m/min or 200 m/min, using hot plates at 90°C, at such stretching ratios that will give stretched yarns having equal strengths and elongations. Then, the stretchability, and the occurrence of filament breakage at the start of and during the stretching operation were examined for each tow.

The conventional tow generally in use (Sample A) generated filament breakage at a high speed such as 160 m/min or 200 m/min, at the start of the stretching operation by the sharp change of the tension, and generated much filament breakage also during the stretching operation. Therefore, it was impossible to obtain fibers of commercial merit.

In the case of the polyester tow (Sample B) representing a higher birefringence than the generally used ones but representing a lower birefringence than the lower limit (0.020) specified in this process, there were also similar troubles in stretching as in the generally used polyester tows. In contrast thereto, the tows composed of the highly oriented unstretched yarns (Samples C to F) were completely exempted from filament breakage at the start of and during the stretching operation even by such high speed stretching, giving high quality stretched yarn, which was faultless and perfect from the commercial viewpoint.

Incorporation of Additives by Cold Drawing

R.T. Guthrie, J.L. Hirshman, S. Littman, E.L. Sukman and P.H. Ravenscroft; U.S. Patent 4,055,702; Oct. 25, 1977; assigned to M & T Chemicals Inc. disclose a process in which additives can be permanently incorporated into melt spun fibers by cold drawing the fibers under conditions that generate a network of interconnecting microvoids within the fiber. The microvoids are formed in the presence of specified liquid or vapor media which fill the microvoid network. The temperature of the medium is below the effective glass transition temperature of the fibers containing the medium. The additive is either present in the medium or is applied to drawn fibers containing the medium.

Figure 10.2 is a schematic diagram of an apparatus suitable for processing textile fibers in accordance with this method.

Undrawn or partially drawn fiber (A) in the form of monofilament, yarn or tow is continuously withdrawn from a spool or other supply (B) and is passed around driven feed roll (C) which determines the linear speed of the undrawn fiber. The fiber then passes into the draw bath (D) and around a driven roll (E) rotating at the same linear speed as the feed roll (C) to ensure that no drawing occurs before the fiber is immersed in the mixture of additive and liquid diluent (F) which can be maintained at any desired temperature by a thermostatically controlled heating and/or cooling device (not shown).

The driven roll (E) can be replaced by one or more devices which will localize drawing of the fiber within the bath. The fiber passes around a guide (G) and a draw roll (H). The draw roll is driven at a higher linear speed than the feed roll. The ratio between the two speeds, equal to the desired draw ratio, is be-

Figure 10.2: Cold Drawing Apparatus

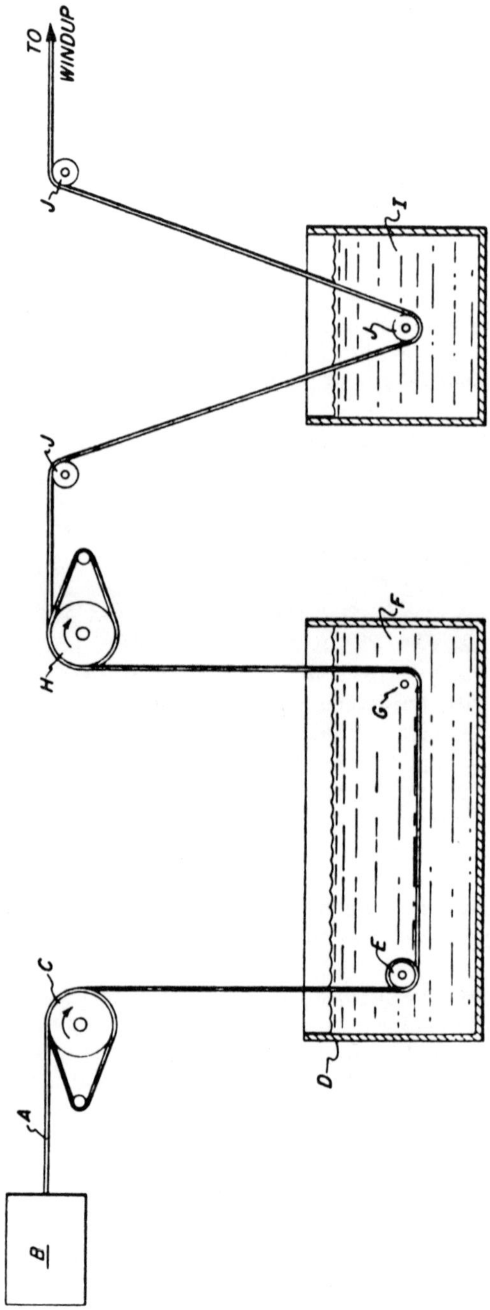

Source: U.S. Patent 4,055,702

tween 1.5:1 and the breaking elongation of the fiber. Preferably the fiber is passed through a wash bath (I) using suitable guides (J) before proceeding to windup or other collection means.

Example: An undrawn, untwisted, semidull yarn of 148 denier, containing 34 polyethylene terephthalate filaments was treated using the apparatus shown schematically in Figure 10.2. The surface speed of the draw roll was 3.5 times that of the feed roll, equivalent to a draw ratio of 3.5:1. The draw bath was 18 inches (45.7 cm) long and contained a methanol solution of the product obtained by reacting the cyclic phosphite $CH_3CH_2C(CH_2O)_3P$ with dimethyl methylphosphonate in an equimolar ratio. The solution contained 200 g of reaction product per liter of solution and was maintained at ambient temperature. The residence time of the fiber in the draw bath was 0.5 second. The drawn yarn was passed through a container of methanol to remove excess treatment compound and then wound on a take-up spool.

A 0.5 g sample of the treated yarn was removed from the spool and scoured by immersing the sample with constant stirring for 15 minutes in perchloroethylene at a temperature of 60°C, after which the sample was treated in a similar manner for 20 minutes using an aqueous solution containing 0.1% by weight of sodium carbonate and 0.1% of an anionic surfactant (Igepal CO 430). The solution was heated to between 70° and 85°C. The yarn sample was then rinsed with water until the washings were clear, rinsed with methanol and dried for 16 hours in an oven heated to 60°C.

Upon analysis the fiber was found to contain 1.59% by weight of phosphorus. A sample of the treated yarn was two-plied and knitted into a number of 4 inch (10.2 cm)-diameter tubes. The tubes were scoured by immersing them for 30 minutes in an aqueous solution containing 1.0 g of a conventional scouring agent and 1.0 g of sodium carbonate per liter of solution. The temperature of the solution was 71°C.

The yarn sample was then rinsed with ambient temperature water until the washings were clear. The scoured tubes were then dyed at 100°C by immersing and agitating them for 1 hour in a 2% solution, based on fabric weight, of Latyl Cerise N, a red dye, in an aqueous solution containing 0.5 and 3.0 g per liter of a conventional dye carrier and 0.1 g per liter of a nonionic surfactant together with sufficient acetic acid to attain a pH of 5.5. A number of the knitted tubes were not dyed but were scoured and laundered 50 times in a domestic washing machine.

The flammability of both the dyed and undyed fabric samples was determined using the Limiting Oxygen Index, which is a measure of the minimum concentration of oxygen in a nitrogen-oxygen mixture that will support combustion. Prior to being tested the knitted tubes were slit open and stitched with fiber glass yarn (Type E12 from Owens-Corning Fiberglas Corp.). Each line of stitching was parallel to the longitudinal axis of the original knitted tube, and contained 7 stitches per inch. The lines of stitching were 0.125 inch (0.318 cm) apart. The stitched fabrics were placed in a frame, the bottom of which was located 2 inches (5 cm) above a bed of glass beads through which a mixture of oxygen and nitrogen was admitted. Various mixtures of the two gases were employed to determine the minimum oxygen content that supported combustion of the sample. The Limiting Oxygen Index (LOI) value was calculated using the formula

$$\frac{C_{oxygen}}{C_{oxygen} + C_{nitrogen}} \times 100$$

The concentration (C) of the gases can be expressed in any convenient units, including flow rates in cubic centimeters per minute.

The LOI value of the scoured, undyed sample was 27.0 and the value for the dyed sample was 26.5. The LOI value for the undyed samples which were laundered 50 times was 27.3.

The distribution of phosphorus within an undyed fiber was determined using electron micrographs of a fiber cross-section which showed that the phosphorus compound employed to treat the fiber had penetrated throughout the entire fiber cross-section.

Two-Step Drawing Followed by Setting for Tire Monofilaments

G. Vock; U.S. Patent 4,043,985; August 23, 1977; assigned to Hoechst AG, Germany describes a process in which polyester monofilaments having very high strength and low elongation at elevated temperature and resistant against tire rubber are obtained from polyesters having an intrinsic viscosity of more than 0.67 and terminal carboxy groups of less than 25 meq/kg by spinning into a water bath, first drawing in a water bath having a temperature of from 70° to 100°C in a ratio of from 1:4.0 to 1:6.0, second drawing at a temperature being a maximum 15°C below the melting point to attain a total drawing ratio of from 1:6.0 to 1:7.5, and subsequent setting at a temperature being a maximum 15°C below the melting point without allowing a shrinkage.

Example: A polyethylene terephthalate raw material having an intrinsic viscosity of 0.81 (measured at 25°C in a mixture of phenol/tetrachloroethane in a weight ratio of 3:2) and a content of terminal carboxy groups of 5.0 meq/kg is spun at 280°C into a water bath by means of a screw extruder, and drawn off at a speed of 10 m/min. The monofilament is drawn at 85°C in a second water bath in a ratio of 1:4.9, and drawn at 260°C (measured in the air space) in an infrared chamber having a length of 7.5 m, in order to achieve a total drawing ratio of 1:6.5.

Subsequently, the monofilament is thermoset in a second infrared chamber having a length of 7.5 m, also at 250°C (measured in the air space), while allowing a shrinkage of 1.5%, thus imparting to the monofilament the desired final diameter of 0.30 mm. The monofilament obtained has the following textile data:

Strength, g/tex	72
Elongation at break, %	10.8
Elongation at the specific load of 27 g/tex, %	3.3
Shrinkage at 160°C, %	5.1
Force of shrinkage at 120°C, g	300
Intrinsic viscosity, dl/g	0.77
Terminal carboxy groups, meq/kg	8.7

Spin Finishes in Polyester Tire Yarn

It has been suggested that certain polyglycol ethers, in particular mixed polyoxyethylated-polyoxypropylated monoethers, are useful as spin finishes in the production of polyester tire yarns. Polyglycol ethers having a relatively low molecular weight of about 300 to 1,000 make the best lubricants; however, their effectiveness is significantly diminished by volatilization loss during high temperature processing of the tire yarn.

R.M. Marshall and K.C. Dardoufas; U.S. Patent 4,019,990; April 26, 1977; assigned to Allied Chemical Corporation have found that the volatilization loss is greatly decreased by including in the finish a small percentage of certain antioxidant compounds. The particular antioxidant compound used has been found to be very critical.

Example 1: This example demonstrates that the finish of the process can be used with excellent results as a spin finish in a conventional spin-draw process for producing polyethylene terephthalate yarn. The polyoxyalkylated monoether used was a condensation product of 50% ethylene oxide and 50% propylene oxide terminated with butyl alcohol and having a molecular weight of about 570 and a viscosity of 100 SUS at 100°F. About 99 parts by weight of this polyoxyalkylated monoether was mixed with about 1 part by weight of 2,2'-methylenebis(4-methyl-6-nonylphenol) to form the finish composition. This finish was applied directly, i.e., without diluting with water, as a spin finish to polyethylene terephthalate tire yarn (approximately 1,300 denier, 192 filaments) by means of a kiss roll prior to drawing the yarn.

After drawing, the yarn was heated to 135°-230°C on relax rolls. Less than 9% of the finish was volatilized during this heating step. The finished yarn was readily twisted and plied into greige cord. The resulting cords were then tensilized by a conventional tensilization procedure.

To show the effect of application rate, the application level ranged from 0.2 to 1.5% based on the weight of fiber. Yarns with finish levels from 0.5 to 1.0% were considered most desirable in view of relatively low finish required and significantly improved compression-extension durability in rubber as tested in a disc fatigue test where the percent strength retention is recorded. The results are shown below.

Percent Finish on Yarn	Disc Fatigue, % Strength Retained
0.2	76
0.5	86
0.8	92
1.0	93
1.5	94

Example 2: The procedure of Example 1 was followed except that the amount of 2,2'-methylene-bis-(4-methyl-6-nonylphenol) (MMNP) in the spin finish was varied to show the effect upon the amount of finish volatilized from the yarn at yarn temperatures up to about 150°C. The results are shown below. The individual standard deviation (σ) of the test was about 1% so that a difference of 3% or more in the amount of finish volatilized is highly significant.

Percent of MMNP in Finish	Percent of Finish Volatilized
3	8
1	9
0.5	15
0	21

From these data, the preferred amount of the 2,2'-methylene-bis-(4-methyl-6-nonylphenol) in the finish is about 1-3% by weight.

Polyglycol Ether Finish for Spindraw of Tire Yarn

R.M. Marshall and K.C. Dardoufas; U.S. Patent 4,054,634; Oct. 18, 1977; assigned to Allied Chemical Corp. provides an improvement in a process for producing polyethylene terephthalate yarn wherein a liquid finish is applied to the yarn, the process involving spinning and drawing steps, the improvement comprising: (a) first applying to the yarn prior to the drawing step a liquid finish composition consisting essentially of a polyalkylene glycol compound which is a mixed polyoxyethylated-polyoxypropylated monoether prepared in accordance with the equation:

$$ROH + x\underset{\diagdown O \diagup}{CH_2CH}CH_3 + y\underset{\diagdown O \diagup}{CH_2CH_2} \longrightarrow RO(C_2H_4O,\ C_3H_6O)_{x+y}H$$

where R is an alkyl group having 1 to 8 carbon atoms, x and y are the number of mols of propylene oxide and ethylene oxide respectively and wherein ethylene oxide comprises 40 to 60% by weight of the combined total of ethylene oxide and propylene oxide and x+y has a value to produce a molecular weight of from 300 to 1,000; and then (b) applying to the yarn after the drawing step a liquid finish composition consisting essentially of about 70 to 95 parts by weight of the mixed polyoxyethylated-polyoxypropylated monoether, about 5 to 30 parts by weight of a silane having the structural formula:

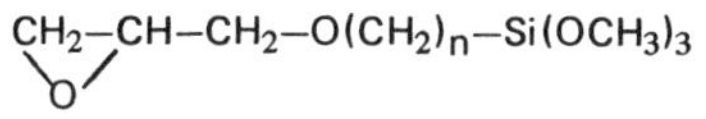

wherein n = 2 to 5, and a sufficient amount of a water-soluble alkaline catalyst to adjust the pH of the finish composition to 8-10.

Example: This example demonstrates that the polyglycol ether finish system can be used with excellent results in a conventional spin-draw process for producing polyethylene terephthalate yarn. The polyglycol ether used was a condensation product of 50% ethylene oxide and 50% propylene oxide terminated with butyl alcohol and having a molecular weight of about 570 and a viscosity of 100 SUS at 100°F. This polyglycol ether was applied directly, i.e., without diluting with water, as a spin finish to polyethylene terephthalate tire yarn (approximately 1,300 denier, 192 filaments) by means of a kiss roll prior to drawing the yarn. About 0.4 weight percent of the spin finish was applied based on the weight of the yarn.

After drawing, the yarn was heated to 135°-230°C on relax rolls. Less than 15% of the finish was volatilized during this heating step. Then, about 0.7% based on the weight of the yarn of a modified polyglycol ether overfinish was applied to the yarn. This overfinish consisted of 40 parts of the same polyglycol ether used in the spin finish, 10 parts of gamma-glycidoxypropyltrimethoxysilane, 50 parts of water, and sufficient amount of 5% aqueous NaOH solution to adjust the pH of the finish solution to about 8.0-8.5.

About 2 parts of the NaOH solution was required. The finished yarn was readily twisted and plied into greige cord. The resulting cord was then treated with a conventional R/F/L adhesive composition (resorcinol, formaldehyde and latex) and tensilized by a conventional tensilization procedure. The conventional isocyanate dip was omitted.

The adhesion between the resulting cord and rubber was tested in a conventional peel adhesion test where the peel force and visual rating of the amount of rubber remaining on the cord is recorded. Results are shown for samples prepared at various molding temperatures.

Molding Temperature, °F	Peel Adhesion, oz	Visual Rating
350	224	1.8
400	272	2.5
450	384	4.2

In a comparative test, a tire yarn prepared as above but without addition of the NaOH as catalyst showed peel adhesion values of only 256 ounces and visual ratings of only 1.5 at a molding temperature of 450°F. This clearly indicates the criticalness of using the alkaline catalyst in the process.

Two-Stage Drawing for Polyamide Monofilaments

S.P. Gauntt; U.S. Patent 4,009,511; March 1, 1977; assigned to E.I. Du Pont de Nemours and Company has developed a process in which heavy denier, polyamide monofils are produced by the coupled steps of spinning, quenching, drawing and winding. Quenching involves passage of the filament through an air gap and a water bath. There are two stages of drawing. In the first stage, pressurized steam is used to deorient surface polymer and as a draw assist.

Example: Polyhexamethylene adipamide having a relative viscosity of 70 is extruded at the rate of 27.2 lb/hr (12.3 kg/hr) through a rectangular spinneret orifice having rounded corners (2.95 x 9.70 mm), passed vertically downward through an air gap for 22 inches (55.9 cm), quenched in water at 30°C for a distance of 134.5 inches (342 cm), passed to a feed roll having a surface speed of 126.3 ypm (115.5 m/min) and then passed through a steam chamber 122 cm in length containing wet steam with 94% quality at 110 psig (7.73 kg/cm^2) and at 173°C (saturation temp) to a first stage draw roll which orientation-stretches the filament 4.0X. The filament is under a relatively low tension of about 5,000 grams between the rolls.

From the first stage draw rolls, the filament makes 3 passes through a 122 cm radiant heating zone at 1050°C to second stage draw rolls which orientation-stretch the steamed and drawn filament an additional 1.43X for a total draw ratio of 5.70X. The filament is wound into a package. The filament has a denier of 3,000 and a substantially flat, ribbon-shaped cross-section with a thickness of 0.338 mm and a width of 0.922 mm. The filament has a straight tenacity of 8.6 gpd and a loop tenacity of 4.1 gpd with 13.3% and 8.9% elongations, respectively, measured as described in U.S. Patent 3,650,884.

Optical examination of the filament cross-section under a microscope shows a visibly apparent surface layer about 3.9 microns thick as a result of the steam treatment. The filament surface has a refractive index of about 1.565. This surface layer has a reduced density compared to the filament core.

Prior to the second stretch, the filament surface was found to have substantially zero birefringence. After the second stage, it was found to have a surface birefringence of about 0.046. Prior to the second stretch, the filament had a straight tenacity of 4.5 gpd and a loop tenacity of 4.7 gpd with 51.7% and 56.6% elongations, respectively.

Another filament was prepared in essentially an identical manner except for the use of superheated steam at 110 psig (7.73 kg/cm^2) and 235°C in the first stage stretch. This produced a filament having a straight tenacity of 9.2 gpd, a loop tenacity of 5.0 gpd, and elongations of 17.1% and 8.6%, respectively. Prior to the second stretch, the filament had a straight tenacity of 4.7 gpd and a loop tenacity of 4.7 gpd with 51.4% and 57.8% elongations, respectively.

Stretched and Interlaced Polyamide-6

H. Kubitzek, G. Hinrichsen and W. Giessler; U.S. Patent 4,035,464; July 12, 1977; assigned to Bayer AG, Germany describe a process for the continuous production of a stretched and interlaced polyamide-6 filament yarn wherein the filaments are interlaced before stretching.

Example: Polyamide-6 granulate with a relative solution viscosity of 2.65 in the form of a 1% solution in m-cresol is melted in a grid head and the resulting melt forced at 280°C through 12-bore spinnerets by means of metering pumps. The bore diameter is 0.25 mm. The ratio between the length of the metering bores to the bore diameter of the spinneret amounts to 2. Air at ambient temperature is blown onto the filaments in the spinning duct, after which the filaments are wetted with an antistatic water-in-oil emulsion below the spinning duct and run off at 1,767 meters per minute by means of a godet with an intermediate roller. The filament yarn is looped once around the take-off unit consisting of the unheated godet and transfer roller.

At the same time, the filament yarn passes through an interlacing jet operated with air under 6 atm pressure which is arranged between the godet and the transfer roller and whose filament duct is 20 mm long and 1.5 mm in diameter. Two opposite air inlet openings 0.88 mm in diameter open into the filament duct. The air inlet bores are perpendicular to the filament duct and are situated in the middle of that duct. Immediately afterwards the filament yarn is stretched by a pair of stretching godets rotating at 3,800 m/min and heated to a surface temperature of 120°C.

After passing 6 times around the stretching unit, the filament yarn, which is stretched to a denier of 67 dtex, is wound at 3,722 m/min by means of a friction winder. The uniform, high-volume filament package obtainable in a yield of almost 100% over long spinning times can be warped with snarl counts of less than 0.1/kg under a uniform take-off tension over the entire width of the package, and further processed into knitted fabrics with an extremely good dye finish. If the tension under which the filament yarn enters the yarn-jet , as measured in the absence of compressed air, amounts to between 6.5 and 7.5 g, the hook-drop value of the twisted filament yarn (according to U.S. Patent 2,985,995) amounts to between 55 and 85 mm.

If, under the conditions mentioned above, the supply of air to the interlacing jet is switched off, or if interlacing is carried out under the same conditions between the stretching pair and the winding unit, it is not possible to obtain useful packages, even by varying the winding speed. Under these conditions, a package of favorable structure is obtained only at stretching pair temperatures above 150°C.

Polyamide-6 Filaments with γ-Modification

H. Kubitzek, K. Feltgen and F-K. Rosendahl; U.S. Patent 4,102,965; July 25,

1978; assigned to Bayer AG, Germany describe a process for the production of continuous polyamide-6-filaments present in the γ-modification both before and after stretching at room temperature by spinning polyamide-6 with a relative solution viscosity of from 2.3 to 3.1, preferably from 2.6 to 2.85, as measured on a 1% solution in m-cresol, wherein the filaments (a) after leaving the spinneret, are cooled by air-blowing and preoriented up to an elongation at break of from 50 to 80%, preferably from 55 to 75%, by being run off, preferably in the absence of godets, at a speed of 3,700 to 4,500 m/min; (b) before winding into package form and during take-off, are prepared with an aqueous preparation oil in such a way that they contain less than 3.0% by weight of water, and (c) before winding into package form are subjected to interlacing.

By adopting this procedure, the polyamide-6-filaments are obtained in the highly oriented γ-modification which remains intact, even after stretching at room temperature. This is certainly the case when, after winding into package form, the filaments do not show any (020)-reflex and when the intensity ratio of the (002) and (200) reflexes is above 1.1, i.e. when I(002):I(200) is greater than 1.1, and in addition when the orientation of the equatorial (200)-reflex, which is defined as the reciprocal value of half the width at half maximum intensity of the azimuthal intensity distribution of the (200)-reflex, is greater than 0.08 [L.E. Alexander, *X-Ray Diffraction Methods in Polymer-Science,* John Wiley and Sons, New York, (1969), pages 241 et seq].

In addition, the filament yarns show increased gloss after stretching at room temperature. For example dtex 44 f 10-filament yarns according to the process containing 0.3% of TiO_2 have G_{45}-gloss values approximately 50 to 150% higher, and gloss levels (h) approximately 50 to 80% higher than filament yarns of the same denier produced from the same chip material by the conventional two-stage process (G_{45}- and h-measurements according to the publication of the Bekleidungsphysiologisches Institut EV, February 1970; Research Project No. 1823, "Investigations into the Development of a Method for Determining the Influence of the Texture of Polyester Textiles on Gloss".

Example: Chips of polyamide-6 with a relative solution viscosity, as measured on a 1% solution in m-cresol, of 2.7 and containing 0.3% of TiO_2, were melted in a screen head and the resulting melt was spun at 280°C at a rate of 18.8 g/min per spinneret from 10-bore spinnerets with bore diameters of 0.2 mm. After they had been cooled to 20°C by blowing with air, the filaments were treated with an aqueous preparation oil in such a way that the filaments contain 2.7% by weight of water and 0.8% by weight of preparation oil, interlaced and wound into package form directly, i.e. in the absence of godets, at 3,905 m/min using a friction winder. In the winding room, the temperature was 19°C and the humidity 49%. The filaments thus produced had the following properties:

Denier, dtex	50.0	
Tensile strength, g/dtex	3.7	
Elongation at break, %	64.0	
(200)Orientation	0.106	} γ-modification
I(002):I(200)	1.45	

The filaments were stretch-textured with a stretching ratio of 1:1.211 on the false-twisting friction principle and processed into ladies stockings on an 8-system hosiery knitting machine. The stockings had a clear stitch pattern and were extremely uniform. The contraction in length between unfinished and finished stockings amounted to 2.4%.

The filaments were then stretched in a ratio of 1:1.26 at room temperature on a stretch-twisting machine. The stretching yield was high. The filaments then had the following properties:

Denier, dtex	43.50	
Tensile strength, g/dtex	4.70	
Elongation at break, %	38.00	
(200)Orientation	0.119	γ-modification
I(002):I(200)	1.34	γ-modification
G_{45}-value	30.9	
h-value	20.9	

For comparison, filament yarns of 10 individual filaments with a gross denier of 136 dtex were produced from the same chips at a winding speed of 804 m/min and stretched in a ratio of 1:3.29 at room temperature. These filaments, in the α-modification, had the following properties:

Denier, dtex	43.00
Tensile strength, g/dtex	4.26
Elongation at break, %	38.00
G_{45}-value	17.3
h-value	13.5

Continuous Melt Spinning and Drawing of Nylon-6

The process and apparatus developed by *T.F. Corbin, Jr., B.B. Eskridge and B.M. Lyon; U.S. Patent 4,042,662; August 16, 1977; assigned to Akzona Inc.* are directed to the continuous spinning and drawing of polyamide yarn wherein an aqueous emulsion spin finish is applied to undrawn yarn and the yarn liveliness on the first or winding godet is significantly reduced by use of a first and substantially unheated godet having an increase in diameter from the first to at least the second yarn wrap prior to drawing and take-up.

Example: Yarn produced by this process has the properties suitable for use as a feeder yarn into, e.g., 2,500 denier carpet yarn. A specific polyamide yarn prepared according to this process utilizing an emulsion spin finish was processed as follows.

Polycaproamide (nylon 6) was melt spun, drawn, and wound continuously as set forth in Figure 10.3. This polymer had a relative viscosity of 2.41 (1% solution of polymer in 90% formic acid at 30°C) and was melt spun at a temperature of 285°C; the spun denier was about 8,300; the filaments comprised 136 in number; and the spun yarn relative viscosity was 2.45.

Figure 10.3: Continuous Melt Spinning and Drawing of Nylon-6 with Reduced Liveliness

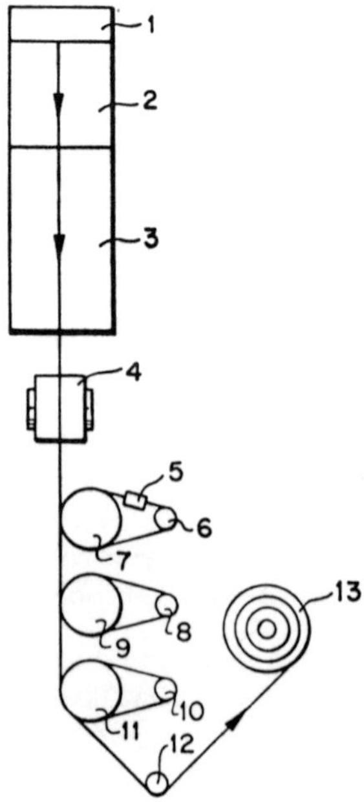

Source: U.S. Patent 4,042,662

From spinneret **1** the yarn was passed through cooling chamber (blow box) **2** and cooled with air. From cooling chamber **2**, the yarn was passed through chimney **3** and treated with 10% emulsion finish applied by stone wheel applicator **4**. The treated yarn was then taken up by stepped winding godet **7** and idler roll **6**, the number of wraps being sufficient to prevent slippage (i.e., 12). The stepped godet **7** was substantially maintained at a temperature of about 28°C and the surface speed (mpm) was 425. The yarn was then passed to draw godet **9** maintained at a temperature of about 50°C and a surface speed (mpm) of 1,610; the yarn then passed over idler roll **8** and then to draw godet **11** and idler roll **10**. Draw godet **11** was maintained at about 80°C and the surface speed (mpm) was 1,610. The yarn was then collected in untwisted condition on takeup package **13**.

The temperature range over the initial winding godet **7** can be from 25° to about 100°C, and that of draw godet **9** can be from about 25° to about 100°C, and that of draw godet **11** can be from about 50° to about 100°C. The surface speed of the draw godet in meters per minute, for example, can be varied within the skill of the art; the draw godets utilized in this example were of the normal straight cylindrical type.

The yarn reflected a moisture content of 3.2, an oil percentage of 1.8, a boiling water shrinkage factor of 12%, a denier of 2,185, a tenacity (gpd) of 3.0, and an elongation at break of 39.

Improving Crystallinity in Nylon-6

B.B. Eskridge, B.M. Lyon and E.H. Boasson; U.S. Patent 4,049,766; Sept. 20, 1977; assigned to Akzona Incorporated disclose an improvement in continuous spin-draw-winding of nylon-6 to produce yarn filaments of mixed beta,alpha-crystalline morphology, the improvement wherein prior to drawing, a spin finish emulsion containing from about 0.1 to about 6% by weight of benzyl alcohol is applied to the filaments to enhance conversion of the beta-hexagonal crystalline form to the alpha-monoclinic crystalline form, and immediately thereafter continuously drawing the nylon-6 yarn filaments thereby imparting improved uniformity of dye acceptance to the drawn filaments.

Example: Unmodified nylon-6 polymer chips having a relative viscosity (RV) of 2.40 and a 220°C melting point were spun to an increase of RV to 2.45 and drawn as 1040/68 yarn, which yarn was treated, by a single stone wheel, with an emulsion spin finish containing from 0 to 6% benzyl alcohol. This treated yarn was then drawn at a total ratio of 3.7 to produce a yarn of 15 denier per filament. The yarn samples were then bulked by combining two ends and bulking same.

It was found that benzyl alcohol has little influence on yarn shrinkage, tenacity, elongation, wet bulk, and crimps per inch. An increase in benzyl alcohol generally corresponds to an increase in alpha-crystalline content. The use of increased benzyl alcohol content results in an increase in yarn density.

An increase in amount of benzyl alcohol as emulsion component results in an increase in alpha-crystallinity, especially within the range of from about 3 to about 6%, and this is reflected in an increase in dye depth and an improvement in uniformity of dye acceptance.

The spin finish was a finish emulsion (8%) consisting of the following components in wt % with from 0 to 6% benzyl alcohol added thereto:

Butyl stearate	77.3
Dioctyl sodium sulfosuccinate	6.5
Polyethylene glycol (g mol wt 600) monolaurate	4.9
Oleyl acid phosphate	8.7
KOH (56% solution prepared from flakes of 90-92% purity)	2.6

The chip relative viscosity was measured in 90% formic acid with a 1% nylon-6 polymer solution. Using x-ray diffraction measurements made in transmission using Ni filtered Cu-K alpha-radiation, the beta/alpha-ratio was formed by taking the intensity of the alpha (002) meridional reflection and dividing it by the intensity of the (020/220) (doublet) equatorial reflection.

Carpet samples of yarn were tufted and routinely dyed a commercial avocado color using disperse dyes. These carpet samples dyed uniformly and evidenced fully acceptable dye depth.

Nylon-66 with Soft Hand

J.E. Bromley, M.M. McNamara and W.T. Mowe; U.S. Patent 4,093,147; June 6, 1978; assigned to Monsanto Company provide nylon 66 yarn having particular stress-strain properties, typically also having a soft, luxuriant hand in fabric form. As compared to conventional nylon 66 with comparable boiling water shrinkage, the yarn exhibits a higher modulus at break, a lower modulus at 10% elongation, a positive stress index, and excellent denier uniformity. The process involves subjecting the yarn, within 0.016 to 0.11 seconds after solidification of the filaments, to a tension of 0.2 to 1.5 grams per final denier and heating the yarn to a temperature between 50° and 250°C long enough to reduce yarn retraction below 1%.

As illustrated in Figure 10.4, molten nylon-66 polymer is metered and extruded from a nonillustrated conventional block through spinneret **22** into quench zone **24** as a plurality of molten streams. The streams are cooled and solidified in zone **24** by a flow of transversely moving air into filaments which constitute yarn **26.**

Figure 10.4: Apparatus for Producing Flat Nylon-66

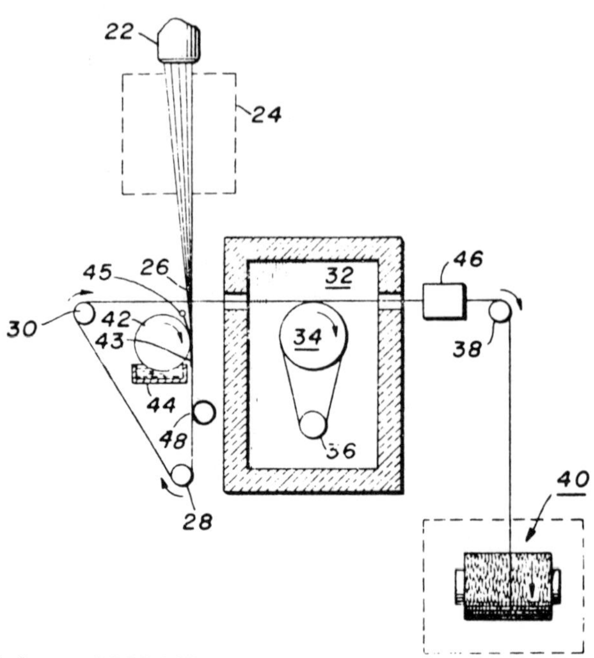

Source: U.S. Patent 4,093,147

Yarn **26** passes in partial wraps around rolls **28** and **30** prior to entering insulated chamber **32**. Driven heated feed roll **34** and its associated skewed separator roll **36** are mounted within chamber **32** for forwarding yarn **26**, which passes in several wraps around rolls **34** and **36** prior to leaving chamber **32**. Yarn **26** next passes in a partial wrap around roll **38** and then downwardly to schematically illustrated yarn winding apparatus **40**.

In the preferred embodiment, spin finish is applied by slowly rotating conventional finish roll **42**, whose lower surface is immersed in liquid finish carried in trough **44**. A conventional gauze finish skirt **43** transfers the finish from roll **42** to yarn **26**, skirt **43** being anchored at **45**. While it is preferred to locate finish roll **42** above roll **28** as illustrated, it may be located between rolls **28** and **30** or at other locations. Optionally, the filaments of yarn **26** may be interlaced or entangled by an interlacing apparatus **46** of any desired design.

Rolls **28**, **30** and **38** may be supported on air bearings, and at least one of rolls **28** and **30** may be driven at a controlled speed for controlling the tension of the yarn entering chamber **32**. Roll **38** may be driven at a controlled speed for adjusting the tension in yarn **26** passing through device **46**, and for adjusting the winding tension.

The table below shows the physical properties of the yarns produced by the process disclosed above (A, B, and C), and compares these properties with those of commercially available yarns having the same nominal denier and the same number of filaments. The data reported are the average of at least five bobbins for all items. Item D is a commercially available nylon 66 premium quality yarn produced by a single-stage-draw coupled process; Item E is a commercially available nylon 66 premium quality yarn believed to be produced by a two-stage-draw coupled process; Item F is a commercially available nylon 66 yarn produced by a two-stage-draw coupled process, and Item G is a commercially available nylon 66 yarn produced by the split process. Items D, E, and F are relaxed yarns, that is, they were heat-treated under appropriate tensions so as to reduce the shrinkage. Item G was not heat-treated and is not a relaxed yarn, as evidenced by the high shrinkage. All items are flat (untextured) yarns.

Item	A	B	C	D	E	F	G
Properties							
Drawn denier, D_O	70	70	68	69	71	71	71
Tenacity, (g/d)	3.8	4.4	3.9	5.3	5.2	4.6	5.0
Elongation, (%)	57	52	55	39	26	38	31
Shrinkage, S (%)	5.5	6.2	6.0	5.0	4.0	6.9	10
Uster unevenness (% U)	0.67	0.57	0.64	0.67	0.69	1.2	1.5
Initial Modulus, M_i (g/d)	19	21	19	25	25	25	28
10% Modulus, M_t (g/d)	8.2	11	9.1	30	23	34	31
Final Modulus, M_f (g/d)	8.2	8.5	8.4	7.4	6.5	4.9	5.2
Modulus ratio (R)	1.0	1.3	1.1	4.2	3.7	7.1	6.1
Stress index (α)	31	23	32	-6	-5	-9	-2

ACRYLICS, POLYOLEFINS, POLYURETHANES, POLYOXADIAZOLES

Highly Shrinkable Acrylic Fibers

U. Reinehr, A. Nogaj and G. Lorenz; U.S. Patent 4,108,845; Aug. 22, 1978; assigned to Bayer AG, Germany describe highly shrinkable fibers or filaments of acrylonitrile polymers or copolymers which have a shrinkability of at least 35% and a fiber strength of at least 2 p/dtex.

The process comprises prestretching the dye-spun polymer in a ratio of up to 1:3.0 at a temperature of from 60° to 100°C; after-stretching in a ratio of up to 1:2.5 at a temperature of from 60° to 90°C, the total stretching ratio of the two stretching stages carried out in an aqueous medium amounting to at least 1:3.0; crimping the stretched and spun polymer while still wet at a temperature of up to 90°C; and drying at a temperature of up to 70°C.

Example: An acrylonitrile copolymer of 93.6% of acrylonitrile, 5.7% of methyl acrylate and 0.7% of sodium methallyl sulfonate was dry-spun by standard methods known in the art. The two, which had an overall denier of 1,200,000, was stretched in a ratio of 1:1.5 in boiling water and was subsequently washed under tension in 3 successive washing baths at 80°C (washing baths 1 and 2) and 50°C (washing bath 3). The two was then after-stretched in a ratio of 1:2.0 at a stretching-bath temperature of 75°C, so that the total stretching ratio amounted to 200%, i.e., to three times the original length of the tow. The rate of travel of the tow after the second stretching stage amounted to 50 m/min.

Individual filaments removed from the tow showed a shrinkage of 45.0% in boiling water. The tow was then treated with an antistatic preparation and crimped in a stuffer box into which steam was sprayed. The shrinkage of a number of individual filaments removed from the crimp tow was determined and gave an average shrinkage value of 44.6% in boiling water. The tow was then cut into staple fibers, dried in a dryer at 30° to 40°C, baled and packaged. The final denier of each individual fiber amounted to 2.4 dtex. The fiber shrinkage of a number of individual filaments amounted to 43.7% in boiling water. Fiber strength was 2.3 p/dtex, elongation at break 23%. Fiber strength and elongation at break were measured with a Statigraph IV(Textechno Company, West Germany). The high-shrinkage fibers were then spun into yarn with yarn counts of 40/1. Yarn constants: tensile strength = 11.5 RKm, elongation at break = 12.5%, satisfactory travel over cards and intersecting gill boxes, density = 1.174 g/cc.

Highly Shrinkable Wet-Spun Acrylic Fibers Fixed with Steam

U. Reinehr, A. Nogaj and T. Herbertz; U.S. Patent 4,067,948; Jan. 10, 1978; assigned to Bayer AG, Germany describe a process for the production of high-shrinkage wet-spun acrylic fibers having good fiber strength and high densities by fixing the unstretched fibers with saturated steam and then stretching in a ratio of from 1:3.5 to 1:5.

Example: An acrylonitrile copolymer of 93.6% of acrylonitrile, 5.7% of methyl acrylate and 0.7% of sodium methallyl sulfonate was wet-spun from dimethyl formamide by conventional methods. The tow with an overall denier of 1,400,000 dtex was washed with water at 50°C, fixed with saturated steam for 7 min at 120°C in a steaming box, drawn in a ratio of 1:3.5 at 75°C, treated with antistatic preparation and moist crimped. The fiber shrinkage of the crimped tow, as measured on a series of individual filaments amounts to 54.7%. The tow was then cut into staple fibers which were dried at 50°C. The final individual fiber denier was 5.4 dtex. The fiber shrinkage of a series of individual filaments amounts to 51.5% in boiling water. Fiber strength: 1.9 p/dtex; density: 1.181 g/cc.

Dry-Spun Polyacrylonitrile Subjected to Hot Drawing and Relaxing

H. Lohwasser; U.S. Patent 4,140,844; Feb. 20, 1979; assigned to Bayer AG, Germany discloses dry-spun polyacrylonitrile filament yarns having a tensile strength of at least 47 cN/tex and an individual filament denier of at most 1.5 dtex and preferably an overall denier of 20 to 145 tex as well as a process for the production thereof by spinning, hot drawing and relaxing, wherein the spun filaments are drawn during spinning to such an extent that, after subsequent hot drawing in a ratio of from 1:6 to 1:10 and relaxation, the individual filaments have a denier of at most 1.6 dtex.

In the following example, contents of N,N-dimethyl formamide (DMF), preparation (oil) and extractable fractions in the filament yarns are expressed in percent by weight, based on the dry mass (PAN). Effective yarn finenesses describe the condition of the material, including DMF and oil. Tensile strength and elongation at break were measured in a Wolpert apparatus.

Example: A 25% solution of polyacrylonitrile in DMF was dry spun at a throughput of 22.2 g of PAN/minute through a spinning jet having 144 bores with a diameter of 0.2 mm, so that a 1,760 dtex filament yarn containing 13% of DMF was obtained for a take-off rate of 126 m/minute. 2.4% of an oil preparation was applied during winding. The spun material (spinning factor 0.82) was drawn in a ratio of 1:9.3 in a drawn-twisting arrangement by a single passage over a heating godet at 147°C, a yoke at 145°C and an unheated take-off godet, the filament yarn being looped several times around each godet.

The drawn filament yarn wound onto cops had a total denier of effectively 204 dtex, a DMF-content of 9.4% and a boiling-induced shrinkage of 16%. A fully shrunk DMF-free filament yarn of dtex 224 f 144 Z 150 was obtained therefrom by after-twisting and steaming under pressure at 125°C in package form. Tensile strength 47.0 cN/tex, elongation at break 18.1%.

When spinning jets having only 96 or 72 bores (spinning factor 0.55 and 0.41, respectively) were used under otherwise the same spinning and after-treatment conditions, the tensile strengths fell to 45.6 and 43.5 cN/tex, respectively.

Acrylonitrile Polymers with Microvoided Core

According to *R.A. Blickenstaff; U.S. Patent 4,094,948; June 13, 1978; assigned to E.I. DuPont de Nemours and Company* substantially single-phase compositions which are substantially partial or complete hydrates of an acrylonitrile polymer containing at least 40% acrylonitrile units can be shaped into filaments and films. The nitrile groups of the polymer are hydrated to a selected extent depending on the particular polymer used. Hydration occurs only within a certain elevated temperature range and under at least autogenous pressure.

Also disclosed are films and filaments having structural features which are spun from the substantially single-phase compositions comprising polymers or copolymers of 80% or more acrylonitrile. The filaments are characterized by a sheath which has a density gradient in which the density increases toward the surface and, after drawing, by a unique luster source ratio related to reflective interfaces. Many of the filaments have surface striations and an optically microvoided core. The filaments contribute good hand and good water absorption properties to fabrics made therefrom.

A dimensionless parameter, termed luster source ratio (LR) is a convenient base for comparison of the drawn filaments with those of the prior art. It is developed by light-scattering measurements as described below, using a "Sofica" light-scattering photometer, Type 701, (Mechrolab, Inc., Mountainview, Calif.) modified slightly to permit (1) reproducible positioning of single, straight filaments or arrays of (randomly twisted) filaments, (2) determination of light scattering patterns of filaments (or multifilament arrays) immersed in either air or a liquid of matching refractive index and (3) improved baffling to minimize the effect of stray light.

The filaments to be tested are boiled off for 30 minutes in a 0.1% "Duponol" WA solution, rinsed in distilled water, dried in a forced-draft oven for 1 hour at 40°C and further conditioned for 16 hours at 21°C, 65% RH.

To measure LR, two to four filaments (depending on denier, not critical) are mounted in a close parallel array in a perforated holder such that the sample may be angularly adjusted relative to the plane which includes the incident light slit.

For this test, the filaments are oriented parallel to the scattering plane. The incident light is a narrow beam of monochromatic green (λ = 545 mμ), unpolarized light. Light scattered by the filament is measured by a photomultiplier and recorded on a graph of log intensity vs angle relative to the incident beam. A maximum intensity is seen at some angle near 45°. The pattern of the ascending and descending slopes around this maximum intensity is characteristic of the particular filament. A horizontal line is drawn on the graph, at one-half the value of the maximum intensity, to intersect the ascending and descending legs of the intensity curve. The distance, in degrees, between these two intersections, termed Full Width, Half Mode (FWHM) is a convenient reference to the luster of a given filament.

The above procedure is repeated after immersion of the sample in a 50/50 mixture of ortho- and para-xylene (selected to approximate the refractive index of the acrylonitrile polymer filament). The ratio of FWHM in air to that in xylene is termed LR. Filaments of this process have an LR of 0.7 or more, suggesting a large dependence of light scattering on the internal (core plus sheath) morphology typical of the products of this process.

Example 1: A copolymer acrylonitrile/methyl acrylate/sodium styrenesulfonate (93.63/6/0.37 weight ratio) having an inherent viscosity of 0.9 is finely ground. Twenty-two parts of water is sprayed onto 100 parts of the copolymer in an open tray. The mixture is placed in a jar, sealed, and rolled for twelve hours. The apparently dry powder is transferred to a cylinder equipped with one 50-mesh (~20 wires/cm) and two 200-mesh (~79 wires/cm) screens and a spinneret with a single hole 0.01 inch long and 0.01 inch in diameter (0.25 x 0.25 mm). A closely fitting Teflon-gasketed, free piston is inserted. The spinneret is then closed by pressing a Teflon pad against its outer face.

The cylinder is then cooled to -10°C and evacuated via a valved side-port between the piston and the spinneret until the water vapor pressure at that temperature is reached. The side-port valve is closed and the cylinder is then heated to 180°C and the spinneret to about 160°C for five minutes. A pressure of 500-600 psig (35-42 kg/cm^2) of nitrogen is applied to the piston, the pad is removed from the spinneret and a continuous filament is wound up at a rate of 45 ypm (41 m/min).

Example 2: A (I.V. = 1.0) copolymer of 93.95% acrylonitrile, 6% vinyl acetate and 0.05% sodium styrenesulfonate is mixed as in Example 1 with water in the ratio 100/22.7; the mix is metered to a 5-zone Werner and Pfleiderer twin-screw extruder which performs the sequential functions of feeding, melting and mixing, the highest temperature therein being 180°C, then to a pump which meters the resulting melt at 550 psig (38.5 kg/cm^2) to a 39-hole 0.007 inch x 0.007 inch (0.18 x 0.18 mm) spinneret having a temperature of 172°C and the resulting white yarn is wound up at 45 ypm (41 m/min). It is drawn to 460% of its spun length in saturated steam at 120°C (28 psig - 1.96 kg/cm^2) and boiled off to yield a product having straight filament tenacity/elongation/modulus/work-to-break values of 3.0/19/74/0.33 (gpd/%/gpd/g-cm-cm^{-1}-den^{-1}). The filament has a denier of 17.

Example 3: One hundred parts of a 1.0 $[\eta]$ copolymer of acrylonitrile/methyl acrylate/sodium styrenesulfonate (93.63/6/0.37 ratio) is mixed with 25 parts water as described in Example 1. The mix is metered to a twin-screw extruder like that employed in Example 2, the highest temperature is 177°C and the molten mass is metered at 700 psig (49 kg/cm^2) to a spinneret having 39 holes of 0.010 inch x 0.007 inch (0.25 mm x 0.18 mm) in length x diameter which is maintained at 175°C.

The filaments issuing from the spinneret are passed through a spinning chamber containing a mixture of air and water vapor at a pressure of 20 psig (1.4 kg/cm^2)

and a temperature of 64°C. The yarn is collected on a bobbin at about 75 ypm (68 m/min). The yarn is then unwound and passed through a tube containing saturated steam at a pressure of about 20 psig (1.4 kg/cm^2), where it is drawn to approximately 600% of its as-spun length and wound up at about 150 ypm (137 m/min). The boiled-off yarn has filament tenacity/elongation/modulus of 3.0/20/68 (g per denier/%/g per denier). The filament has a denier of 11.8. Microscopic examination of one filament cross section shows a sheath on the filament of a thickness amounting to 28% of the filament radius; the remaining structure comprises a core which contains multitudinous voids. The average density of the filament is 1.083 g/cc. Microscopic examination of the filament surface shows deep longitudinal striations.

The resulting structure shows a water absorption rate of 0.095 cc/sec as determined by a modification of the method of E.H. Buras, Jr. et al, "Measurement and Theory of Absorbency of Cotton Fabrics," *Tex. Res. Journal,* Vol. 20, April 1950, pp. 239-240, the modification being to use as sample a yarn wrapped on a form rather than the wadding or fabric of Buras et al. The wicking rate of a commercial dry-spun acrylic fiber in the same range of denier is about 0.06 cc/sec in this test.

A sample of the drawn filament of this example is treated by a cross-section dyeing test. The filaments are placed in the dye bath for 40 seconds at a temperature of 180°F (82°C). The core of the filament is sharply differentiated from the sheath by being dyed to a distinctly darker color.

A sample of the drawn filaments is mounted in epoxy resin and 0.2 μ sections are cut with a diamond knife microtome. Pictures taken in the electron microscope at 16,120X magnification show that the core voids have been partially closed in one direction by the action of the diamond knife, but the diameters parallel to the knife edge measure about 0.2 μ.

Example 4: Apparatus such as described in Example 3 is used to prepare a 90 dpf (as spun) yarn from a melt comprising 100 parts 1.0 I.V. polymer having the composition of that employed in that example and 25 parts water. The conditions are as follows: highest extruder temperature: 175°C; pressure: 1,050 psig (74 kg/cm^2); spinneret: 38 holes 0.36 x 0.18 mm (length x diameter); spinneret temperature: 177°C; spinning chamber: 46 cm long, 26 psig (1.8 kg/cm^2) pressure, 66°C temperature in a mixture of air and water vapor and windup: 89 m/min.

The yarn is drawn to 800% of its as-spun length in an atmosphere of saturated steam at 20 psig (1.4 kg/cm^2). After boil-off in water, the yarn has a straight tenacity/elongation/initial modulus of 4.0 gpd/25%/66 gpd and corresponding loop properties of 0.86/2.1 and 58. It is found to have a luster source ratio (LR) of 2.43. At a magnification of 25,500X in a transmission electron microscope, a cross-section reveals 140 x 10^5 voids/mm^2 in the core, typical voids being 0.07 to 0.08 micron in diameter, the largest being about 0.31 micron and the smallest about 0.04 micron in the overall cross-section. This filament has a sheath of about 4 μ thickness.

Antisoiling Flame-Retardant Acrylics Involving Hot-Wet Relaxation

A. Maranci; U.S. Patent 4,081,498; March 28, 1978; assigned to American Cyanamid Company describes acrylic fibers of a fiber-forming first acrylonitrile polymer containing at least 50% acrylonitrile, a flame retardant amount of a halogen-containing vinyl monomer, and any balance of a halogen-free vinyl monomer having heterogeneously dispersed therein a small amount of an incompatible, halogen-free, second acrylonitrile polymer containing at least 70% acrylonitrile and one or more halogen-free vinyl monomers. The fibers are prepared by wet-spinning an intimate mixture of the polymers separately dissolved in aqueous inorganic solutions of the same salt following conventional procedures but including a hot-wet relaxation of the stretched wet-gel filaments prior to drying.

Although in certain prior art procedures wet gel relaxation is carried out prior to drying the filaments, most procedures do not have the criticality associated with the particular order of steps as does the process. The combination of blending small portions of the second polymer solution with the first polymer solution followed by conventional wet spinning including the wet-gel relaxation of the stretched filaments prior to drying provides flame-retardant acrylic fibers of high luster and desirably low apparent soiling properties, the latter being distinguished by low light transmission properties.

Example 1 (Comparative): A spinning solution was prepared containing 10% of a fiber-forming polymer of composition 81.1% acrylonitrile, 9.2% methyl methacrylate, and 9.7% vinylidene chloride in 90% of an aqueous solution of 46% sodium thiocyanate. The solution had a viscosity of 34 poises at 28°C and was extruded through a spinnerette having 10 orifices, each of 200 microns diameter, into an aqueous 12% sodium thiocyanate solution maintained at -2°C, to form filaments. The filaments were continuously withdrawn from the bath, stretched at a stretch ratio of 2, washed with water, and drawn a second time in water at 99°C so as to provide a cumulative stretch ratio of 12 and a denier of 9.6.

The stretched filaments were dried in a free-to-relax state at 127°C dry bulb and 60°C wet bulb and then further relaxed in saturated steam at 130°C. The filaments obtained had a high degree of luster and a high degree of light transmission, indicating that the filaments had an undesirably high level of apparent soiling tendencies.

Example 2 (Comparative): A portion of the wet-stretched filaments of Example 1, prior to drying were first exposed to saturated steam at 110°C in a free-to-relax state for 10 minutes and thereafter dried at 100°C dry bulb and 36°C wet bulb. The filaments obtained were highly delustered but had a very low light transmission, thus indicating that the apparent soiling tendencies were low but the luster value of 9.2 was unsatisfactory.

Example 3: A solution of 11.2% of a polymer having a composition of 89.3%

acrylonitrile and 10.7% methyl methacrylate, was prepared in 88.8% of an aqueous solution of 40% sodium thiocyanate. Five parts of this solution were mixed with 95 parts of the solution used in Example 1. The mixed solution was turbid, thus indicating incompatibility between the two solutions. The mixed solution was spinnable, however, and stretched wet gel filaments were made following the procedure of Example 1.

The wet-stretched filaments were then exposed to saturated steam and dried as in Example 2. The filaments obtained had a high degree of luster, 19.8%, and a low degree of light transmission, indicating low apparent soiling tendencies.

Spontaneously Crimping Composite Fibers of Polyacrylonitrile

H. Lohwasser, A. Nagaj and H. Wieden; U.S. Patent 4,013,753; March 22, 1977; assigned to Bayer AG, Germany describe a process for the production of spontaneously crimping composite fibers of acrylonitrile polymers wherein the dry-spun-filaments which still contain solvent are stretched to at least three times their original length in a bath of hot water which contains from 12 to 30% by weight of dimethyl formamide. The stretching operation may be carried out in one or two stages.

Example 1 (Comparative): Polymers of the composition (A) 93.4% acrylonitrile, 5.6% methyl acrylate, 1.0% methacryloylaminobenzene-benzenedisulfonimide; and (B) 89.3% acrylonitrile, 9.8% vinyl acetate, 0.9% methacryloylaminobenzene-benzenedisulfonimide; were spun side-by-side, in a 50:50 ratio to produce composite fibers with a solvent content of about 18% dimethylformamide. A fiber cable with a total mass per length of 88 g/m was stretched by a total ratio of 1:3.6 in two stages in fresh water at 98°C, passed through a finishing bath and dried under tension at 120°C. The cable was mechanically crimped and cut up into staple fibers and steamed at 110°C under normal pressure. The residual solvent content was then 1.5%, the remaining boiling shrinkage 2.4%. The fiber had a titer of 4.8 dtex, a tensile strength of 2.5 g/dtex and an elongation on tearing of 50%. It developed 8.9 crimp arcs per cm after boiling and drying at 80°C.

Example 2: The abovedescribed procedure was modified by inserting an additional stage between the two stretching sections. The dimethylformamide contents in the preliminary stretching vat, additional vat and main stretching vat were 24.1%, 12.6% and 14.5% respectively. The fibers contained 3.9% dimethylformamide after drying and 1.3% after steaming. The residual boiling shrinkage was 0.4%, and the development of crimp 7.0 crimp arcs per cm. The titer of the fibers was 5.0 dtex, the tensile strength 2.8 g/dtex and the elongation on tearing 44%.

Knitting samples were prepared from fibers treated as described in Examples 1 and 2 by worsted spinning and hand dyeing. The sample from Example 2 had a boiling shrinkage of 4.8% in the raw yard and a more open, softer hand and stronger gloss than that of Example 1 which had a boiling shrinkage of 8.5% in the yarn.

Polyolefin-Reinforced Rubber Products

A. Bonnefon and J. Benard; U.S. Patent 4,005,054; Jan. 25, 1977 and A. Bonnefon, A. Coupard and M. Maffre; U.S. Patent 4,123,476; Oct. 31, 1978; both assigned to Pneumatiques, Caoutchouc Manufacture et Plastiques Kleber-Colombes, France disclose a flexible, homogeneous rubbery product comprising a matrix of a rubber selected from the group consisting of natural and synthetic elastomer in which there is uniformly dispersed at least 5% by weight based on the weight of the elastomer of a crystalline polyolefin whose molecular weight is equal to or greater than 500,000. A substantial proportion of the polyolefin is in the condition or form of fine particles which are drawn out in the form of elongated fibrils having a diameter smaller than 5 microns. This product exhibits a very high resistance to elongation in at least one direction.

When the mixing or blending operation of the rubber and polyolefin is started in an internal mixer such as a Bambury mixer and continued on a mill, or when it is carried out entirely on a mill, there is obtained a preferential orientation of the polyolefin fibrils in that the polyolefin particles are more or less pushed or forced in the direction of the drawing or stretching out of the mixture between the cylinders. The same orientation occurs when the mixture, having been made with another type of mixing apparatus, is thereafter treated at a temperature close to or higher than the melting temperature of the polyolefin, with an apparatus that draws out the mixture, like a mill, a calender, or an extruder. The products obtained from these mixtures may have a high degree of anisotropy, i.e., a longitudinal modulus of elasticity much higher than the transverse modulus of elasticity with respect to the direction of orientation of the fibrils.

Example: Two identical batches or series are prepared of a rubbery mixture on the basis of a styrene-butadiene rubber, SBR, and of a polyethylene having a high molecular weight GUR Hostalen (polyethylene having an average molecular weight of 1,000,000 measured by viscosimetry, in the form of a fine powder having an average granulometry in the order of 100 microns with a size range between 20 and 400 microns, a melting point of 138°C, and a density of 0.945). The composition is as follows:

	Parts by Weight
SBR	100
Polyethylene	30
Phenyl-beta-naphthylamine (antioxidant)	1
Zinc oxide	3.5
Stearic acid	2

The mixing or blending operation is carried out on a mill and the polyethylene, in powder form, is incorporated at the beginning when the rubber has a temperature of about 60° to 80°C, which facilitates the dispersion. The mixture is then processed for 10 minutes approximately at a temperature of 156°C.

In the course of the blending of one of the batches, 4 parts of sulfur are added thereto about 3 minutes before the end of the operation. In the course of the blending or mixing of the other batch, 2 parts of zinc diethyldithiocarbamate accelerator are added thereto.

Each batch of the mixture is extracted from the mixer in the form of thin leaves or sheets which are cooled under tension so as to prevent their becoming longitudinally contracted, and thereafter alternate sheets of one and the other series of the mixture are vulcanized together and under pressure of 8 bars at 120°C for 35 minutes. The properties of the product obtained are as follows:

	Longitudinal Direction	Transverse Direction
Modulus of elasticity at 1% elongation (kg/cm^2)	185	–
Modulus of elasticity at 100% elongation (kg/cm^2)	–	40
Resistance to breaking (kg/cm^2)	1.043	136
Breaking elongation, %	5.5	360
Dynamic modulus (kg/cm^2)	11.525*	152**
Loss angle, °	3.9	6.2

*Tests by extension with a pretension of 3%, a frequency of 7.5 Hz and an amplitude of 0.5% of the dynamic deformation (the value indicated represents the inclination of the straight stress or traction line).

**Tests by extension with a pretension of 15%, a frequency of 7.5 Hz and an amplitude of 5% of the dynamic deformation (the value indicated represents the inclination of the stress or traction curve drawn back to the origin).

These properties show notably an extremely elevated anisotropy ratio in the order of 100, i.e., a very high modulus in the longitudinal direction, a much less elevated modulus in the transverse direction. The product obtained is comparable in this respect to a cord layer or ply consisting of parallel textile cord embedded in a layer of ordinary vulcanized rubber. The loss angle, which is small, indicates a weak hysteresis of the product.

Elastic Hard Fibers

According to *H. Toba, K. Ohata and N. Tsukane; U.S. Patent 4,006,208; Feb. 1, 1977; assigned to Daicel, Ltd., Japan* an elastic hard fiber composed mainly of polyisobutylene oxide and having an elastic recovery ratio of at least 70% from 50% extension and a work recovery ratio of at least 70% from 5% extension, is prepared by extruding molten polyisobutylene oxide at a temperature of from 175°C up to the decomposition temperature thereof, cooling the extrudate rapidly to a temperature of -20° to 70°C, and spinning it at a draw ratio of 50 to 1,000.

Example 1: Molten polyisobutylene oxide having an inherent viscosity of 2.5 and containing 0.8% by weight of tetrakis-[methylene-3-(3,5-di-tertiary-butyl-4-hydroxyphenyl)propionate]-methane and 0.2% by weight of 4-4'-thiobis-(3-methyl-6-tertiary-butylphenol) was melt-extruded at 210°C from nozzles of an extrusion molding machine through orifices having a diameter of 1 mm at a flow speed of about 0.185 m/min. The extruded filaments were rapidly cooled in a flowing air bath maintained at 10°C and disposed at a distance of 5 cm below the nozzles. The filaments were wound at a rate of 50 m/min by means of a winding apparatus disposed at a distance of 3 m from the spinning nozzles. The filaments were stretched at the spinning nozzles and also were stretched 2-3 times their original length during the solidification passage from the nozzles to the winding apparatus, this stretch ratio corresponding to a draw ratio of 270.

The thus-obtained elastic hard fibers had a size of 250 denier. The elastic recovery ratio from 50% extension was 92% as the initial elastic recovery ratio and rose to 96.5% after 10 hours. The fibers were also characterized by a work recovery ratio from 5% extension of 92%, a breaking strength of 0.91 g/d, an elongation at break of 120% and a stress at 50% extension of 0.69 g/d. The fiber density was 1.0468 and the birefringence was 30.

Example 2 (Comparative): The same polyisobutylene oxide composition as used in Example 1 was melt-extruded, employing the same apparatus, but without conducting the special cooling and the filaments were wound on a winding apparatus spaced 3 m from the nozzles. The draw ratio during this operation was 1,200. Although the resulting fibers had a breaking strength of 4.3 g/d, they underwent plastic deformation when stressed and did not exhibit a significant elastic recovery after extension. The fiber density was 1.0685.

Drawing Polyurethane Filaments

In the method described by *T.V. Peters; U.S. Patent 4,002,711; Jan. 11, 1977;* a substantially linear polyurethane filament is elongated a number of times its original length and heated at a temperature above its fiber stick temperature for a sufficient time so that the filament permanently retains that elongation and does not recover any substantial amount of the length to which it has been elongated.

As used herein, fiber stick temperature means the temperature at which a fiber will stick to a heated brass block when held against the surface of the block for 5 seconds with a standard 200 gram brass weight.

As shown in Figure 10.5, the method may be carried out by passing a number of identical, wet-spun elastomeric polyurethane filaments from a creel to form a beam. The beam is transferred to a second position (adjacent the inlet drive roll in Figure 10.5), and the filaments are passed from the beam, controlled to feed the filaments at a specified speed by the inlet drive roll, to feed rolls. The feed rolls are set at a higher speed than the drive roll so as to elongate each filament several times its former length, preferably no more than two times. The filaments are then passed to a zone of heated cans (which may be heated by steam, hot oil, infrared or other means), operating at the same surface speed as the drive rolls, where each filament is heated to a temperature above the fiber stick temperature for a period of time.

Figure 10.5: Drawing Polyurethane Filaments

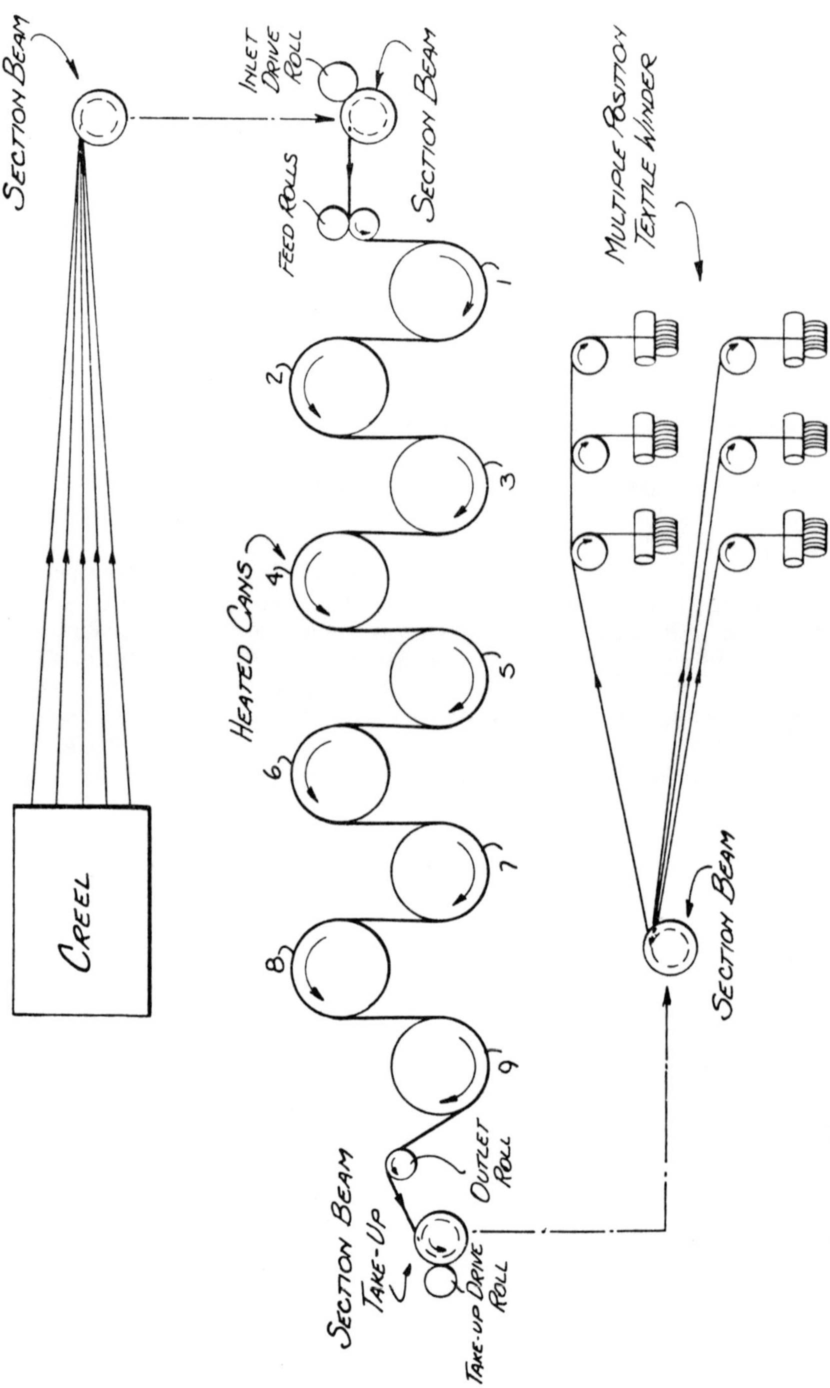

Source: U.S. Patent 4,002,711

This time is of sufficient length so that the filament remains permanently deformed at that elongation and does not recover any substantial portion of the length to which it has been elongated. The filaments are then taken up on a take-up beam at the speed of the feed rolls, and the take-up beam is transferred to a holder (not shown) where the filaments are passed to a textile winder and wound up on packages ready for commercial use.

Example 1: Approximately 120 parts of a 2,000 molecular weight polyester glycol comprising the reaction production of a 9/1 molar mixture of ethylene glycol and propylene glycol with adipic acid was reacted with 28 parts of methylene-bis(4-phenylisocyanate) and 1.6 parts of tolylene diisocyanate at a temperature of 90°C for 60 minutes to form an isocyanate-terminated urethane prepolymer.

The above prepolymer was diluted with 49 parts of DMF and slowly added to a rapidly agitated solution comprising 1,860 parts of dimethyl formamide (DMF), 2.4 parts of methyl-imino-bis-propyl-amine, 9.4 parts of ethylene diamine and 1.0 parts of diethanolamine until a viscosity of 120 poises was attained.

After the addition of pigments and antioxidants, the above solution was extruded through a spinnerette containing 60 holes, each hole having a diameter of 0.0065 inches, into a bath containing 20% DMF in water. The resulting filaments were continuously removed from this bath and extracted countercurrently in a series of ten extraction baths. The wet filaments were then thoroughly dried by passage through a three-pass dryer (at a residence time of 40 seconds in each pass, the first pass being at 150°C, the second at 175°C and the last at 200°C). A textile lubricant was added to the filaments, and they were subsequently taken up on a winder.

Example 2: A spandex filament, produced according to the process of Example 1, which had a nominal denier of 420 (417 actual), an elongation of 660% and a modulus (stress) at 300% elongation of 0.296 gram per denier and had a fiber stick temperature of 178°C, was passed through the apparatus shown in Figure 10.5.

The single filament was taken off a feed package by means of a drive roll, passed through a comb guide and then through an additional set of drive rolls. The speeds of the first drive roll and the second set of drive rolls were adjusted to produce an elongation in the filament of approximately 3.5 times. The filament was then passed ten times in contact with about one-quarter of the surface of a heated 6 inch roll. The temperature to which the filament was exposed was approximately 225°C for a time of 2.6 seconds. The filament was then taken up on a take-up roll operating at a surface speed of 90 fpm. The resulting filament was 118 denier, had an elongation of 690% and a 300% modulus of 0.351 gram per denier. The filament from that drawing operation was then subjected to an identical treatment to produce a filament of 34 denier, an elongation of 720% and a 300% modulus of 0.355 gram per denier.

Drawing Polyoxadiazole Filaments

D.L. Brydon; U.S. Patent 4,035,465; July 12, 1977; assigned to Imperial Chemical Industries Limited, England discloses a method of making a drawn filament from a hot drawable polyphenylene-1,3,4-oxadiazole having p-phenylene radicals or p- and m-phenylene radicals, with substituted bromine, wherein a solution of the polymer is dry jet wet spun into a coagulant, the filament washed and dried, and drawn at a temperature of at least 350°C at a draw ratio of at least 6:1. The polymer has as characteristic units

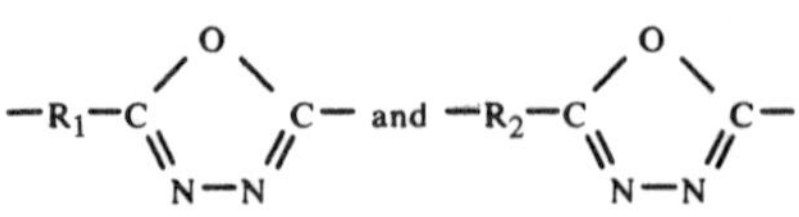

in which R_1 is a p-phenylene radical and R_2 is a m-phenylene radical.

Preferred polymers are based on terephthalic acid alone, or preferably on mixtures of terephthalic acid and isophthalic acid, which are polymerized with a hydrazine salt, especially hydrazine sulfate, in oleum of from 5 to 65% strength, preferably 20 to 40% strength. Bromination may be effected by adding bromine to the polymer in solution in oleum, or by using brominated acid(s), or a mixture of brominated acids and unbrominated acids.The acids are, desirably, dissolved in the oleum before addition of the hydrazine derivative.

The preferred polymer has a ratio of R_1 radicals to R_2 radicals in the range 95/5 to 70/30, because in this range the drawn filament has extremely good mechanical properties. Most desirably, the bromine content of the polymer is at least 15% by weight.

Example 1: Terephthalic acid (166.2 parts) was dissolved in oleum (2,320 parts) of 35% strength at 85°C and then bromine (68.6 parts) was added dropwise over 4 hours to the stirred, heated solution which was protected from atmospheric moisture by a silica drying tube on top of a water condenser. After a further hour, when the bromine had reacted with the terephthalic acid in the heated stirred solution, the source of heat was removed and the reaction flask was flushed for a short time with compressed air to remove any residual traces of bromine.

Isophthalic acid (166.2 parts) and hydrazine sulfate (273.3 parts) were added to the solution of brominated terephthalic acid at room temperature and this reaction mixture was then heated for 4 hours at 130°C with stirring to give a viscous solution.

The resultant polymer solution was diluted to a polymer concentration of 8% by weight with 100% sulfuric acid at 100°C, and the diluted solution was used as a spinning dope. Polymer isolated from the dope had a bromine content of 16% and an inherent viscosity of 2.38.

Following degassing the dope was dry-jet wet spun at 110°C through a ten-hole spinneret (each hole being 200 μ diameter) with a one centimeter air gap into water at 12°C and wound up with a spin stretch factor of 1.01. The spinning temperature was chosen to be 10°C higher than the extrusion flow instability temperature. The spun filaments were washed free of sulfuric acid and dried. The inherent viscosity of the polymer of the filaments was 2.12.

The filaments were drawn between feed rolls and draw rolls while passing over and in contact with a grooved hot plate 46 cm in length. The temperature of maximum draw ratio was found to be 430°C and the maximum draw ratio was 18X. At 430°C, with a feed speed of 6 m/min and a draw ratio of 18X, the resulting draw filaments had the properties given in the table below.

Example 2 (Comparative): In this example the bromine is substituted in the 5-position in the m-phenylene radicals only, i.e., not in a position ortho to the heterocyclic ring.

The bromination, polymerization and spinning dope preparation were as described in Example 1, except that the isophthalic acid was dissolved in oleum and brominated, and the terephthalic acid, together with hydrazine sulfate, subsequently added. Polymer isolated from the spinning dope had a bromine content of 14.25% and an inherent viscosity of 2.14.

The spinning, washing and drying steps were as described in Example 1, except that the spinning temperature was 150°C and the spin stretch factor was 0.99. The inherent viscosity of the polymer of the spun filaments was 2.14.

The filaments were drawn between feed rolls and draw rolls while passing over and in contact with a grooved hot plate 46 cm in length. The temperature of maximum draw ratio was found to be 540°C and the maximum draw ratio was 5X. Drawing under these conditions, at a feed speed of 6 m/min, the resulting drawn filaments had the properties given in the table.

Example	1	2
Tenacity (g/decitex)	9.75	2.38
% Extension	6	11.8
Initial modulus (g/decitex/100% extension)	240	70

Dyeing

For many years, the textile industry has sought ways for producing yarns from continuous filaments such that the yarns have the characteristics of a spun yarn comprised of staple. Prior to the advent of synthetic filaments, all yarns were produced from staple products. Synthetic filaments, however, are made as continuous filaments and, in order to provide the desirable effects of staple products, a vast proportion of synthetic filament production is cut into staple length fibers. Such fibers are then twisted into yarns, called spun yarns.

Spun yarns have a particularly desirable characteristic of being somewhat fuzzy or hairy along their length giving them the desirable attributes of softness and cover and, when produced into fabrics, the ability to produce low density, porous, permeable and comfortable materials.

Continuous filament yarns also have many desirable attributes but they also have their limitations, particularly in respect to bulk, cover and comfort factors. Nevertheless, continuous filament yarns have replaced spun yarns in many end uses. Of course, it is obvious that if a continuous filament yarn could be made into a spun-like yarn, the otherwise expensive steps of cutting continuous fibers into staple followed by carding, coning and twisting into roving, followed by drafting and twisting further into yarns could be eliminated.

Many attempts have been made to accomplish this feat but various limitations in the resulting product have kept such continuous filament yarns from being complete replacements for spun yarns. In particular, previous methods, such as the very popular false twist texturing method for crimping continuous filament yarns to produce bulk and cover, have had their limitation in that the yarns always end up having a rather synthetic feel and look. This is probably due to the lack of the fuzzy and hairy projections which are present in spun yarns.

Another attribute staple yarns have is the ability to blend different fibers of different dyeability, such as polyester and wool or polyester and cotton, to produce a heather effect when dyed. Many suggestions have been made for combining filament yarns but such previous attempts have failed to effect an adequate blending of the fibers such that the yarns would equal the appearance of staple in fabric form. The lack of a total intimate blending results in an undesirable moire or plaiting appearance in the resulting fabric. The material in this chapter deals with improvements in the dyeing process for synthetic filaments, and with the resulting products.

POLYESTERS

Two Differently Dyeable Yarns Resulting in Heather Appearance

J.R. Talbot; U.S. Patent 4,164,117; August 14, 1979; assigned to Fiber Industries, Inc. provides a method for producing ingrain spun-like yarn simulating a blended staple yarn wherein continuous filament yarn is produced which has characteristics similar to blended staple spun yarns. Fabrics made from the yarns described herein provide a fine grained heather appearance without the normal moire or plaiting effect seen in previous ingrain continuous filament yarn fabrics.

The unusual continous filament yarn described herein is made by a texturing process which involves the combination of two or more differently dyeable continuous synthetic yarns which are false twist textured followed by overfeeding to an air bulking means.

The preferred method utilizes false twist texturing of synthetic filaments such as polyester, nylon, cellulose acetate or cellulose triacetate and mixtures thereof wherein two differently dyeable yarns are combined with an air bulking jet interposed to act on the false twist textured yarn while it has a high residual torque and subsequently decaying the torque.

Referring to Figure 11.1, a typical draw texturing schematic is shown wherein feed yarns **8** and **9** are withdrawn from packages **12** and **13** and combined to form yarn **10** prior to passing over feed roll **15** and across heat setting zone **19** and through twisting means **21**. Twisting means **21** rotates yarn **10** to a highly twisted state wherein the twist backs onto heating means **19** wherein the twist is set.

As the yarn is drawn through the twisting means **21** by draw rolls **17**, it is untwisted. The untwisted yarn is then fed through bulking means **22** in a substantial overfeed which is determined by the different speeds between draw rolls **17** and takeup rolls **27**. Preferably, prior to reaching takeup roll **27**, the yarn is heat set by heat means **23**. As the yarn passes from takeup rolls **27**, it is taken up on package **25** in the conventional manner.

Figure 11.1: Draw Texturing Schematic

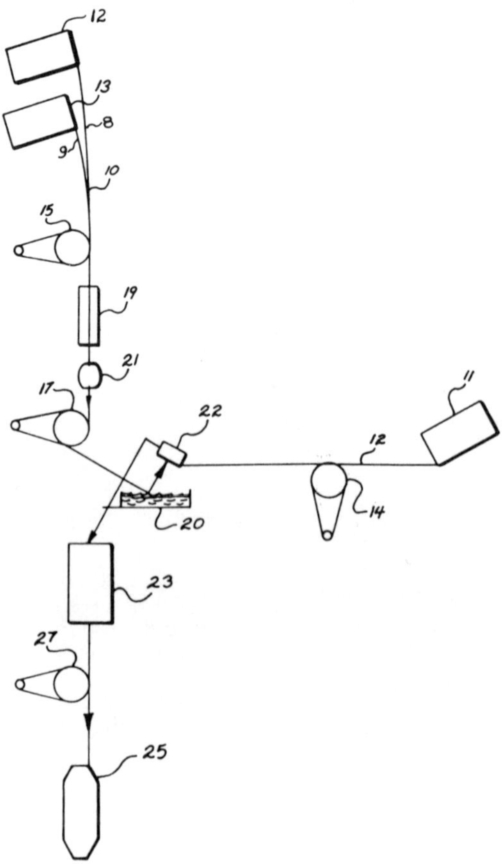

Source: U.S. Patent 4,164,117

Example: Ingrain yarns are made utilizing the apparatus as schematically shown in Figure 11.1. Various feedstocks are utilized as set forth hereinafter with the processing speeds and conditions as follows for the various feed yarns set forth hereinafter. Two or more feedstocks are combined prior to the feed roll, which is operated at 585 ft/min, passed across the first heater set at 200°C through the false twist twister to the draw roll which is operated at 1,000 ft/min, thus effecting a draw ratio of 1.71. The drawn yarn is then fed through a water bath to a texturing jet operated at 90 pounds per square inch gauge (psig) and an air flow of 3.4 standard cubic feet per minute and hence through a second heater operated at 260°C to the relax roll operating at 756 ft/min, representing a second heater overfeed of 32.2%. The yarn is taken up on a package at a takeup speed of 810 ft/min.

The feedstock for the process is 290 denier, 60 filament partially orientated polyethylene terephthalate round cross section plied with 70 denier, 9 filament, partially orientated polyethylene terephthalate yarn containing 2 weight percent of 5-sodiosulfoisophthalic acid, a cationic dye enhancing additive. The resulting ingrain yarn has potentially different dyeability due to the cationic ingredient and has more than 5 kinks per centimeter. This yarn comprises approximately 19% cationic dyeable polyester and 81% disperse dyeable polyester, which, when constructed into fabric and dyed, results in a fabric with a wool-like hand and feel and a fine grain heather appearance.

Difference in Dye Uptake Produced by Incomplete Drawing

D.E. Bosley; U.S. Patent 4,026,098; May 31, 1977; assigned to E.I. Du Pont de Nemours and Company has developed a process in which yarn of polyester filaments having a random distribution along the filament length of thick and thin sections, which differ in dye uptake, is produced by incompletely drawing a spin-oriented yarn, preferably one prepared by melt-spinning polyethylene terephthalate into yarn while withdrawing the yarn from the spinneret at a take-off speed of 3,000 to 4,000 yards per minute. The spin-oriented yarn is drawn on a hot draw pin of about 1.3 to 6 cm in diameter, using a draw ratio which is 82 to 95% of the draw ratio needed to fully draw the yarn. The incompletely drawn yarn can be false-twist textured under conditions conventionally used for fully drawn polyester yarn. The textured yarn provides attractive dyed fabrics due to darker shades of color in the thick filament sections.

Example: A 245 denier polyester yarn is prepared by melt-spinning at 288°C polyethylene terephthalate of 22 relative viscosity, using a spinneret having 34 round orifices (each orifice 9 mils diameter, 12 mils deep) and winding the filaments at 3,400 yd/min (3,110 m/min). The yarn has a tenacity of 2.35 gpd, a break elongation of 142% and boil-off shrinkage of 55%. This yarn is drawn 1.706X in a separate step on a Whitin RK drawwinder at 454 yd/min wind-up, the yarn taking 1 wrap around a 1.6" diameter draw pin heated to 90°C.

The yarn contains filaments having thick and thin sections. Each thick section is about ¼ to ½ inch (6 to 12 mm) long and has a cross section area which is about 1.8X that of adjacent sections along the length of the filament. The drawn yarn is textured on an ARCT FTFL 440B false-twist texturing machine using the following conditions: 0% 2nd overfeed, +14% third overfeed, 0% supplementary overfeed, and -4% takeup overfeed; 60 turns per inch (2,360 turns/meter); 247,323 rpm spindle speed; 210°C Ist heater; 230°C 2nd heater; and a cooling zone stringup. No melting occurs in the texturing zone. The yarn has good bulk.

The textured yarn is knit into a circular tubing on a Lawson FAK machine and dyed with Latyl Blue FLW (Disperse Blue 27) dye. The resulting fabric showed 20 to 30 deep-dyed streaks per square inch (3 to 5 per cm^2) each streak being about 0.1 to 0.2 inch (2 to 5 mm) long. Microscopic examination of the yarns in the deep dyed sections showed that there were thin filament sections that dyed lighter and thick filament sections that dyed deeper. The fabric is attractive.

Fluid-Induced Crystallization of Portions of Spin-Oriented Yarn

In a process developed by *D.E. Bosley; U.S. Patent 4,096,222; June 20, 1978; assigned to E.I. Du Pont de Nemours and Company,* polyester yarn having portions which differ in dye uptake, arranged along its length in a predetermined pattern, is produced by treating portions of a spin-oriented yarn with a fluid to induce crystallization and then drawing the yarn to provide a product having a break elongation appropriate for the intended use. The spin-oriented yarn is preferably prepared by melt-spinning polyethylene terephthalate into yarn while withdrawing the yarn from the spinneret at a take-off speed of 3,000 to 4,000 yd/min. Suitable treating fluids include acetone or methylene chloride at room temperature and water or air at elevated temperatures. The treated yarn can be draw-textured on false-twist texturing machines having heaters at temperatures above 200°C. The treated portions of the yarn dye deeper than the untreated portions to provide attractive effects in dyed fabrics.

Example: A 235 denier spin-oriented polyester yarn is prepared by melt spinning at 284°C, polyethylene terephthalate (RV = 20) using a spinneret having 34 round orifices (each orifice 0.28 mm wide, 0.51 mm deep) and winding the filaments at 3,107 mpm (3,398 ypm). The yarn is interlaced during its travel to the windup as shown in U.S. Patent 2,985,995 to a pin count (the length of yarn in inches that passes by probe 18 of U.S. Patent 3,290,932 before the probe is deflected about 1 mm; a force of about 8 g is required to deflect the probe) of 40 cm. The yarn has a birefringence of 0.038, a tenacity of 2.2 gpd, a break elongation of 120%, and a boil-off shrinkage of 55%.

The yarn is wound on a conventional 6 cm diameter, 28 cm long cylindrical paper tube to form a package.The yarn is wound on the package at a helix angle of 7°. The package weighs 4.5 kg, has an outside diameter of 17 cm and a length of 24.5 cm. The package is dipped into a bath of acetone to a depth of 0.5 cm for 2 seconds. It is then withdrawn from the acetone and allowed to dry in air. The treated yarn sections (deep-dying sections) have a force-to-draw of about 50 grams and the other sections have a force-to-draw of about 40 grams.

The treated spin-oriented yarn is draw-textured using a draw ratio of 1.60 on a 440 ARCT false-twist-texturing machine modified to draw the yarn in the texturing zone, using the following conditions: 59.5 m/min feed speed, 14% second overfeed, 0% supplementary overfeed, -4% takeup overfeed, 2,598 turns per meter twist; 247,323 rpm spindle speed; 210°C first heater; 230°C second heater; and a cooling zone stringup. No melting occurs in the texturing zone. The textured yarn is knit into a single-feed jersey on a Lawson-Hemphill FAK knitter and dyed with Latyl Blue FLW. The resulting fabric has an attractive variegated appearance.

Filaments of Different Melting Points and Different Dye Affinity

A twist-free texturing feed yarn of drawn thermoplastic filaments is prepared by *M.E. Mirhej; U.S. Patent 4,155,214; May 22, 1979; assigned to E.I. Du Pont de Nemours and Company* from a bundle (A) of filaments having a MP 255°-275°C and a bundle (B) of filaments having a melting point above 230°C which is 15° to 39°C lower than that of the filaments of bundle (A). The bundles differ in coloration or affinity for dyestuffs. The feed yarn is false-twist textured at a temperature 53° to 78°C lower than the melting point of the filaments of bundle (A) to produce a textured yarn which provides a heather of high contrast and low directionality (no color streaks) in fabric made from the yarn.

Filament compositions may be polyethylene terephthalate or poly[ethylene terephthalate/5-(sodium sulfo)isophthalate] or polyhexamethylene adipamide for bundle (A); and poly(ethylene terephthalate/glutarate) or poly(ethylene terephthalate/adipate) or poly(ethylene/2,2-dimethyl-1,3-propylene terephthalate) for bundle (B).

Contrast refers to the ease with which the eye can distinguish between different colors in a fabric. At maximum filament intermingling, the different colors will appear to blend and provide a substantially uniform color. U.S. Patent 3,593,513 disclosed a method of measuring the degree of filament intermingling (DFI). For the filament counts disclosed a DFI value of about 40% indicates no intermingling of the two types of filaments and 100% indicates the maximum obtainable filament intermingling. A high contrast heather corresponds to a DFI value of about 40 to 60%.

Example: A 70 denier bundle (A) of 17 drawn filaments of polyethylene terephthalate is combined with a 70 denier bundle (B) of 17 drawn filaments of poly(ethylene terephthalate/glutarate) having a mol ratio of 88.5 terephthalate: 11.5 glutarate and a MP of 238°C. The filaments of bundle (A) have a MP of 273°C and are colored black as extruded by incorporation of carbon black in the melt of polyethylene terephthalate. The two bundles are combined to form texturing feed yarns which are substantially free from twist or intermingling of the two types of filaments. The DFI value of the yarns is about 45%.

The feed yarns are textured on a Leesona 955 false-twist texturing machine operated at 300 ft/min (91.4 m/min) and 60 turns per inch (2,360 turns/meter) of inserted twist, using texturing temperatures of 190°, 205°, 210° and 220°C. Single-knit tubings are prepared from the yarns of a Model 68 Lawson-Hemphill Fiber Analysis Knitter fitted with a 3.5 inch (8.9 cm) head and operated at a feed rate setting of 4.5 (8:1 gear ratio). A high contrast heather with good crimp and good bulk is provided by the yarns textured at temperatures of 190°, 205° and 210°C.

At a texturing temperature of 220°C sufficient filament fusion occurs to render the fabric harsh and boardy. At 190°C the heather effect is undesirably directional. At 205°C reduced directionality results, and a twist entrapment frequency of 1 per 40 inches (100 cm) is observed. At 210°C a very fine, nondirectional, high contrast heather results. A twist entrapment frequency of 1 per 4 inches (10 cm) is observed. The DFI value of the textured yarns is about 50%.

Thick and Thin Sections with Differing Dyeability

According to *K. Nakagawa, Y. Otaki, M. Mineo, Y. Aratani, Y. Tatsuoka and H. Kusakabe; U.S. Patent 4,084,622; April 18, 1978; assigned to Toray Industries Inc., Japan* a textured polyester yarn having improved variation in dyeability along its length is composed of a plurality of individual filaments each having thick and thin sections randomly distributed along the filament axis, the yarn having an Uster evenness value of 1.0 to 10.0%, a T index of 3 to 30 and a variation in reflected light intensity in a continuous dyeing test of ±0.15 to ±0.80 and the yarn preferably having not-untwisted portions containing genuine twists of a fixed direction intermittently retained at random along its length.

The yarn is produced by false twisting a polyester yarn composed of a plurality of individual filaments each having variation in cross-sectional area along its length, in which the thick sections have a birefringence of 15 to 80 x 10^{-3} and the thin sections have a birefringence of 90 to 200 x 10^{-3}, at a temperature of not lower than 180°C and under a twisting tension of 0.05 to 0.8 g/den with a variation with respect to the average twisting tension of ±5 to ±20%.

Example: A thick and thin yarn was prepared by nonuniformly drawing at a draw ratio of 1.38 a polyethylene terephthalate undrawn yarn composed of 48 filaments having a birefringence of 41.6 x 10^{-3}. The individual filaments of the obtained thick and thin yarn had thick sections of a birefringence of 42 x 10^{-3} and thin sections of a birefringence of 108 x 10^{-3}, the ratio in cross-sectional area of the thickest sections to the thinnest sections was 1.7, the average fineness of the obtained yarn was 155 denier and the Uster evenness value of the yarn was 11.0% (determined by ASTM method D 1425-67).

The thick and thin yarn was then false twisted on a false-twisting machine at a speed of 400 m/min and at a temperature of 210°C. The false-twisting machine employed was of the type wherein a hot plate type heater having a length of 1.5 m and an external friction type twister capable of feeding the yarn were provided between a feed roller having a nip with an apron type roller and a delivery roller of the same construction as the feed roller. The twisting tension to which the yarn was subjected was 36 grams, i.e., 0.23 g/den, and the variation in twisting tension was ±3.5 grams, which corresponded to ±9.7% with respect to the average twisting tension. The number of twists imparted was 2,350 T/M and the broken filament count observed in the processed yarn was 6 per 2,000 meters.

The ratio of the cross-sectional area of the thickest sections to that of the thinnest sections of the individual filaments composing the false twisted yarn was about 1.7, the Uster evenness value of the yarn was 2.5% and the T index of the yarn was 11. To determine the T index, an Uster evenness curve is drawn on a standard chart marked off from 0 to 100% so that the portion of the curve corresponding to the thinnest sections of the yarn coincides with the 0% base line. Then, the area between the curve and the 0% base line with respect to a portion corresponding to a yarn length of 8 meters is determined. The T index of 1 corresponds to the area of 0.333 cm^2.

It is important that the yarn have a variation ($\Delta L/\bar{L}$) in reflected light intensity in a continuous dyeing test of preferably about ±0.20 to ±0.60. Where the variation in reflected light intensity is less than ±0.15, the yarn has too small a difference in color shades when dyed and where the variation in reflected light intensity is more than ±0.80, the yarn has too large a difference in color shades. Thus, in either case, the yarn can not produce a knitted or woven fabric capable of having a desirable color tone. The variation in dyeability of the yarn of this example was a $\Delta L/\bar{L}$ of ±0.33 and an $\bar{L}$ of 24% and thus, the yarn was proved to have sufficient variation in dyeability.

A fabric woven from this yarn was dyed using a dye having tendency to result in uneven dyeing. The dyed fabric had a favorable color tone of a sprinkly colored effect.

High Speed Spinning

According to *B.H. Knox; U.S. Patent 4,156,071; May 22, 1979; assigned to E.I. Du Pont de Nemours and Company* polyethylene terephthalate flat yarns and tows having properties closer to cellulose acetate than to conventional polyethylene terephthalate are prepared directly by spinning at speeds of about 4,000 m/min. Among the useful physical properties are a modulus of 30 to 65, which is relatively unaffected by boiling, low boil-off shrinkage, no need for heat setting, low shrinkage tension, large crystal size, and low amorphous orientation which is measured by a value termed "amorphous modulus." The flat yarns may be used, for example, in textile fabrics, without drawing, and may be modified by air-jet texturing. The tows may be converted into staple fiber.

The dyeability of various yarns is compared herein by measuring their disperse dye rate, DDR, which is defined hereby as the initial slope of a plot of percent dye in filament by weight versus the square root of dyeing time, and is a measure of a dye diffusion coefficient (if corrected for difference in surface to volume ratio). The values of the disperse dye rate are normalized to a round filament of 2.25 denier per filament having a density of 1.335 g/cm^3, i.e. of an amorphous 70-34 round filament yarn after being boiled, as a relative disperse dye rate, RDDR, defined by the relation:

$$RDDR = DDR\ \text{measured}\quad [(dpf/2.25)(1.335/\rho)(100/100\text{-}S)]^{1/2}$$

where ρ is the polymer density, dpf is the filament denier, and S is the boil-off shrinkage. The RDDR value is more or less independent of the surface to volume ratio of the dyed filaments, and reflects differences in filamentary structure affecting dye diffusion.

The modulus (M) and other tensile properties herein are measured on an Instron Tester TTB using 1" x 1" (about 2.5 x 2.5 cm) flat-faced jaw clamps (Instron Engineering Corporation) with a twister head made by Alfred Suter Company, using a 10 inch sample length and two turns of twist per inch (about 25 cm length and 8 turns twist per 10 cm) at a 60% per minute rate of extension at 65% relative humidity and 70°F (21°C).

The amorphous modulus (M_A) correlates with amorphous orientation, and is calculated using a normalized value of M (modulus of as-produced yarn)

$$M_n = (0.65/[\eta])^{0.3}M$$

from which is subtracted a normalized crystallinity factor or "X value"

$$[530(\rho-1.335)(0.65/[\eta])^{0.3}]$$

which is between 5 and 25. The range of 28 to 38 g/den for the amorphous modulus (M_A) provides suitable tactile aesthetics, similar to those of cellulose acetate filaments of similar denier. A low amorphous modulus is one of the factors related to improved dyeability (as measured by RDDR). Preferred yarns have a relatively low amorphous modulus within this range, preferably less than 36.5, especially less than 35 g/den.

The shrinkage values herein are generally boil-off shrinkages (S) and are measured by suspending a weight from a length of yarn to produce a 0.1 g/den load on the yarn and measuring its length (L_o). The weight is then removed and the yarn is immersed in boiling water for 30 minutes. The yarn is then removed, loaded again with the same weight, and its new length recorded (L_f). The percent shrinkage (S) is calculated by using the formula:

$$\text{Shrinkage (\%)} = 100\ (L_o-L_f)/L_o$$

A low shrinkage is highly desirable for most textile purposes.

The crystal size (CS) is estimated from the Scherrer formula $CS = K\lambda/\beta\cos\theta$ where K is taken to be unity; λ is 1.5418 A, the wavelength of CuK_α x-rays; θ is the Bragg angle of diffraction; β is line broadening corrected for instrumental broadening by $\beta^2 = B^2-b^2$ where B is the observed broadening and b is the instrumental broadening as measured on a ZnO pattern assuming infinitely large crystallites (all angle measurements in radians). B is measured on a photographic film pattern of a sample using the diffraction arc at $2\theta = 17.5°$ (the 010 diffraction), and is measured radially along the equator, i.e., at its maximum intensity.

The filaments preferably have crystal sizes that are related to the fiber density by the relation $CS \geqslant 1430(\rho-1.335)$A, and are preferably greater than about 50 A, especially greater than 60 A. Generally, the larger the crystal size, the better the tensile properties, about 90 A being a maximum that is likely to be attainable in practice.

Example: Polyethylene terephthalate of intrinsic viscosity 0.66 is spun on apparatus essentially as described above and illustrated in Figure 11.2 to form a 68 filament flat yarn of 1.02 denier per filament (round cross section) at a windup speed of 4,500 yd/min (4,115 m/min), using a spinneret block at 298°C and a pack pressure of 3,500 psig through spinneret capillaries of diameter (D) 9 mils and of length (L) 50 mils, the emerging filaments being protected by a hollow tube of length about 2 inches and then subjected to a radially-inwardly-directed flow of air at room temperature at a rate of 25 scfm. The solidified yarn contacts a finish roll, the finish being as described in Example 1 of U.S. Patent 3,859,122, and the yarn is interlaced and wound up, without any drawing step.

Figure 11.2: Schematic of High Speed Spinning Process

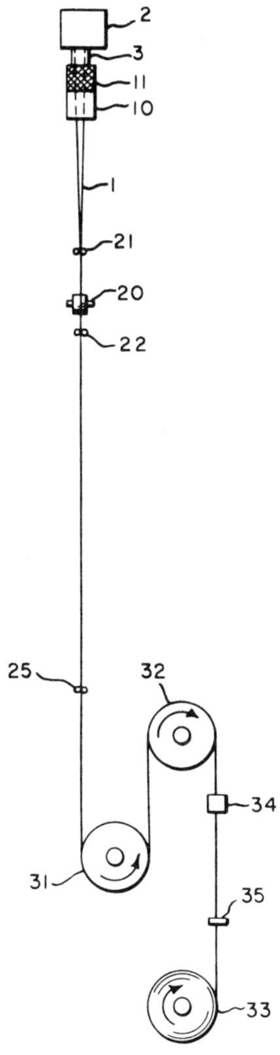

Source: U.S. Patent 4,156,071

It will be noted that the dyeability is good (RDDR of 0.1), the amorphous modulus (M_A) being 32.4 g/den, the modulus (M) is 51.4 g/den and the boil-off shrinkage is only 3.6%. The thermal stability is excellent as shown by a dry heat shrinkage after boil-off (S_2) of only 0.3%. The modulus after boil-off (M_2) is 54.5 g/den, so the difference ΔM between M and M_2 is only 3 g/den. The X value (difference M_n - M_A) is 19 g/den, i.e., between 5 and 25. The shrinkage modulus (M_S) is 3.22 g/den. The crystal size (CS) is 71 and the density of the polymer (ρ) is 1.3707, so CS $>$ 1430(ρ - 1.335), i.e., CS $>$ 50. The birefringence (Δ) is 0.0883.

CELLULOSICS

Regenerated Cellulosics Modified with Poly-2-Alkyl-2-Oxazoline

S.L. Miller and Y.J. Dickert; U.S. Patent 4,113,674; September 12, 1978; assigned to The Dow Chemical Company have developed a process in which modified, regenerated cellulosics having good receptivity to acid dyes are prepared by introducing a poly-2-alkyl-2-oxazoline, such as poly-2-ethyl-2-oxazoline, having a weight average molecular weight of at least about 10,000 into an aged, cellulose xanthate solution and forming modified, regenerated cellulosics therefrom. These cellulosics exhibit rapid and quantitative acid-dye absorption and good dye fastness.

Example: Three viscose rayon samples were prepared. The first sample was without a modifier (dye receptor). The second sample contained as a modifier 2.5% poly-2-ethyl-2-oxazoline (PEO), 92,000 molecular weight. The third sample contained as a modifier 2.5% PEO, 325,000 molecular weight. The PEO-modified samples were prepared by introducing a 20 weight percent aqueous solution of the modifier into a cellulose xanthate solution and forming 720 filament, 1,100 denier yarn therefrom.

A 6 foot length of each sample was placed in a 250 ml glass beaker and dried at 70°C for 6 hours in a constant-temperature oven. The samples were adjusted so that their weights were within 3.6 mg of each other and then wound around a 3 inch (o.d.) 10 mesh stainless steel screen cylinder with uniform tension being achieved throughout winding. The screen-bound samples were then immersed in a 3 liter beaker containing a prepared dye bath at ambient temperature. The prepared dye bath comprised Irgalan Blue FBL, a dye (0.1 g); Alkanol DN (Du Pont) (0.575 g), and $Na_3PO_4 \cdot 12H_2O$ (3,306 g) diluted to 2,300 ml volume with deionized water.

The continuously agitated dye bath was at ambient temperature at the instant of immersion and then gradually heated to boiling over a 40-45 min period and then maintained thereat for 1 hr. The screen-bound fibers were then removed and soaked in deionized water for 30 min. The dyed and washed fibers were then contacted with two 20 ml portions from a 40 ml mixture of dimethylformamide (30 ml) and deionized water (10 ml) to extract the dye therefrom. The dye extracts were then compared to standard dye concentration samples by using a UV-visible spectrophotometer:

Sample	Fiber	Fiber Weight (g)	Equivalent* Dye Uptake (g)
1	Unmodified rayon	0.1290	0
2	92,000 MW 2.5% PEO rayon	0.1254	6.20×10^{-4}
3	325,000 MW 2.5% PEO rayon	0.1270	7.76×10^{-4}

*Computed by converting absorbance to g/l dye concentration and then multiplying by 0.04 (conversion factor for liters to 40 ml).

Uniformly Dyed Water-Swellable Cellulosic Fibers

J. Dehnert, G. Lamm, W. Juenemann and G. Meyer; U.S. Patent 4,150,943; April 24, 1979; assigned to BASF AG, Germany have produced uniformly dyed water-swellable cellulosic fibers which are fast to washing, dry cleaning, sublimation and light by contacting water-swellable cellulose fibers sequentially or concomitantly with water, ethylene glycol or a derivative thereof and, while the fibers are still swollen, with an essentially water-insoluble dye of the formula

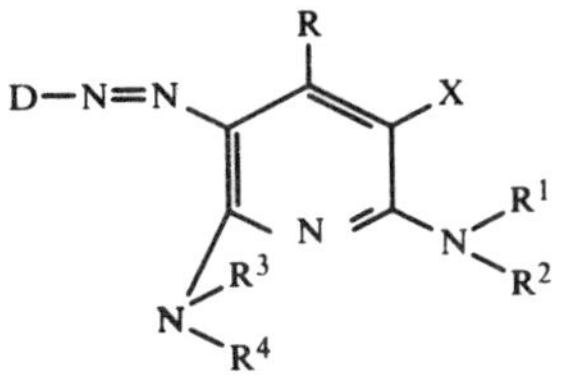

in which

- D is the radical of a diazo component;
- R is hydrogen, C_{1-7} alkyl, phenyl or phenyl bearing methyl as a substituent;
- X is cyano, carbamoyl or substituted carbamoyl;
- R_1 is hydrogen or an aliphatic, cycloaliphatic, araliphatic or aromatic radical;
- R_2 is hydrogen or an aliphatic, cycloaliphatic or araliphatic radical;
- R_3 and R_4 independently of one another have the same meanings as R_1 and R_2; and
- R_1—N—R_2 and R_3—N—R_4 independently of one another are saturated heterocyclic radicals.

Example: Cotton cloth is printed by a rotary screen printing method with an ink consisting of 10 parts of the dye of the formula:

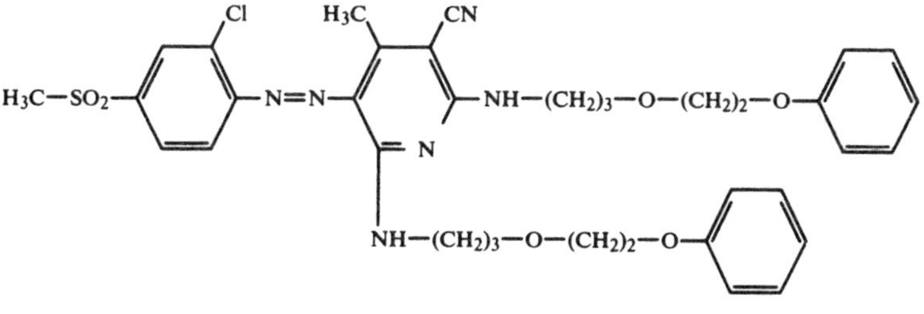

100 parts of polyethylene oxide of a molecular weight of 300 and 790 parts of a 3% alginate thickening and the print is dried at 100°C. The print is then treated for one minute at 200°C with hot air, rinsed cold, soaped at the boil, again rinsed cold and dried. A light-fast and wash-fast yellow-orange print is obtained on a white ground.

In a related patent, *J. Dehnert and G. Dunkelmann; U.S. Patent 4,157,893; June 12, 1979; assigned to BASF AG, Germany* describe dyes of the formula

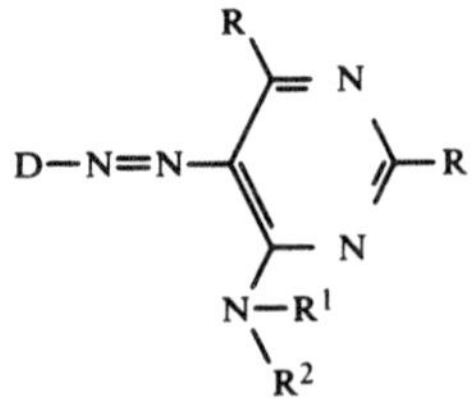

in which one R is R^1, OR^1, SR^1 or NR^1R^2 the other R is NR^1R^2. R^1 and R^2 independently of one another are hydrogen or an aliphatic, cycloaliphatic, araliphatic or aromatic radical, NR^1R^2 may also be a saturated heterocyclic radical, and D is the radical of a diazo component. Cotton and polyester-cotton fabrics dyed with the above exhibit very good fastness properties.

ACRYLICS

The processes which are known and usually used for obtaining acrylic fibers comprise passing a solution, called collodion, of homopolymers or copolymers of acrylonitrile in a suitable solvent through spinnerets pierced with very tiny holes usually having a diameter of approximately 5/100 of a millimeter and then eliminating the solvent to coagulate the polymer in the form of fibers or filaments.

Acrylic fibers may be colored by adding coloring matters to the collodion before spinning. If these matters have affinity for polymers, the operation is called "spin dyeing". To spin dye or color acrylic fibers, coloring matters soluble in the collodion or pigments completely insoluble in the collodion have been used.

Dyestuffs of intermediate solubility cannot be suitable because very often, in this case, the equilibrium between the soluble and insoluble fractions is unstable and its development considerably disturbs the course of spinning.

As is well known, serious problems arise from using pigments. One of these problems is that very few pigments are in fact completely insoluble in the spinning solvents or the coagulation bath. Another problem arises from the necessity of effecting very fine dispersion of the pigment in the collodion. These problems are avoided by the use of dyestuffs which are soluble in the collodion, the most frequently recommended ones being cationic dyestuffs or their salts.

Nitrate of a Cationic Dyestuff

F.R.D. Manier and P.L.A. Caullet; U.S. Patent 4,100,131; July 11, 1978; assigned to Produits Chimiques Ugine Kuhlmann, France provide a process for the spin dyeing of acrylic fiber obtained by spinning a collodion of a homopolymer or copolymer of acrylonitrile, in which the coloring agent comprises at least one nitrate of a cationic dyestuff.

Example 1: A stock solution is prepared in a container provided with a mixing device, this stock solution containing: 10 parts by weight of 4'-amino-3'-chloro-3-phenylazo-1,2-dimethylindazolium nitrate, 85 parts by weight of dimethylformamide and 5 parts of a compound obtained by copolymerizing 94% by weight of acrylonitrile and 6% by weight of methyl methacrylate.

With the aid of an arrangement currently used for the spin coloration of chemical fibers, comprising a metering device and another homogenizer for continuously mixing two liquids having different rates of flow and similar viscosity, 2 parts of the stock solution of dyestuff and 100 parts of a collodion comprising 80% of dimethylformamide and 20% of the same copolymer as that entering the composition of the stock solution of the dyestuff are homogenized.

This collodion which is thus colored is then extruded through a spinneret immersed in a coagulation bath containing 60% of dimethylformamide and 40% of water. The filaments thus obtained are drawn out and washed and then, according to conventional spinning conditions, they are subjected to one or several fixing treatments before being oiled and dried. They have a very bright and very intense scarlet shade. These filaments are dyed by 1% of the cationic dyestuff nitrate.

It is found that the coagulation bath and the baths of the subsequent treatments remain practically colorless, even after a prolonged spinning period. If the formate, acetate or phosphate of the same dyestuff cation are used, then a change of shade takes place. With the formate, bleeding occurs. The phosphate shows a lack of solubility.

Example 2: Under the same conditions as in Example 1, 10 parts of N-(trimethylammonio-ethyl)-N-ethyl-4'-amino-2'-methyl-4-nitro-2-chloroazobenzene nitrate are used. Fibers are obtained which have a very intense and very bright red coloration.

If the chloride of this same dyestuff cation is used, it is necessary to use 15 parts of dyestuff to obtain the same intensity.

Transparent Solutions Containing Anhydro or Carbinol Bases

M. Hähnke, T. Papenfuhs and W. Teige; U.S. Patent 4,020,037; April 26, 1977; assigned to Hoechst AG, Germany disclose a process for spin dyeing of acid-modified polymers or copolymers of acrylonitrile with basic dyestuffs wherein the spinning solutions of the polymers contain the basic dyestuffs in form of their free dyestuff bases as anhydro or carbinol bases. The process is conducted with completely transparent spinning solutions, avoiding clogging and mechanical abrasion of the spinning nozzles, and yields filaments and fibers which do not show matting effects but have a high tinctorial strength and good and improved fastness to rubbing, to light and to wet processing. As dyestuffs used there may especially be cited those which belong to the triarylmethane, indoldiarylmethane, azine, oxazine, thiazine, acridine or xanthene series. Such dyestuffs may be converted from their salts by reaction with alkalis such as for example, sodium hydroxide solution, potassium hydroxide solution, sodium carbonate or ammonia or other comparable strong organic amines into the carbinol or anhydro bases.

Example 1: 2.8 parts of a 10% solution of a dyestuff base corresponding to the formula

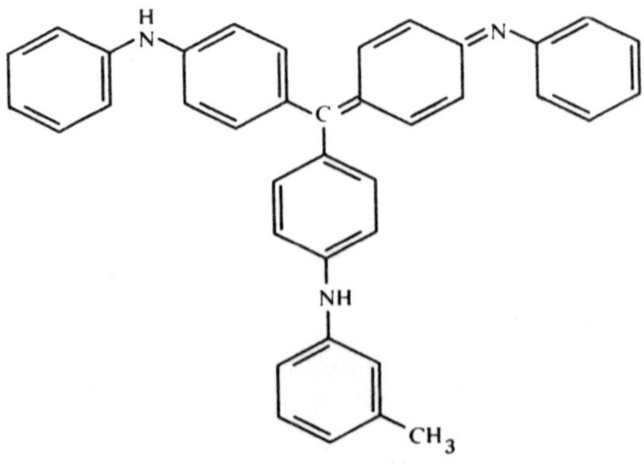

in dimethylformamide were added immediately before the spinning to 100 parts of a 28% solution of an acid-modified polyacrylonitrile in dimethylformamide by means of adequate dosing pumps. By means of a mechanically or statically operating mixer the solution to be spun was homogenized. Afterwards the solution was spun according to a dry-spinning process generally known and used in the art.

The abovementioned dyestuff base was obtained after two hours' boiling for a mixture consisting of 1 part of its hydrochloric acid compound, of an equivalent amount of sodium hydroxide and 5 parts of ethanol; and subsequently pouring the mixture into 10 parts of an ice water mixture and separating the precipitated dyestuff base by filtration.

The so-obtained fiber or filament material presents a brilliant reddish blue shade of a high color depth (intensity) and excels additionally by its very good fastnesses, especially to wet-processing, to rubbing, to sublimation and to cross-dyeing.

Example 2: 1.5 parts of a dyestuff base were dissolved in 770 parts of dimethylformamide at room temperature and under stirring. To this solution 300 parts of an acid-modified polyacrylonitrile were added which was dissolved under heating up to 80°C and under constant stirring. The dyestuff base had the following formula:

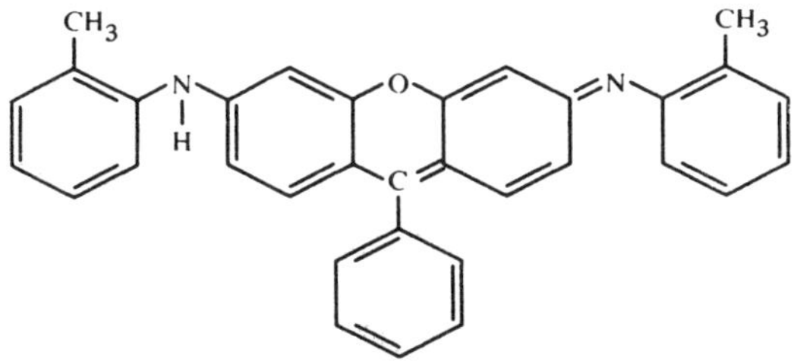

The so obtained red-violet spinning solution was filtered and spun according to a dry-spinning process generally known and used in the art.

The so obtained filament material presents an intense red-violet shade of high tinctorial strength and of excellent fastness properties especially of a very high fastness to light as well as of fastness to wet processing, to sublimation, to rubbing and to cross-dyeing.

Carbon Black Pigments Added Prior to Spinning and Dyeing

U. Reinehr, H.-D. Braun and R. Dippelhofer; U.S. Patent 4,087,494; May 2, 1978; Bayer AG, Germany describe a process for the production of dyed filaments or fibers of acrylonitrile polymers, wherein carbon black pigments are added to the polymer before the spinning process and the filaments containing carbon black pigments are dyed. By this process it is possible to save considerable amounts of dyestuff. The dyes used in the examples below are:

Dye A

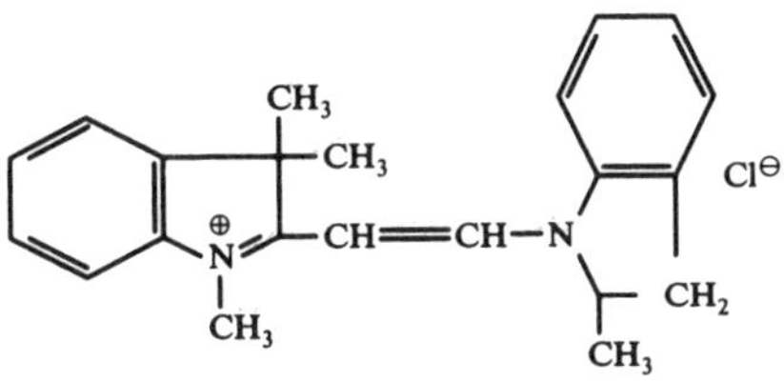

Dye B

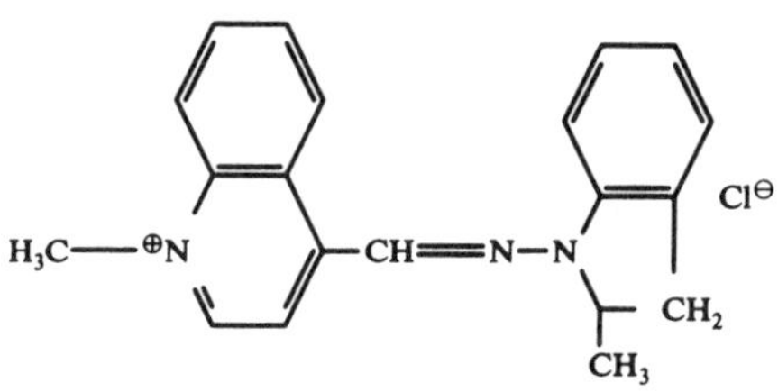

Dye C

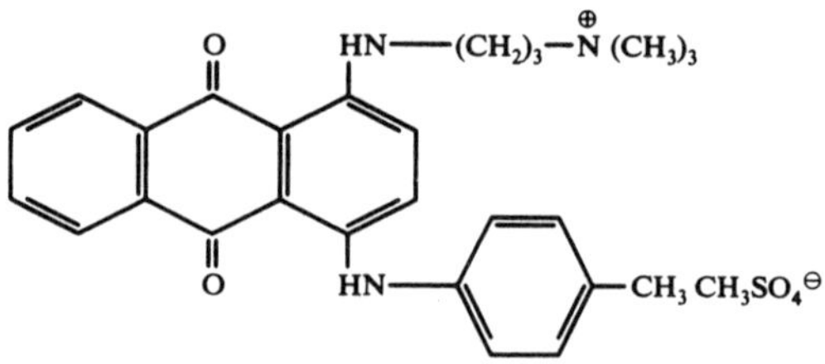

Example 1: An acrylonitrile copolymer of 91.1% of acrylonitrile, 5.5% of methyl acrylate and 3.4% of sodium methallyl sulfonate was dissolved in dimethyl formamide (DMF), followed by the addition with stirring of a 10% by weight solution of "Ravenschwarz 30" in DMF in such a quantity that the addition of carbon black pigment made up 0.01% by weight, based on the solids content of the polymer. The filtered spinning solution, which had a final concentration of approximately 29% by weight, was spun by the dry-spinning process. The spun filaments are subsequently further processed into a tow with a total denier of 960,000.

The tow is first passed through a tank filled with water at room temperature for conditioning and removing the solvent. The tank is provided with a pair of pressure rollers at its inlet and outlet ends. The moist, parallelized tow which is about 150 mm wide and has a moisture content of 65% is then squeezed out through a pair of pressure rollers (to give a residual moisture content of 40 to 42%) and passed under tension through a dryer. Depending upon the rate of travel of the tow, the temperature of the dryer is regulated in such a way that the tow leaves the dryer with a residual moisture content of approximately 5%.

For example, the temperature of the dryer is adjusted to 110°C for a rate of travel of the tow of 7 m/min and a drying capacity equivalent to 16 meters of tow with an overall denier of 960,000 dtex. The unstretched tow is then dyed at room temperature in a dye trough with a dye mixture of 0.1 g/l of a dye of formula A, 0.15 g/l of a dye of formula B and 1.1 g/l of a dye of formula C, in order to obtain a silver-grey dye finish. The residence time in the dye trough is about 3 to 4 seconds. The uptake of the dye solution amounts to 120%. The dyed tow is then fixed under saturated steam conditions at 102° to 103°C in a festoon-type steamer. The residence time in the steamer amounts to about 2 minutes for a rate of travel of the tow of 7 m/min.

The fixed, dyed tow is then stretched in a ratio of 1:3.6 in boiling water, washed and brightened. The dyed tow, which shows a shrinkage of 23 to 24% after stretching, is then subjected to a dry shrinkage treatment in a drying unit comprising 14 perforated drums. Shrinkage is carried out in stages of about 15% and about 5%, so that the tow leaves the drying unit with approximately 3 to 4% residual boiling-induced shrinkage. The rate of travel of the tow is approximately 25 m/min and the residence time approximately 2 min. The temperature at the input end of the drying unit is between 130° and 140°C and, at its output end, between 60° and 70°C. The tow dyed silver-grey is subsequently crimped and deposited in cartons.

Fiber yarns with a denier of 3.3 dtex produced from the dyed tow show thoroughly dyed dumbbell forms in the fiber cross section without any peripheral dyeing. Milliequivalents of acid groups/kg of polymer = 198. Fastness to light = 6.7; fastness to washing and perspiration = 5; fastness to rubbing in dry and wet form = 5; fastness to overdyeing: bleeding against polyacrylonitrile, neutral and in acetic acid = 5.

Fastness to light was determined in accordance with DIN 54 004, pages 68-72. Fastness to washing at 40°C was determined in accordance with DIN 54 014, pages 88-89. Fastness to perspiration was determined in accordance with DIN 54 020, pages 96-97. Fastness to rubbing (dry and wet) was determined in accordance with DIN 54 021, page 98. Fastness to overdyeing was determined in accordance with DIN 54 049, pages 139-140.

Example 2: 100 grams of dry polyacrylonitrile tow from Example 1 with an individual denier of 3.3 dtex and a tow weight of approximately 27 g/m are padded with the dye mixture (for types and concentrations of the dyes used, see Example 1) through squeezing rollers, and subsequently weighed. After padding, the material of Example 1 weighs 220 grams; in other words the dye absorption amounts to 120%.

POLYAMIDES

Crimping Fibers While Wet and Maintaining Moisture Level Before Dyeing

R.L. Provost; U.S. Patent 4,144,023; March 13, 1979; assigned to E.I. DuPont de Nemours has developed a process in which high strength, high modulus aromatic polyamide fibers are dyed by crimping the fibers while wet to at least 10 crimps per inch and maintaining at least 15% by weight moisture based on the dry fibers at all times before dyeing. K/S is a measure of apparent dye depth (visual color intensity) calculated from the equation:

$$K/S = (100-R)^2/200R$$

where R is the percent light (of wavelength corresponding to that of maximum absorption) reflected from the sample. In practice, R is measured on a carded pad of staple fibers weighing 2.5 to 3 grams in a "Colormaster Model V Differential Reflectometer" (Meeco Instruments) modified to permit rotation of the sample during the reading using an appropriate filter. The depth of color of the dyed fiber is approximately proportional to the K/S value, i.e., the larger the K/S value is, the deeper the shade of the fiber. A difference of 10% in K/S values is easily seen as a shade difference between fibers.

Example 1: A tow of continuous poly(p-phenylene terephthalamide) filaments (1.5 den/filament, tenacity/elongation initial modulus = 24 gpd/4.3%/540 gpd) obtained according to the process of U.S. Patent 3,767,756, containing 7% moisture is passed through a 21°C water bath and then into the sparger (steaming tube) of a steam stuffer-box crimper of the general type shown in U.S. Patent

2,747,233 and crimped to give 12 crimps per inch (4.7/cm). The crimped fibers are stored in sealed plastic bags and then scoured with aqueous detergent for 30 minutes at 82°C, drained, and rinsed with hot water. The wet fibers are dyed at 121°C under pressure for two hours in a 15/1 dye bath prepared from 4.4% by weight of Color Index Basic Green 6 (Astrazon Olive Green BL-200) and 0.6% by weight Color Index Basic Orange 22 (Astrazon Orange R 48046) (both based on the dry fiber weight). The dyes are pasted in water containing a small amount of glacial acetic acid, and the pH is adjusted to 2.0 to 2.5 with formic acid. The bath is cooled, the fibers are rinsed with hot water and scoured 30 min at 82°C, rinsed and dried. The dyed samples are cut to 3.8 cm staple, carded into pads and K/S values determined.

Example 2: Example 1 is repeated except additionally the crimped tow is sprayed with water as it leaves the crimper.

Example 3: Example 2 is repeated except the starting tow is flooded with water and piddled into a container and left in the sealed container for at least 6 hours before crimping.

Example 4: Example 1 is repeated except that the tow is not passed through the 21°C water bath before entering the crimper. The tow is crimped to 10 crimps per inch before dyeing.

The examples are summarized in the table. When the fibers of Example 3 are dried before dyeing, a K/S value of 3.9 or less is obtained.

The addition of sodium sulfate to the dye bath of Example 1 at a rate of 25 g/l of bath, and the addition of a steaming step of the dyed sample at 30 psig for 30 minutes before scouring increases the dye pick-up and retention so that smaller amounts of dye yield the same K/S values as in Example 1.

			. . . Crimped Fiber	
Example	Precrimp Treatment	Postcrimp Treatment	Moisture Before Dyeing (%)	Dyed K/S
1	Water bath	None	15	6.0
2	Water bath	Water spray	42	6.9
3	Water soaked	Water spray	32	8.3
4	None	None	8-10	3.9

Fixing Agent

A fixing agent for improving the fastness of dyes on polyamide fibers is disclosed by *A. Kobayashi, Y. Shimohiro, K. Shimizu and Y. Murakami; U.S. Patent 4,147,512; April 3, 1979; assigned to Dainippon Pharmaceutical Co., Ltd., Japan*. The fixing agent is a condensation product of 4,4'-dihydroxydiphenylsulfone, 4,4'-dihydroxydiphenylsulfonesulfonic acid and a lower aliphatic aldehyde, which has a mean molecular weight of about 5,000 to 30,000 and a ratio of carbon atom content and sulfur atom content of 3.0:1 to 4.8:1, or its metal salt, and when polyamide fibers are treated with the condensation product or its metal salt, or the condensation product in the presence of a salt of metal, at a temperature of 95° to 120°C, the fastness properties, especially wet-fastness properties of dyeings on polyamide fibers are remarkably improved.

Example 1: To 300 parts of acetic anhydride was added with agitation 250 parts of 4,4'-dihydroxydiphenylsulfone. Thereto was added with agitation 80 parts of a 98% by weight sulfuric acid and the mixture was heated at 120°C for 2 hours. The resultant was then mixed with 132 parts of 25% by weight of formalin and heated at 100°C for 4 hours to give 760 parts of a solution of the condensation product.

To the solution was added 2,300 parts of ethyl acetate and 985 parts of water. The mixture was agitated and allowed to stand so as to separate into three layers. After the separation was completed, the lower layer was taken out to give 540 parts of an aqueous solution (solid content: 35% by weight/volume).

The aqueous solution was passed through a column packed with 300 parts of Sephadex G-25, followed by the successive passing of 500 parts of water. The eluate was concentrated to dryness to give 126 parts of a powdery product.

The obtained powdery product has a mean molecular weight of about 8,000 and a ratio of carbon atom content and sulfur atom content of 4.5:1. The molecular weight was determined according to the gel filtration method employing Sephadex G-25 as a packing and acid-dyestuffs, C.I. acid orange 7, C.I. acid red 88 and C.I. acid red 114 as standard reagents. The ratio of carbon atom content and sulfur atom content was determined by the elementary analysis.

Example 2: 80 denier "Woolly" nylon yarn (made of nylon 6 fiber) was dipped at 20°C into a dyeing bath containing 2% owf of C.I. acid blue 129 and 3% owf of acetic acid (48% by weight) at a liquor-to-goods-ratio of 50:1. At the same time, the fixing agent obtained in Example 1 was added to the dyeing bath so as to reach to 2.5% owf. The dyeing bath was gradually heated to reach to a temperature of 100°C and maintained at the same temperature for 30 minutes. The treated yarn was then rinsed and dried.

The fastnesses to washing, to hot water, to perspiration and to water of the dyed nylon yarn after steam-set at 120°C for 60 seconds were determined according to the methods provided in JIS L 0844 A No. 2, JIS L 0845 No. 2, JIS L 0848 A and JIS L 0846 A, respectively.

The numerical values are graduations on the gray scale (provided in JIS L 0805) for determining a degree of staining on white cloths which are stained by the dyestuff fallen out from the dyed nylon yarns. The numerical value 5 shows the lowest staining and the numerical value 1 shows the highest staining. The yarn of this example had the following fastness values: washing, 4-5; hot water, 4-5; perspiration, 5; and water, 5.

ASSOCIATED APPARATUS

Dye Solid with Predetermined Surface Area

B.M. Phillips and D.R. Gregory; U.S. Patent 4,045,529; August 30, 1977; assigned to Eastman Kodak Company describe a process for making producer-colored fibers, yarns, films and related products by extruding from a source, such as from an extruder, a flow of polymer to a spinneret placing in the path of flow between the polymer flow and the spinneret a dye solid or pigment solid having a predetermined shaped surface area, and flowing the polymer into contact against the exposed surface of the dye solid or pigment solid, and by such contact dissolving at a predetermined rate a predetermined portion of the dye solid or pigment solid into the polymer flow and coloring the polymer flow for subsequent extrusion from the spinneret as colored filaments.

In reference to Figures 11.3a and 11.3b, the process may be practiced by providing space within a spinneret pack assembly **10** to receive a dissolvable dye cake **12.** The spinneret pack assembly comprises an outer pack wall **14,** a sand filter **16** disposed upon a support screen **18,** a breaker plate type support **20** defining a flow channel **22** for polymer flowing through the spinneret pack assembly and a spinneret plate **24** defining a plurality of orifices **26** through which polymer is extruded into filaments. A series of stationary mixer elements **18** may be disposed within the flow channel **22** for achieving a thorough mixing of the polymer with the dye, which is dissolved from the dye cake by the polymer flowing into contact with the exposed surfaces of the dye cake. Reference may be made to U.S. Patents 3,577,308, 3,182,965, and 3,806,097 as to the nature and operation of mixer elements in general. In brief, however, it is sufficient to say that the flow through each of the mixer elements is divided and redivided as many times as desired by the number of mixer elements positioned in the spinneret pack assembly and then recombined in a homogeneous colored flow for extrusion through the spinneret plate orifices **26.**

The dissolvable dye cake **12** has been designed so as to present a renewable constant surface area continuously exposed to the polymer flow as the dye cake gradually becomes dissolved. For purposes of explaining this constant surface area, Figure 11.3b shows the dye cake **12** as a circular solid **30** surrounded by an annular solid **32.** As the outer circumference of the circular solid **30** becomes decreased by dissolution into the polymer flow, the inner circumference of the annular solid **32** becomes increased by such dissolution in such manner that the same amount of surface area continues to be exposed to the polymer flow until the moment shortly before the dye cake becomes essentially dissolved or before collapse of one or the other solids **30, 32.** In order to assure that such collapse is deferred as long as possible, a series of supporting wires or support members **34** (Figure 11.3b) are positioned within the dye cake **12** for this purpose.

As previously mentioned, the dye that is employed must be compatible with the polymer in concern. The dye must also be capable of withstanding the temperatures that occur in the spinneret pack assembly. For example, in the use of a

Figure 11.3: Dissolving Dyes into Polymer Flow

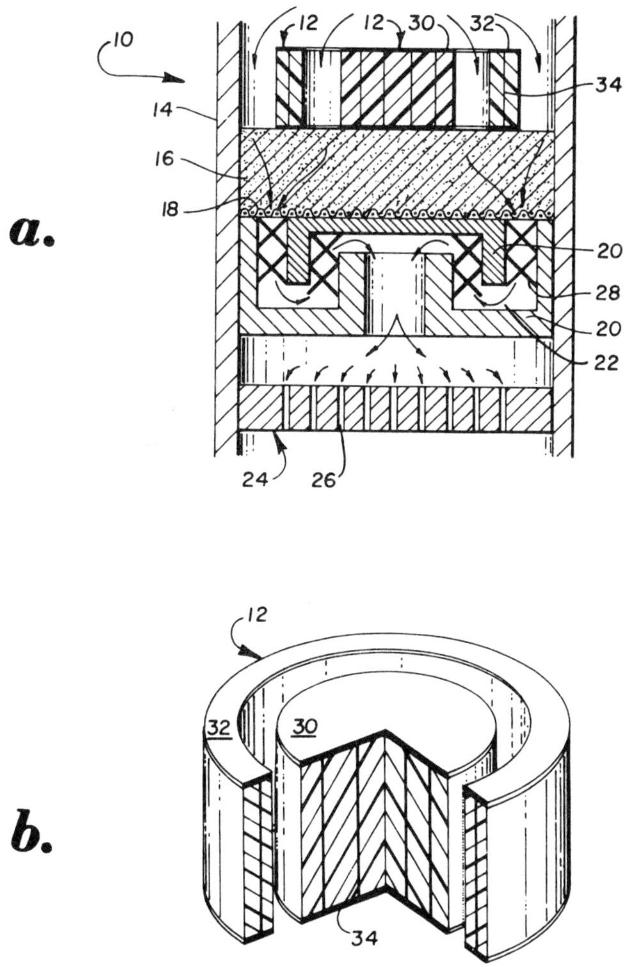

(a) Spinneret assembly showing dye cake, sand filter, mixer, and spinneret plate

(b) Enlarged isometric view, partly broken away and in cross section, of the dye cake

Source: U.S. Patent 4,045,529

polyester polymer such as poly(ethylene terephthalate), a blue dye from 4-(4'-beta-hydroxyethylanilino)-5-nitro-1,8-dihydroxyanthraquinone may be used. Also, a yellow dye from 4-[4'-(phenylazo)phenylazo] phenol may be used. These dyes may be used in the form of cakes within the spinneret pack assembly.

Example: A sand cavity of a spinneret pack assembly was half filled with 40/60 mesh sand, and the remainder of the cavity was filled with a granular blue dye of 5/10 mesh size. The dye had a melting point of 310°C and dissolved in molten poly(ethylene terephthalate) polymer of 0.60 I.V. (inherent viscosity) at 290°C at a rate sufficient to provide the desired concentration in the finished polyester fiber. It was observed that the particle size and the amount of dye could be changed to effect filtration as well as coloring. It was also observed in this example that the extruded polyester fibers contained decreasing amounts of dye with time because of reduction in bed depth and surface area of the dye granules with time.

It was also found, however, that a preformed dye cake, such as that disclosed in Figures 11.3a and 11.3b, could be designed such that for constant dissolution rates the surface area of the dye cake would remain constant with time. Ideas from solid racket propellant technology can be utilized in designs other than that shown in Figure 11.3b. This eliminates the potential problem of decreasing dye concentration with time, which occurred in the example of the dye granules in one-half of the sand cavity, as mentioned above.

The size of the dye cake is limited by the space provided within the spinneret pack assembly, therefore, arbitrarily it is decided that in reference to the dye cake shown in Figure 11.3b, the overall or total radius R_t, of the dye cake shall be 2.5 inches; the radius R_1 of the circular solid **30** shall be 1.125 inches; the radius R_2 to the inside diameter of the annular solid **32** shall be 1.375 inches and the gap width (between the two solids) shall be 0.25 inch.

The dye dissolution rate for this particular cake of 4-[4'-(phenylazo)phenylazo]-phenol is 5.04×10^{-4} lb dye/hr/in^2. Thus, the exposed dye cake area required for 1% dye dissolution is 9.6 lb/120 hr/area = 5.04×10^{-4}. Therefore, area = 9.6 lb/(120 hr)(5×10^{-4} lb/hr/in^2) = 158.6 in^2.

The length of the required cake is

$$L = \frac{158.6 \text{ in}^2}{2\pi(R_1 + R_2)} = \frac{158.6 \text{ in}^2}{2(3.14)(2.5 \text{ in})} = 10.1 \text{ in}$$

therefore the volume of the dye cake is

$$V = \pi(R_T^2 - R_2^2)L + \pi R_1^2 L$$

$$= 3.14(2.5^2 - 1.375^2)10.1 + 3.14(1.125)^2 10.1 = 178 \text{ in}^3$$

In summary of the above calculations: the dye dissolution rate must be determined for the particular dye and type of cake required; the shade desired is selected; the time of run must be selected; and then the length of the particular dye cake can be determined.

Wet Treatment of Endless Strands of Textile Material

M. Schuierer; U.S. Patent 4,165,548; August 28, 1979; assigned to Bruckner Apparatebau GmbH, Germany has developed a process in which an endless strand of textile material is initially dyed by transporting it spirally through successive compartments of a vat and simultaneously and separately circulating a homogeneous dye liquor through the compartments individually, and then the textile material is rinsed while continuing to transport it through the successive compartments, by causing a rinsing liquor to flow successively through the compartments in countercurrent to the movement of the textile material.

A wet-treatment apparatus is shown in Figures 11.4a, 11.4b, and 11.4c. This wet-treatment apparatus includes a wet-treatment vat **1** through which the textile material **2** to be treated with liquid is spirally transported in the form of an endless strand in the direction of the arrow **3**. The elongated vat **1** may be symmetrical in cross section, as shown in Figure 11.4b, and may comprise substantially straight, opposite housing side walls **4**, **5**, an arcuate base **6** and a correspondingly shaped cover **7**, all the above mentioned housing sections preferably merging smoothly with one another.

The cross section of the vat may of course also assume other forms, for example a cross-sectional form characterized by an inclined base at one end. The ends of the wet-treatment vat **1** are essentially formed by straight, flat walls **8, 9**.

According to one important aspect, the interior of the wet-treatment vat **1** is longitudinally divided into a number of wet-treatment compartments **10, 11, 12, 13, 14** and **15** (i.e., into six wet-treatment compartments in this case) by partitions **16, 17, 18, 19** and **20** which are sealingly built, preferably welded, into the wet-treatment vat **1** and which extend upwards over only part of the internal height of the vat from the base **6** thereof.

As can be seen from Figure 11.4c, the partitions **16** to **20** are arranged substantially at right angles to the longitudinal axis **21** of the vat, being separated at substantially equal intervals from one another and from the adjacent side walls **8** and **9** of the vat **1** and extending parellel thereto, so that in this case all the wet-treatment compartments **10** to **15** are equal in size.

Above the individual wet-treatment compartments **10** to **15**, there are guide means and transfer means for the textile material **2** so that the textile material can be guided and transferred from one compartment to the following compartment, the last compartment **15** through which the textile material is transported (**3**) communicating with the first compartment **10**, for example through a de-

Figure 11.4: Wet Treatment of Endless Strands of Textile Material

a.

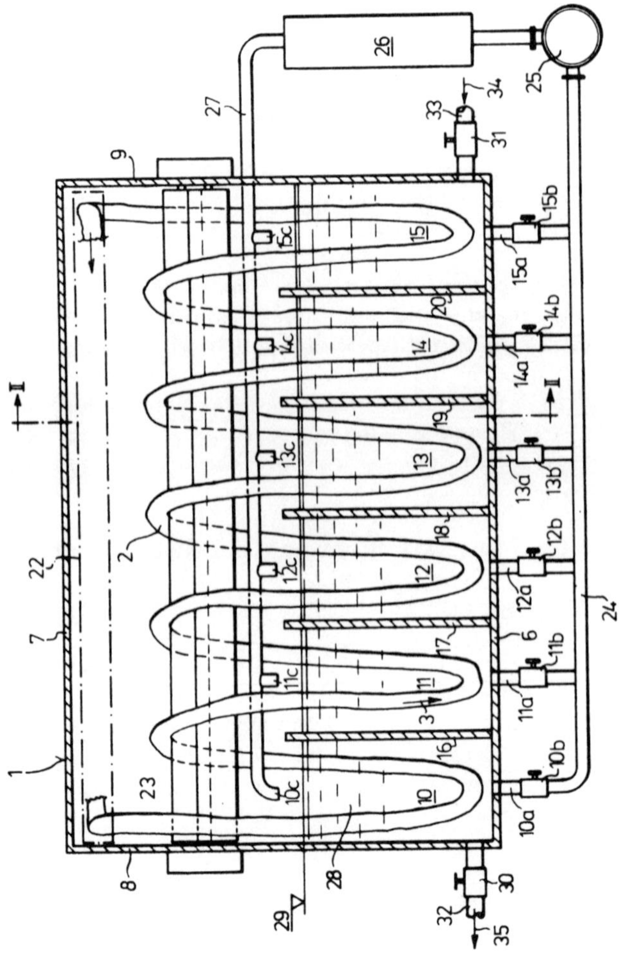

Source: U.S. Patent 4,165,548

(continued)

Figure 11.4: (continued)

b.

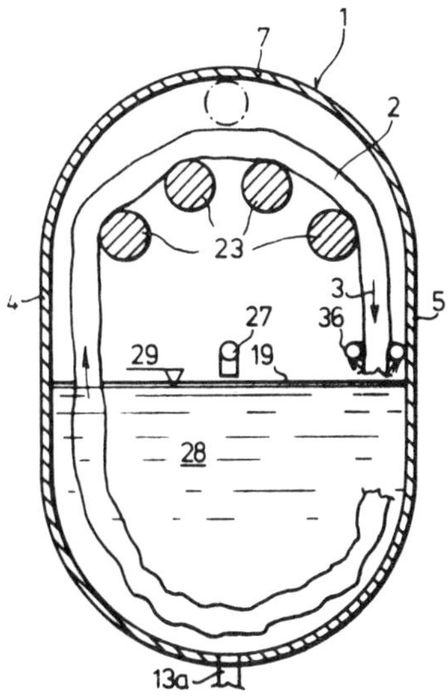

c.

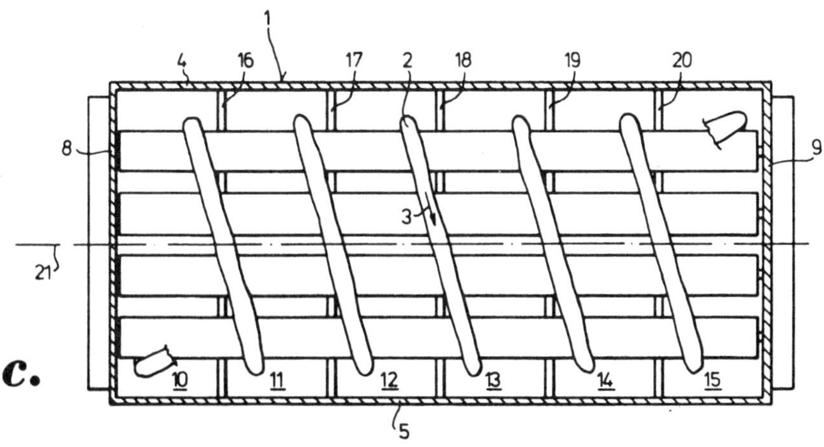

(a) A highly simplified longitudinal section

(b) A section through the wet treatment vat of the apparatus shown (on the line **II—II**) of Figure 11.4a

(c) A horizontal section through the upper portion of the wet-treatment vat shown in Figure 11.4a

Source: U.S. Patent 4,165,548

livery passage **22** shown as dash-dot lines (see Figure 11.4a), so that the textile material **2** can be returned from one end of the wet-treatment vat **1** to the opposite end, particularly during the dyeing process.

In the embodiment shown, the guide and transfer means are formed by several guide rollers **23**, at least some of which are also used for mechanically transporting the textile material. In addition, at least one of these guide rollers could of course also be in the form of a standard winch. The number and size of the guide rollers and winches will generally be determined by the type of textile material to be treated and by the width of the vat **1**.

In the embodiment illustrated in Figure 11.4b, the endless textile material is guided through the individual wet-treatment compartments **10** to **15** freely dependent (it could also be temporarily deposited in these compartments). In some cases, however, it would also be possible to use additional guide means in the form of guide rollers.

Whereas, in the embodiment shown in Figures 11.4a to 11.4c, only one side of the textile material lies on the guide rollers **23**, it would also be possible to associate a counter roller (not shown) with at least one guide roller in order to obtain greater adhesion and hence to ensure positive transport of the textile material. By providing a counter roller of the type in question, as known per se, it would also be possible mechanically to remove surplus water by squeezing from the section of textile material ascending from the associated wet-treatment compartment.

In this embodiment, the guide and transport rollers **23** for the textile material preferably extend over the entire length of the wet-treatment vat, so that they can be driven by a common drive (not shown).

For each of the wet-treatment compartments **10** to **15**, the base of the vat comprises a pipe connection **10a** to **15a** serving as outlets for the dye liquor, a shut off valve **10b, 11b, 12b, 13b, 14b,** and **15b** being provided in each of these pipe connections. All the pipe connections **10a** to **15a** communicate through a collecting pipe **24** with a common delivery pump **25** for the dye liquor, which at its output or delivery end is connected to a common heat exchanger **26**. From the heat exchanger, a distributing pipe **27** extends longitudinally through the wet-treatment vat above the individual wet-treatment compartments **10** to **15**. The distributing pipe has outlets **10c** to **15c** for each of the wet-treatment compartments **10** to **15**, these outlets representing the inlets for the dye liquor into the compartments.

Accordingly, each wet-treatment compartment **10** to **15** may be individually supplied with dye liquor and the dye liquor may be separately run off from each of the wet-treatment compartments at their lower ends. In the individual wet-treatment compartments **10** to **15**, a bath of dye liquor **28** is maintained at a suitable level **29** which lies below the upper edge of the partitions **16** to **20** (for example known level indicators, not shown, may be used for this purpose).

In this way, each individual wet-treatment compartment **10** to **15** is continuously supplied with fresh dye liquor from a common source so that the particular bath composition may always be kept homogeneous throughout.

Since, in this embodiment, the partitions **16** to **20** extend over only a certain part of the height of the vat from its base, the individual wet-treatment compartments **11** to **15** communicate freely with one another through the upper vat space, so that the spirally transported textile material can be guided and transferred without interference in the upper part of the vat.

It can also be seen from Figure 11.4a that pipe connections **32, 33** closeable by valves **30, 31** are provided on the end walls **8** and **9** of the wet-treatment vat (preferably near its base) for the introduction and removal of rinsing liquor. In the present case, it is assumed that the textile material is removed from the vat after the wet treatment, for example in the vicinity of the right-hand end wall **9** of the vat. In this case, the rinsing liquor is delivered through the pipe connection **33** provided on the end wall **9**, as indicated by the arrow **34**, while the spent rinsing liquor is removed through the pipe connection **32** provided on the opposite end wall **8**, as indicated by the arrow **35**.

In addition, changeover means (not shown in detail) are provided, by which, manually or automatically, the dye liquor is run off from the individual wet-treatment compartments **10** to **15** on completion of the dyeing process, the shut-off valves **10b** to **15b** are closed and the supply of dye liquor to the individual wet-treatment compartments is shut off, while the means for the removal of rinsing liquor through the pipe connections **32** and **33** is switched on.

Finally, this embodiment may comprise means for additionally introducing dye liquor onto the textile material above the wet-treatment compartments (or even only over a few certain wet-treatment compartments) in the region preceding the point at which the textile material re-enters a wet-treatment compartment. As shown in Figure 11.4b, these means for additionally introducing dye liquor may be formed by a spray system **36** which comprises spray nozzles and which is arranged in the region between the dye liquor level **29** and the last guide roller **23** over which the textile material is transported (arrow **3**). If desired, the spray system **36** may also be substantially annular or partly annular in configuration, so that it may also serve as another guide element for the textile material.

The preceding description of the apparatus also provides a clear idea of the process. In this process, the textile material is introduced in the usual way into the wet-treatment vat in which it is first dyed and then rinsed, i.e., in one and the same vat. During the dyeing process, the textile material sewn endlessly together in strand form is transported spirally in the direction of the arrow **3** through the adjoining, separate wet-treatment compartments **10** to **15**. This operation may be repeatedly carried out in the required manner by the textile material being returned from the last wet-treatment compartment **15** of the vat through the delivery passage **22** to the first wet-treatment compartment **10**. While the textile material is being transported in this way through the vat or successively through its wet-treatment compartments **10** to **15**, the wet treatment compartments are kept filled with dye liquor **28** to the required level **29**, the dye liquor of each individual wet-treatment compartment **10** to **15** being circulated through the separate inlets and outlets **10c** to **15c** and **10a** to **15a** for the dye liquor by the common delivery pump **25** and the common heat exchanger **26** so that the composition of the dye bath is always kept homogeneous.

As already mentioned, the dye liquor is run off from the individual wet-treatment compartments and the supply of dye liquor is switched off on completion of the dyeing process, and rinsing liquor or rinsing water is delivered through the wet-treatment vat via the pipe connection **33** in countercurrent to the direction in which the textile material is transported, as indicated by the arrow **3**. Accordingly, the rinsing liquor is initially pumped into the wet-treatment compartment **15** from below, flows over the upper edge of the partition **20** into the following compartment **14** and so on until it is run off from the opposite compartment **10** through the pipe connection **32**.

This rinsing operation may be carried out by initially transporting the textile material, still in the form of an endless strand as in the dyeing process, through the wet-treatment vat, then severing the strand of textile material after a reasonable time and removing it from the vat in the region of its end wall **9**, or, in the case of material which is fairly easy to rinse, by severing the strand of textile material at the beginning of the rinsing process, transporting it spirally towards the end wall **9** and removing it in the region of the end wall.

As already mentioned in connection with the transport system for the textile material, surplus water may be removed from the section of textile material ascending from a wet-treatment compartment (above the dye liquor level **29**), not only mechanically (by squeezing) but also by means of a compressed-air nozzle system.

For particularly intensive dyeing, it can be of advantage additionally to supply dye liquor to the textile material before it re-enters the corresponding wet-treatment compartment, which in the embodiment shown in Figure 11.4b may be done by means of the spray system **36**.

Reduction of Frosting Effect

In the continuous dyeing of textile fabrics made of staple fibers, an effect is found in most cases which is labelled "frosting effect". This is a markedly reduced dyeing of the individual fiber ends (pile ends) protruding from the fabric surface, which effect is caused above all by a dyestuff migration in the intermediate drying of the fabrics. The "frosting" may occur on fabrics made of synthetic fibers as well as on those made of natural fibers or fiber mixtures. The frosting effect is seen most clearly in the case of polyester fibers and spun rayon, and the problem of frosting is faced in practice predominantly in the continuous dyeing of mixed fabrics made of polyester fibers and spun rayon.

W. Birke, H.-U. von der Eltz and F. Schön; U.S. Patent 4,178,705; December 18, 1979; assigned to Hoechst AG, Germany disclose a process for reducing the frosting effect in the dyeing of textile fabrics made of staple fibers according to a continuous operation with intermediate drying, which comprises flattening the fiber ends protruding from the material web, between impregnating and drying the textile material, by the action of mechanical means onto the surface of the material web and holding them tight to the web surface by ironing, and a device for carrying out the process.

Figure 11.5: Apparatus for Reducing Frosting Effect

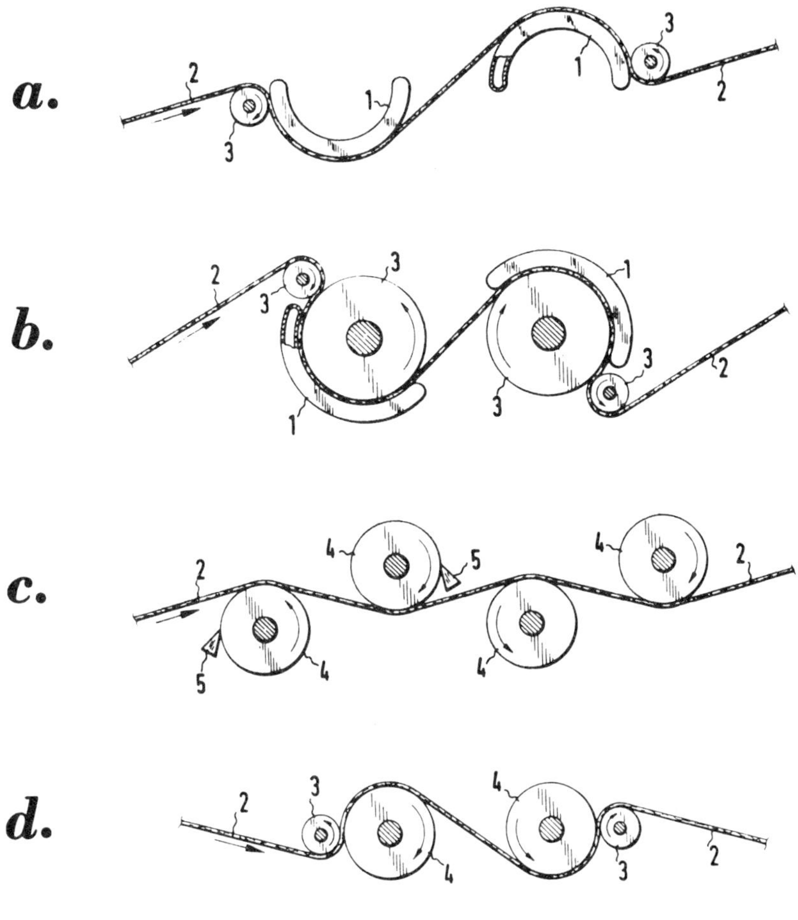

(a)(b) Side-view embodiments of apparatus having stationary curved surfaces

(c)(d) Side-view embodiments of apparatus having rotating rollers

Source: U.S. Patent 4,178,705

In its simplest variant, the device according to Figures 11.5a and 11.5b consists of stationary elements, especially vaulted plates **1**, which may also be designed as oval or cylinder-shaped hollow parts (not shown) and over which the material web **2** is guided in a sliding manner. Guide elements **3** for the material, for example guide rollers, ensure that the textile material in its open width remains in a sliding contact with the surface of **1**.

As movable sliding elements there are mentioned rotating rollers **4** in the first place. An arrangement of this kind will suitably comprise at least two rollers. Depending on the quality of the goods and the desired effect, the material may be guided round the rollers **4** with less (Figure 11.5c) or more (Figure 11.5d) contact.

Apart from special fabrics and/or specially desired oneside effects, the device has been designed in such a manner that both sides of the fabric may be treated in the same way, i.e., the vaulted surfaces and/or rollers are arranged on either side of the material web.

The treatment of textile materials with very high deposits of liquid is effected preferably by means of movable devices instead of stationary ones. In this case it may also be advantageous if doctor blades **5** or similar appliances are installed at the rotating rollers, which will strip off a liquid film possibly adhering to the roller surfaces.

Polymer Recovery

REGENERATED CELLULOSE

Filtration of Cellulosic Solutions

Cellulosic solutions are filtered by a process developed by *E.M. LaPolla, C.F. Murphy and A. Sarkisian; U.S. Patent 4,018,681; April 19, 1977; assigned to International Telephone and Telegraph Corporation,* enabling virtual complete recovery and recycling of the undissolved cellulosic components of the solutions. The process comprises filtering the solution through a bed of particulate material until the build-up of undissolved material impedes the further flow of solution through the bed, mixing the undissolved material and particulate bed with a solvent to partially dissolve the undissolved material, which is highly concentrated relative to its original concentration before filtration, adding a solubilizing reactant to the concentrated partial solution of undissolved material to complete the solution of the cellulosic residues, and separating and recycling the resulting solution.

Example 1: A stainless steel powder bed was prepared of 1¾ inch depth and five inches diameter. The bottom one inch of the bed was of -40 to +70 U.S. standard mesh size. The top ¾ inch of the bed was -20 to +40 mesh size stainless steel powder. The filtration was carried out in a sealed pressure vessel using compressed air at 40 psig to maintain a constant feed pressure. The viscose solution had a composition of 9% regenerable cellulose (based on total solution), 6.0% NaOH (based on total solution) and 28% CS_2 (based on cellulose content of solution).

After 235 kg of viscose solution passed through the system with only partial plugging (i.e., the filtration rate had dropped about 10 to 30%), the bed was digested in about one liter of 9.0% NaOH containing a CS_2 addition of 5.0 cc per 100 cc of NaOH solution amounting to about 250% CS_2 based on the estimated amount of undissolved cellulose trapped in the bed. After a one-hour treatment, the digestion liquor was decanted from the bed and used as part of the make-up caustic for the next batch of viscose to be filtered through the same stainless steel bed.

After decantation, the bed was remounted and prepared for reuse. The cycle was repeated four times, each filtration cycle consisting of two batches totaling 200 to 240 kg (190 to 227 liters) of viscose. The second and fourth cycles used 2 kg of digested washings from the previous cycle.

In order to determine plugging values, separate, but identical, solutions were filtered through a stainless steel bed and through a standard viscose filter medium of one layer of double faced 2 ounce cotton and one supporting layer of 6 ounces canton flannel. The filtered solutions were then refiltered through the standard medium until plugging of the filter occurred. The plugging values obtained for the particulate bed filtration were at least equal to, and generally better than, the conventional filter medium. Moreover, filtration efficiency was not decreased as a result of the recycle procedure of reusing the particulate bed or of reclaiming the undissolved cellulosic particles and introducing them back into the viscose stream as part of the make-up caustic.

Example 2: The ultimate test of filtration quality is its effect on the spinning of viscose fibers. A series of 4-liter pots of viscose was taken from a single 30-gallon batch and each was processed with a different filtration method, i.e., one was filtered through nylon felt/cotton batting/cotton flannel, another through cotton batting/cotton flannel, and a third was filtered through a powdered stainless steel bed of –40 to +70 mesh as in Example 1.

In each case, the viscose solution had a composition of a type used for tire yarn, 7% regenerable cellulose (on total solution), 6% NaOH (on total solution) and 34% CS_2 (on cellulose). The results indicated that the fatigue life, cord strength and iron pick-up of the sample filtered through the stainless steel bed, in accordance with Example 1, were equal to or better than that of control samples filtered in conventional fashion nylon felt/cotton batting/cotton flannel or cotton batting/cotton flannel. And, of course, such results were achieved with substantially no waste effluent.

It has been found that the process utilizing a stainless steel powder bed of 5 inches diameter and 1¾ inch depth, achieves an average flow rate at 40 psig operating pressure of 19.2 gal/ft^2/hr. This flow rate is over four times the flow rate through conventional viscose filter presses (2.2-4.4 gal/ft^2/hr as reported in the literature) for operating pressures several times greater, 100-150 psig.

The process thus provides a number of advantages over conventional filtration processes including longer operating cycles, less maintenance and shut-down time, avoidance of pollution by recovery of entrapped viscose and undissolved gels and capability of full automation of the filtration bed thus eliminating filter disposal and reducing cost of maintenance of filter media. It also makes possible the potential use of economical pulp grades in compositions which cannot be tolerated by conventional filter systems.

Formation of Soluble Cellulose Nitrite Ester

A regenerated cellulosic product is produced by *N.A. Portnoy; U.S. Patent 4,044,090; August 23, 1977; assigned to International Telephone and Telegraph Corporation* by nitrosating and dissolving a cellulosic material in a solution comprising a dialkylacylamide solvent and adding from 2 to 25% by weight, based on the weight of the solution, of a tertiary amine to form a soluble cellulose nitrite ester, contacting the cellulose ester with a coagulant therefore, stretching the coagulated cellulose ester prior to complete regeneration thereof and then completing the regeneration of the cellulose to produce a regenerated cellulosic product.

Example 1: A cellulose solution was prepared from 160 parts of aged prehydrolyzed kraft pulp (DP 485), 300 parts liquid N_2O_4 and 1,540 parts dimethyl formamide (DMF) keeping the temperature 10°-30°C, preferably 20°-25°C. To this was added 450 parts (18.4%) of pyridine keeping a solution temperature of 20°-40°C, preferably 20°-30°C, to make a clear, honey colored solution. This was deaerated and vertically spun using a glass 300 hole (0.0025 inch) spinnerette. The fiber tow was passed through a primary bath at 22°C of 900 parts by volume DMF, 60 parts by volume water, 120 parts by volume 70% HNO_3 and 200 parts by weight $Ca(NO_3)_2 \cdot 4H_2O$. The secondary bath was water at 22°C. The tertiary bath was 5% aqueous HNO_3 at 22°C.

Following the final bath, the fibers were treated with water and 5% sodium bicarbonate, both at room temperature for 10-20 minutes and then finished with 0.3% of a lubricating finish prior to air drying. Godet stretch was 57%. (As used in this and the following examples, godet stretch is the difference between the circumferential speed of the initial and final godet, divided by the speed of the initial godet, times 100.) The fiber properties are shown in the table below.

Example 2: The same procedure as in Example 1 was followed except that 290 parts (12.6%) pyridine was added to the cellulose solution. The primary bath was 850 ml methanol containing 25 ml of 70% aqueous HNO_3 at 22°C. The secondary bath was water at 22°C and the tertiary bath was an aqueous 5% HNO_3 solution at 22°C. Godet stretch was 64%. Fiber properties are shown in the table.

Example 3: The same procedure was used as in Example 1 except that pyridine was added to the primary bath rather than the cellulose solution. The primary bath was composed of 30 parts of pyridine and 850 parts of methanol. Secondary bath was water at 22°C, tertiary bath was 3% aqueous solution of HNO_3 at 22°C. Godet stretch was 83%. Fiber properties are shown below.

	Example 1	Example 2	Example 3
Denier	1.43	1.39	1.15
Tenacity, g/den			
Conditioned	2.22	2.31	3.23
Wet	0.82	1.25	1.75
Elongation, %			
Conditioned	10.97	11.01	5.88
Wet	27.47	27.66	12.63
Wet tenacity, at 5% elongation, g/den	0.23	0.44	0.80
Toughness, g/d*	24.35	25.43	18.99

*Conditioned tenacity x cond. elongation (work to rupture).

The table shows that both the addition of pyridine to the cellulose solution improves elongation and toughness over comparable solutions without pyridine and that the use of pyridine in the spin bath (rather than in the cellulose solution) does not produce an equivalent improvement.

Improved Resistance to Alkaline Solubility

Regenerated cellulosic articles are produced by *R.B. Hammer, A.F. Turbak, N.A. Portnoy and A.C. West; U.S. Patent 4,076,932; Feb. 28, 1978; assigned to International Telephone and Telegraph Corporation* by nitrosating and dissolving cellulose in a dialkylacylamide solvent, such as dimethylformamide, and regenerating the dissolved cellulose by passing into a coagulant bath. The coagulant comprises water or a 1 to 4 carbon atom aliphatic monohydric alcohol and a soluble reactive base, the latter in an amount ranging from 6 to 25% by weight of the regenerant. The process produces regenerated cellulose products of improved resistance to alkaline solubility.

Example 1: A bleached, prehydrolyzed kraft pulp was alkaline aged to a degree of polymerization (DP) level of about 450, neutralized with 10% acetic acid, washed with water, acetone and air dried. The pulp was cut with a rotary cutter to hasten dissolution.

An 8/15/77 cellulose/N_2O_4/DMF solution was prepared by charging 160 parts of the alkali aged prehydrolyzed kraft pulp (DP 450) and 1,540 parts of dimethylformamide (DMF) into a 2-liter four neck resin reaction flask equipped with a stainless-steel mechanical stirrer, thermometer, and a 250 ml equalizing pressure addition funnel. The resulting slurry was stirred and cooled to below +20°C, preferably between -5°C and +10°C, while 300 parts of liquid nitrogen tetroxide (N_2O_4) was added dropwise over ca 60 minutes. The temperature of the resulting exothermic reaction was maintained below 20°C preferably in the range previously specified during N_2O_4 addition and for the duration of the remaining dissolution process.

All solutions were observed microscopically to be free of gels and unreacted fibers. The solutions were then filtered through a 90 mm diameter, polypropylene, pan-cake, in-line filter during spinning. A 300 hole glass spinnerette having 0.0025 inch holes was employed for all spin trials. Viscosities were measured by a Brookfield Viscometer and found to be in the range of 8-16,000 cp at 22°C. The solutions were deaerated under vacuum prior to spinning.

The solutions were promptly spun into a water primary regeneration bath at 20°C and the resulting fibers passed vertically to a primary godet, then through a water secondary bath at 13°C to a secondary godet, whose speed could be altered to produce desired stretch conditions. The fibers were treated with 90°C 6% $NaHCO_3$, 50°C water, 50°C 0.3% finish, centrifuged and oven dried at 100°C. Godet stretch was 50-60%. (As used in this and the following examples, godet stretch is the difference between the circumferential speed of the initial and final godet, divided by the speed of the initial godet, times 100.)

Examples 2 through 7: Example 1 was repeated to produce a series of fiber samples, varying only the composition of the primary bath to include a progressively larger amount of NaOH. $S_{6.5}$ solubility tests were then conducted on each of the samples.

The $S_{6.5}$ solubility tests were conducted by adding a fiber sample to a 6.5% NaOH solution at 20°C. The mixture was vigorously shaken to disperse the sample in the caustic and allowed to steep for a total time of 1 hour from addition. The solution was then filtered with a slight vacuum and a 5 ml aliquot of the filtrate was oxidized by the addition of 10 ml of 0.5N potassium dichromate and 25 ml of concentrated H_2SO_4. After stirring the filtrate was permitted to stand for five minutes and 75 ml of distilled water and 2 to 3 drops of ferroin indicator (0.025 M ortho-phenanthroline ferrous sulfate) were added. The oxidized filtrate was then titrated to the endpoint with ferrous ammonium sulfate (FAS). The $S_{6.5}$ is calculated by the following formula

$$\begin{array}{c}\text{\% Solubility}\\ \text{in 6.5\% NaOH}\end{array} = \frac{34.23\,(B-T)}{(S)(\text{Sample Wt})}$$

in which B is the volume in milliliters of FAS required to titrate a blank sample of 0.5N dichromate, T is the volume in milliliters of FAS required to titrate the test sample and S is the volume in milliliters of FAS required for standardization against dichromate. Sample weight is oven dried. The results of the $S_{6.5}$ solubility tests are set forth in the following table.

Example	Primary Bath Composition	Percent $S_{6.5}$
1	Water	28.5
2	3% Aqueous NaOH	21.3
3	6% Aqueous NaOH	15.6
4	8% Aqueous NaOH	17.7
5	10% Aqueous NaOH	6.8
6	15% Aqueous NaOH	4.5
7	16% Aqueous NaOH	2.7

Examples 8 and 9: Additional fibers were prepared as in Example 1, but using a methanol coagulant-regenerant as the primary bath. In Example 8, the primary bath contained methanol alone; in Example 9, the primary bath contained methanol and 10.2% by weight of sodium methoxide, prepared by the addition of pure sodium to methanol. The $S_{6.5}$ solubility of fibers prepared from the pure methanol primary bath was 28.2%. The $S_{6.5}$ solubility of fibers prepared from the 10.2% CH_3ONa/CH_3OH primary bath was 10.9%.

These results demonstrate quite clearly that the addition of a soluble reactive base to the regenerant solution produces a regenerated article of improved resistance to alkaline solubility. Similar improvement is obtained with the use of two, three and four carbon aliphatic monohydric alcohols to which the required amount of a soluble reactive base has been added.

Fibers with a Crenulated Shape

Shaped cellulosic fibers are produced by *A.F. Turbak, R.B. Hammer, N.A. Portnoy and A.C. West; U.S. Patent 4,076,933; Feb. 28, 1978; assigned to International Telephone and Telegraph Corporation* by nitrosating and dissolving a cellulosic material in an aprotic solvent selected from the group consisting of a dialkylacylamide and dimethylsulfoxide and regenerating the dissolved cellulose by contacting the cellulose solution with a regenerant. The regenerant is butanol, pentanol or benzyl alcohol where the solvent is a dialkylacylamide and propanol, butanol or benzyl alcohol where the solvent is dimethylsulfoxide. The fibers have a generally crenulated shape as regenerated and require no special spinnerette designs nor special spinning conditions.

Example 1: A prehydrolyzed kraft pulp having a degree of polymerization (DP) of 1,050 was alkaline aged to a DP level of 450, neutralized, washed, dried, then fluffed and defibered by means of a rotary cutter. An 8/25/67 cellulose solution containing 8 parts of cellulose, 25 parts of N_2O_4 and 67 parts of DMF was prepared by charging 160 grams of the alkali aged prehydrolyzed kraft pulp (DP 450) and 1,340 grams of dimethylformamide (DMF) into a two-liter four neck resin reaction flask equipped with a stainless-steel mechanical stirrer, thermometer, and a 250 ml equalizing pressure addition funnel.

The resulting slurry was stirred and 500 grams of liquid nitrogen tetroxide (N_2O_4) was added dropwise over ca 60 minute time period. The solution was deaerated and vertically spun using a glass 300 hole (0.0025 inch) spinnerette. The fiber tow was passed through a primary bath of isoamyl alcohol (2-methyl-butanol-4) at 15°C, then through a secondary bath of water at 22°C. The fiber tow was cut from the take up godet, cut into staple fiber lengths, neutralized with bicarbonate, washed with water, finished with a 0.3% lubricating finish, air dried and carded. Godet stretch, as defined in the preceding patent, was 60%. The fibers produced had a trilobal cross-section. The fibers had tenacities of 2.74 g/den (conditioned) and 1.23 g/den (wet), elongations of 6.94% (conditioned) and 10.15% (wet) for a 0.90 denier filament of 0.54 g/den wet modulus.

Example 2: The process of Example 1 was repeated using however a primary bath coagulant of benzyl alcohol at 15°C. The fibers were largely quadrilobal. Properties were as follows.

Tenacity	(conditioned)	1.94 g/den
	(wet)	0.97 g/den
Elongation	(conditioned)	7.68 %
	(wet)	12.17 %
Wet modulus		0.41 g/den
Denier		0.96

Reduction of Formaldehyde Evolution

In the process developed by *A.B. Auerbach, J.P. Thelman and A.F. Turbak; U.S. Patent 4,120,836; Oct. 17, 1978; assigned to International Telephone and Telegraph Corporation,* the evolution of formaldehyde from shaped cellulose acetate articles containing methylolated melamine or guanamine polymers is reduced by adding to the solutions from which the cellulose acetate article is regenerated, from 1 to 20% by weight based on the weight of the methylolated polymer of a cyclic urea compound.

Example 1: A solution was prepared by dissolving 896 grams of cellulose acetate (800 oven dry, 39.7% acetyl) in 3,224 ml of acetone. Total solution water content was adjusted to 2%. Mixing was accomplished using a helical mixer over a 16-hour period. To this solution 100 grams of a methyl capped methylol melamine polymer and 20 grams of ethylene urea were added in 100 ml of acetone. The capped methylol polymer was prepared from 3.2 mols of formaldehyde for each mol of melamine with approximately 2 mols of methanol used in capping. The polymer has a degree of polymerization of 2, an approximate molecular weight of 380, is 72% alkylated with methyl groups and contains 28% free methylol groups.

After addition of the polymer and ethylene urea, the resulting solution was stirred for an additional 4 hours prior to a double stage pressure filtration. The solution was then deaerated on standing for 12 hours. The resulting acetone spin dope was then spun through a 40 hole (0.06 mm) spinnerette at a head pressure of 300-400 psig with a head temperature of 60°-70°C. Jetting rate was approximately 30 ml/min with a takeup of 1,030 linear fpm. This yielded a stretch of 14%. Cell jacket temperature was maintained at 110°C during spinning. The resultant 175 denier acetate yarn was twisted to 2.5 turns/inch and evaluated. Properties of this yarn are shown in the following table.

	Example 1	Control*
Formaldehyde, ppm	30	240
Tenacity, g/den**		
Dry	1.34	1.38
Wet	0.74	0.79
Elongation, %**		
Dry	21.3	22.4

(continued)

	Example 1	Control*
Wet	27.6	22.5
Fabric strength after bleaching, kilopascals***	305	245

*Identically prepared, no ethylene urea additive.
**Following heat set for 30 min. at 175°C.
***AATCC method 92-1974.

Example 2: A series of spinning solutions were prepared containing 27.7% cellulose acetate flake in acetone and 10%, based on the weight of total solids, of the capped methylol melamine polymer of Example 1. A control solution was prepared without additive. The remaining solution samples contained 2% by weight of the total solids of ethylene urea or of a known formaldehyde scavenger. The resulting 175 denier yarns were then evaluated, as in the foregoing table for formaldehyde evolution by AATCC Method 112-1975. The results are set forth below.

Additive	Formaldehyde (ppm)
Ethylene urea	30
Phthalimide	175
Urea	160
Polyurethane*	180
Isopropyl carbamate	165
Control	240

*Estane 5716.

These data clearly demonstrate the superiority of ethylene urea as compared with other scavengers of the type used in resin coating applications.

Highly Absorbent Filaments with Collapsed Hollow Structure

C.R. Woodings; U.S. Patent 4,129,679; Dec. 12, 1978; assigned to Courtaulds Limited, England describes highly absorbent regenerated cellulose filaments having collapsed hollow structure, a multilimbed cross-section and a water imbibition of 170 to 345%. They are made by a viscose process using sodium carbonate as an inflating agent.

Example: A viscose comprising 7.85% cellulose, 5.23% caustic soda and 2.55% sodium carbonate, and with a salt figure of 3.8 and a ball fall viscosity of 35 seconds at 18°C, was extruded into an aqueous spin bath comprising 13.5% sulfuric acid, 0.8% zinc sulfate, 24.0% sodium sulfate and 61.7% water to form a tow of filaments having an average count of 1.7 dtex.

The temperature of the spin bath was 70°C and the immersion path length was 24 inches. The spinning speed was 30 meters per minute. The spun tow of filaments was stretched by 30% in a 2% aqueous sulfuric acid stretch bath at a temperature of 95°C and then the filaments were washed.

The filaments of the tow were predominantly of multilimbed cross-section specified with a small proportion of filaments having the round cross-section of conventional regenerated cellulose filaments.

The filaments had an average water imbibition of 170%. The water imbibition was measured by taking a 1 gram sample of the fiber, soaking it in water at a temperature of 20°C for 15 minutes, centrifuging it at a force of 1,000 g for 5 minutes, weighing it, drying it at a temperature of 110°C for 2.5 hours, and finally reweighing it. The percentage water imbibition is then equal to:

$$\frac{\text{weight of wet fiber} - \text{weight of dry fiber}}{\text{weight of dry fiber}} \times 100$$

Alloy Rayon Fibers

F.R. Smith; U.S. Patent 4,179,416; Dec. 18, 1979; assigned to Avtex Fibers Inc. discloses alloy fibers having high fluid-holding capacity, and a method for making the same, the alloy fibers being comprised of a matrix of regenerated cellulose having polyvinylpyrrolidone dispersed therein. The polyvinylpyrrolidone may be present in combination with an anionic polymer.

Example 1: Using conventional rayon spinning equipment, aqueous solutions of polyvinylpyrrolidone, designated as K-60 (GAF Corp.) and having an average molecular weight of about 160,000 and K-value of 50-62, were separately injected by a metering pump into a viscose stream during its passage through a blender and the blend thereafter extruded. During this the blend was subjected to high mechanical shearing. The viscose composition was 9.0% cellulose, 6.0% sodium hydroxide and 32% (based upon the weight of the cellulose) carbon disulfide. The viscose ball fall was 56 and its common salt test was 7.

The mixtures of viscose and polyvinylpyrrolidone were extruded through a 720 hole spinneret into an aqueous spinning bath consisting of 7.5% by weight of sulfuric acid, 18% by weight of sodium sulfate, and 3.5% by weight of zinc sulfate. After passage through the spinning bath, the resulting continuous tow was washed with water, desulfurized with an aqueous solution of sodium hydrosulfide, washed with water, acidified with an aqueous HCl solution, and again washed with water. The still wet multifilament tow was cut into staple fibers and, without any further treatment, dried.

The fluid-holding capacity of sample fibers, made with various approximate proportions (tabulated below) of cellulose and polyvinylpyrrolidone in the spinning solution, was determined using the following test procedure.

Sample staple fibers were carded or otherwise well opened and then conditioned at 75°F and 58% relative humidity. Two grams of such alloy fibers were placed in a one-inch diameter die, pressed to a thickness of 0.127 inch, and maintained in this condition for one minute. This compressed pellet of fibers was removed from the die and placed on a porous plate of a Buchner funnel. The upper surface of the pellet was then engaged with a plunger which was mounted for free vertical movement, the plunger having a diameter of one inch and a weight of 2.4 pounds.

The funnel stem was connected by a flexible hose to a dropping bottle from which water was introduced into the funnel to wet the pellet of fibers. Control over the water flow was exercised by the position of the dropping bottle. After an immersion period of two minutes, the water was permitted to drain from the fiber pellet for three minutes, after which the still wet pellet was removed from the funnel and weighed. One-half of the weight of water in the sample pellet is a measure of the fluid-holding capacity of the fibers, expressed in cc/g. The test results of sample fibers, as described above, are set forth below.

Sample	Cellulose	Polyvinyl-pyrrolidone	Fluid-Holding Capacity cc/g	% Water Retention*
A	100	0	3.06	105
B	95	5	3.16	112
C	90	10	3.52	121
D	80	20	4.15	145
E	70	30	4.69	186
F	65	35	4.68	178
G	60	40	4.65	190

*Percent water retained by the loose mass of fibers after centrifuging the same at 1000 *g* for 3.5 minutes.

Example 2: A 20% aqueous solution of polyvinylpyrrolidone, designated as K-90 (GAF Corp.) and having an average molecular weight of 360,000 and a K-value of 80-100, was injected into a viscose having a composition as described in Example 1, after which the mixture was extruded as a continuous tow and processed as described above. The relative proportions of cellulose and polyvinylpyrrolidone in the spinning solution were 83:17. The resulting fibers had a fluid-holding capacity (tested as in Example 1) which was 28% higher than conventional regenerated cellulose fibers.

POLYESTERS AND OTHER POLYMERS

Polyester Polymer Recovery

N.C. Sidebotham, P.D. Shoemaker and C.W. Young, III; U.S. Patent 4,064,079; Dec. 20, 1977; assigned to Monsanto Company have developed a process in which polyester fibers are dissolved and solid linear polyester polymer recovered for use in the production of new linear polyester products, particularly polyester fibers or filaments, by the process of dissolving the fibers in a nondepolymerizing dissolution solvent for polyester characterized by inclusion of carbocyclic rings in its structural formula under dissolution conditions for polyester polymer; thereafter quenching the polyester solution; and separating the solid polyester from the solution.

Figure 12.1 is a flow sheet of a preferred embodiment. Polyester waste fabric **1** may include dyestuff, finish, knitting oils, etc.; and there may be contamination in the form of other fibers, including cotton, rayon, wool, nylon, acrylics, and the like. Scraps of paper or wire may also be in the feedstock. The waste fabric is loaded into washer-dissolver **2**, which is an agitation-spin-dry device, preferably having a wire-mesh covered rotatable bowl to prevent solid matter from being pumped out with liquids, this machine being similar to or identical with a conventional washing machine.

Recycled naphthalene, at a temperature of about 165°C, is pumped from quench solvent flasher **11** into washer-dissolver **2**. With agitation, more than half of the dye, knitting oil, finish, etc., is removed, and any moisture in the polyester is evaporated. The waste fabric load is spun out, and the spent naphthalene is recycled through flasher **3** to provide dye-free naphthalene for reuse in the dissolution step. Residual dyestuff, permanent press resins, etc., are discarded. The relatively dye-free naphthalene is recycled at about 190°-210°C into washer-dissolver **2** causing immediate dissolution of the polyester. The solution of polyester in hot naphthalene is pumped out of washer-dissolver **2** through in-line filter **4** thereafter to be mixed in two-fluid nozzle **5** with the quenching medium which is dimethyl formamide (DMF) and dropped as a precipitate into slurry tank **6**. Alternately, the two-fluid nozzle **5** may be by-passed and the quenching medium pumped directly into slurry tank **6** while the hot solution of polyester in naphthalene is sprayed into slurry tank **6**.

Washer-dissolver **2**, at this point contains only solid material which was not carried off in solution. This may include any paper, metal scraps, cotton, rayon, wool, nylon or acrylic fibers which were present in the waste fabric. This material is removed before a new load is placed in washer-dissolver **2**.

Figure 12.1: Polyester Polymer Recovery

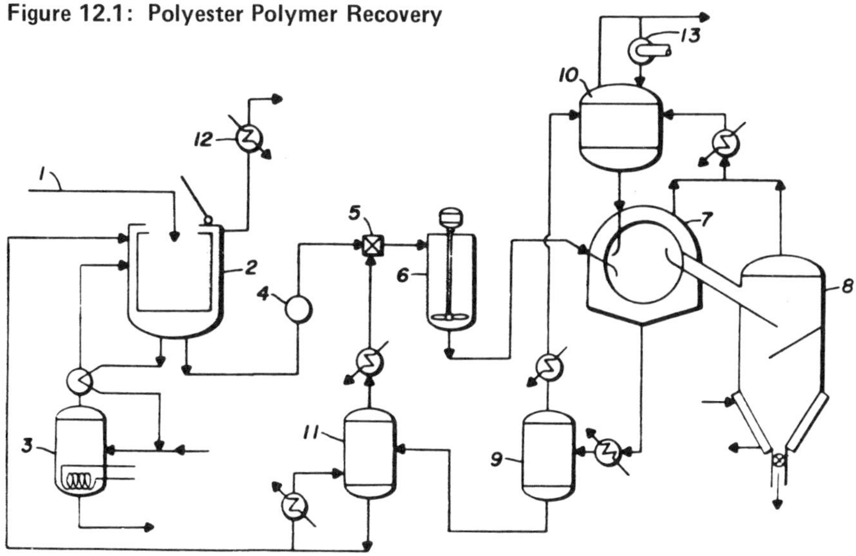

Source: U.S. Patent 4,064,079

The low temperature DMF injection at two-fluid nozzle **5** has caused a shock-quenching effect and thereby precipitated finely divided polyester powder which forms a suspension in the solution of DMF and naphthalene. Some residual dyestuff is also dissolved in the solution.

The polyester slurry is then fed to batch centrifugal filter **7** in which, during the first centrifuge cycle, a polyester filter cake is formed. The slurry feed is then interrupted and the filter cake sprayed with dichloromethane to rinse off the DMF, naphthalene and any residual dye. The white filter cake is then discharged to drying bin **8** where it is dried under mild heat conditions (above 40°C) to produce a crumb-like polyester material. Intrinsic viscosity of the polyester crumb product is equal to the intrinsic viscosity of normal polyester spinning feedstock. No water or air is present in the drying bin, the vaporization of the dichloromethane generating its own continuous inert gas purge. Under such conditions, the material may then be fed directly from drying bin **8** to a bypass-vented extruder (not shown) and thereafter spun into filament.

In addition to the dye purge flasher **3**, recovery systems for the naphthalene, the quench solvent, and the rinse solvent are integral with, although not essential to, this process. Included in the material fed to centrifuge **7** is the naphthalene polyester dissolution solvent, the DMF quenching medium, and the dichloromethane rinse solvent. This combination of solvents is fed from centrifuge **7** to rinse solvent flasher **9** wherein dichloromethane is vaporized at about 40°C and fed to rinse solvent tank **10** for reuse. The DMF/naphthalene/residual dye solution is fed from rinse solvent flasher **9** to quench solvent flasher **11** where DMF is vaporized at about 153°C for use in quenching the polyester/naphthalene solution. The residue from quench solvent flasher **11** is fed to washer-dissolver **2** for use in the preliminary dye removal step. Washer-dissolver **2** and rinse solvent tank **10** are equipped with condensers **12** and **13** respectively, thereby to recover overhead losses.

Example 1: A brown commercially available double knit polyester (polyethylene terephthalate) sample was dissolved at a 10% polymer concentration in naphthalene, and the solution was poured (hot) into an excess of dimethyl formamide at 140°C, producing a dilute slurry of white polyester in a colored solution. The slurry was allowed to cool and was thereafter filtered. The solids were rinsed twice with dimethyl formamide, and then with water. After drying, the granular solids were an off-white color. The heat of the dimethyl formamide in this example apparently detracted from the quenching effect.

Example 2: The same hot polyester naphthalene solution was poured into an excess of dimethyl formamide at room temperature, then filtered and rinsed with dimethyl formamide and water. After drying the powdery solid was white.

Example 3: To illustrate the significance of a quench as opposed to slow precipitation from a product quality point of view, commercially purchased dyed polyester rags were dissolved at 17.25% polymer concentration in naphthalene. Half of this solution was poured at 200°C directly into acetone. The other half was cooled slowly to room temperature. The slowly cooled material yielded a pale blue powder having an intrinsic viscosity of 0.52-0.54. The quenched product was a flocculent white product having an intrinsic viscosity of 0.52-0.53. Apparently slow cooling traps dyestuff within the polyester powder particles.

As previously stated polyethylene terephthalate is the preferred polymer. However, copolymers are workable as are other linear and quasi-linear polymers.

Polyester Recovery from Dyed Fibers

In a process by *N.C. Sidebotham, P.D. Shoemaker and C.W. Young, III; U.S. Patent 4,137,393; Jan. 30, 1979; assigned to Monsanto Company,* polyester polymer is recovered from polyester fibers (in the form of filaments, yarns or fabrics and the like, both dyed and undyed), and used in the production of new undyed fibers, films, and other polymer products by the process of stripping the dye from the dyed polyester fibers and dissolving the polyester fibers in a solvent in a separate but functionally interrelated (with the dye stripping step) step; separating the dissolved fibers from any undissolved impurities, and thereafter separating the solvent from the polyester without precipitating the polyester by evaporating the solvent.

Example 1: In all examples, the feedstock was first prewashed in naphthalene at 165°C wherever dye stripping was included as a step in the procedure. Polyester feedstock was then dissolved in naphthalene at about 30% polyester and at a solutioning temperature of about 210°C. The solution was then filtered to remove any insoluble contaminants such as nylon, acrylics, paper, cotton, wool, silk, rayon, acetate, glass or metals. The polyester/naphthalene solution was then boiled in an oil-heated three-neck boiling flask, fitted with a thermometer, an inert gas purge nozzle, and a short air-cooled condenser connected to a two-neck flask, which functioned as a naphthalene collector. A small purge of inert gas was used to sweep naphthalene vapors into the condenser. Boiling was first accomplished at a temperature of about 218°C as the first heating oil temperature was subsequently increased to about 270°C.

When the rate of boiling declined, a vacuum was slowly applied to the apparatus through the second neck of the naphthalene collector. A vacuum regulator was used to control the vacuum pull-down rate on the boiling solution to minimize foaming, splashing, and possible freezing of the solution. When solvent removal was completed, the vacuum was replaced with dry nitrogen and the flask was removed from the hot oil. The inert blanket was maintained as the polyester cooled and froze. The flask was wrapped with cloth as it cooled, because expansion of the solidifying polyester always broke the flask. After cooling was complete, the polyester product was recovered from the broken flask for analysis and subsequent spinning to fiber.

Example 2: Mixed dyed polyester waste fabric was subjected to the naphthalene dye stripping and dissolution procedures, with the result that although the initial intrinsic viscosity could not be ascertained, recovered polymer of comparable fiber viscosity was obtained which appeared to be reasonably pure based on color observation. These recovered polymer samples were of a quality which were suitable for melt spinning into synthetic fibers.

The spinning procedure was to place chunks of recovered polyester in a laboratory autoclave; pressurize several times with dry nitrogen; place a vacuum (20-40 mm Hg) on the autoclave; heat the autoclave until the polyester melts; bleed off the vacuum with dry nitrogen; remove the agitator shaft and put on a spinneret pack; apply dry nitrogen pressure to form filaments from the spinnerets; and take up bobbins of the fiber on a Lessona Type 955 winder. Spun bobbins were subsequently drawn over a hot pin at 105°C at several known draw ratios and physical properties of the drawn fiber were measured.

Only crude indications of spinnability and drawability were obtained due in large part to the lack of denier control. The intrinsic viscosity of the recovered prewashed polyester mixed rag feedstock was 0.6867. The spun yarn (10 filaments with conventional fiber finish) was successfully drawn over a hot pin at 105°C, with drawn fiber properties as indicated below.

Draw Ratio	Drawn Denier	Tenacity g/den	Elongation %
4.78	85.5	3.88	15.37
5.00	83.0	4.33	10.62
5.50	77.5	3.43	7.23

Upgraded Aromatic Polyester

A.H. Gerber; U.S. Patent 4,055,534; Oct. 25, 1977; assigned to Horizons Inc. describes a process for upgrading polyesters derived from aromatic dicarboxylic acids which comprises reaction of the polyester material (both virgin and scrap) in an aprotic polar solvent with any of the following:

(a) diisocyanates represented by the general formula $R(NCO)_2$ with difunctional reagents in which the functionality is selected from –OH, –NH and/or –COOH; and

(b) isocyanate-terminated polymers alone or with the above difunctional reagents.

The process also includes stabilization of the products obtained from the above, with organic phosphites. The polyester products, whether or not stabilized, are found to be upgraded with respect to both molecular weight and processability as compared with the starting material and are readily recoverable in powder form which is most desirable for subsequent use.

Example 1: Poly(ethylene terephthalate) powder (29 g) was added with stirring, to a solution (130°C) of sulfolane (150 ml), toluene (70 ml) and hydroxyl-terminated poly(tetramethylene ether) (4.5 g, 4.3 mmol). The mixture was gradually heated to 170°C while distilling off toluene. To the polyester solution was added dimeryl diisocyanate, (36 carbon cycloalkylene diisocyanate DDI 1410, General Mills Chemical Co.) (3.6 g, 6.0 mmol). The reaction mixture was kept for 2 hours at 170°±2°C and then for 1 hour at 180°C. The mixture was allowed to cool to 70°C at which time the precipitated mass was broken up and stirred well with methanol (250 ml). The polymer was filtered, washed well with methanol, and ultimately dried under high vacuum for 2 hours at 170°-180°C. Thirty-five and one-half grams of product was recovered. The product showed significant softening at 250°C, melted at 255°C and had a relative viscosity of 2.0.

When the above experiment was repeated without the poly(tetramethylene ether) and using 0.90 g (1.5 mmol) dimeryl diisocyanate, 29.3 g product of relative viscosity 1.81 was obtained. When 4,4'-methylene dianiline (0.40 g, 2.0 mmol) was prereacted with polyester and poly(tetramethylene ether) as above and subsequently reacted with dimeryl diisocyanate (8.0 mmol) a product of relative viscosity 1.86 was obtained.

Example 2: A stirred solution of sulfolane (300 ml), toluene (75 ml), and hydroxyl terminated poly(tetramethylene ether) (6.0 g, 6.0 mmol) was heated under nitrogen to 130°C while distilling off toluene. Poly(ethylene terephthalate) powder (58 g) was added and the mixture heated to 175°C and then maintained for 3¾ hours. An aliquot showed the polyester to have a relative viscosity of 1.45. Anhydrous zinc acetate (0.1 g) and 4,4'-dicyclohexylmethane diisocyanate (2.36 g, 9.0 mmol) were added and the resulting solution was then maintained for 3 hours at 175°C. The reaction mixture was then cooled, stirred with methanol (700 ml) and filtered. The polymer was twice washed well with methanol and vacuum dried at 175°-185°C. Sixty-two grams of product of relative viscosity 1.80 was recovered.

Example 3: A solution of sulfolane (175 ml), benzene (50 ml) and hydroxyl terminated poly(tetramethylene ether) (16.5 g, 8.09 mmol) was azeotroped and, at 110°C, 4,4'-dicyclohexylmethane diisocyanate (2.51 g, 9.59 mmol) and dibutyltin dilaurate (35 mg) added. The solution was stirred for 1 hour at 110°C, then poly(ethylene terephthalate) (33 g) was added and heating continued for 1.5 hours more at 160°C. Hot xylene (150 ml) was added and the reaction mixture was stirred well with methanol. The reaction mixture was filtered and the solid product was washed well with methanol and then high vacuum dried to afford 39.5 g of solid product which completely melted at 255°C and had a relative viscosity of 2.17. Percent nitrogen found was 0.50 (calculated, 0.52). Extraction of the polymer with methanol, tetrahydrofuran or benzene led to essentially no change in nitrogen analysis.

Extracting Granular Polyamides

It is known that polyamides which are processed into shaped bodies or fibers are treated in granular form with extraction means in order to remove monomers or low mol weight components. Practically, water or low mol weight alcohols having 1 to 4 carbon atoms and also benzol and toluol are used as extracting means. Generally therefore those substances are used as extracting means which dissolve the monomers or low mol weight components, but do not dissolve the polyamide itself. Furthermore, the extraction means causes no alteration in the property profile of the polyamide, such as yellowing or a lessening of mechanical qualities.

It is also known how to completely dissolve the polyamide in the preparation of powdered polyamides at high temperatures, and how to obtain the polyamide powder therefrom by precipitation or cooling. However, such a process results in higher mol weight components also remaining in the solvent besides the monomers and low mol weight ones, so that such powders differ in their property profiles from the polyamides obtained by conventional extraction.

The extraction is a relatively long process because the extracted granular polyamides must be freed of the extracting means, then dried and again dissolved in another solvent for the preparation of a powder. A grinding process is of no help in this respect because it does not lead to powder polyamides which are useful in every application.

Surprisingly, the monomers and low mol weight components are selectively dissolved and the mechanical properties of the polyamide are not disadvantageously modified in the process described by *S. Sridhar and H. Hedden; U.S. Patent 4,046,746; Sept. 6, 1977; assigned to Chemische Werke Huls AG, Germany*.

The method provides for extracting granular polyamides having at least 10 aliphatically bound carbon atoms per carbonamide group and a process for preparing polyamide powders from the extracted granular polyamides. The granular polyamides are extracted with dimethyl formamide at temperatures between about 70° and 100°C. The extracted granular polyamides are then dissolved in dimethyl formamide and the polyamide powders are formed by cooling the solution.

Example 1: Granular polylauryllactam of a grain size of about 2-3 mm in diameter is treated in a tumbling drier at 90°C with twice the amount by weight of dimethyl formamide. The extract is removed after 8 hours, the granulate is washed with a single amount by weight of fresh dimethyl formamide and dried at a temperature of 90°C and a pressure of 50 torrs (mm Hg). The single extraction reduces the proportion of extractable substance from 1.2 to 0.3% by weight.

If, on the other hand, a double amount of weight of methanol is used at 65°C, the extraction process must be carried out twice. It will then last 10 hours. The proportion of extractable substances then drops merely from 1.2 to 0.5%.

Example 2: A mixed polymer consisting of 80% by weight of lauryllactam and 20% by weight of caprolactam is treated at 85°C with dimethyl formamide under the conditions of Example 1. The grain size of the mixed polyamide used amounts to 2-3 mm. After 6 hours, the extractible proportion has dropped from 1.8 to the constant value of 0.35% by weight.

Purification of Chitin

Chitin is an aminocellulose derivative that occurs widely in nature, for example, in the cell walls of fungi, bovine cartilage, cuttlefish bone and the hard shell of insects and crustaceans. The waste from shrimp, lobster and crab seafood industries contains 10-15% chitin and is a potentially important source of chitin, although the isolation and purification of the chitin, associated therein with mineral components, protein and other ingredients, presents considerable difficulty.

More specifically, chitin is a mucopolysaccharide, believed to be poly-N-acetyl-D-glucosamine, with an empirical formula of $(C_8H_{13}O_5N)_n$ in which n may be any number into the thousands range, but is commonly in the area of 100-10,000. Chitin is a generally intractable material, soluble only in strong mineral acids, lithium thiocyanate solutions, and other special concentrated salt solutions, most of which cause disintegration or rapid degradation with loss in molecular weight or hydrolysis of the acetyl groups or both.

P.R. Austin, U.S. Patent 4,062,921; Dec. 13, 1977; assigned to University of Delaware discloses solvents for chitin comprising dimethylacetamide, N-methylpyrrolidone or mixtures of these in combination with a minor proportion of lithium chloride, and their use in the purification of chitin for regeneration in the form of films and fibers.

Example: To a solution of one part of lithium chloride in 20 parts of dimethylacetamide is added 0.6 parts of chitin (red crab) and the mixture stirred at room temperature for one hour. Then 20 parts of N-methylpyrrolidone is added and the mixture stirred for 2 hours. The resulting solution is filtered through felt and the filtrate spread on a glass plate at a thickness of 1 mm. The wet film is dried for one hour in a current of air at room temperature, and then placed in excess acetone for 16 hours to extract solvent from the film. The film is extracted for another 24 hours with fresh acetone, and finally rinsed again with fresh acetone and dried in air at room temperature for 30 minutes. The film is cut into small strips and are cold drawn to produce a 66% increase in length. The drawn film is strong and pliable.

Company Index

The company names listed below are given exactly as they appear in the patents, despite name changes, mergers and acquisitions which have, at times, resulted in the revision of a company name.

U.S. Patent Number Index

Inventor Index

Notice

Nothing contained in this Review shall be construed to constitute a permission or recommendation to practice any invention covered by any patent without a license from the patent owners. Further, neither the author nor the publisher assumes any liability with respect to the use of, or for damages resulting from the use of, any information, apparatus, method or process described in this Review.